Environment Conscious Manufacturing

Environment
Conscious
Manufacturing

Edited by

Surendra M. Gupta
A. J. D. (Fred) Lambert

CRC Press
Taylor & Francis Group
Boca Raton   London   New York

CRC Press is an imprint of the
Taylor & Francis Group, an **informa** business

CRC Press
Taylor & Francis Group
6000 Broken Sound Parkway NW, Suite 300
Boca Raton, FL 33487-2742

© 2008 by Taylor & Francis Group, LLC
CRC Press is an imprint of Taylor & Francis Group, an Informa business

First issued in paperback 2019

No claim to original U.S. Government works

ISBN-13: 978-0-367-45279-7 (pbk)
ISBN-13: 978-0-8493-3552-5 (hbk)

Library of Congress Cataloging-in-Publication Data

Environment conscious manufacturing / editors Surendra M. Gupta and A.J.D. Lambert.
   p. cm.
   Includes bibliographical references and index.
   ISBN 978-0-8493-3552-5 (alk. paper)
   1. Manufacturing processes--Environmental aspects. 2. Engineering design--Environmental aspects. 3. Industries--Environmental aspects. 4. Waste minimization. I. Gupta, Surendra M. II. Lambert, A.J.D. III. Title.

TS155.7.E57 2008
670.28'6--dc22   2007020180

Visit the Taylor & Francis Web site at
http://www.taylorandfrancis.com

and the CRC Press Web site at
http://www.crcpress.com

Dedicated to Our Families

Sharda, Monica and Neil
SMG

Teuntje, Henrike, Florian and Frans
AJDL

Contents

Preface

Environment conscious manufacturing (ECM) is an emerging discipline that is concerned with developing methods for manufacturing new products from conceptual design to final delivery, and ultimately to end-of-life disposal, that satisfy environmental standards and requirements.

The environment and global warming are receiving increasing attention these days. The Academy Award winning documentary, "An Inconvenient Truth,"* released in 2006 and presented by former U.S. Vice President Al Gore, has sent a warning signal to the masses to sharpen their awareness about global warming. At the same time, most industrialized nations are facing serious repercussions from the rapid technological development that has taken place in the past few decades. In recent years, environmental awareness and recycling regulations have been putting pressure on many manufacturers and consumers, forcing them to produce and dispose of products in an environmentally responsible manner. Government regulations are becoming more persuasive, and thus many manufacturers are under pressure to use recycled materials whenever possible. Occasionally, manufacturers are even required to take care of the products at the ends of their useful lives. This regulation has created a need to design products that are environment friendly, as well as easy to disassemble and recycle. Hence, there is more than ever a need to develop algorithms, models, heuristics, and software for addressing designing, recycling, and other issues (such as the economic viability, logistics, disassembly, recycling, and remanufacturing) for an ever-increasing number of products produced and discarded.

This text provides a comprehensive coverage of this discipline, exploring topics such as industrial metabolism, product design for the environment, design of reverse and closed-loop supply chains, disassembly modeling, and case studies in ECM. Students, academicians, scholars, consultants, and practitioners worldwide would benefit from this text. It is our hope that this book will inspire further research in ECM and motivate new researchers to get interested in this all too important field of study.

The book is organized into 15 chapters. The first chapter, by Lambert, presents an introduction to the basic concepts of industrial ecology including its historical roots. The author discusses the concepts of industrial metabolism and integrates them with the concepts of reverse logistics. The second chapter by Giudice, takes the life cycle approach to designing the product for the environment, considering all phases of the life cycle, from definition of product requirements to its disposal. In the third chapter, Vadde et al. describe sensor-embedded products in the context of product life cycle management.

---

* "An Inconvenient Truth," starring Al Gore and Billy West and directed by Davis Guggenheim. Studio: Paramount. Available on DVD.

The effectiveness of the embedded sensors is modeled using simulation and measured in terms of the average downtime, average maintenance cost, average disassembly cost, and average life cycle cost. It is shown that embedding sensors in computers provides a beneficial result.

The fourth chapter by Pochampally et al. illustrates how various quantitative techniques can be employed in the design phase of reverse and closed-loop supply chains to address a variety of decision-making problems. The decision-making problems addressed include the selection of economically used products, collection centers, recovery facilities, production facilities, second-hand markets, and new products; the optimal transportation of goods; the evaluation of marketing strategy; and the futurity of used products. In the subsequent chapter (Chapter 5), Jalil et al. assert that, in closed-loop supply chains, uncertainty could be managed by using information. The authors illustrate their point by considering the case of CopyMagic.

Chapters 6 through 10 address various issues associated with disassembly, which is the first step in product recovery. Product recovery seeks to obtain materials and parts from old or outdated products through reuse, remanufacturing, or recycling to minimize the amount of waste sent to landfills. Disassembly is defined as the methodical extraction of valuable parts or subassemblies and materials from discarded products through a series of operations. Chapter 6, by McGovern and Gupta, deals with disassembly line balancing. Since a disassembly line is the best choice for automated disassembly, it is essential that the disassembly line is designed and balanced to work efficiently. The disassembly line balancing problem seeks a disassembly sequence that is feasible, minimizes the number of workstations, minimizes total idle time, and ensures similar idle times at each workstation as well as addresses other disassembly-specific concerns. As finding the optimal balance is computationally prohibitive due to exponential growth, the chapter presents several metaheuristics algorithms that are easy to implement to solve the problem. Chapter 7, by Udomsawat and Gupta, presents a variant of the Toyota Production System (known as the multikanban system) that is implemented in a disassembly line. The authors' investigation reveals that the multikanban system is effective in controlling the system's inventory while providing a decent customer service level. In Chapter 8, Tripathi et al. suggest several random search techniques that can be used to solve a disassembly sequencing problem. Chapter 9, by Tang and Zhou, presents an overview of two models for uncertainty management in disassembly process planning, owing to human intervention. The first model mathematically represents the influence of human factors on disassembly, while the second incorporates fuzzy learning strategy into the disassembly process. In Chapter 10, Inderfurth and Langella discuss the planning of disassembly for the remanufacture-to-order systems. The authors argue that linear programming models can be used to plan disassembly for simple situations when yields are known. Their model, first discussed in the context of a single period, is subsequently relaxed to accommodate multiple periods, which necessitate the use of heuristic methodologies. The authors suggest the use of recourse models to incorporate the randomness to accommodate the uncertain yields of disassembly.

Chapter 11, by Topcu et al., highlights the issues arising from the design of facility and storage space in the context of remanufacturing. The authors argue that the number of usable parts or candidates for remanufacture retrieved from returned products varies significantly, causing fluctuations in inventory capacity and configuration requirements. Therefore, remanufacturing requires storage designs that not only minimize warehousing space and inventory-holding costs but also facilitate effective coordination of facilities planning and remanufacturing decisions. The authors use two mathematical models to illustrate their views. In Chapter 12, Mukherjee and Mondal highlight the current status of Indian remanufacturing. They point to the lack of such economic activities in the country and examine the reasons for such absence through empirical investigation. Chapter 13, by Nakashima, proposes a Markov model to evaluate and optimize environment-conscious manufacturing systems with stochastic variability stemming from customer demand, recovery rate, and disposal rate. The model can be used to calculate the total expected cost per period. In Chapter 14, Dhanda and Peters report that while the developed countries are guilty of discarding a majority of electronic products in the waste stream, most of the waste is actually shipped overseas to Asian countries. The authors analyze the reasons for such behavior and suggest some solutions to this problem. In the final chapter, Sarkis et al. point out that the adoption of environment conscious manufacturing practices needs to overcome a variety of barriers. The authors propose the use of interpretive structural modeling to help investigate, analyze, and overcome these barriers. The chapter provides an overview of the major barriers and an illustrative example with initial insights.

This text would not have been possible without the devotion and the commitment of the contributing authors. They have been very patient in preparing their manuscripts. We would also like to express our appreciation to Taylor & Francis and its staff for providing seamless support in making it possible to complete this timely and important manuscript.

Surendra M. Gupta

A.J.D. Lambert

Editors

Dr. Surendra M. Gupta, PE, is a professor of mechanical and industrial engineering and director of the Laboratory for Responsible Manufacturing at Northeastern University in Boston, Massachusetts. He received his BE in electronics engineering from Birla Institute of Technology and Science (India), an MBA from Bryant University, and an MSIE and PhD in industrial engineering from Purdue University. Dr. Gupta's research interests are in the areas of production or manufacturing systems and operations research. He is primarily interested in environment conscious manufacturing, manufacturing of electronic products, MRP, JIT, and queueing theory. He has authored or coauthored about 350 technical papers that have been published in prestigious journals, books, and conference proceedings. His publications have been cited by thousands of researchers all over the world in journals, proceedings, books, and dissertations. He has traveled to all seven continents, that is, Africa, Antarctica, Asia, Australia, Europe, North America, and South America and presented his work at international conferences there (except Antarctica). He is currently the area editor of environmental issues for *Computers and Industrial Engineering*, the associate editor for *International Journal of Agile Systems and Management*, and an editorial board member of a variety of journals. He has also served as a conference chair, track chair, and member of technical committees of a variety of international conferences. Dr. Gupta has been elected to the memberships of several honor societies and is listed in various *Who's Who* publications. He is a registered professional engineer in the state of Massachusetts and a member of ASEE, DSI, IIE, INFORMS, and POMS. Dr. Gupta is a recipient of the Outstanding Research Award and the Outstanding Industrial Engineering Professor Award (in recognition of teaching excellence) from Northeastern University. His recent activities can be viewed at http://www1.coe.neu.edu/~smgupta/, and he can be reached by e-mail at gupta@neu.edu.

Dr. A.J.D. (Fred) Lambert, is an assistant professor of industrial ecology in the Department of Technology Management at the University of Technology at Eindhoven, the Netherlands. He received his BS in electrical engineering from the Technical College at Vlissingen and his MSc in technical physics and his PhD in theoretical plasma physics from the University of Technology at Eindhoven, all in the Netherlands. In addition to the University of Technology at Eindhoven (the Netherlands), Dr. Lambert has participated in research projects at the Philips Company, the University of Greifswald (Germany), the University of Trieste (Italy), the Technical University at Lausanne (Switzerland), and the FOM Institute for Plasma Physics at Rijnhuizen (the Netherlands). Dr. Lambert has published papers on several topics, including nuclear fusion, nonequilibrium thermodynamics, MHD power generation, energy systems modeling, process integration, materials flow modeling, and, more recently, on disassembly sequencing. He has published more than 40 research papers in various scientific journals and has contributed to numerous books, conference proceedings, and professional papers. Dr. Lambert teaches undergraduate students energy efficiency and the managerial aspects of reuse and recycling; he supervises group projects on industrial ecology–related topics. He coaches graduate students on topics such as energy and waste management in industry, and sustainable energy resources. His recent activities can be viewed at http://w3.tm.tue.nl/en/subdepartments/aw/ (technology studies). Dr. Lambert can be reached by e-mail at A.J.D.Lambert@tue.nl.

Contributors

Shubham Agrawal Department of Manufacturing Engineering, National Institute of Foundry and Forge Technology (NIFFT), Ranchi, India

James C. Benneyan Department of Mechanical and Industrial Engineering, Northeastern University, Boston, Massachusetts, USA

Thomas P. Cullinane Department of Mechanical and Industrial Engineering, Northeastern University, Boston, Massachusetts, USA

K. Kathy Dhanda Department of Management, De Paul University, Chicago, Illinois, USA

Fabio Giudice Department of Industrial and Mechanical Engineering, University of Catania, Catania, Italy

Surendra M. Gupta Laboratory for Responsible Manufacturing, Department of Mechanical and Industrial Engineering, Northeastern University, Boston, Massachusetts, USA

Mohd. Asif Hasan University Polytechnic, Aligarh Muslim University, Aligarh, India

Karl Inderfurth Faculty of Economics and Management, Otto-von-Guericke-University Magdeburg, Germany

Muhammad N. Jalil Department of Decision and Information Management, RSM Erasmus University, Rotterdam, the Netherlands

Sagar V. Kamarthi Laboratory for Responsible Manufacturing, Department of Mechanical and Industrial Engineering, Northeastern University, Boston, Massachusetts, USA

Harold Krikke Department of Organization and Strategy, Tilburg University, Tilburg, the Netherlands

A.J.D. Lambert Faculty of Technology Management, Eindhoven University of Technology, Eindhoven, the Netherlands

Ian M. Langella   Faculty of Economics and Management, Otto-von-Guericke-University Magdeburg, Germany

Seamus M. McGovern   National Transportation Systems Center, U.S. DOT Research and Innovative Technology, Cambridge, Massachusetts, USA

Sandeep Mondal   Department of Management Studies, Indian School of Mines University, Dhanbad, India

Kampan Mukherjee   Department of Management Studies, Indian School of Mines University, Dhanbad, India

Kenichi Nakashima   Department of Technology Management, Osaka Institute of Technology, Osaka, Japan

Satish Nukala   Laboratory for Responsible Manufacturing, Department of Mechanical and Industrial Engineering, Northeastern University, Boston, Massachusetts, USA

Adrian Peters   Neohapsis, Chicago, Illinois, USA

Kishore K. Pochampally   School of Business, Southern New Hampshire University, Manchester, New Hampshire, USA

Joseph Sarkis   Graduate School of Management, Clark University, Worcester, Massachusetts, USA

Ravi Shankar   Department of Management Studies, Indian Institute of Technology Delhi, New Delhi, India

Ying Tang   Electrical and Computer Engineering, Rowan University, Glassboro, New Jersey, USA

M.K. Tiwari   Department of Industrial Engineering and Management, Indian Institute of Technology, Kharagpur, India

Aysegul Topcu   Department of Mechanical and Industrial Engineering, Northeastern University, Boston, Massachusetts, USA

Mukul Tripathi   Department of Metallurgy and Material Engineering, National Institute of Foundry and Forge Technology (NIFFT), Ranchi, India

Gun Udomsawat   Laboratory for Responsible Manufacturing, Department of Mechanical and Industrial Engineering, Northeastern University, Boston, Massachusetts, USA

Srikanth Vadde Laboratory for Responsible Manufacturing, Department of Mechanical and Industrial Engineering, Northeastern University, Boston, Massachusetts, USA

Ibrahim Zeid Department of Mechanical and Industrial Engineering, Northeastern University, Boston, Massachusetts, USA

MengChu Zhou Electrical and Computer Engineering, New Jersey Institute of Technology, Newark, New Jersey, USA

Rob A. Zuidwijk Department of Decision and Information Management, RSM Erasmus University, Rotterdam, the Netherlands

1

Industrial Metabolism: Roots and Basic Principles

A.J.D. Lambert

CONTENTS

1.1   Introduction

Terms such as "life cycle management," "industrial ecology," "industrial metabolism," and "industrial symbiosis" are common in the literature on environmentally conscious production, but the precise definition of these concepts and the distinction between them varies with authors. For clarity on these topics, we will present a brief review of the work done by authors who first proposed these disciplines of study and related concepts. We also intend to explain how industrial ecology is embedded within the development of related disciplines. This approach is relevant because a multidisciplinary approach is followed in implementing industrial ecology.

Because of its multidisciplinary character, the field that is covered by industrial ecology has been gradually extended to domains beyond its original concepts. Therefore, it might be useful to redefine industrial ecology while reconsidering the original definitions of this topic. The essential aspect in which industrial ecology differs from other approaches is in the comparison between the industrial system and the natural ecological system. This comparison is not only of academic interest, but it is considered a tool for the redesign of the industrial system on various levels of aggregation aimed at achieving increased sustainability. Most of the industrial ecologists advocate the closing of the materials cycles as an essential path toward this end. The method of quantitative materials—and energy flow analysis—also called industrial metabolism, is the principal instrument for achieving this goal. Therefore, this chapter describes the basic concepts of industrial metabolism.

1.2   Roots of Industrial Ecology

1.2.1   Malthusianism and Neo-Malthusianism

Industrial ecology deals with the impact of human activities, specifically economic activities, on the environment. In the course of history, the ancient Greeks had observed this impact. It involves

- depletion of natural resources;
- impact on human well-being, including health risks and dangers posed by toxins and infectious diseases; and
- impact on the ecosystem caused by degradation of natural habitat and emission of ecotoxic substances.

The fact that the earth's resources are finite was noted several centuries ago. T.R. Malthus, who published his principal work, *An Essay on the Principle of Population*, in 1798, was one of the most well-known early proponents of this idea. He postulated that food production increased in an arithmetic progression whereas population increased exponentially. This situation would inevitably lead humankind to a catastrophe. Although Malthus's pessimistic theory, soon called Malthusianism, fueled nonscientific and selfish political positions, he was among the first to use mathematical concepts to consider the impact of human activities on the earth.

Architect Buckminster Fuller (1963) promoted a more mature way of dealing with the earth's natural resources. He introduced the concept of "spaceship Earth," undoubtedly inspired by emerging space programs that produced the first photographs of the earth as an isolated, brilliant, bluish sphere floating in space, although the metaphor of the earth as a ship had been used earlier. Fuller advocated an integrated worldview and supported the idea of global cooperation. He also coined the term "synergetics." As an architect, he supported the idea of ecodesign, also called design for the environment. Kenneth E. Boulding (1966), an economist and a follower of Fuller's philosophy, introduced concepts such as evolutionary economics and the constraint imposed by the second law of thermodynamics, which is the law of increasing entropy, or disorder. Like Fuller, he also advocated a global approach. The need for such an approach became extremely relevant with the discovery of adverse impacts of human activity on the global environment. These impacts included depletion of the ozone layer and the reinforced greenhouse effect. His position is clear from his comment on mainstream economics: "Anyone who believes exponential growth can go on forever in a finite world is either a madman or an economist." This way of thinking, with regard to the restrictions imposed by the laws of nature, is referred to as neo-Malthusianism.

The same applies to more recent and sophisticated world models such as that found in the report of the Club of Rome (Meadows et al., 1972). Overpopulation and overconsumption form the crux of this report, which is based on Forrester's (1971) dynamic modeling. Forrester's view was based on the modeling of complex systems via a set of coupled differential equations, with additional constraints that referred, for instance, to stocks and capacities. He designed dynamic models of the world as a whole as well as of cities and industries. The approach of the Club of Rome has been criticized, however, for its lack of essential regulatory mechanisms. Price mechanisms, substitution, and technological change could surely delay the inevitable collapse of the industrial system predicted by these models. Apart from this, the Club of Rome model assumed hard constraints on the quantity of resources, but in practice these are soft constraints. For instance, oil reserves, although finite, manifest themselves quite differently from a discrete "subterranean lake of oil" that is gradually going to be emptied. In practice, every additional quantity of oil can be made available only at an increased expense. This constraint relaxes consumption in a smooth way rather than abruptly, offering some tolerance for substitution and other conservation measures. Despite the criticism leveled at it, the report has exerted a considerable impact on humankind's attitude toward the environment. It also stimulated the scientific community to design more sophisticated models of the interaction between human activities and the environment.

1.2.2 Environmental and Ecological Economics

Environmental economics, a branch of macroeconomics that is close to mainstream economics, has made an important contribution to the understanding of the interaction between human activities and the environment (see, e.g., Freeman et al., 1973). The perspective, of course, has been financial rather than physical. Environmental economics primarily aims at the quantification of tariffs, taxes, and costs or benefits that are related to specific environment-related measures. This branch of economics also deals with internalization of external costs, for instance, those related to the impact of carbon dioxide (CO_2) emission on the environment and quantification of substitution effects. Environmental economics usually uses highly aggregated models that do not include process characteristics and physical laws.

Ecological economics connects economic theory with physical laws, particularly mass conservation, energy conservation, and an increase of disorder. It is called the second law economics, as the second law of thermodynamics plays a crucial role in its formulation (see Boulding, 1966). This branch of economic science considers matter and energy, rather than labor and capital, as basic production factors of any economic system. Ecological economics attempts to design strategies that result in sustainable development. Some early works in this field include those by Georgescu-Roegen (1971, 1976) and Dasmann et al. (1973). The concepts have been compiled and elaborated by, among others, Ruth (1993), who studied the application of renewable energy sources in macroeconomic models.

The macroeconomic instrument of input–output analysis (IO analysis), which was formerly developed by Leontief (1953) as an instrument for the analysis of interaction between different sectors of the economy based upon national accounts, has also been extended to deal with materials and energy. This concept is called physical IO analysis. One of the principal advantages of IO analysis is that it considers the mutual interdependence of industrial branches, with respect to materials and energy, through matrices that can easily be mathematically manipulated. Material and energy flows are usually derived from flows of money, which are at the basis of the original approach of IO analysis. Waste IO and energy IO are frequently used examples of physical IO analysis, which has become a field of research in its own right.

Economists also deserve credit for the first attempts at making material balances for countries, or for even larger systems. For instance, Kneese et al. (1974) was the first to estimate the mass that is affected by the U.S. population. Termed total materials requirement (TMR), this amounted to about 15 ton $\times$ cap^{-1} $\times$ year^{-1}. The TMR involves fossil fuels, metal ores, minerals, excavation, biomass, anthropogenic erosion, and imports; it refers to the complete economy. The concept of determining the TMR is connected to the idea of dematerialization. It implies that a service is available at the expense of a minimum amount of materials. Herman et al. (1990) and Wernick et al. (1996) have contributed to this topic.

A related tool is the gross energy requirement (GER), which refers to the total use of fossil resources or primary energy during the life cycle of a product or material (IFIAS, 1973). This tool was developed in the wake of the 1973 oil crisis. The method of quantitative life cycle assessment (LCA) is an extension of this method. Boustead and Hancock (1979) have discussed systems analysis focused on energy use in the process industries. Data on energy use in industry and agriculture have been compiled in various handbooks, for example, those by Carroll (1980), Pimentel (1980), and Brown et al. (1985). Since then, attention has shifted to a more comprehensive method of chain analysis, life cycle assessment (LCA), which will be briefly discussed in the next subsection.

1.2.3 Quantitative Life Cycle Assessment

At the micro level, we observe the tendency in management science toward environmentally conscious production, which is typically at the company or product level. It includes organizational tools such as environmental management systems, and systematic analysis methods, of which LCA is the most well-developed and standardized tool. LCA is aimed at quantifying various selected aspects of the environmental impact of a product or a material in the course of its complete life cycle.

Originally designed as a tool to quantitatively compare products such as packaging, LCA was given a scientific basis by the work of Guinée et al. (1993a,b). A drawback of LCA is that it does not include a complete mass balance but is confined to a restricted set of environmental impact classes. Another principal weakness of LCA is that it does not deal in a natural way with recycling and exchange of materials between various production chains. This problem, which is related to the definition of systems boundaries, is discussed by Tillman et al. (1994) and others. Management tools for environment-conscious production are now standardized through the ISO 14000 series. LCA is represented in the ISO 14040 series, and the GER is one aspect of the LCA.

Today, various software packages that support LCA are available. These packages include extended databases on the environmental impact of multiple products and processes, which are to be combined by the modeler to chain modes.

A combination of quantitative LCA and physical IO analyses is advocated by many authors, starting with Heijungs (1994). Contributions by Suh (2004) are also worth mentioning.

A considerable number of quantitatively oriented scientific papers on general LCA, case studies, and the combination of LCA and IO analyses have been published.

1.2.4 Depletion of Materials and Energy

Environmental consciousness has emerged primarily from concern for human well-being in addition to concern for the salvation of habitat and biodiversity. In society, this consciousness was initially confined to hygienic measures and the avoidance of environmental nuisances such as noise, stench, and dust. Efforts toward the reduction of hazardous emissions into the atmosphere, surface water, and soil, called pollution prevention, did not exist before the 1960s and the 1970s. Interest in these topics was stimulated by several industrial accidents, the Bhopal gas leak disaster (1984) being the most noteworthy.

The practice of reuse and recycling is not new, but in the past, this practice was dictated by economic rather than environmental reasons. Depletion of material and energy resources, emphasized by the Club of Rome's report, received increasing attention after the 1973 oil crisis, also called the energy crisis. Although this event had a political overtone, the industrialized world woke up to a harsh reality—its dependence on crude oil from fairly unstable or unreliable countries such as Saudi Arabia, Iraq, and Iran. The crisis resulted in growing attention toward energy use in the production of commodities. Possibilities for energy conservation in industry were investigated and implemented. Apart from these measures, the crisis resulted in measures aimed at diversification of primary energy resources to reduce the risks of a sudden break in supply. However, the nuclear option that was popular in those days became unfeasible for quite some time in the aftermath of the 1979 Three Mile Island accident and the 1986 Chernobyl disaster.

Obviously, possible future events that are comparable with the oil crisis would have a significant impact on the world economy because every commodity that is produced has a considerable energy component in its costs. Ore mining, in particular, is of interest here because of an increasing scarcity of energy resources and decreasing ore grades. In the wake of the oil crisis, Chapman and Roberts (1983) quantitatively studied this type of mechanism. The introduction of renewable energy resources such as solar, wind, hydro, and biomass was initially propagated by ecologically oriented non-governmental organizations (NGOs). Later, however, such resources became increasingly driven forward by the call for diversification and the tenacious priority for economic growth without any emphasis on dematerialization. When introduced on a massive scale some of these renewable resources, particularly hydroelectricity and biomass from primary crops, appear extremely harmful in spite of the fact that they are marketed as "green" or "sustainable." In addition, renewable energy is restricted by the amount of power that is available.

The alarming message is that the demand, which is generated by the unrestrained growth of consumption and intensified by extensive publicity around extravagant life styles, can never be met through environment-friendly technologies. In Southeast Asia, rapid economic growth and a large population have created an additional demand on materials and industry.

1.3 Pioneers in Industrial Ecology

1.3.1 Pollution Prevention

The term "industrial ecology" was coined by Evan (1974), who used it first in a seminar held in 1973 on the chemical industry and the environment. His paper was published at a time when environmental policy focused mainly on pollution prevention in the process industries. The author did not present the integrated view that emerged later in industrial ecology; instead, he reflected on the discussion by the process industries about the introduction of pollution prevention technologies, particularly on the socioeconomic consequences of these measures. He made an inventory of the costs that would accrue from prevention of pollution, the increase in price of products due to these costs, and the benefits of the emergence of a new branch of industry for the production of pollution prevention equipment. Apart from these economic issues, he discussed the effects on income, income distribution, employment, and the working environment. Thus, Evan's perspective on environmental economics had a strong focus on the social sciences as well.

Evan advocated a multidisciplinary approach that present-day industrial ecology is already taking, although along rudimentary lines. He defined industrial ecology as

> an interdisciplinary systems approach to environmental problems arising from industrial activities, i.e., the production, consumption and disposal of manufactured products and their raw material and energy inputs, as well as from related mining, agricultural, transportation and construction processes.

He discussed six aspects of industrial ecology: technology, environment, natural resources, biomedical aspects, institutional and legal matters, and socioeconomic aspects.

As we will focus on industrial metabolism, we highlight the first three aspects as follows:

- *Technology*—Evan stressed the role of technology, mainly pollution prevention techniques such as removal and fixation of pollutants, alternative processes and materials, waste disposal, and recycling.
- *Environment*—Evan proposed research on "the interaction of effluents and wastes with the environmental media and the dynamics of self-regeneration." He also included the work environment, which is usually related to safety issues.
- *Natural resources*—Evan advocated the optimal use of raw materials, energy, water, and natural sites.

Fortunately, many of Evan's recommendations have been implemented. Today, environmental legislation exists, end-of-pipe technologies such as filters and water purification units have been introduced, and processes have been modified, for instance, the replacement of batch processes by more controllable continuous processes.

1.3.2   Restructuring Global Management

The concept of industrial ecology took on a new dimension with the publication of a paper by Frosch and Gallopoulos (1989) in a special issue of *Scientific American* entitled "Managing Planet Earth." This issue focused on the global impact of human issues such as population growth, food shortages, freshwater scarcity, and the constant need for energy. Frosch and Gallopoulos unfolded an ideal view of how a sustainable industrial system should be organized.

> The industrial ecosystem would function as an analogue of biological ecosystems.... An ideal industrial ecosystem may never be attained in practice, but both manufacturers and consumers must change their habits to approach it more closely if the industrialized world is to maintain its standard of living—and the developing nations are to raise theirs to a similar level—without adversely affecting the environment.

The latter statement is closely related to the concept of sustainable development as explained in the Brundtland report (Brundtland, 1987). In this report sustainable development is described as "a development that meets the needs of the present generation without compromising the ability of future generations to meet their needs." A weakness of this definition is, when used in its restricted form, "human needs" are not clearly explained and the intrinsic value of the ecosystem itself is not highlighted. Because the Brundtland report focuses on development economy, the definition of sustainability is extended by formulating a compulsory condition that involves meeting "the essential needs of the world's poor, to which overriding priority should be given." A further essential aspect is "preservation and protection of diverse ecosystems." The Brundtland report builds on, among others, the ideas of the physicist and economist J. Tinbergen, who investigated the requirements of the industrial system in coping with population growth. Thus, he represented neo-Malthusian thought.

In contrast with Evan's work, which is mainly restricted to pollution prevention in the process industries, the work of Frosch and Gallopoulos has an emphasis on global issues, although it is also strongly focused on one of the many subjects mentioned by Evan, namely recycling. This is expressed as an incentive in their paper: "Wastes from one industrial process can serve as the raw materials for another, thereby reducing the impact of industry on the environment." The authors illustrate this point with the existing cycles for plastics, ferrous metals, and noble metals of the platinum group. Here, the authors observe that the recycling rate of plastics is poor, but that of ferrous metals is considerable, although much of it is still discharged. The rate for the platinum group of metals is nearly perfect. Historical, technological, and, most importantly, economic factors account for these differences.

1.3.3   Zooming in on Industrial Metabolism

Ayres and Simonis (1994), Ayres and Ayres (1996), and Graedel and Allenby (1995) have presented more recent studies on industrial ecology.

Ayres and Simonis (1994) introduced the concept of industrial metabolism. They define this as

> the whole integrated collection of physical processes that convert raw materials, plus labor, into finished products and wastes in a (more or less) steady-state condition.… The stabilizing controls of the system are provided by its human component (… The system is stabilized, at least in its decentralized competitive market form, by balancing the supply and demand for both products and labor through the price mechanism.

This definition guides us on how to model the technosystem, through which this system of processes is controlled. Because a decentralized competitive market does not exist to its full extent, other mechanisms such as legislative regulation must also play a role in stabilizing this. Regulation is the result of policy on different levels, which is subjected to multiple actors, each with his own interests. The study of this process is beyond the scope of industrial metabolism, which focuses on physically oriented methods. The definition's addition, "in a (more or less) steady-state condition" must also be put into perspective, since the technosystem can only be considered a more or less steady state when a relatively short period of time is considered. In practice, the model is then frozen over a time interval of typically one year. This is usual when materials and energy balances are studied in which short-term fluctuations are averaged out. Apart from this, statistical data are typically available with yearly intervals.

In work by Ayres and Ayres (1996), the term "industrial ecology" has been reused. Again, the focus is on closing of the material cycles. The one-liner, "closing the loops," is typically based on industrial ecology thinking. Dematerialization, material substitution, recycling, and waste-mining are considered the principal strategies to be implemented. Waste-mining refers to the recovery of low-value by-products from waste. As an example, the possible use of coal-ash for metals recovery is mentioned. The authors discuss flows of various solid materials. Ultimately, they introduce the concept of industrial ecosystems, referred to now by the term "industrial symbiosis." This concept involves the mutual exchange of process by-products among different industrial plants that share the same site.

In a report from Tibbs (1993) that focused on management, industrial ecology is explained as follows:

> Industrial ecology involves designing industrial infrastructures as if they were a series of interlocking manmade ecosystems interfacing

with the natural global ecosystem. Industrial ecology takes the pattern of the natural environment as a model for solving environmental problems, creating a new paradigm for the industrial system in the process.

This description does not specify the principal aspects of the natural environment, but the absence of waste production has been considered the most important feature of the ecosystem.

Next, we discuss the work of Graedel and Allenby (1995), which focuses on the manufacturing industries. The authors employ a comprehensive definition of industrial ecology that highlights some of its essential aspects.

> Industrial ecology is the means by which humanity can deliberately and rationally approach and maintain a desirable carrying capacity, given continued economic, cultural, and technological evolution.... The concept requires that an industrial system be viewed not in isolation from its surrounding systems, but in concert with them. It is a systems view in which one seeks to optimize the total materials cycle from virgin material, to finished material, to component, to product, to obsolete product, and to ultimate disposal. Factors to be optimized include resources, energy, and capital.

The first part of this definition bears a strong similarity to that of sustainability, which we explained earlier. The second part emphasizes a systems approach, a life cycle approach, and the use of optimization techniques. The authors discuss the link between ecosystems and industrial systems and present data on global cycles of, for instance, chlorine, nitrogen, and cadmium. They also discuss the application of tools such as LCA for evaluating and improving products and production processes, inclusive of those of complex products. Much of the discussion is in the form of case studies in their work.

The work by Socolow et al. (1994) is a compilation of previously published works on industrial ecology. A subsequent textbook on industrial ecology has been compiled by Lowe et al. (1997). It contains various topics and well-known cases, but it also discusses the ambiguities in the definitions of industrial ecology. We will summarize here the similarities in the approaches of various authors, which are clearly analyzed in the book:

- Industrial ecology is a systems approach.
- It focuses on the interaction of the technosystem and the ecosystem.
- It studies the flows of materials through the technosystem and in interaction with the ecosystem.
- It seeks transformation from a linear to a closed-loop system.

Our focus is on industrial metabolism, and we note that the characteristics mentioned so far are in common with industrial metabolism. However, one similarity remains unmentioned by Lowe et al., namely: industrial ecology studies the complete physical life cycle of a product.

Additional characteristics, mentioned below, that the authors on industrial ecology state in common do not refer to industrial metabolism but to the more extended field of industrial ecology. This field is interdisciplinary, and this perspective agrees with that of Evan's, discussed earlier:

- Enables short-term innovations with awareness of their long-term impacts
- Balances environmental protection with business feasibility
- Offers an objective for coordinating design of public policy

The differences in approach of the various authors are also summarized in this book. They are as follows:

- The level of aggregation in time and complexity of the system considered range from incremental changes in companies to drastic changes that are affecting the complete global industrial system.
- The value of modeling the industrial system according to the ecosystem's principles is questioned by several authors.
- Some studies invoke nothing more than a materials flows analysis.
- Many industrial ecology studies are restricted to the industries that produce physical products, while others advocate extension to service industries, public policy, and consumer behavior.
- Ecological concerns about biodiversity, carrying capacity, and so on are often not considered.
- Some authors emphasize technological change, while others focus on institutional change.

Other differences can be found in the more or less far-reaching character of the improvement options, which range from gradual improvement of efficiency in existing production processes to the introduction of completely new processes based on renewable materials and energy.

In his paper on industrial ecology, Den Hond (2000) cites the nine attributes of industrial ecology listed by Garner and Keoleian (1995). Briefly, these attributes are as follows:

- A system's view on the interactions between industrial and ecological systems
- The study of material and energy flows and their transformations into products, byproducts, and waste (industrial metabolism)
- A multidisciplinary approach
- An orientation toward the future
- A change from linear to cyclical processes
- An effort to reduce environmental impact from industrial activities

- An emphasis on harmonious integration of industrial systems in the ecosystem
- Industrial systems mimicking the ecological system
- Identification and comparison of industrial and natural system hierarchies

The author emphasizes that two different viewpoints exist in industrial ecology:

- Describing material and energy flows
- Controlling material and energy flows through traditional governance instruments such as hierarchy, the market, and the law

The first mentioned viewpoint coincides with the concept of industrial metabolism. The author discusses the establishment of eco-industrial parks and dematerialization as important tools for achieving the ideal of an environmentally benign production system.

Martin (2001) has explained clearly the role of reverse logistics in industrial ecology. He presents the "six Is" as basic elements of industrial ecology, which are mainly used for environmental management purposes. These are given below.

- Information, which is an inventory of used and wasted materials
- Incentives, which mean enabling creativity in pollution prevention and product redesign
- Investment, which includes the location of production plants in the vicinity of others that can apply the residuals
- Integration, which implies materials exchange in an eco-industrial park
- Interaction, which is the establishment of strategic alliances
- Infrastructure, which embodies the principles of reverse logistics

Finally, we want to briefly discuss some aspects described in the book by Bourg and Erkman (2003). This book contains a brief history of industrial metabolism, many case studies, and a concluding chapter entitled "Perspectives in Industrial Ecology". It is reiterated there that no standard definition of industrial ecology exists. According to the author, the following are the main characteristics of industrial ecology (see also Erkman, 1997):

- It is a systematic, comprehensive, and integrated view of all the components of the industrial economy and their relationship with the biosphere.
- It focuses on materials flows within and outside the industrial system.

- It considers industrial dynamics, which is long-term evolution of key technologies, as a crucial element to achieve a transition from an unsustainable to a viable industrial system.

The concept of industrial metabolism is considered the basic methodology of industrial ecology. It is defined in Bourg and Erkman's book as "a descriptive and analytical methodology based on the conservation of mass." This is a rather unsatisfactory and incomplete definition, as it does not mention energy, for instance, as another crucial physical factor.

1.3.4   Industrial Symbiosis

Apart from the confusion caused by different definitions given in the preceding subsections, another source of confusion arises from considering the concept of industrial symbiosis as similar to industrial ecology. See, for instance, Ehrenfeld and Gertler (1997). To avoid confusion, we therefore will define industrial symbiosis as follows:

> Industrial symbiosis is the theory and practice of interchanging residual material and energy flows between companies on a site or in a region.

Methods of industrial ecology and even of industrial metabolism can be used to study industrial symbiosis, but it is not identical to any of these, as it refers to a restricted interpretation and does not include the complete life cycle of a product. Although the concept of eco-industrial parks has been introduced within the framework of industrial symbiosis, only a few scientific papers have been published as yet on this topic.

1.3.5   Industrial Ecology as a Mature Science

1.3.5.1   *Scientific Results*

Today, many scientific journals carry papers on industrial ecology. The following are the most important among them and regularly publish on subjects related to industrial ecology:

- *Journal of Cleaner Production* (31)
- *Journal of Industrial Ecology* (27)
- *Ecological Economics* (22)
- *Environmental Science and Technology* (12)
- *Resources Conservation and Recycling* (10)

This ranking of journals is according to the number of papers published in this field (in parentheses), as compiled by ISI Web of Science. An additional important scientific journal on industrial ecology, although not listed by ISI, is *Progress in Industrial Ecology*.

Scientific papers with an emphasis on recycling, in addition to a detailed discussion of specific recycling technologies for plastics, paper, water, and batteries, can be found, for instance, in the following journals:

- *Resources, Conservation and Recycling* (359)
- *Waste Management and Research* (141)
- *Waste Management* (119)

In addition, scientific papers on LCA can be found in the *International Journal of Life Cycle Assessment*. In the period from 1988 to 2006, over 1000 papers have been published in this field.

Since 2001, the International Society for Industrial Ecology (ISIE) is active in this field. Its former president, B. Allenby, has written some of the previously discussed books on industrial ecology. The society has an extensive domain of interests, including the fields of eco-design and environment-conscious manufacturing. The society publishes the *Journal of Industrial Ecology*.

A number of dedicated graduate courses on industrial ecology are available today, and some management tools inspired by industrial ecology perspectives are in use in the industry.

1.3.5.2 Criticism and Further Development

In spite of all its merits, the current theory and practice of industrial ecology has a a noticeable weakness—it sometimes appears as a theory of everything. Unfortunately, it is not a unified theory on the relationship between economic activities and the ecosystem. To date there is no coherent body of knowledge on this topic. In fact, no clear distinction exists even between theory and practice. Although the promise of industrial ecology includes a systematic and integrated approach, the result so far is an addition of completely diverging approaches. While industrial ecology originally referred to a great challenge, which included the restructuring of the global manufacturing system, its approach as shown in the literature remains rather eclectic, as it embodies a set of poorly coherent ideas on environmentally benign production and consumption. No general theory has been developed so far and even terminology in the field is far from coherent. The literature in this field is a rather diffuse body of knowledge that extends from global challenges to detailed management and product design instructions. Fortunately, multiple elements that can form a coherent body of knowledge are found throughout the literature. To reinforce these, we must first classify the topics that are embraced by the concept of industrial ecology and then identify the core topic of industrial ecology that distinguishes it from other approaches. This core topic involves the analysis of anthropogenic physical flows, or industrial metabolism.

1.4 Structuring Industrial Ecology

1.4.1 Classification of Subdomains

To provide structure to the vast domain of industrial ecology, we need to break the subject down into the following subdomains:

- Environmental policy, which includes
  - environmental economics, as specified earlier and
  - legislation, consumer behavior, and related topics.
- Environmentally conscious production, which includes
  - environmental management;
  - environmental assessment, such as LCA and GER; and
  - environmental design or eco-design. As the latter topic includes both product and process design, it also embraces the wide field of environmental technology, which includes recycling techniques, end-of-pipe technology, and reverse logistics.
- Physical flow analysis (industrial metabolism), which includes the following areas of analysis:
  - Global
  - Regional or urban
  - Site-oriented
  - Branch of industry-oriented
  - Product-oriented
  - Material-oriented
  - Service-oriented

All of these aspects are discussed to varying extents in the literature.

1.4.2 Back to Basics

To avoid confusion, we will restrict the concept of industrial ecology to the comparative study of human economic activities and the ecosystem. This concept is based on that of ecology, which originates from biology. It is defined as follows:

Ecology is the science of the interaction between living organisms as well as the interaction between these organisms and their abiotic environment. It is a branch of biology, first explicitly mentioned by the biologist Ernst Haeckel (1866).

Three aspects can be discerned here:

1. The presence of evolutionary mechanisms, which are self-organization, adaptation, and competition.
2. The interrelationship between organisms as shown by the food pyramid, the food chain, and symbiosis.
3. The relationship between biotic and abiotic environments as shown by energy sources based on renewables and no waste production.

Material and energy flows within the living system and the physical interaction with the abiotic world are studied as a part of ecology. Therefore, ecology is also related to geophysics and geochemistry.

When comparing human economic activities with ecology, it is evident that evolutionary mechanisms are present in the cultural, scientific, and technological development of humankind. Self-organization is undoubtedly at work here, but the dynamics of this kind of evolutionary processes are mostly considered beyond the scope of industrial ecology. Economic development and the diffusion of environmentally conscious technologies are also subjects for study in related disciplines rather than in industrial ecology. Fundamental differences in biological evolution are found in the absence of species and even of discrete organisms, and also in the absence of blueprints such as genomes. Although a company can be compared with an organism, fundamental differences such as the absence of reproduction are present.

The interrelationship between companies or production processes is evident. Companies usually do not eat each other, apart from mergers and competition issues, which are not studied by industrial ecology. In particular, the symbiosis aspect is of interest, which refers to the art of living together and is aimed at mutual benefits. Such benefits are exactly why companies interact with each other and with consumers as well. The term "industrial symbiosis," however, must be reserved to mutual interchange of by-products between companies that reside on a well-defined site (eco-industrial park) or region (ecoregion). Although other complex relationships between companies such as those that are present in supply chains can also be considered as a kind of symbiosis, these are usually not covered by industrial ecology. Interrelationship on a lower level of aggregation such as the mutual exchange of by-products between production processes, typically within the same company, is called process integration.

The relationship between the biotic and the abiotic environments in ecology is replaced in industrial ecology by the relationship between the industrial system and its biotic and abiotic environments.

1.4.3 Limiting to the Core Domain: Industrial Metabolism

We observed that many topics can be included under industrial ecology. We started by restricting ourselves to the comparison of the ecosystem and the industrial system. We will call the latter "technosystem" from now on. This definition excludes typical environmental management topics from industrial ecology. If confined to the study of physical flows and their transformation processes within the technosystem and the exchange of these flows with the ecosystem, we are dealing with industrial metabolism. We will apply the following definition:

> Industrial metabolism is a modeling method aimed at the investigation of the industrial system, the mutual interchange of physical flows between its subsystems, and the interchange of physical flows between the industrial system and its natural environment.

- It is an integrated view.
- It is based on systems theory.
- It can be applied to various levels of aggregation, varying from process-oriented to global.
- It has an emphasis on physical flows, which are material and energy flows.
- It is quantitatively oriented.
- It is life cycle-oriented.

The purpose of industrial metabolism is to understand the relationship between the system and its environment, with a focus on the extraction from and the discharge to the environment of physical flows. The domain of application follows from the fact that this method acts as a decision and an evaluation aid with respect to measures aimed at minimizing the environmental impact caused by human economic activity.

Industrial metabolism thus embodies a reductionist view, which is inherent to modeling, because a model is a simplified mapping of real conditions, aimed at investigating a restricted set of properties. The goal of minimizing the environmental impact is achieved by minimizing both the quantity and the toxicity of physical flows entering and leaving the technosystem. Apart from simplification of the system itself, many interactions between the technosystem and its environment are externalized without considering the way these come about. These include constraints set by regulation, ecological issues, demand, availability of resources, and costs of materials and technologies; these factors are introduced in the model as external parameters.

1.5 Basics of Industrial Metabolism

1.5.1 Systems Theory

Systems theory starts with defining the supersystem or the universe that is all inclusive. As a subset of this universe, a system is defined as a set of objects and their mutual relationships. An object can be considered a "black box"; its internal structure is not studied in detail. If the black box is opened, the object appears to be a system in itself. It is also called a subsystem as it is a part of the original system.

Systems are confined by a systems boundary that externalizes all the features not belonging to the system, as defined by the systems designer. This boundary is not necessarily a geographic boundary. Rather, it excludes those objects that are different from those sought after in the model. By this description, the systems boundary is a rather abstract concept. Apart from defining the system of interest, one also often specifies those parts of the universe that are closely related to the system under consideration. These parts together are called the environment of the system. Relationships crossing the systems boundary thus represent the interaction of the system with its environment.

Systems exhibit an intrinsically hierarchical structure because it is theoretically possible to decompose the subsystems downward to ever more detailed structures. This process is called disaggregation. It is also possible to combine a system and its environment to arrive at a new, extended system, which in turn has an environment, and so on. This process is called aggregation. A global system has a high level of aggregation. Hierarchy in the sense of the occurrence of higher level objects that control lower level ones, such as occurs in hierarchical organizations, is in general not present in industrial metabolism because only physical flows are considered. To be considered a system, a certain extent of internal cohesion must be present among objects, which means that the objects of the system are more or less strongly mutually interacting.

Although systems are of different types, we will restrict ourselves to those systems in which the objects are processes and the relationships are physical flows. These flows are directed. This implies that the system can be represented by a directed graph in which the nodes are the objects and the arcs are the physical flows. The objects can be considered multi-input multi-output processes. The presence of processes presupposes transformation of the physical flows. If the transformation is restricted to storage, we call the node a reservoir. A reservoir might have finite capacity, which is in contrast to its use in thermodynamics, where a reservoir is considered infinite.

1.5.2 Ecosystem

Although focused on the technosystem, we begin by considering the ecosystem since it makes up the environment of the technosystem. It is connected with the technosystem through the exchange of physical flows. If physical flows are transferred from the ecosystem into the technosystem, this process is called extraction. Physical flows that leave the technosystem and enter the ecosystem are said to be discharged.

A study of industrial ecology starts with a consideration of the ecosystem. Thus, first Earth is assumed as a system, with the solar system as its environment. The planet experiences negligible materials exchange but experiences massive energy exchange with respect to the surrounding interplanetary space. The latter takes place mainly through solar radiation and re-radiation from Earth to interplanetary space. Within the system multiple energy and materials cycles, both long term and short term, can be distinguished. These include the water cycle and the carbon cycle. In many of these cycles, the ecosystem plays a crucial role. This ecosystem consists of the biosphere, which includes all living and dead biomass, the soil, the hydrosphere (seawater and freshwater), and the atmosphere. Confusion may arise about the concept of the biosphere, as some scientists use the term when referring to the complete ecosystem. We use the term biosphere in a restricted sense only, thus referring to the aggregate amount of biomass on Earth. We notice that, for instance, the lithosphere or upper crust exchanges materials with the ecosystem through erosion and sedimentation and through phenomena such as volcanism.

In the technosystem, the economic activities of humankind, although essentially belonging to the ecosphere, are placed separately from the ecosystem. The technosystem and its environment, together with the principal material flows between these two, are depicted in Figure 1.1.

In Figure 1.1, we discern the nodes, which are reservoirs, and the arcs, which represent interactions.

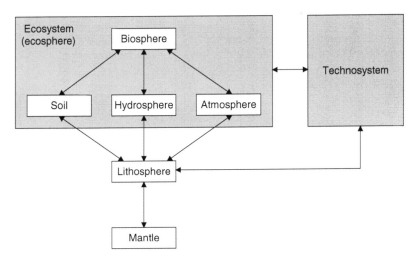

FIGURE 1.1
The environment of the technosystem.

With regard to the reservoirs, the following quantities are important:

- *Abundance* represents the averaged share of specific substances in the reservoir. The substance can be an element, a compound such as a molecule or a mineral, or a material such as sand or gravel. With regard to the lithosphere, seawater, and the atmosphere, elemental abundances are given. Apart from a few light elements, most of the elements are marginally present. Fortunately, there exist natural concentration mechanisms that gradually locally enhance abundances, thus creating exploitable ore bodies and other exploitable reserves.

- *Time of residence* refers to the average time that a substance is present in the reservoir.

In this model, *interactions* are expressed as physical flows, which are characterized by their composition and their flow rate, expressed, for instance, in ton/year.

Cycles of materials can be discerned as well. Because driving forces depend on the kind of material, all the various materials cycles have different characteristics. The cycles are driven by both abiotic and biotic mechanisms. Atmospheric nitrogen, for instance, is fixed by both lightning and bacteria and subsequently transformed into inorganic compounds, such as ammonium, and organic compounds, such as amino acids and proteins.

In addition to materials cycles, energy cycles must be considered. Energy can be present as chemical energy, tangible heat (related to temperature), latent heat (e.g., evaporation heat), kinetic energy such as wind, and potential energy such as water residing at a definite height above sea level.

The reservoirs are not always considered inert because transformations other than storage will proceed inside them. These include transport phenomena as well as chemical and physical transformation.

In many ecological studies, the unaffected ecosystem is considered static in the short term. In this case, the system is in some equilibrium state, which means that no or marginal accumulation takes place. Anthropogenic activities impose both quantitative and qualitative changes of the flows, which usually result in a—sometimes considerable—shift in the equilibrium in the short or medium term. An example of this process is the gradual shift in carbon dioxide concentration in the atmosphere due to combustion of fossil fuels.

In contrast with a graph representation, some authors apply a method that resembles a Sankey diagram as is frequently utilized in energy analysis. Although such a representation sometimes provides a clear picture, it lacks discrete nodes that represent transformation processes. For this reason, such a model cannot easily be modified. Therefore, we prefer the graph representation as depicted in Figure 1.1, which is a basis for mathematical modeling of various systems.

Figure 1.2 shows an example of a flow diagram that is visualized as a Sankey diagram. Originally, it was part of a much larger diagram that depicted the

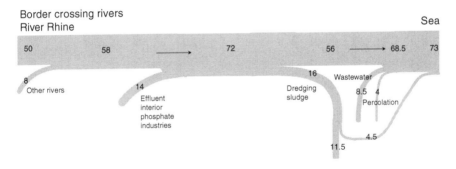

FIGURE 1.2
Flow diagram of phosphate load in Dutch surface water, 1980 (Unit: 10^6 kg P/year; data: Central Statistics Bureau of the Netherlands).

complete phosphorus flow in the Netherlands due to human activities. It can be seen that nodes are failing here.

1.5.3 Technosystem

With Figure 1.1 in mind, we begin by considering the global technosystem as a black box, which represents a single transformation system that extracts materials from the ecosystem and the lithosphere and that discharges the transformed materials back into the ecosystem. Next, we discuss the different features of this system in more detail. After opening the black box, we discern a set of transformation processes with their mutual relationships. This system is represented by a directed graph similar to that in Figure 1.1. The most primitive organization of the technosystem is a linear chain of transformation processes, starting with extraction of materials from the ecosystem or the lithosphere, and ultimately discharging the wasted materials back into the ecosystem. A more advanced organization of the technosystem includes materials cycles aimed at minimizing the incoming and the outgoing materials flows.

First, we will discuss and classify the transformation processes in the technosystem in greater detail. Transformation implies change, and the condition for any change is the availability of energy. Although energy is conserved, it will also be transformed in the course of the process.

Transformation can take place as follows:

- With respect to time: in storage and accumulation
- With respect to place: in internal and external transportation
- With respect to quality.

A typology of transformation processes with respect to quality is

- nuclear transformation;
- chemical transformation; and
- physical transformation, with or without phase transition.

Another typology based on change in coherence is

- coherence increase: joining, mixing, and pressing;
- coherence decrease: separation, demixing, and loosening; and
- coherence remains unchanged.

With regard to energy transformation, the aggregate quality of the involved energy flows always decreases, which means that its properties are approaching closer to those of its environment. This can be seen, for example, when observing temperature. Quality of energy is usually expressed in its ability to perform a certain amount of mechanical work, which is called exergy. Because of energy transformation the aggregate exergy decreases, which is a consequence of the second law of thermodynamics. The same law also affects the quality of materials such as their purity and concentration. Quality of materials and energy reflects itself in the scope of utilization. Exergy loss can result in energy that is hardly applicable, for instance, because its energy density is too low or its temperature is too close to the ambient temperature. In this case, we are dealing with residual heat. Although energy can be converted into mass, and vice versa, this effect is only substantial in nuclear transformations, which are uncommon in the majority of production processes. The same principle implies that materials conservation is valid for every individual chemical element, because elements can only be converted into each other via nuclear processes.

In static modeling, materials and energy conservation result in node equations that express that the sum of the masses of the incoming flows equals that of the outgoing flows for every node. In the case of dynamic modeling, storage effects must be accounted for. In this case, the yearly accumulation of mass equals the difference in mass of aggregate incoming and outgoing flows. Conservation laws result in a set of node equations.

The interrelationships between processes are maintained by product flows. These flows are characterized like those in the ecosystem, namely, by quantity, composition, and, when required, by physical quantities such as temperature. Product flows are related to transport processes, particularly when external transport between different sites is considered. Ancillaries needed for transport are usually assigned to the product flows instead of considering transport as a separate process in the chain. The complete graph of processes and product flows is called a product–process graph. Essentially, this graph is a process flow diagram on an aggregated level.

1.5.4 Production and Consumption: Linear System

1.5.4.1 Production Process

A production process is a process in which raw materials or semifinished products are transformed into products. These in turn can be considered semifinished products from the viewpoint of the next producer in the production chain. The aim of any production process is to increase the aggregate value of the outgoing flows with respect to the aggregate value of the incoming flows.

Value is a concept that has two aspects:

- *Practical value,* which refers to the usefulness of the product for the user
- *Market value,* which refers to the yield that one obtains when purchasing the product in the market

Mechanisms that determine the price of products are usually externalized in industrial metabolism studies. The price of some residual products might be negative.

Every production process needs ancillaries such as energy carriers, water, and lubricating oil. A production process not only creates products but also creates unintended residuals. If these residuals find a useful application, they are called by-products; in other cases these are called waste, which can be solid waste, effluent, or emission. The waste is called process waste because it is due to the process rather than the product. Not only materials flows but also energy flows must be considered. Energy is not only applied as an ancillary, but it is also incorporated in many raw materials and products such as those in the petrochemical and plastics industries and, of course, in the energy industries that consider energy transformation as their key activity.

The aim of the production process must be achieved by producing valuable products with a minimum of waste. Therefore, a trade-off of the benefits of the products and the costs of the process, with those of the input materials included, must be made.

A model of a general production process is depicted in Figure 1.3a,b.

Production processes in the process industries are usually segregated into unit operations. This approach has been advocated by Little (1915), although the original idea is attributed to Hausbrand (1893), see, for a more recent discussion, McCabe and Smith (1976). One must keep in mind that there already

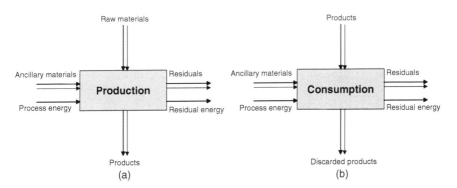

FIGURE 1.3
Production and consumption processes.

existed about 2000 different synthesis processes in the year 1915. Nowadays, about 50 unit operations are discerned that bring structure to the still increasing number of such processes. See, for example, McCabe and Smith (1976).

Alting and Boothroyd (1982) attempted classification of the discrete production processes. The DIN-8580 standardization series is partly based on Kienzle's work.

1.5.4.2   Consumption Processes

Consumption processes have many features in common with production processes, although some essential differences exist between them. For instance, a general consumption process has two flows of residues: process waste and product waste. These flows result from the aim of the consumption process, which involves providing services to the consumer instead of creating value. In the course of this process, the value of the product decreases until the product ultimately must be discarded. The utility of a car, for instance, is transportation. The discarded car makes up the product waste, and the emissions, spent oil, and so on, make up the process waste. Gasoline, lubricant, and the like are the ancillaries. It is arbitrary whether we consider packaging waste as product or process waste, as, for instance, in the case of food consumption.

The consumption process is depicted in Figure 1.3b. Not every consumption process generates product waste; many consumption processes completely destroy the original product during consumption. This applies, for instance, to food and fuel. Investment goods can be considered consumption goods to some extent, depending on the point of view that has been taken up.

1.5.4.3   Linear Product–Process Chain

The combination of processes is organized as shown in Figure 1.4, which represents a linear materials flow in the technosystem. The flow consists of the basic processes: extraction, production, consumption, and discharge. The way of a product through these processes is called the physical product life cycle, in contrast to the conceptual product life cycle, which starts with the design and ends with the outdating of a product. Although still far removed from the principles advocated in industrial metabolism, the scheme in Figure 1.4 is the backbone of any analysis of the technosystem or its subsystems.

1.5.5   Vertical Disaggregation

Next, we discuss the vertical disaggregation of a linear system. Subsequently, the graph will be extended so that it includes waste upgrading processes, and, finally, cycles are introduced.

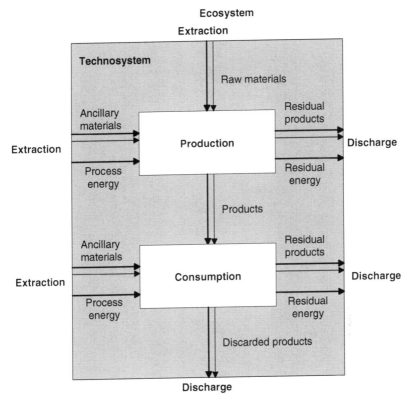

FIGURE 1.4
Simple linear technosystem.

1.5.5.1 Vertical Disaggregation of the Production Process

Production processes can be vertically disaggregated, which results in a subdivision of production processes into various categories. Most important is the subdivision in materials production, which takes place in the process industries. Next comes component production, which takes place in the manufacturing industries, and then comes assembly, which is the process that ultimately combines the components to arrive at a complex product. Materials production transforms the intrinsic properties and component production the extrinsic properties of the product. Although the emphasis on the different transformation types varies from product to product, this classification is typical for the most advanced product categories, which include complex mechanical, electrical, and electronic products. We introduced this distinction in the graph in Figure 1.5. All of these types of production processes generate process waste. The emphasis in Figure 1.5, however, is on the product waste that emerges when a product is discarded. Therefore, the process waste flows are not included in the graph.

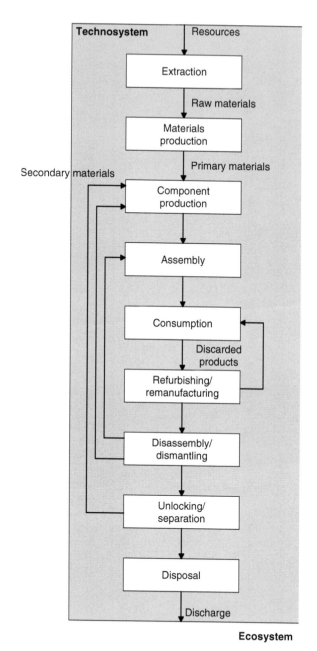

FIGURE 1.5
Disaggregated product–process chain with recycling simplified.

1.5.5.2 Upgrading Processes

Even in a linearly organized system, one will treat the waste, prior to its discharge, in such a way that its environmental impact decreases to an acceptable level. This treatment typically refers to end-of-pipe measures such as filters aimed at pollution prevention, and to some preparation of waste, for instance, through volume reduction and containment. These kinds of upgrading processes can be considered analogous to production processes.

Closing the cycle, however, also requires upgrading processes aimed at transforming waste to a useful product. Processes of this kind are aimed at maintaining the product's value as much as possible. This can be achieved by passing through subsequent stages of a reverse production chain. For several reasons, such a chain of processes will not be an exact reverted copy of the production chain in which the product was produced because the discarded product is in a different state from the new one:

- The product can be outdated.
- The product can be damaged, contaminated, or modified.
- The products of a specific type are dispersed in place and in time, which impairs the viability of recollection.
- The value of a discarded product is significantly lower than a new one, which restricts the viability of multiple reverse production steps.
- A lot of production processes are irreversible.

With these considerations in mind, upgrading of residuals is only partly modeled analogous to production.

The hierarchy of the final use of the discarded product, after recovery, is as follows:

- *Repair,* aimed at reuse of the complete product
- *Refurbishing,* aimed at reuse of the product in a different configuration
- *Reuse,* aimed at reuse of modules or components, which can be in an "as new" product or in a spare part
- *Recycling,* aimed at reuse of materials
- *Energetic recycling,* aimed at reuse of the energy content
- *Disposal,* aimed at environmentally benign discharge

It is evident that every subsequent type of recovery results in a further degradation of the product's original value.

The following upgrading processes are frequently applied to meet the above-mentioned purposes:

- Recollection, testing, cleaning, and sorting
- Disassembly and subsequent reassembly

- Disassembly, which is nondestructive removal of modules and components
- Dismantling, which is destructive removal or separation of components
- Unlocking by shredding, grinding, milling, and so on
- Separation of chunks that are created by shredding
- Upgrading for improvement of material properties
- Incineration, aimed at energy recovery and volume reduction
- Final disposal

The products that are obtained by these upgrading processes are called secondary products, in contrast to primary products. In the case of materials, primary or virgin materials are acquired by extraction. Figure 1.5 presents the extended product life cycle chain, which is the basis for materials flow analysis through the technosystem. Practice is more complicated because multiple production systems that are connected with each other exist. Distinguishing between those various systems is called horizontal disaggregation.

Even when reuse and recycling loops are optimized, we must account for losses in the system such as recollection rates that are lower than unity and restricted selectivity of separation processes. In addition, upgrading processes require energy and ancillaries and generate process waste as well, which is not different from other processes. Moreover, it is not always possible to apply recovery processes that guarantee secondary materials of the same quality as the original primary materials. Although recycling of metals can often be done without loss of quality, this situation is not viable for products such as plastics, rubber, glass, textiles, and printed circuit boards. Furthermore, recycling processes are at the expense of energy and materials, which means that a trade-off always has to be made. When the recycling process leaves us with a significant loss of quality with respect to the original materials, we call this downcycling or cascading.

1.6   Conclusion

In this chapter, we presented a brief introduction of industrial ecology. We discussed the different approaches that have been practiced by the key authors who developed this field of interest. Because the concept of industrial ecology has often been interpreted in a rather wide sense, we proposed to confine it to the comparison of the system of human economic activities and the ecosystem and to mimic the essential sustainability aspects of the ecosystem by the technosystem. The fields of industrial symbiosis and industrial metabolism, which form a part of industrial ecology, have been redefined.

Because the actual literature on industrial ecology is dominated by environmental management issues, LCA studies, and case descriptions, there remains a need for a revival of the art of modeling materials and energy flows in the technosystem according to a unified and integrated concept.

As a contribution to this model, the basic concepts of industrial metabolism have been reconsidered and discussed. This approach has also been followed to integrate the basic concepts of reverse logistics within this picture. With this study, we hope to add to the development of industrial metabolism and to the wider field of industrial ecology.

References

Alting, L. and Boothroyd, G., 1982, *Manufacturing Engineering Processes*. New York: Marcel Dekker.

Ayres, R.U. and Ayres, L.W., 1996, *Industrial Ecology: Towards Closing the Materials Cycle*. Cheltenham, UK: Edward Elgar.

Ayres, R.U. and Simonis, U.E. (eds.), 1994, *Industrial Metabolism*. Tokyo: United Nations University Press.

Boulding, K.E., 1966, The economics of the coming spaceship Earth. In: Jarrett, H. (eds.), *Environmental Quality in a Growing Economy*. Baltimore, MD: John Hopkins Press, pp. 3–14.

Bourg, D. and Erkman, S. (eds.), 2003, *Perspectives on Industrial Ecology*. Sheffield, UK: Greenleaf Publishing.

Boustead, I. and Hancock, G.F., 1979, *Handbook of Industrial Energy Analysis*. Chichester, UK: Ellis Horwood.

Brown, H.L., Hamel, B.B. and Hedman, B.A., 1985, *Energy Analysis of 108 Industrial Processes*. Atlanta: Fairmont Press.

Brundtland, G.H., 1987, *Our Common Future*. Oxford, UK: Oxford University Press.

Buckminster Fuller, R., 1963, *Operating Manual for Spaceship Earth*. New York: Dutton & Co.

Carroll, L. (ed.), 1980, *Industrial Energy Use Data Book*. Oak Ridge, TN: Oak Ridge Associated Universities.

Chapman, P.F. and Roberts, F., 1983, *Metal Resources and Energy*. London: Butterworth.

Dasmann, R.F., Freeman, P.H. and Milton, J.P., 1973, *Ecological Principles for Economic Development*. London: Wiley.

Den Hond, F., 2000, Industrial ecology: a review. *Regional Environmental Change* **1**(2), 60–69.

Ehrenfeld, J. and Gertler, N., 1997, Industrial ecology in practice: the evolution of interdependence at Kalundborg. *Journal of Industrial Ecology* **1**(1), 67–79.

Erkman, S., 1997, Industrial ecology: a historical view. *Journal of Cleaner Production* **5**(1–2), 1–10.

Erkman, S., 2003, Perspectives on industrial ecology. In: Bourg, D., and Erkman, S. (eds.), *Perspectives on Industrial Ecology*. Sheffield UK: Greenleaf Publishing, pp. 338–342.

Evan, H.Z., 1974, Socio-economic and labour aspects of pollution control in the chemical industries. *Journal for International Labour Review* **110**(3), 219–233.

Forrester, J.W., 1971, *World Dynamics*. Cambridge: Wright-Allen-Press.

Freeman, A.M., Haveman, R.H. and Kneese, A.V., 1973, *The Economics of Environmental Policy*. London: Wiley.

Frosch, R.A. and Gallopoulos, N.E., 1989, Strategies for manufacturing. *Scientific American* **261**(September), 144–152.

Garner, A. and Keoleian, G.A., 1995, *Industrial Ecology: An Introduction*. Ann Arbor, MI: University of Michigan National Pollution Prevention Center for Higher Education.

Georgescu-Roegen, N., 1971, *The Entropy Law and the Economic Process*. Cambridge, MA: Harvard University Press.

Georgescu-Roegen, N., 1976, *Energy and Economic Myths: Institutional and Analytical Economic Essays*. New York: Pergamon Press.

Graedel, T.E. and Allenby, B.R., 1995, *Industrial Ecology*. Englewood Cliffs, NJ: Prentice Hall.

Guinée, J.B., Udo de Haes, H.A. and Huppes, G., 1993a, Quantitative life cycle assessment of products 1: goal definition and inventory. *Journal of Cleaner Production* **1**(1), 3–13.

Guinée, J.B., Heijungs, R., Udo de Haes, H.A. and Huppes, G., 1993b, Quantitative life cycle assessment of products 2: classification, valuation and improvement analysis. *Journal of Cleaner Production* **1**(2), 81–91.

Heijungs, R., 1994, A generic method for the identification of options for cleaner products. *Ecological Economics* **10**, 69–81.

Herman, R., Ardekani, S.A. and Ausubel, J.H., 1990, Dematerialization. *Technological Forecasting and Social Change* **38**, 333–347.

IFIAS, 1973, *Report of the International Federation of Institutes for Advanced Study*, Workshop no 6 on Energy Analysis. Guldsmedshyttan, Sweden.

Kneese, A.V., Ayres, R.U. and d'Arge, R.C., 1974, Economics and the environment: a materials balance approach. In: Wolozin, H. (ed.), *The Economics of Pollution*. Morristown: General Learning Press.

Leontief, W., 1953, *Studies in the Structure of the American Economy: Theoretical and Empirical Explorations in Input–Output Analysis*. Oxford, UK: Oxford University Press.

Lowe, E.A., Warren, J.L. and Moran, S.R., 1997, *Discovering Industrial Ecology: An Executive Briefing and Sourcebook*. Columbus, OH: Battelle Press.

Martin, M., 2001, Implementing the industrial ecology approach with reverse logistics. In: Sarkis, J. (ed.), *Greener Manufacturing and Operations: From Design to Delivery and Back*. Sheffield, UK: Greenleaf Publishing.

McCabe, W.L. and Smith, J.C., 1976, *Unit Operations of Chemical Engineering*. London: McGrawHill.

Meadows, D.H., Meadows, D.L. and Randers, J., 1972, *The Limits to Growth: A Report for the Club of Rome's Project on the Predicament of Mankind*. New York: Universe books.

Pimentel, D., 1980, *Handbook of Energy Utilization in Agriculture*. Boca Raton, FL: CRC Press.

Ruth, M., 1993, *Integrating Economics, Ecology and Thermodynamics*. Dordrecht: Kluwer Academic Publishers.

Socolow, R.H., Andrews, C. and Berkhout, F., 1994, *Industrial Ecology and Global Change*. Cambridge, UK: Cambridge University Press.

Suh, S., 2004, Materials and energy flows in industry and ecosystem networks: LCA, IO analysis, materials flow analysis, ecological network flow analysis, and their combinations for industrial ecology. PhD thesis, Leiden University.

Tibbs, H., 1993, *Industrial Ecology: An Environmental Agenda for Industry*. Emmeryville, CA: Global Business Network.

Tillman, A., Ekvall, T., Baumann, H. and Rydberg, T., 1994, Choice of system boundaries in life cycle assessment. *Journal of Cleaner Production* **2**(1), 21–29.

Wernick, I.K., Herman, R., Govind, S. and Ausubel, J.H., 1996, Materialization and dematerialization: measures and trends. *Daedalus* **125**(3), 171–198.

2

Product Design for the Environment: The Life Cycle Perspective and a Methodological Framework for the Design Process*

Fabio Giudice

CONTENTS

---

* The main contents of this chapter were previously published in a book by F. Giudice, G. La Rosa, and A. Risitano, entitled *Product Design for the Environment: A Life Cycle Approach* (CRC/ Taylor & Francis, 2006).

2.1   Introduction

In recent years, considerable innovation has gone into product design and management. The aim of innovation is to reduce time taken and resources used in design, production, distribution, and disposal of products with elevated and diverse performance requirements. Methodological approaches have evolved to enable designers harmonize the various specifications of functionality, safety, quality, reliability, and cost, aimed at achieving a broader spectrum of performances.

Environmental awareness is another wave that has simultaneously swept the production process. Over the past decades, this has resulted in strategies to promote environment-friendly production, integrating environmental concerns with product standards. In product development, these new requirements involve a shift away from a conventional approach to an innovative approach. Specifically, it means considerations beyond the sale of the product to the end of the product's useful life and to its retirement.

Thus, environmental requirements must lead to innovations toward a successful and "sustainable" product design. A design approach directed at the systematic reduction or elimination of the environmental impacts implicated in the life cycle of a product, from the extraction of raw materials to product disposal, can help. This methodology, known as design for environment, is based on evaluating the potential impacts throughout the design process.

The new design challenges require a systematic, integrated, and simultaneous intervention into a product and its correlated processes. This is based on the new methods known as concurrent engineering (CE) and design for X (DFX). These methodological approaches, although starting from different premises, tend to embrace the life cycle approach. It takes a holistic view of the product and its life cycle, where the latter is no longer seen as a series of independent technological processes, but as a complex system set in its environmental and sociotechnological context.

This view takes shape in life cycle design, a design intervention that uses tools and methods to reconcile the evolution of the product, from conception to retirement, with a wide range of design requirements (functional performance, ease of production, requirements of use, servicing and maintainability, and environmental aspects). All the phases of a product's life cycle (development, production, distribution, use, maintenance, disposal, and recovery) are considered in the context of the entire design process, from concept definition to detailed design.

2.2 Product Design and the Environment

Design essentially consists of molding material and energy flows to meet the needs of humankind. Therefore, the reach of this instrument clearly extends to the management of environmental concerns.

Analysis of factors influencing the environmental efficiency of industrial systems enables the identification of contexts most appropriate for a design intervention directed at environmental protection. In particular, the importance of process and product design on the efficiency of working, using, recollecting, and recycling of materials and other resources is emphasized. Due to its great potential, therefore, design is one of the most influential factors in the development of sustainable production systems and products.

2.2.1 Origin and Evolution

Despite a clear perception of the leading role design plays in resource transformation and consumption, a full understanding of its potential and responsibility toward the environmental question has been slow to arrive.

Although the necessary influence of socioecological systems on technical design was observed in the early 1960s (Asimow 1962), the transition from a "design for needs" to a "design for environment" occurred only in the early 1970s (Madge 1993). The environmental question and its revolutionary effects on the structure of conventional design were recognized.

In the field of design, there already existed an explicit reference to the potential of using biological systems as virtuous models for systems developed by humankind and to the consequent opportunities of reusing, repairing, and recycling artifacts (Papanek 1971). This idea of a possible ecological metaphor (the transposition of the organizational principles of ecological systems into economic–industrial systems) applied the basic concepts of industrial ecology that emerged later (Ayres 1989; Allenby 1992; Jelinski et al. 1992).

Starting out from different viewpoints, both economic and social, other authors also emphasized the design phase. This included optimizing production systems; promoting maximum well-being with the least possible consumption of resources (Schumacher 1973); and spreading a correct perception of the environmental question among consumers. These factors were fundamental to promoting an industrial production directed at limiting the obsolescence of products and at encouraging their recycling (O'Riordan 1976).

Over the following decade, these first initiatives to revise conventional design paradigms were consolidated. This second phase was perceived as a turning point from an industrial to a postindustrial design perspective (Cross 1981). Based on the nonsustainability of a development exclusively oriented toward economic expansion, these new design paradigms drew inspiration from alternative models of development and projected toward new concepts such as sane humane ecological future (Robertson 1980). As opposed to specialized industrial products with limited functionality and durability, postindustrial design focused on multifunctional products that were repairable and durable and designed to be socially responsive and ecosustainable. Conventional product requirements regarding functionality and cost were integrated with new requisites: energy efficiency, durability, recyclability, and appeal to consumers sensitive to environmental issues (Elkington et al. 1988). At the same time, the postindustrial design perspective emphasized that this extension in product requirements must not be seen as a disadvantage by the manufacturer: environmentally compatible products can not only be economically competitive but also innovative and particularly attractive to the consumer (Elkington and Burke 1987).

Thus, in the early 1990s, a comprehensive view of the effects of environmental issues on the design activity existed. This view covered diverse areas and was clarified by the results of the first experiences (Mackenzie 1991).

Considerations regarding the technical aspects of environment-friendly designs emerged in the first half of the 1980s (Overby 1979; Lund 1984). The early 1990s witnessed a phase of greater understanding of new needs to safeguard resources. This resulted in the clear objective of integrating environmental demands in traditional design procedures (Overby 1990;

Navin-Chandra 1991; OTA 1992) and culminated in design for environment (DFE), green design (GD), environmentally conscious design (ECD), and ecodesign (Ashley 1993; Allenby 1994; Dowie 1994; Fiksel 1996; Billatos and Basaly 1997; Zhang et al. 1997; Brezet and van Hemel 1997; Graedel and Allenby 1998; Gungor and Gupta 1999). This new approach aimed at minimizing the impact, on the environment, of products that were already in the design phase.

2.2.2 Design for Environment: Definition and Approach

The implementation of the major principles of environmental protection in industrial practice requires the direct involvement of the product design and development process, as a vector of dissemination and integration of the new environmental needs. DFE originated to facilitate this strategic role. Its definition, which initially was not univocal, has evolved over the last decade. First presented as a design approach for the reduction of industrial waste and the optimization of the use of materials (OTA 1992), DFE subsequently acquired a more appropriate dimension. It can be understood more completely as

> a design process that must be considered for conserving and reusing the earth's scarce resources; where energy and material consumption is optimized, minimal waste is generated and output waste streams from any process can be used as the raw materials (inputs) of another (Billatos and Basaly 1997).

Ultimately, DFE can be defined as a methodology for the systematic reduction or elimination of environmental impacts implicated in the life cycle of a product, from the extraction of raw materials to product disposal. This methodology is based on an evaluation of the potential impacts throughout the design process. In addition, DFE is characterized by two other aspects, as highlighted in Figure 2.1:

- Dual level of intervention, involving both products and processes
- Proactive intervention, based on the presupposition of the greater efficacy of intervening early in the product development process

A similar concept, green engineering design, brought these factors to design practice. It suggests a two-part approach (Navin-Chandra 1991):

- Evaluation of designs to assess their environmental compatibility using a spectrum of indices and measures (green indicators)
- Analysis of the relationship between design decisions and green indicators

Particular attention is given to the use of indicators quantifying the environmental benefit of design choices. These indicators make up a part of environmental

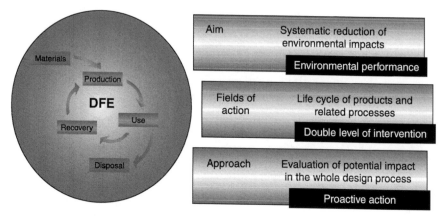

FIGURE 2.1
Objectives and characteristics of DFE.

metrics, which can generally be considered the algorithmic interpretations of levels of performance within an environmental criterion, that is, an attribute of the product found to be significant in determining the environmental performance of alternative product designs (Veroutis and Fava 1996).

Obviously, any type of design intervention directed at environmental protection cannot be separated from the requirements of product functionality, performance, reliability, quality, and cost. Having guaranteed these, environmental demands must evolve from simple constraints to new opportunities and incentives for innovation.

2.2.2.1 Reduction of the Environmental Impact and Design Approaches

The central theme unifying the various studies on DFE is reduction of the environmental impact of a product across its entire life cycle, from design to disposal (Coulter et al. 1995). The concept of "reduction of the environmental impact" is not, however, limited to the simple quantification and minimization of direct impacts on the ecosystem. Its implications are wider, extending to the optimization of the environmental performance, which includes the following aspects:

- Reduction of scrap and waste, allowing more efficient use of resources and decrease in the volumes of refuse; reduction in the impact associated with the management of waste materials
- Optimal management of materials, including the correct use of materials on the basis of the performance required, their recovery at the end of the product's life, and the reduction of toxic or polluting materials
- Optimization of production processes by the planning of processes that are energy efficient and result in limited emissions

- Improvement of the product, in particular its behavior during the phase of use, to reduce the consumption of resources or the need for additional resources during its operation

These objectives of making a product more environment-friendly can take form by following two different approaches to the problem on the basis of the circumstances they address. In the case of intervention into a preexisting production cycle, a first evaluation of the negative impact of the product's life cycle is followed by research into the interventions most effective in reducing that impact. These interventions primarily involve implementation of a higher level of technology.

In contrast, intervention during the development of a new product can focus on the complete sustainability of the life cycle, with the ideal objective of obtaining a product manufactured, used, and retired without giving rise to significant environmental impacts. This can be achieved through a design intervention, which has precisely the characteristics of DFE. Substantial benefits can only be obtained by taking into account the entire life cycle of a product.

2.2.2.2 Areas of Intervention: Product and Process Design

Although we will focus on product design and development, it should again be emphasized that, in general, DFE covers the design of both products and processes. For this reason, a distinction is frequently made between environmentally conscious product design and environmentally conscious process design. The latter is also known as environmentally conscious manufacturing (Zhang et al. 1997).

The influence of process design on the environmental question takes different forms and gives different results than those of product design. Processes are not confined to a specific application; the same processes can be part of the life cycles of many different products. They transform raw materials into products and therefore define the flows of all material resources as well as energy flows. They are largely responsible for the outflows from the industrial ecosystem they belong to. From this point of view, their optimization has greater strategic importance than that of the life cycle of a specific product. Further, once implemented in a production system, a process cannot generally undergo significant modification in the short term without heavy costs.

Product design has greater margins for flexibility. This advantage is counterbalanced by the much wider field of investigation required, which inevitably includes problems of process. In fact, an exhaustive analysis of the environmental performance of a product within its life cycle also includes the choice and the planning of the processes needed to perform each phase of the life cycle itself.

The industry–environment interaction is treated by two distinct groups of designers: those who design processes and those who design products.

DFE vision requires that these two groups direct their efforts in the same direction and harmonize the results obtained in their different areas of competence.

2.2.3   DFE Implementation and General Guidelines

DFE can be implemented in design practice through three successive phases. This applies to a product, a process, or each single flow of resources that is the subject of environmental improvement. The phases are as follows (Allenby 1994):

- Scoping involves defining the target of the intervention (product, process, or resource flow) and determining the depth of analysis.
- Data gathering means acquiring and evaluating significant environmental data.
- Data translation involves transforming the results from the preliminary data analysis into tools (from simple guidelines and design procedures to more sophisticated software systems assisting the design team to apply environmental data in the design process).

In practice, the second and third phases are implemented using two instrument typologies:

- Tools aiding the analysis of the life cycle (life cycle assessment and environmental accounting), allowing the acquisition, elaboration, and interpretation of environmental data
- Tools aiding the design or redesign (design for environmentally conscious manufacturing, use, end-of-life)

These tools are based on a wide-ranging series of suggestions and guidelines for the designer (OTA 1992; Fiksel 1996; Billatos and Basaly 1997), which can be summarized as follows:

- Reducing the use of materials, using recycled and recyclable materials, and reducing toxic or polluting materials
- Maximizing the number of replaceable or recyclable components
- Reducing emissions and waste in production processes
- Increasing energy efficiency in phases of production and use
- Increasing reliability and maintainability of the system
- Facilitating the exploitation of materials and the recovery of resources by planning the disassembly of components
- Extending the product's useful life
- Planning strategies for the recovery of resources at end-of-life; facilitating reuse, remanufacturing and recycling; and reducing waste

- Controlling and limiting the economic costs incurred by design interventions aimed at improving the environmental performance of the product
- Respecting current legal constraints and evaluating future regulations in preparation

Applying these general guidelines in relation to the main phases of the product's life cycle (from production to retirement), it is possible to obtain useful information and to explore the whole set of environmental opportunities for an eco-efficient intervention in the product design and development process.

2.3   Life Cycle Concept and Environmental Protection

As underlined in the previous section, the most significant benefits of DFE can only be obtained if the product's entire life cycle, including other phases together with those specific to development and production, is considered at the design stage. Products must be designed and developed in relation to all these phases, in accordance with a design intervention based on a life cycle approach. This approach is understood as a systemic approach "from the cradle to the grave," the only approach able to provide a complete environmental profile of products (Alting and Jorgensen 1993; Keoleian and Menerey 1993). Only this extended view can guarantee that the design intervention will identify the environmental criticalities of the product and reduce them efficiently without simply moving the impacts from one phase of the life cycle to another.

2.3.1   Life Cycle and the Product-System Concept

The concept of product life cycle has different connotations in different contexts. In the management of product development, it means the spectrum of phases from need recognition and design development to production. It may extend to any possible support services for the product but usually does not include the retirement and disposal phases.

This limited view of the life cycle has its origins in a statement of the problem conditioned by the competencies and direct interests of different actors involved in the life of manufactured goods. This view leads to a fragmentation of the life cycle according to the main actors: manufacturer (design, production, and distribution); consumer (use); and third actor, defined on the basis of the product typology (retirement and disposal). It is clear therefore that the managerial concept of "life cycle" springs from the interests of the manufacturer and does not usually include postproduct phases.

Interaction among all the actors involved influences the environmental performance of a product across its entire life cycle. Therefore, an effective approach to the environmental problem must be considered in the context of the entire society. It consists of a complex system of actors, including the government, manufacturers, consumers, and recyclers (Sun et al. 2003). This system is also characterized by complex dynamics since the various actors interact through the application of reciprocal pressures dependent on political, economic, and cultural factors (Young et al. 1997).

Therefore, from a comprehensive perspective not limited by a specific actor's point of view, the life cycle of a product is well represented by its main phases of need recognition, design development, production, distribution, use, and disposal, as suggested by other authors (Alting 1993). This perspective, in contrast with the limited view of the environmental question held by the single actor "manufacturer," imposes a sort of "social planner's view" (Heiskanen 2002).

The aspects considered above can be summarized in a holistic vision of the product and its life cycle. The latter is no longer thought of as a series of independent processes expressed exclusively by their technological aspects but rather as a complex system set in its environmental and sociotechnological contexts (Zust and Caduff 1997). It is then possible to speak of product-system, which in its most complete sense includes the product, understood as integral with its life cycle, within the environmental, social, and technological contexts in which the life cycle evolves. The environmental point of view places the actions of all actors involved in the product-system, in the context of the global ecosphere, which includes the biosphere (i.e., all living organisms) and the geosphere (all lands and waters).

2.3.2 Product Life Cycle and Environmental Impact

Analyzing the main aspects related to the concept of "reduction of the environmental impact" previously introduced (Section 2.2.2.1), it is clear that the environmental assessment of the product-system must be oriented toward a view of the life cycle of a product associated with its physical reality, excluding the conception and development phases, and focusing on the interaction between the ecosphere and all the processes involved in the product's life, from production to disposal. From this perspective, the product becomes "a transient embodiment of material and energy occurring in the course of material and energy process flows of the industrial system" (Frosch 1994), and the life cycle is understood as a set of activities or processes of transformation, each requiring an input of flows of resources (quantities of materials and energy) and generating an output of flows of by-products and emissions.

Finally, the environmental impact of the product-system is the consequence of the exchange of substances, materials, and energy between the life cycle processes and the ecosphere. The different environmental effects produced can be summarized in three main typologies (Guinée et al. 1993):

- *Depletion.* The impoverishment of resources, imputable to all the resources taken from the ecosphere and used as input in the product-system

- *Pollution*. All the phenomena of emission and waste, caused by the output of the product-system into the ecosphere
- *Disturbances*. All the phenomena of variation in environmental structures due to the interaction of the product-system with the ecosphere

In detail, the main impacts of life cycle can be summarized as follows:

- Consumption of material resources and saturation of waste disposal sites
- Consumption of energy resources and loss of energy content of products dumped as waste
- Combined direct and indirect emissions of the entire product-system

As regards the first aspect, the impact can be quantified by an analysis of the distribution of the volumes of material in question across the entire life cycle. The energy and emission aspects require a more complete approach that takes into account the energy and emission contents of the resources and of the final products.

Refer to the scheme of an elementary transformation process shown in Figure 2.2. Each typology of resource introduced (materials and energy) is characterized in terms of both energy and emission content, and a distinction is made between direct and indirect emissions.

The energy and emission contents of a material resource are, respectively, understood as follows:

- The energy expended to produce the material resource
- All the emissions correlated with its production

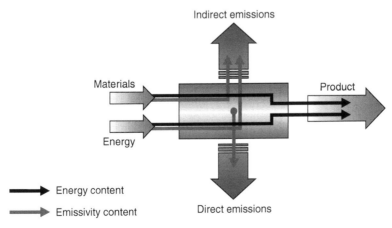

FIGURE 2.2
Scheme for the identification of the environmental impact of a process.

The energy and emission contents of an energy resource are, respectively, understood as follows:

- The sum of energy expended to produce this energy resource in the form in which it is used in the process
- The sum of emissions correlated with its production

The distinction between direct and indirect emissions is, respectively, understood as follows:

- The sum of characteristic emissions of the process itself (dependent on the materials, the type of process, and the product of this process)
- The sum of the emissions correlated with the production of the resources used by the process, corresponding therefore to the emission content of the resources

According to this statement, again referring to Figure 2.2, it is possible to say that

- the sum of the direct and indirect emissions quantifies the total emissivity that can be associated with the process and therefore with the final product, and
- the sum of the energy contents of the materials and of the energy consumed by the process quantifies the energy content of the final product and expresses the consumption of energy resources associable with it

The conceptual schematization proposed offers a detailed definition of the environmental impact of an elementary process; in theory, it is easily extended to the product-system by considering all the processes that make up its life cycle.

2.3.3   Product Life Cycle Modeling

Generally, in modeling a product, process, or system, the object to be represented is reduced to an abstraction. This reduction involves simplifying the functional mechanisms and limiting the information under consideration, with the aim of simulating its behavior and estimating a wide range of attributes (performance level, quality, reliability, cost etc.).

In the present context, the product must be understood as product-system, characterized by flows of resources transformed through the various processes making up the life cycle and by interactions with the ecosphere. The model of the life cycle must then be a fundamentally physical model and represent a system with accurately predefined boundaries. Everything that falls outside these boundaries constitutes the environment in which the system operates and with which it interacts through flows of resources, energy, and information. The product-system can be further broken down into subsystems and elementary activities that interact according to a functional structure to achieve the functionality of the original system.

With this general approach, "the task for life cycle modeling is to construct an appropriate framework in which the system architecture (hierarchy) and structure (connections) can be first represented and then evaluated consistently and rigorously" (Tipnis 1998).

Prediction and planning are possible only through adequate modeling of a product's life cycle. The modeling of a system generally tends to reduce its complexity, although there is a consequent loss of information. Such simplification, however, becomes necessary in environmental evaluations because of the high complexity of the real systems. This aspect is clearly highlighted in the ISO 14040 series of standards (International Standards Organization 1997), which explicitly use product-system modeling for the purposes of evaluating environmental impacts. They suggest some fundamental stratagems in the construction of the model such as breaking down the product-system into subsystems and defining elementary unit processes that perform specific functions and necessitate resource flows in input and produce flows in output.

With this clearly physical technical-based approach, the behavior of the model can be described and simulated using mathematical models of limited complexity. These models refer to the analysis of a system with static, linear behavior.

The complexity would be markedly greater if the life cycle were treated from a sociotechnological perspective, since in this case, the system would be characterized by dynamic, nonlinear behavior. As generally proposed in the literature, such considerations justify the choice of the physical-technological viewpoint in modeling the life cycle of the product-system (Vigon et al. 1993; Keoleian and Menerey 1993; Billatos and Basaly 1997; Hundal 2002).

2.3.3.1 Modeling by Elementary Activity

In modeling the life cycle with these premises , the entire product-system is subdivided into elementary functions (Zust and Caduff 1997; Hundal 2002). These are also represented by activity models (Navin-Chandra 1991) that summarize the elementary processes characterizing the main phases of the life cycle.

In other words, modeling by activity (activity modeling) consists of defining a set of single activities that make up a complex system. These activities can be the transformation, handling, generation, use, or disposal of material resources, energy, data, or information (Tipnis 1998).

Appropriate activity modeling first requires a clear definition of the primary objective, which is to be attained using the model, and of the initial viewpoint from which the model will be developed. In fact, both factors are necessary in defining the boundaries of the system to be modeled and in structuring the model, which must be broken down into subsystems, sequences, and operating units and processes, in relation to the aims and the viewpoint.

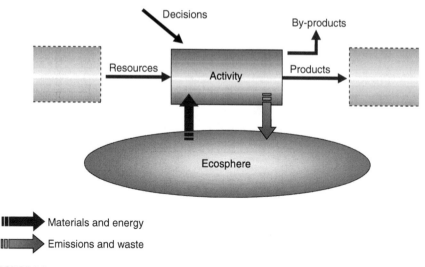

FIGURE 2.3
Reference model for elementary activity.

Having defined these factors on the basis of the environmental requisites, it is possible now to apply activity modeling to the product-system and its life cycle. The reference activity model is therefore of the type shown in Figure 2.3. It is characterized by input flows of physical resources, output flows, and a possible input flow of information when there is a margin of choice on how the activity is to be performed.

For the input flows, given that they are physical resources and can consist of materials and forms of energy, it is possible to distinguish between resources produced by preceding activities and resources coming directly from the ecosphere.

For the output flows, consisting of products of the activity, it is possible to distinguish among true main products, secondary by-products, and various types of emissions into the ecosphere. Having defined the reference model for elementary activity, the product's life cycle is now translated into a system model by the following procedure (Zust and Caduff 1997):

- Defining the boundaries of the system
- Identifying the elementary processes and functionalities
- Identifying and quantifying the connections between elementary activities
- Evaluating any possible changes in the activities and connections over time

Using the system model developed through activity, it is possible to perform simulations of the life cycle and to interpret the results of these simulations.

The reference model of Figure 2.3 can be read in different ways, based on the kind of environmental evaluation to be undertaken.

As introduced in Section 2.3.2, in terms of the physical-chemical exchanges of technological processes with the ecosphere, a product's environmental impact can be principally expressed as

- consumption of material resources and saturation of waste disposal sites;
- consumption of energy resources and loss of energy content of products disposed off as waste; and
- combined direct and indirect emissions of the entire product-system.

In the first case, a quantification of the impact can be based on an analysis of the distribution of the volumes of materials under consideration, in the context of the entire life cycle. The energy and emission aspects instead require a more complete approach, which considers the energy and emission contents of the resources and final products. For a complete environmental analysis, therefore, the reference model for elementary activity can be read as in Figure 2.2.

However, when the aim is to develop a life cycle model that supports only the analysis of the material resources under consideration, a more simplified reading of reference activity is required, such as that represented in Figure 2.4.

This representation takes only the flows of material into account. The resources fueling the activity constitute the input, and the product of the activity and any possible discards and waste make up the output. The input is of the following types:

- Primary (virgin) resources, coming directly from the ecosphere
- Secondary (recycled) resources

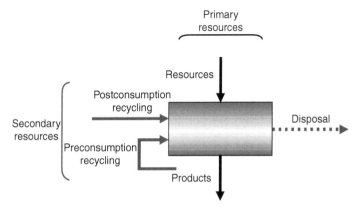

FIGURE 2.4
Activity model for analysis of material resources flows.

The latter can in turn be divided into

- preconsumption secondary resources, that is, originating from discards and waste generated by the activity itself and
- postconsumption secondary resources, that is, originating from the recycled product after its use and retirement.

2.3.3.2 Reference Model for Product Life Cycle

The various considerations introduced in the preceding sections, particularly those related to product life cycle, the appropriateness of considering the physical life cycle in environmental analysis, and the basic principles of modeling for elementary activities, are interpreted by the general life cycle model introduced below. It can be considered a reference model. All the processes of transformation of resources involved in the product's entire physical life cycle are grouped according to the following main phases (Manzini and Vezzoli 1998; Sánchez 1998):

- Preproduction, where materials and semifinished pieces are prepared for the production of components
- Production, involving the transformation of materials, production of components, product assembly, and finishing
- Distribution, comprising the packing and transport of the finished product
- Use, including the use of the product for its intended function and any possible servicing operations also
- Retirement, corresponding to the end of the product's useful life, can consist of various options, from product reuse to disposal as waste.

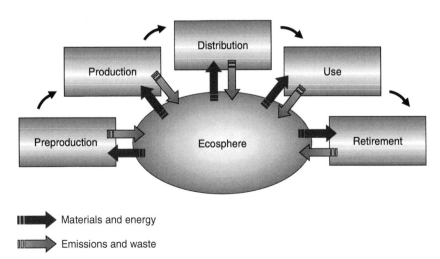

FIGURE 2.5
Phases of physical life cycle and their interactions with the ecosphere.

Each of these phases interacts with the ecosphere, since it is fueled by input flows of material and energy and produces not only by-products or intermediate products that fuel the successive phase but also emissions and waste (Figure 2.5).

By developing each main phase according to the different primary activities the phase encompasses, it is possible to obtain a vision of a product's entire physical life cycle and of the resource flows that characterize it. In Figure 2.6 the flows of material resources are shown based on the simplified activity model of Figure 2.4. The first phase of preproduction consists of the production of materials and semifinished pieces required for the subsequent

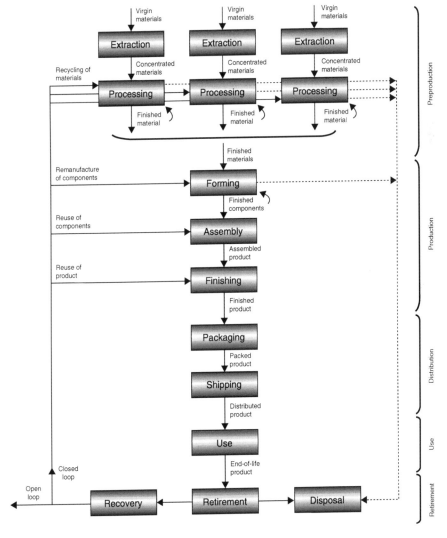

FIGURE 2.6
Complete physical life cycle of product and flows of material resources.

production of components. It includes, therefore, the production phases of all the materials, which will result in the final product. After the product is manufactured, distributed, and used, it arrives at the final phase of retirement and disposal.

Figure 2.6 shows the division of all these phases according to their primary activities. It provides an overview of all the waste flows generated during the cycle, which in the model proposed, are principally due to the phases of processing the various materials and of forming the components, together with the disposal of the product.

Figure 2.6 also offers a complete picture of all the options for recovery of the product at the end of its life and of how the recovery flows can be distributed within the same life cycle that generated them, providing the post-consumption secondary resources for various activities. Alternatively, the flows can be directed outside the cycle. In fact, it is necessary to distinguish between the two typologies of recovery flows:

- *Internal recovery (closed loop)*. The resources recovered reenter the life cycle of the same product, which generated the flows, replacing the input of virgin resources. This can occur by directly reusing the product at the end of its useful life, reusing some parts, or reusing other parts after appropriate reprocessing (remanufacturing), or by recycling materials. From the viewpoint of the environmental consequences, these recovery processes lead to an increase in the expenditures and emissions for the treatment and possible transport of these volumes before they reenter in the cycle. They also lead to a decrease in the consumption of materials in general, due to the partial reduction in the input of virgin materials and a reduction in the volumes disposed off as waste.

- *External recovery (open loop)*. At the end of the product's life, some of its parts are directed to the production processes of other materials or products external to the cycle under examination. This can result in recovering part of the energy content of materials to be eliminated, saving virgin materials in other production cycles, and obtaining financial benefits through the sale of materials for recycling.

2.4   Integrated Product Development Based on the Life Cycle Approach

DFE must act as a bridge between production planning and development on one hand, and the environmental management of the production itself on the other. To fulfill this role, the design activity must generate a statement based on some key features: a product life cycle orientation, the balancing of a wide range of requirements, and the simultaneous and integrated structure of the design intervention. Only on the basis of these premises it is possible to conceive a process of product development that furthers the sustainability of the product life cycle. The objective is to obtain a product whose manufacture, use, and disposal have the least environmental impact.

Today, while many environmental aspects of industrial production are the subject of wide-ranging discussions, manufacturing companies continue to encounter difficulty in achieving an environmentally sustainable production. One crucial problem is that the principles and the methods for the environment-friendly design of products have not yet been integrated into design and managerial practice (Gutowski et al. 2005). Therefore, the success factors in product design remain limited to those of quality and development costs, that is, to factors that have an impact on the business environment.

2.4.1 Role of Life Cycle Approach Instruments and Techniques

The life cycle approach can provide a qualitative leap in product development, "making the product fit its natural environment as much as it fits the business environment" (Krishnan and Ulrich 2001). This affirmation is based on observations of factors obstructing environmentally oriented product development in the practice of manufacturing companies (Ries et al. 1999). These factors include

- poor understanding of the environmental impacts of products;
- cost-oriented statement of the product development process; and
- lack of a homogenous and efficient distribution within the context of the entire development process, of the approach directed at the environmental requirements of products.

The first factor is historically linked to the need of producers to address principally those aspects regarding the impact on production sites (consumption of resources and generation of emissions and waste) not directly attributable to products and limited to the context of the production phase alone. The result has been a lack of primary information that could serve as the basis for a strategy aimed at improving the environmental quality of products life cycle. This problem can be resolved by adopting the techniques used in life cycle assessment (LCA). It is a well-known method of analysis, which enables quantification of the environmental effects associated with a process by means of the identification and quantification of the resources used and the emissions and waste generated as well as the assessment of the impact caused by the use of these resources and the emissions produced (Fava et al. 1991; Rebitzer et al. 2004). This typical instrument of the life cycle approach to the environmental question can be particularly useful in its simplified form (streamlined LCA). This form overcomes the disadvantages of very detailed analysis to be undertaken in the preliminary phases of product development (Todd and Curran 1999).

The second factor originates from a dated, "defensive" approach to the environmental question. This approach views the environment as a restrictive and, generally, troublesome constraint without appreciating its potential value. This factor becomes particularly significant when one considers the weight that the functions of cost planning and marketing have in the product development process, and hence the obstacle that the deficiency of an accurate economic analysis and a perception of a product's "environmental value" can constitute for the diffusion of ecocompatible design. Also, in this case, life cycle-oriented techniques can come to the rescue. These techniques include life cycle cost analysis (LCCA) (Fabrycky and Blanchard 1991; Asiedu and Gu 1998) and environmental accounting, together with the other techniques integrating economic and environmental analyses of the life cycle (Gale and Stokoe 2001; Kumaran et al. 2001).

The absence of a complete distribution, within the entire development process, of a homogenous, environmentally oriented approach, is one of the crucial factors in terms of achieving a complete and affective environmentally-oriented approach to design. It has often been observed that this problem principally occurs in the preliminary phases of product development (Bhamra et al. 1999; Ries et al. 1999), where there is a scarcity of methods and tools oriented toward environmental aspects. Anyway it should be noted more generally that design practice lacks an organic approach to environmental aspects in the entire development process despite the desirability of such an approach at the theoretical level.

In this case also, the life cycle approach represented by life cycle design (LCD) or life cycle engineering (Alting 1993; Ishii 1995; Molina et al. 1998; Wanyama et al. 2003) in the design dimension, can constitute an effective basis for the integration of environmental aspects into product development. According to this novel approach, design choices are not to be guided exclusively by functional requirements and by considerations of cost effectiveness. Designers must also take into consideration an ever broader spectrum of features requested to the products with regard to each phase of their whole life cycle (production, distribution, use, maintenance, and disposal).

2.4.2 Basic Outline to Implement an Integrated Product Development Process

When LCD is expressly oriented toward environmental requirements (Keoleian and Menerey 1993), it can become an example of an environmentally oriented approach to the design process, and provides a reference methodological statement to achieve the complete and integrated distribution of environmental aspects within the development process.

The specification of the design objectives and strategies plays a crucial role in this respect. DFX can also play a vital role. It provides the tools and techniques for a design directed at specific product requisites, so that environmental requirements can also be included among the other requirements (the role of DFX approach with regard to the environmental aspects in product design will be presented in detail in Section 2.7).

The proposed analysis is summarized in Figure 2.7. It shows the instruments with which the life cycle approach can help overcome the factors impeding the implementation of environmentally oriented product development in company practice. The same figure shows another important obstructive factor, the cross-functional character of both design practice and environmental aspects. This factor is linked to the multidisciplinary nature of the competencies required and to the transversal nature of the correlated activities, with respect to the principal functions of the company (design, production, and marketing). This issue is well known in the context of organization and planning of the product design and development process, and CE was conceived precisely to face these necessities in design practice (Kusiak and Wang 1993; Prasad 1996). By implementing the organizational structures of CE, environmental aspects can be integrated into product development.

To sum up, and again referring to Figure 2.7, the full integration of environmental aspects into product development must occur at two different and complementary levels:

- *External integration.* Concerns the relationship between the product development process and the factors external to the design team, which must be taken into consideration, that is, customer and market demands, production constraints, and environmental requirements. This integration, as shown in Figure 2.7, is obtained by adopting the life cycle approach and using its tools.

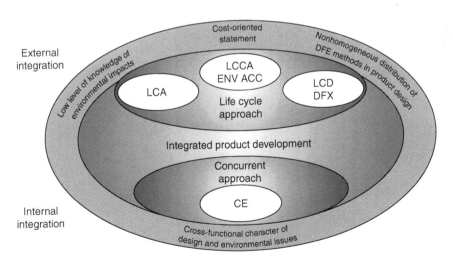

FIGURE 2.7
Implementation of integrated product development oriented toward environmental aspects.

- *Internal integration.* Concerns the relationship between the internal functions and competencies of the design team. This integration, necessary to best manage the cross-functional character of design practice and of the environmental aspects, is obtained through the simultaneous and concurrent approach to product development.

Having achieved the integration on this dual level, it is finally possible to speak of integrated product development (IPD), understood in its most complete sense and including environmental aspects. IPD can resemble the general concept of improving the design solution in response to consumer demands and market opportunities (Wang 1997). The life cycle approach extends this perspective, placing alongside the needs of the consumer, those of all the other actors involved in the various phases of the product's life cycle (Prudhomme et al. 2003). The further extension of the concept underlying IPD, to include a response to the needs of the environment, thus constitutes the fundamental premise for achieving an integrated product design, which also takes into account environmental requirements.

2.5 Orientation Toward the Environmental Requirements in the Product Design and Development Process

When we consider the physical dimension of a concrete industrial product with engineering content, the design activity is a process of transforming resources (cognitive, human, economic, and material). This transformation is aimed at translating a set of functionality requirements into the description of a physical solution (product or system) satisfying these requirements.

Although the terms "product design" and "product development" are sometimes considered interchangeable, they are commonly used as complementary terms, giving rise to the expression "product design and development." This evidences a possible distinction between the specific activity of design, as explained earlier, and a more extended activity. Although this activity includes design, it encompasses a wider arena, beginning with the identification of a need or market opportunity and concluding with the start of product manufacture.

2.5.1 Product Design and Development Process

The process of product development involves a sequence of activities that must be performed to ideate, design, and introduce a product into the market (Ulrich and Eppinger 2000).

Although no single model exists for this process, which can define a schematic pathway common to the vast typology of possible applications, it is possible to identify the main elements the great variety of possible product development processes have in common. The identification and understanding of these shared factors can enable a descriptive summarization of the main activities involved and a reference modeling for the comprehension and management of the entire process.

To describe the product development process, we take recourse to the models available in the literature. According to a traditional viewpoint, the product development process is an essentially sequential process and can be summarized into the main stages as shown in Figure 2.8, combining the suggestions of several authors (Dieter 2000; Ulrich and Eppinger 2000; Ullman 2003):

- *Need identification.* This phase consists of acquiring information on the needs of the customer, identifying the competing products in the market, and evaluating the most appropriate strategy (improvement of a preexisting product or development of new technologies). To underline the close ties this phase has with knowledge of the market and the opportunities afforded by technological innovation, the parallel activities of market analysis and research and development are also included in the figure.

- *Project definition.* This is the phase where the project is approved, and it constitutes the true and proper beginning of product development. It summarizes the company strategies, market reality, and technological developments in what is called the "project mission statement," which describes the market goal of the product, company objectives, and main constraints of the project.

- *Development process planning.* This phase involves planning the entire design and development process through the decomposition, planning, and distribution of activities, the definition and distribution of resources (temporal, financial, and human), and the acquisition and distribution of information.

- *Product design.* This phase includes the design-related activities, from the definition of product requirements and concept generation, to the translation of the latter into a producible system. This phase is in turn divided into a subprocess, the design process, which will be discussed later.

- *Postdesign planning.* This generic title is used to indicate the specific phase regarding the planning of the production-consumption cycle. For some authors, this planning is limited solely to production needs. In this case, it involves the definition and complete planning of product manufacture, from the sequence of machining the components, the preparation of tools and machinery, to planning the assembly. Some authors add to production planning, the necessities of distribution, use, and retirement of the product (Dieter 2000).

- *Prototyping and testing.* This phase requires the development of product prototypes, which are then tested to evaluate how well the proposed solution satisfies the prescribed requisites. The performance levels and the reliability of this solution are also noted. Clearly, this phase has a preeminent influence on interventions to improve the product and on the evolution of the design solution.

- *Production ramp-up.* This phase of starting up production completes the product development process. It involves manufacturing the product using the planned production system (this is not the case in the prototyping phase, where the product is created in another way). The aim is to verify the suitability of the real production process, to resolve any problems, and to identify any remaining defects in the final product. Ramp-up is followed by a transitional phase toward true and proper production and the definitive launch of the product into the market.

As shown in Figure 2.8, feedback processes guide the improvement of the final product. These feedback processes transmit information across the entire development process, from the postdesign planning and production phases, to the product design phase (and, if necessary, also to the preceding phase of development process planning). This mechanism for improving the final result, based on feedback assessments and corrections, has its origins in the first studies theorizing on the design process (Asimow 1962) and has always been considered its engine (Dieter 2000). It interprets the role of true and proper evolutionary mechanisms, leading to the final solution.

Once distributed and marketed, the product comes to the final user. This actor is closely linked to the initial phase of the product development process, that of need identification, because the user interacts with the market and technological innovation, influencing and being influenced by them.

2.5.1.1 *Product Design Subprocess*

As shown in Figure 2.9, within the product development process, the design phase is divided into a subprocess (the product design process), which transforms the set of functional specifications and product requirements into the detailed description of the constructional system interpreting them. This transformation is achieved through the use of various types of resources (cognitive, human, economic, and material), which serve to fuel the main phases of the design subprocess. These phases consist of: (i) a preliminary phase of specifying the problem and defining the product requirements; (ii) phases of design at different levels (concept, system, detail); and (iii) phases of assessment, making up iterative cycles of analysis–synthesis–evaluation (Pahl and Beitz 1996; Dieter 2000, Ulrich and Eppinger 2000; Ullman 2003).

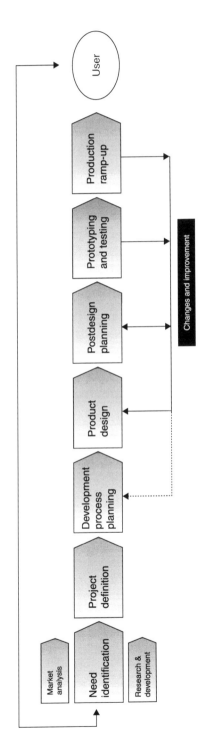

FIGURE 2.8
Product development process.

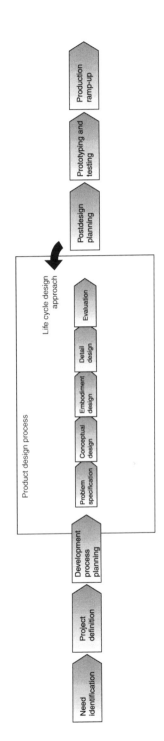

FIGURE 2.9
Product design phases in the development process.

The first four phases are considered in greater detail:

- *Problem specification.* In this phase, all the information related to the project in question is elaborated to define the requisites that must characterize the product. Information describing the needs to be satisfied, consumer requirements, market conditions, and company strategies must be clarified, if necessary, integrated and used to generate the specifications guiding the design phases.

- *Conceptual design.* Having defined the project specifications, it is necessary to develop ideas for the creation of a product with the desired requisites. In this phase, product is understood in the abstract sense, as a set of attributes, which must be embodied in the product concept. The embodiment of attributes in the product concept is achieved through a first step of generating ideas (concept generation) and the second step of assessment and selection (concept evaluation) (Dieter 2000; Ullman 2003).

- *Embodiment design.* Having identified the most appropriate concept, the next phase consists of a preliminary interpretation of the design idea in a physical system. The concepts formulated in the previous phases are developed, their feasibility is verified, and finally they are translated into a general product layout, defining subsystems and functional components. In this phase, the physical elements are combined to achieve the required functionality, and the product architecture thus takes shape. This phase also involves a preliminary study of the shape of the components and a first selection of materials.

- *Detail design.* The layout developed in the previous phases must be translated into geometric models and detailed designs. This requires the application of methods and tools, aiding a correct definition of the design details. The choice of materials, study of the shapes, definition of the geometry of components and assemblies, the development of the assembly sequences, and definition of the junction systems must all be guided by the entire range of product requirements (performance, economic, environmental, etc.). To complete this phase, some authors provide for the comprehensive planning of the production process (Ulrich and Eppinger 2000), while others suggest including instructions for production, assembly, shipping, and use in the final documentation (Pahl and Beitz 1996).

The subsequent evaluation phase has the purpose of evaluating the degree to which the proposed solution corresponds to the design specifications defined in the problem specification phase and of guiding modifications and improvements that can make the process evolve toward the definitive solution, that is, the product that best satisfies the desired requisites. With this aim, it is necessary to undertake analyses of the critical aspects of the design; studies to predict the way in which the chosen solution will behave over time in relation to environmental factors (socioeconomic conditions, consumer tastes, competing offers, and availability of raw materials) and technological factors (technological progress and deterioration in performance) and verification programs (modeling and initial prototyping).

2.5.2 Integration of the Environmental Aspects

Referring to the entire product design and development process reported in Figure 2.9, it is possible to say that the full and homogeneous integration of environmental aspects is obtained through a series of interventions, varying according to the phases of the development process:

- In the preliminary phases (project definition, development process planning, and problem specification), this integration is achieved through the extension of the factors conditioning the preliminary structuring of the project and the definition of product specifications and requisites. These aspects, together with consumer requirements and market opportunities, will also include environmental necessities, and the latter are given their due weight in defining company policies and strategies. A set of data, not exclusively environmental, regarding the expected life cycle of the product is added to the information input of the design process.

- The statement of the design-related phases (i.e., those making up the product design subprocess) must be guided by appropriate strategies of approach toward the environmental aspects of the product's life cycle. This particular consideration will be analyzed in detail later.

- In the main phases of the design process, beginning with conceptual design, and particularly the phases of embodiment and detail design, the statement of the design intervention must be directed at harmonizing the ever wider range of design requirements, as envisaged by LCD. On the basis of this statement, the various specifications can be achieved using the DFX tools, each addressing a specific typology of product requisite and giving appropriate emphasis to those requisites oriented toward environmental requirements (this will be introduced in Section 2.7).

- The postdesign planning, which in the general scheme of the product development process follows the product design phases, must be integrated with them, as is established in the concurrent approach to design. This integration must be performed according to the presuppositions introduced in Section 2.5.1, that is, extension of postdesign planning to cover the entire life cycle, including the production, distribution, use, and retirement of the product. It is precisely in relation to the planning of the production-consumption-disposal cycle that the most appropriate DFX tools are introduced.

2.6   Implementing the Environmental
Strategies for the Product Life Cycle

Design strategies play an essential role in the life cycle approach. They enable translation of the environmental requisites demanded of the product into design practice. It should therefore be underlined that the environmental strategies most appropriate and effective for a specific design problem must be carefully chosen only after the objectives of the project have been accurately translated into product requirements (Keoleian and Menerey 1993).

In general, strategies oriented toward the environmental efficiency of the life cycle can be defined on the basis of the main aspects of the product's impact on the environment. These impacts are ascribable to exchanges with the ecosphere of the physical–chemical flows involved in the technological processes making up the life cycle (see Section 2.3.2). Numerous environmental strategies are directed at reducing this wide spectrum of impacts (Keoleian and Menerey 1993; Hanssen 1995; Fiksel 1996; Bhander et al. 2003). These strategies can be distinguished on the basis of the phase of the life cycle in which they are intended to intervene, as shown in Table 2.1. This table reports some important environmental strategies.

2.6.1   Strategies for Environmental Efficiency

Given that the environmental efficiency of a product directly depends on its design, it is of fundamental importance that any strategy that is to be followed should take into account the main design parameters (Whitmer et al. 1995). However, not all the strategies reported in Table 2.1 are true and proper design strategies. In fact, some of the strategies involve interventions that are not directly linked to design choices.

By summarizing the various strategies presented in the table, we can conclude that a design intervention intended to take account of a product's behavior, in environmental terms, during its life cycle must aim to optimize the distribution of the flows of resources and minimize the emissions. In other words, the design intervention must

- reduce the volumes of materials used and extend their life span;
- close the cycles of resource flows through recovery interventions; and
- minimize the emissions and energy consumption in production, use, and disposal.

To fully achieve these conditions, it is necessary to intervene in the two separate areas of product design and process design (see Section 2.2.2.2). The latter is not directly relevant to the objectives of this treatment, although it is of primary importance and often considered to be intimately linked to the product development process. Here, in fact, attention is focused more on the design of the product understood as a set of material components designed in such a way that they constitute a functional system which satisfies certain

TABLE 2.1

Environmental Strategies and Life Cycle Phases

Life Cycle Phases	Environmental Strategies
Preproduction	Reducing the use of raw materials
	Choosing plentiful raw materials
	Reducing toxic substances
	Increasing the energy efficiency of processes
	Reducing discards and waste
	Increasing flows of recovery and recycling
Production	Reducing the intensive use of materials
	Using materials with low impact
	Reducing the use of toxic materials
	Using recycled and recyclable materials
	Using materials on the basis of their required duration
	Selecting processes with low impact and high energy efficiency
	Selecting processes with high technological efficiency
	Reducing discards and waste
Distribution	Planning the most energy-efficient shipping
	Reducing the emissions of transport
	Using containment systems for toxic or dangerous materials
	Reducing packaging
	Using packaging with low environmental impact
	Reusing packaging
Use	Using products under the intended conditions
	Planning and execution of servicing interventions (diagnostics, maintenance, repair)
	Reducing energy consumption and emissions during use
Retirement	Facilitating product disassembly at end-of-life
	Analyzing the condition of materials and their residual life
	Planning the recovery of components at end of use
	Planning material recycling at end of use
	Reducing volumes for disposal

requisites. This is the product-entity dimension that is directly linked to the choices made in the design-related phases of the development process (conceptual, embodiment, detail design), whose parameters are ascribable to the product's physical dimension: materials, shape and geometrical properties of the components, system architecture, interconnections, and junctions.

The physical dimension of the product-entity is, in its life cycle, expressed by the flows of material resources. This, therefore, leads us back to the first of the three main aspects of a product's impact on the environment, that of the employment and consumption of material resources. Figure 2.6 gives an overview of the product life cycle and the resource flows characterizing it.

This partial view of the environmental problem may seem limited, but in reality it is very wide ranging. The only possibility that is completely ignored is the intervention into the various technological processes constituting the life cycle. It does not exclude the possibility of taking into account the other two aspects of impact (energy consumption and product-system emissions) in environmental evaluations. The contributions of the energy and emission contents of the materials under consideration to the environmental impact, are clearly ascribable to the volumes of the material flows. The contributions due to the energy fueling the processes and the direct emissions from the processes to the impact can also be generally ascribed to the volumes processed, or to the specific process parameters dependent on the physical properties of the materials or on the geometrical properties of components. Thus, these impacts can be managed through the choices of product design. The design choices, on the basis of the definition of the materials and of the main geometric parameters, also condition the choice of the processes and the way in which they are performed.

The environmental performance of the life cycle can be improved by applying two main types of strategy. These strategies focus on the material flows, and therefore on the physical dimension of the product-entity. Summarized in Figure 2.10, the two types of strategies are as follows:

- Useful life extension strategies, directed at extending the product's useful life and thereby conferring increased value on the materials used and on all the other resources employed in its manufacture (product maintenance, repair, upgrading, and adaptation)

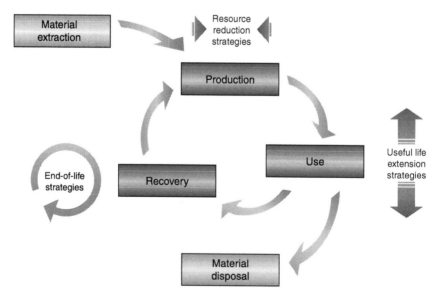

FIGURE 2.10
Environmental strategies for the life cycle.

- End-of-life strategies, directed at recovering the product at the end of its useful life, closing the cycle of materials, and recovering, at least in part, the resources used in its manufacture (reusing systems and components and recycling materials in the primary production cycle or in external cycles).

These strategies must be taken into consideration during the design phase, but they do not come into effect until the product has been manufactured. As shown in Figure 2.10, however, a third, important type of environmental strategy, known as resource reduction strategies, becomes effective during the production phase. These strategies, which are again associated with the product's material dimension, include all the interventions and choices that would reduce the consumption of material and energy resources to be used in the manufacture of the product. Thus, in general terms, these strategies constitute a wide spectrum of expedients that cover not only product design but also production process planning. They may also include radical strategies such as "dematerialization." This strategy deals with thinking of products as "means to perform functions," and to achieve the corresponding economic advantage. From this point of view, products could be replaced by services that meet the same needs requested to products themselves. By substituting a product with a service, the product is "dematerialized," because its functions are performed by the service (which is immaterial), instead of material entities (exactly as common products are). This allows to reduce the quantity of materials necessary to achieve an economic function (Wernick et al. 1997), promote the evolution from the sale of products to the sale of services, and therefore more properly collocate in the realm of business strategies.

The first two typologies of environmental strategies, useful life extension and end-of-life recovery, favor an increased intensity of resource use in the manufacture of the product, thereby improving resource exploitation. In contrast, some strategies aim at reducing the resources used in production.

In the sections that follow, attention will be focused on the two strategies for improving resources exploitation, which will be described briefly, and correlated to main factors which condition their real practicability. Subsequently, it will be shown how these strategies can be implemented in a methodological framework for product design, outlining the full integration of environmental aspects.

2.6.1.1 *Useful Life Extension Strategies*

Extending the product's useful life, that is, the period of time over which the product is used while guaranteeing the required operating standards, results in saving energy and material resources upstream and a reduction in waste downstream of the use phase. With this intervention, it is possible to satisfy the same demand with fewer product units. A product's useful life may be extended through four intervention typologies:

- *Maintenance.* Includes periodic and preventative checking operations as well as monitoring and diagnostic interventions and programmed substitution of parts subject to wear; maintenance also includes ordinary cleaning operations

- *Repair.* Essentially consists of the removal and substitution of damaged parts to reestablish the operational condition and level of performance required of the product
- *Upgrade and adaptation.* They are similar strategies, in that both are motivated by the phenomena of technological and cultural obsolescence and by changes in the conditions of the working environment and in the exigencies of the user. They differ in intervention typology since upgrading involves the substitution or addition of components, while adaptation involves a reconfiguration of the main components of the product.

2.6.1.2 End-of-Life Strategies

Recovery interventions at the end of the product's useful life cause closure of the life cycle, as shown in Figure 2.10. The consequent environmental benefits are a decrease in the raw materials entering the cycle, because they are in part substituted by recovered resources; recovery of energy and material resources used in production, and therefore a better exploitation of their use; and a decrease in the waste flows.

As suggested by several authors (Dowie 1994; Ishii et al. 1994; Navin-Chandra 1994), the strategies for the recovery of resources at the end-of-life can be grouped according to their different recovery levels. In general, the three main recovery levels are direct reuse, reuse of parts, and recycling of materials. For each of these levels, there is a corresponding but different potential of environmental benefit, which depends on the level of incidence of the recovery flows in the life cycle (as is evident in the reference model of Figure 2.6):

- *Direct reuse.* At the end of use, the product can be directly reused, possibly after checking and repair, with consequent savings in energy consumption, any possible emissions, costs relative to the production and assembly of components, and the volumes of virgin materials consumed.

- *Reuse of parts.* Components that have not undergone excessive deterioration during use can be recovered, possibly after regeneration through intermediate processes, as components for reassembly. There will be consequent savings in energy, possible emissions, costs relative to the process of producing the parts, and the volumes of virgin materials consumed.

- *Recycling materials.* The materials of parts that cannot be reused can be recycled by the recovery processes included in the materials' own life cycles, or they can be treated and used in external production cycles to manufacture products with inferior characteristics.

2.6.2 Product Durability and Preliminary Identification of Optimal Strategies

The application of strategies for the extension of the useful life and for the recovery at end-of-life is, in general, conditioned by a wide range of factors, which determine its effectiveness (Rose et al. 1998; van Nes et al. 1999). The evaluation of these factors is essential for a correct implementation of these strategies in product development. External factors conditioning the life expectation of a product are of particular importance (Woodward 1997):

- *Functional life.* Period of time for which need for the product is predicted to last
- *Technological life.* Period of time, which ends when the technological obsolescence is so great that the product must be replaced by another product based on superior technology
- *Economic life.* Period of time, which ends when the economic obsolescence is such that the product must be replaced by another product characterized by analogous performance but costing less
- *Social and legal life.* Period of time, which ends when changes in the desires of the consumer or in normative standards require the product to be replaced.

All these external factors must be placed alongside a final, internal factor, physical life, that is, the period of time for which the product is expected to last physically, maintaining its functional performance. This is, therefore, the factor that is directly linked to the main design choices (architecture, materials, shapes, and geometrical properties).

In the design phase, the possibility of providing for the extension of the product's useful life and for the reuse of its parts depends precisely on the length of its physical life. This aspect in turn is strictly bound to the durability of its components (Giudice et al. 2003) and is understood as the capacity to maintain the required functional performance. However, this property should not be maximized indiscriminately, since, for example, in product sectors with a high level of technological innovation (and therefore with rapid obsolescence), excessive duration has a negative environmental value. It guarantees a useless extension of product and components' life, resulting in a greater consumption of resources in the production phase.

To improve resource exploitation by means of both types of environmental strategies previously introduced, a preliminary evaluation of external factors is necessary. These factors are linked to the market reality, regulatory standards, company policies, technological innovation, and esthetic–cultural conditioning and vary widely with product typology. Having quantified the main external factors, it is possible to identify the strategies most appropriate for varying the durability of the product and that of its components.

A series of significant evaluations can be made by comparing the physical life with the replacement life, defined as the period of time for which the product is effectively usable, comparable to the period of time it is present in the market up to its definitive replacement, and thus incorporating all the external conditioning considered earlier. Physical life represents the predicted duration of a product's full efficiency, its potential lifespan, while replacement life represents its effective lifespan, conditioned by factors such as technological and economic obsolescence and the other external factors.

On the basis of this comparison, it is first of all possible to draw a distinction between two types of useful life extension strategies:

- Maintenance, repair, and, more generally, service operations constitute strategies intervening in the physical life.
- Upgradation and adaptation of the product constitute strategies intervening in the replacement life.

The graph in Figure 2.11 shows that depending on the product typology the favorable conditions for extending the useful life of the product, through the two different types of strategy, can be identified. For the first type, these conditions correspond to a long replacement life, indicating the possibility that the product may be used for a long time, and a short physical life, revealing the contrasting brief duration of some components. Therefore, the

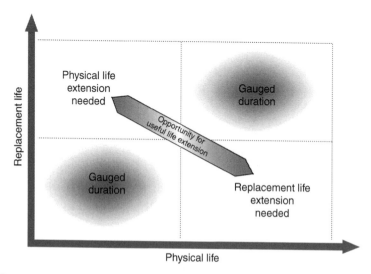

FIGURE 2.11
Preliminary identification of optimal strategies: extension of product useful life.

entire system possesses limited capacity to guarantee the required performance. For the second type, these conditions, conversely, correspond to a short replacement life and long physical life. These conditions not only indicate the inappropriateness of planning maintenance and service interventions pointlessly prolonging the product's life but also reveal a poor design, which is unsuited to the predicted short span of effective use (this is the case, for example, with overdimensioning, or using excessively high-performance materials). This highly ineffective situation can be remedied by the upgrading or adaptation of the product.

The other areas of the graph indicate conditions of equilibrium between the two factors, representing the result of good design, where the design choices were such that the physical duration of the system was calibrated on its expected, effective useful life. This particular condition is generally referred to as a condition of environmental efficiency, where the resources used in manufacturing the product are gauged on the basis of the effective exigencies, avoiding overdimensioning and consequent pointless wastage.

In an analogous way, the graph in Figure 2.12 shows the conditions favoring the different recovery strategies, from low-level recovery (recycling of materials) to a higher level recovery (reuse of the entire product):

- Area 1 represents the condition in which the product, whose useful life and presence in the market is expected to be short, is composed of rapidly deteriorating components. This represents a condition of

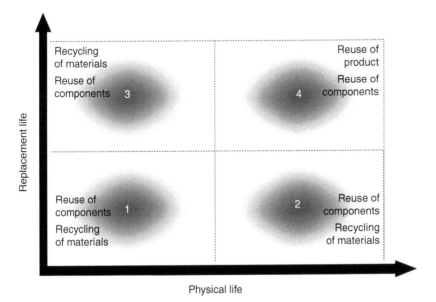

FIGURE 2.12
Preliminary identification of optimal strategies: recovery at product end-of-life.

gauged duration and is therefore, in principle, ecoefficient. At the end-of-life, any integral components are potentially reusable in other products, but the most probable recovery strategy is that of recycling the materials where possible.

- Area 2 represents the condition in which the product, whose useful life and presence in the market is again expected to be short, is composed of long-lasting and functionally efficient components. At the end-of-life, the product may still be fully efficient, but as a result of external factors cannot easily be reused because it is obsolete. Many of the components can potentially be reused as spare parts or in other products. The most probable recovery strategy is again the recycling of materials where possible.

- Area 3 represents the condition in which the product, whose useful life and market presence is expected to be long, is composed of rapidly deteriorating components, so that its performance is not long-lasting. At the end-of-life, components that are still efficient can be reused in the manufacture of a product of the same type. For the remaining components, only the recycling of materials is possible.

- Area 4 represents the condition in which the product, whose useful life and market presence is again expected to be long, is composed of long-lasting and functionally efficient components. Like area 1, this represents, in principle, an ecoefficient condition of gauged duration. If, at the end-of-life, the entire product is fully efficient and if the length of the replacement life is such to allow it, it is possible to directly reuse the entire product. This would, in theory, represent the most efficient strategy for environmental protection, unless the use of the product involved a significant environmental impact, which could be avoided by using a new, more efficient product instead. An alternative to direct reuse is to reuse some of the components of the product. Finally, it is always possible to resort to recycling of the materials.

2.6.3 Introduction of the Environmental Strategies into the Design Process

The environmental strategies for improving the life cycle of a product, introduced earlier and grouped according to the two typologies proposed, can lead to true and proper design strategies. These strategies can guide the designer in the choices that must be made at the different levels of design development. Table 2.2 summarizes some important design strategies, which are classified in relation to the main design parameters. The latter are distinguished according to whether they concern the system design (characteristics of the architecture, particularly layout and relationships between components) or the detailed design of components (materials, shape, and

TABLE 2.2

Design Parameters, Design Strategies, and Environmental Strategies

Design Level	Design Parameters	Design Strategies	Environmental Strategies					
			Useful Life Extension			End-of-Life Recovery		
			(ES1)	(ES2)	(ES3)	(ES4)	(ES5)	(ES6)
System	Layout	Minimize number of components	✓	✓			✓	✓
		Optimize modularity	✓	✓	✓		✓	✓
		Design multifunctional and upgradable components			✓	✓	✓	
		Plan accessibility to components	✓	✓			✓	✓
	Relations between components	Reduce number of connections	✓	✓	✓		✓	✓
		Reduce variety of connecting elements	✓	✓	✓		✓	✓
		Increase ease of disassembly	✓	✓	✓		✓	✓
Component	Materials	Reduce unsustainable and hazardous materials	✓	✓				✓
		Increase biodegradable and low-impact materials		✓				✓
		Reduce material variety						✓
		Increase material compatibility and recyclability						✓
		Specify and label materials						✓
	Shape	Optimize performance, resistance, and reliability	✓		✓	✓	✓	
		Design for easy removal	✓	✓	✓		✓	✓
	Dimensions	Reduce mass	✓	✓	✓		✓	
		Optimize performance, resistance, and reliability	✓		✓	✓	✓	
		Design for easy removal	✓	✓	✓		✓	✓

Note: ES1—maintenance; ES2—repair; ES3—upgrading and adaptation; ES4—direct reuse; ES5—reuse of parts; ES6—recycling materials.

geometric parameters). The table also shows the direct correlations between each design strategy proposed and the environmental strategies it can support.

This information makes it possible to outline a preliminary methodological statement, which would allow the integration of environmental aspects into design practice. The statement is schematized in Figure 2.13 and can be summarized as follows:

- Definition of the environmental requirements to be attained
- Choice of the environmental strategies most appropriate to the desired requisites
- Identification (by means of Table 2.2) of the design strategies that can help further the chosen environmental strategies
- Definition (by means of Table 2.2) of the design parameters to be used in interventions at the two design levels (system and component design).

The result of this preliminary statement, consisting of the overview of a set of design parameters on which to implement the environmental strategies and achieve the desired requisites, can be extremely useful in the management of conflicts between the strategies themselves. Often, design interventions directed at different environmental objectives are mutually opposing (Luttropp and Karlsson 2001). Moreover, this overview can constitute a first step toward clarifying the links between the choices directed at environmental aspects and those inspired by conventional design criteria.

2.6.4 Equilibrium between Conventional Criteria and the Approach to Environmental Efficiency

It is vital to achieve an equilibrium between necessities of different nature (performance, economic and environmental), for the complete and efficient integration of environmental aspects into design practice. This equilibrium is required because of the very nature of the environmental problem and of the design intervention oriented toward environmental necessities. Radical strategies, potentially conflicting with traditional product requisites (performance, producibility, and cost), are required to manage these necessities in a truly effective manner. The design strategies introduced earlier (Table 2.2) are based on a problem-focused approach to the design process, that is, an approach that initially concentrates on the problem to be resolved and subsequently arrives at the solution to the problem (Maffin 1998). This statement, frequently adopted in reference methodology frameworks for the product design process, requires considerable freedom in the preliminary phase of structuring the project, and in effect, it is only possible in the case in which there is the opportunity to ideate and develop a new and highly innovative product. More commonly, a product-oriented approach is usually adopted instead, that is, an approach that analyzes preexisting products and subsequently adapts them to any new necessities. This second type of approach is conditioned by experience acquired previously and expressed in practice through the use of general guidelines and rules of the thumb. This latter characteristic, in particular, evidences the inadequacy of a

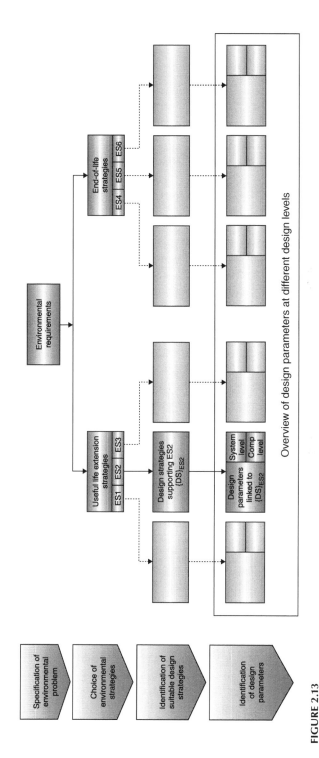

FIGURE 2.13
Introduction of environmental strategies into the design process: methodological statement.

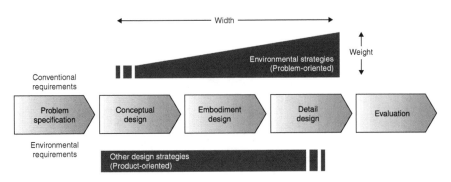

FIGURE 2.14
Equilibrium between product-oriented and problem-oriented approaches.

product-oriented approach as an aid in efficient environmental design intervention, that cannot easily be translated into effective guidelines. Generally, a more radical approach is required.

Environmental strategies can be reconciled with conventional product requirements by adopting a middle path between these two contrasting approaches. Figure 2.14 shows how a conciliatory approach can be implemented by operating on the weight given to the environmental strategies when applied to the design process. The environmental requirements must be clearly defined in the preliminary phase of problem specification, together with all the other product requisites. However, the application of the product-oriented approach, particularly in the early design phases, can guarantee the attainment of the conventional requirements, through the exercise of experience and established rules. The application of environmental strategies, requiring a problem-oriented approach, can be regulated according to the weight to be given to the environmental requirements. Regulation can be achieved by varying both the extent of the field of application within the design process and the weight given to these strategies in the various phases of design.

2.7   Role of Design for X with Regard to the Environmental Strategies

Recently, there has been a growing interest in DFX. This is particularly true for cases in which a generic industrial product must pass through all the phases making up its life and where design strongly influences the product's performance in each phase. In DFX, X stands for a different property of the product that characterizes it in relation to one or more phases of its life cycle (Gatenby and Foo 1990). DFX is an approach to product design that is directed at the maximization of the requisites demanded of the product (functionality and performance, manufacturability, quality, reliability, serviceability, safety, user friendliness, and environmental friendliness) and at the minimization of costs (Bralla 1996; Huang 1996). Since these requisites characterize the behavior of the product over its entire life cycle, these premises have led to the identification and classification of criteria for design oriented at the production, assembly, use, and recovery of the product. The objective is to develop an aid to design as complete and flexible as possible. Therefore, DFX is a design system consisting of tools and techniques aiding decision-making, conducting different types of analysis, and, through appropriate metrics, quantifying the performance of the design choices. These tools can assume different forms, from a set of guidelines to a detailed procedure using analytical models, sometimes implemented in software programs (Herrmann et al. 2004).

As discussed in the previous sections, a design intervention for harmonizing the range of requirements, including those requirements concerning a product's environmental performance during its life cycle, can be achieved through LCD.

The same premises that have led to the introduction and development of DFX harmonize the components of the DFX system with the life cycle approach to design. These components can be operational tools for LCD, each specific for a particular objective-property of the product and characterizing it in relation to one or more phases of its life cycle.

The designer can translate the product requirements and concept into detailed design solutions by incorporating the most appropriate DFX tools. Each tool is oriented toward a specific typology of product requisite and implemented primarily in the embodiment and detail design phases.

2.7.1 DFX Tools Supporting the Environmental Strategies

Although DFE is sometimes understood as belonging to the DFX system, it is more appropriate to consider it as a design approach. It is not, therefore, a true and proper operational design tool, but a design philosophy implying a profound change in the way in which industry relates to the environmental question (Allenby 1994). As a design approach, it requires operational tools, which embody its premises and objectives. Some of the DFX system tools can perform this role effectively. It is interesting to note that, by their very nature, these tools are based on the problem-oriented approach. This confirms that these tools are particularly suited to the environmental aspects of product design (see Section 2.6.4).

Of the different DFX typologies, several are of particular interest in relation to the two intervention strategies for the environmental efficiency of the life cycle:

- Those typologies directed at facilitating the continued functionality of the product during the phase of use, as they favor the extension of its useful life. In this case, one speaks of design for maintainability and design for serviceability (Makino et al. 1989; Gershenson and Ishii 1993; Klement 1993; Dewhurst and Abbatiello 1996; Kusiak and Lee 1997). The latter, taking into consideration the necessities associated with the whole set of servicing operations (diagnosis, maintenance, and repair), usually also encompasses the former.

- Those oriented at the planning of processes at the end-of-life, as they are directed at reducing the impact of disposal and at recovering the resources. In this case, one speaks in general terms of design for product retirement or recovery (Ishii et al. 1994; Navin-Chandra 1994; Zhang et al. 1997; Gungor and Gupta 1999) or, more specifically, of design for remanufacturing (Shu and Flowers 1993; Amezquita et al. 1995; Bras and McIntosh 1999), and design for recycling (Burke et al. 1992; Beitz 1993; Kriwet et al. 1995), depending on whether greater emphasis is placed on the reuse of components or on the recycling of materials.

In both cases, the tools can intervene directly in the most significant design parameters linked to the product architecture and to the characteristics of the components, exactly as required by the statement proposed in Section 2.6.3.

The same can also be said for a third typology of DFX, known as design for disassembly. This typology is directed at the design and planning of the disassembly of constructional systems. While design for disassembly is frequently oriented toward interventions of recovery at the end of a product's useful life (Simon 1991; Jovane et al. 1993; Li et al. 1995; Harjula et al. 1996; Srinivasan et al. 1997), to the extent where it is sometimes considered an integral part of design for recovery and recycling (Beitz 1993; Ishii et al. 1994; Navin-Chandra 1994; Pnueli and Zussman 1997), it cuts across the two environmental strategies, extension of useful life and recovery at end-of-life (Boothroyd and Alting 1992; Ishii et al. 1993; Sodhi et al. 2004; Lambert and Gupta 2005). Both strategies are, in fact, advantaged by an intervention directed at achieving a specific and important product characteristic: the ease with which the product can be disassembled. This aspect is further confirmed by authors who integrate design for disassembly with the requirements of servicing operations (Eubanks and Ishii 1993; Vujosevic et al. 1995).

The DFX tools supporting environmental strategies must clearly assume a determining role in the versatile system of instruments for an integrated design to achieve the integration of environmental aspects into product design. This role must be based on a wide variety of product requisites that constitute the DFX system. Their relationship and integration with the other tools of the DFX system (design for manufacturability, design for assembly, design for variety, design for robustness and quality, and design for reliability) thus becomes a crucial factor in obtaining a final solution. This solution would succeed in interpreting the different requisites and in balancing the various design strategies. Some interesting studies on the links between design for environment and design for manufacturability, one of the more widely used DFX techniques, have reported encouraging results about their complementarities and the possibility of conducting interventions for mutual improvement in relation to their respective objectives (Ufford 1996; Rounds and Cooper 2002).

2.7.2 Integrating the DFX Tools in the Product Development Process

Specificity is a very interesting aspect of DFX tools. This aspect enables the decomposition of a design problem that is already very wide ranging and segmented in its conventional dimension and is further complicated by environmental requirements. Each DFX technique is characterized by methods, procedures, and models and enables the elaboration of specific data through appropriate analytical functions. A suitable set of DFX tools can enable specific parts of the problem to be treated separately, so that each part is managed by those members of the design team who are most skilled in that area.

This approach is based on the decomposition of the problem and the design intervention itself. It is the cornerstone of modern methods of product design and can constitute an effective resource for achieving the integration of environmental and traditional necessities (Jackson et al. 1997). At the same time, it is important not to overlook the negative effect that excessively specific and separate design actions may have on the design process (Bras 1997). These actions may delay or even block convergence on a final, balanced solution. The integrated and simultaneous structuring of the design intervention, recommended earlier (Section 2.4), can help remedy this dangerous tendency. To fulfill environmental requirements through the use of the most appropriate DFX techniques, it is necessary to adopt a product development model of the design-centered type (Yazdani and Holmes 1999). As showed in Figure 2.15, this approach to design is characterized by a structure tending toward the concurrent model while maintaining, in part, the sequential dimension of some phases and giving particular emphasis to the vast range

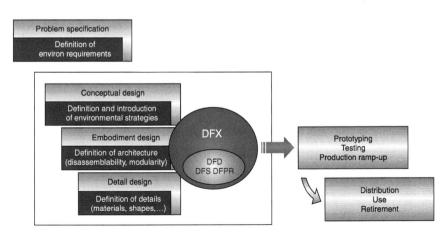

FIGURE 2.15
Integration of environmental aspects into product design using design-centered model and DFX tools.

of requisites demanded of the product in relation to the various phases of its life cycle. The principle of simultaneity is applied to the design-related phases, thus underlining the importance of a high level of information-sharing in the design analysis. The increased effectiveness of design choices is obtained by reinforcing the primary design phases themselves (conceptual, embodiment, detail design) and introducing a series of theoretical and analytical tools and techniques that differ according to product requirements and are applied at different design levels.

With the design-centered model, therefore, it is possible to introduce and prioritize in product development a flexible system of integrated design methodologies and tools. Each of these methodologies and tools is targeted at the attainment of a particular product requisite, such as the DFX system.

Figure 2.15 summarizes the integration of the environmental aspects. The design-related phases are preceded by the phase of preliminary definition of the product's fundamental requirements (problem specification). At this level, as noted earlier, the objectives of the development process are defined together with their correlated problems and the design exigencies and constraints. It is, therefore, the phase in which the environmental objectives are defined. The project variables correlated with these objectives are identified, and the problems resulting from their fulfillment are analyzed. The importance of this preliminary stage has already been underlined: the most appropriate and effective environmental strategies can be chosen judiciously only after the design objectives have been accurately translated into product requirements. It is clear that the life cycle approach must be adopted in formulating the environmental requirements. This approach involves, *de facto*, an added dimension that enlarges the domain of the design objectives.

The problem specification phase is followed by the main design phases, not in a rigidly sequential order but quite simultaneously:

- *Conceptual design.* After defining the design objectives, which are translated into product requirements, it is necessary to develop design ideas that can enable achievement of these objectives. In this phase, certain general proposals are defined (through the initial description of the functions and the principal characteristics of the product), and these proposals are then evaluated to determine which of them best meets the intended target. The earlier an intervention oriented toward environmental aspects is implemented in the development process, the greater is its effectiveness. This is the stage in which it is necessary to introduce those environmental strategies that are held on the basis of the considerations discussed in Section 2.6. These strategies are deemed to be the most appropriate for the chosen objectives.

- *Embodiment design.* This phase involves arriving at a more complete description of the product idea. The concepts formulated in the preceding phases are developed, their feasibility is verified, and they are finally translated into an approximation of the product layout by defining subsystems and functional components. In this phase, DFX techniques supporting the environmental strategies are introduced. This introduction is performed by applying methods and tools aiding a correct definition of the product architecture, with particular regard to the properties of modularity and disassemblability that can favor interventions of both servicing and recovery.

- *Detail design.* In this phase, the product architecture is translated into a complete and detailed solution. The DFX techniques supporting the environmental strategies are introduced again, applying methods and tools aiding a correct definition of the design details. The choice of materials, study of the shape, and definition of components' geometry and junction systems, must all be guided not only by performance and economic requirements but also by the environmental targets.

The design phases, where the most appropriate tools of the DFX system are applied (in Figure 2.15: design for disassembly—DFD; design for serviceability—DFS; and design for product retirement—DFPR), are followed by the last phases of product development: prototyping, testing, and production ramp-up. However, the definitive verification of the environmental performance of the finished product, which can only be a first, approximate assessment if based on the various appositely developed metrics, requires that the entire life cycle runs its course through the phases of distribution, use, and disposal. Only then, is it possible to evaluate the effective environmental performance of the product and to identify the factors to be considered in any future redesign.

2.8 Standards, Regulations, and the Environmental Efficiency of Products

This section will introduce some of the most significant regulatory initiatives for incorporating environmental considerations into product development. These initiatives intend to reinforce and redirect environmental policies. The ultimate aim is to evoke the primacy of ecological production in companies and integrate environmental requirements into product standards.

These initiatives closely follow the main motivating factors that could drive manufacturing companies to adopt policies and instruments aimed at environmental protection (Fiksel 1996; Bras 1997):

- Introduction of standards for the management of environmental questions in product development and production

- Promotion of certification of the environmental efficiency of products
- Legislation for extending the manufacturer's responsibility beyond the commercialization of products

2.8.1 Environmental Standards

Creating standards that encourage a preventive approach to environmental problems is, perhaps, the most effective way to promote the principles of environmental efficiency in design intervention.

Standards have the great advantage of promoting the importance of design in industrial production and product development. Today, designers normally possess knowledge of and refer to the relevant national and international norms. Thus, the inclusion of environmental principles within these standards can guarantee an immediate response.

Various national and international regulatory bodies have initiated research into instruments of standardization aimed at containing and reducing the environmental impact of industrial production. The International Standard Organization, through the ISO 14000 series, first began to standardize the implementation of environmental management systems and subsequently considered the instruments and procedures, such as environmental audit, environmental labeling, and LCA, which serve to introduce environmental variables into company management. Among the standards of the 14000 series, we highlight those that expressly address product life cycle and DFE:

- *ISO 14040 series.* Reference standards for LCA (International Standards Organization 1997) provide guidelines on the principles and conduct of environmental analysis of the product's life cycle.
- *ISO/TR 14062.* The technical report for integrating environmental aspects into product design and development (International Standards Organization 2002) provides guidelines for the improvement of environmental performance of the product development process.

2.8.1.1 *International Standards for Life Cycle Assessment*

Introduced in the early 1990s as "an objective process to evaluate the environmental burdens associated with a product or activity by identifying and quantifying energy and materials used and wastes released to the environment, and to evaluate and implement opportunities to affect environmental improvements" (Fava et al. 1991), LCA rapidly became a well-known technique to evaluate the environmental impact of products and processes.

Recognition of the validity and utility of this methodology led to international standardization through the publication, from 1997 onward, of the ISO 14040 series of norms. The detailed definition of LCA evidences the implicit intention of the standardization to delineate a clear reference methodology. LCA is considered a technique for assessing the environmental impacts associated with a product, "compiling an inventory of relevant inputs and outputs of a product-system; evaluating the potential environmental impacts associated with those inputs and outputs; interpreting the results of the inventory analysis and impact assessment phases in relation to the objectives of the study" (International Standards Organization 1997). The group of ISO 14040 standards describe in detail the general criteria and underlying methodological framework for the conduction of the main phases making up a complete LCA. Table 2.3 shows the main headings of these standards and a summary of their specific contents. The lower section of the table reports references and brief information on the new editions, only just published, which replace the previous standards.

2.8.1.2 Toward an International Standard for Integrating Environmental Aspects into Product Design and Development

The growing sensibility of consumers and manufacturers toward environmental issues associated with industrial production and toward the consequent need to integrate environmental aspects into the product design and development process led, in 1998, to the technical committee, ISO/TC 2007—Environmental Management, setting up a working group to study the specific theme of "design for environment." The proceedings, which lasted 4 years, resulted in the technical report, ISO/TR 14062 "Environmental Management—Integrating Environmental Aspects into Product Design and Development" (International Standards Organization 2002). It has the aim of providing those personnel who are directly involved in the design and development phases, regardless of the typology or size of the company they work in, with a systematic program for predicting and identifying the possible effects their future products could have on the environment and for taking effective decisions during the conception and development of these products to improve their environmental performance.

The main issues treated in the technical report completely embrace the problems associated with creating an environment-friendly product design and with the methodological statement introduced in this chapter.

The evolution of this first report toward a standard, which can be integrated with the other environmental management norms of the ISO 14000 series, would not only favor the realization of more complete and integrated environmental management systems but could also clearly stimulate the implementation of DFE in design practice. This aspect is confirmed by the interest generated by the technical report in the context of some specific regulatory standards. This is in the case of the electromechanical and electronic components production sector, where it is noted that the new editions of the International Electrotechnical Commission (IEC) Guide 109 "Environmental Aspects–Inclusion in Electrotechnical Products Standards" (International Electrotechnical Commission 2003), and European Computer Manufacturers

TABLE 2.3

ISO International Standards for LCA

Designation Document Type Year	Title	Contents
ISO 14040:1997 International Standard 1997	Environmental management— life cycle assessment— principles and framework	General framework, principles, and requirements for conducting and reporting LCA studies
ISO 14041:1998 International Standard 1998	Environmental management— life cycle assessment—goal and scope definition and inventory analysis	Requirements and procedures necessary for the compilation and preparation of the definition of goal and scope for LCA, and for performing, interpreting, and reporting a life cycle inventory (LCI) analysis
ISO 14042:2000 International Standard 2000	Environmental management— life cycle assessment—life cycle impact assessment	General framework for the life cycle impact assessment (LCIA) phase of LCA Key features and inherent limitations of LCIA Requirements for conducting the LCIA phase Relationship to the other LCA phases
ISO 14043:2000 International Standard 2000	Environmental management— life cycle assessment—life cycle interpretation	Requirements and recommendations for conducting the life cycle interpretation phase in LCA or LCI studies
ISO 14040:2006 International Standard 2006	Environmental management— life cycle assessment— principles and framework	Overview of the practice, applications, and limitations of LCA to a broad range of potential users
ISO 14044:2006 International Standard 2006	Environmental management— life cycle assessment— requirements and guidelines	Preparation of, conduct of, and critical review of LCI Guidance on LCIA phase and on the interpretation of LCA results, as well as the nature and quality of the data collected

Association Standard ECMA-341 "Environmental Design Considerations for ITC (Information and Communication Technology) and CE (consumer electronic) Products" (Ecma International 2004), make explicit reference to ISO/TR 14062 and adopt its key concepts, general statement, and design principles.

2.8.2   Product Certification and Labeling

Innovations in products and production technologies and their promotion through the mechanisms of market competition can also be stimulated through environmental certification and labeling. In product certification, different procedures have been developed worldwide for the assigning of environmental quality labels. Germany introduced the "Blauer Engel" program in 1977, making it the first country to implement a national ecolabeling program. Subsequently the Scandinavian countries introduced the "Nordic Swan," and the Netherlands adopted a national ecolabel named "Stichting Milieukeur." In 1992, the European Union issued Regulation 880/92, the community system for assigning a seal of ecological quality to some product typologies, subsequently defined in specific directives. This regulation was revised in 2000 (EC1980/2000).

In other countries too, similar environmental labeling initiatives were undertaken: "Green Seal" and "Energy Star" in the United States, "Environmental Choice" in Canada, "EcoMark" in Japan, and "Ecomark" in India. (EPA 1998).

The objectives underlying these regulations can be summarized as follows:

- Creating a mechanism of voluntary adhesion to promote the market presence of more "environmentally friendly" products
- Indicating to the consumer the more environmentally favorable products among those present in the market.

Also the ISO 14000 standards has specific areas regarding the ecological labeling of products. The ISO 14020 series (International Standards Organization 2000) addresses a range of different approaches to environmental labels and declarations, including self-declared environmental claims, ecolabels, and a possible scheme of environmental declarations for products.

2.8.3   Extension of Manufacturer Responsibility

Extended producer responsibility (EPR) is an underlying principle of more recent environmental regulations. EPR is an environmental policy approach in which the producer's responsibility for a product is extended to the post-consumer stage of the product's life cycle. The focus is on product-systems rather than on production facilities (Davis et al. 1997). This focus can be translated into restrictions at different levels, from the obligation of producers to take on the costs of disposal and, in some cases, the organization of calling in their products after use, to the explicit demand for product requisites such as disassemblability and recyclability.

The principle of EPR relies, for its implementation, upon the life cycle concept. Opportunities to prevent pollution and reduce resource use, in each stage of the product life cycle, through changes in process technology and product design need to be identified. Therefore, this type of regulatory action aims at stimulating the redesign of some categories of products to obtain a reduction in their environmental impact, encouraging producers to adopt an integrated approach for the development and management of ecocompatible products, and introducing an extended vision of the problem to cover the entire life cycle of products. This extended view, typical of the DFE approach, includes a wide range of aspects (energy consumption, use of materials, component duration, reuse of components, and recycling of materials).

Since its introduction, EPR has become a characteristic of regulations for various production sectors, constituting a valid stimulus for a process of innovation directed at environmental sustainability. For example, the concept of producer responsibility for the disposal of products at the end of their useful life has recently been embodied into the European Union directives on two important industrial sectors: vehicles and electrical and electronic equipments. These directives have a great impact on product design, because they encourage alternatives to hazardous substances and design for disassembly and recycling (Kumar and Fullenkamp 2005).

2.9   Summary

For a complete analysis directed at evaluating and reducing the environmental impact of a product, it is necessary to consider, together with the phases of development and production, those of use, recovery, and treatment of the retired product. Further, all these phases must be understood not in relation to the specific actors involved but from a wider perspective, going beyond their direct competencies.

It is possible therefore to speak of product-system, where, in its most complete sense, the product is considered integral with its life cycle and within the environmental, technological, economic, and social contexts in which this life cycle develops. From the viewpoint of environmental analysis, this system is characterized by physical flows of resources transformed through the various processes making up the life cycle and by interactions with the ecosphere, which result in the impact the product-system has on the environment.

It is necessary to operate an integrated and simultaneous product design for environmental requirements to become factors of innovation in the development of a sustainable product. This process must take into account a growing range of specifications and requisites. The design intervention must be structured according to the principles of DFE and LCD. The main phases of a product's life cycle must be taken into consideration, starting from the definition of the design problem and the development of product requirements, which will subsequently be translated first into the product concept and then into the detailed solution. An accurate definition of the environmental objectives and of the consequent product requirements can help in identifying the most suitable environmental strategies that could enable an improvement in the exploitation of the resources used in product manufacturing, and also in identifying the design strategies that could put them in concrete form. These strategies can be supported by different typologies of techniques, completing the versatile and structured system of design tools for product requisites known as DFX. Appropriately introduced into an integrated development process which assimilates the life cycle approach and its proper methodologies, the tools and techniques for environmental requirements can help the designer to make decisions regarding the most important design parameters strongly influencing the final solution, and confronting themselves with the vast system of instruments oriented at conventional product requisites.

Acknowledgments

The author thanks Guido La Rosa and Antonino Risitano for their encouragement to undertake the research activity on this theme, and for their contribution.

References

Allenby, B.R., Achieving sustainable development through industrial ecology. *International Environmental Affairs*, 1992, 4(1), 56–68.

Allenby, B.R., Integrating environment and technology: Design for environment. In *The Greening of Industrial Ecosystems*, B.R. Allenby and D.J. Richards (eds.), pp. 137–148, 1994, National Academy Press, Washington, D.C.

Alting, L., Life-cycle design of products: A new opportunity for manufacturing enterprises. In *Concurrent Engineering: Automation, Tools and Techniques*, A. Kusiak (ed.), pp. 1–17, 1993, John Wiley & Sons, New York.

Alting, L. and Jorgensen, J., The life cycle concept as a basis for sustainable industrial production. *Annals of the CIRP*, 1993, 42(1), 163–167.

Amezquita, T., Hammond, R. and Bras, B., Design for remanufacturing, in ICED 95: 10th International Conference on Engineering Design, 1995, pp. 1060–1065.

Ashley, S., Designing for the environment. *Mechanical Engineering*, 1993, 15(3), 53–55.

Asiedu, Y. and Gu, P., Product life cycle cost analysis: State of the art review. *International Journal of Production Research*, 1998, 36(4), 883–908.

Asimow, M., *Introduction to Design*, 1962, Prentice Hall, Englewood Cliffs, NJ.

Ayres, R.U., Industrial metabolism. In *Technology and Environment*, J.H. Ausubel and H.E. Sladovich (eds.), pp. 23–49, 1989, National Academy Press, Washington, D.C.

Beitz, W., Designing for ease of recycling. *Journal of Engineering Design*, 1993, 4(1), 12–23.

Bhamra, T.A., Evans, S., McAloone, T.C., Simon, M., Poole, S. and Sweatman, A. Integrating environmental decisions into the product development process—Part 1: The early stages, in EcoDesign'99: 1st International Symposium on Environmentally Conscious Design and Inverse Manufacturing, 1999, pp. 329–333.

Bhander, G.S., Hauschild, M. and McAloone, T., Implementing life cycle assessment in product development. *Environmental Progress*, 2003, 22(4), 255–267.

Billatos, S.B. and Basaly, N.A., *Green Technology and Design for the Environment*, 1997, Taylor & Francis, Washington, D.C.

Boothroyd, G. and Alting, L., Design for assembly and disassembly. *Annals of the CIRP*, 1992, 41(2), 625–636.

Bralla, J.G., *Design for Excellence*, 1996, McGraw-Hill, New York.

Bras, B., Incorporating environmental issues in product design and realization. *Industry and Environment*, 1997, 20(1–2), 7–13.

Bras, B. and McIntosh, M.W., Product, process, and organizational design for remanufacture: An overview of research. *Robotics and Computer Integrated Manufacturing*, 1999, 15, 167–178.

Brezet, H. and van Hemel, C., *Ecodesign: A Promising Approach to Sustainable Production and Consumption*, 1997, UNEP United Nations Environment Programme, Paris.

Burke, D., Beiter, K. and Ishii, K., Life-cycle design for recyclability, in ASME Design Theory and Methodology Conference, 1992, pp. 325–332.

Coulter, S., Bras, B., and Foley, C., A lexicon of green engineering terms, in ICED 95: 10th International Conference on Engineering Design, 1995, pp. 1–7.

Cross, N., The coming of post-industrial design. *Design Studies*, 1981, 2(1), 3–7.

Davis, G.A., Wilt, C.A., Dillon, P.S. and Fishbein, B.K. *Extended Product Responsibility: A New Principle for Product-Oriented Pollution Prevention*, 1997, US Environmental Protection Agency, Office of Solid Waste, Washington, D.C.

Dewhurst, P. and Abbatiello, N., Design for serviceability. In *Design for X: Concurrent Engineering Imperatives*, G.Q. Huang (ed.), pp. 298–317, 1996, Chapman & Hall, London.

Dieter, G.E., *Engineering Design: A Materials and Processing Approach*, 2000, McGraw-Hill, Singapore.

Dowie, T., Green design. *World Class Design to Manufacture*, 1994, 1(4), 32–38.

EC1980/2000, Regulation (EC) no 1980/2000 of the European Parliament and of the Council on revised community eco-label award scheme. *Official Journal of the European Communities*, 2000, L 237, 21/9/2000, 1–12.

ECMA International, ECMA-341, Environmental design considerations for ICT and CE products, 2004.

Elkington, J. and Burke, T., *The Green Capitalists: How Industry Can Make Money and Protect the Environment*, 1987, Victor Gollancz, London.

Elkington, J., Burke, T. and Hailes, J., *Green Pages: The Business of Saving the World*, 1988, Routledge: London.

EPA, Environmental labeling: Issues, policies, and practices worldwide, 1998, EPA 742-R-98-009, US Environmental Protection Agency, Office of Prevention, Pesticides and Toxic Substances, Washington, D.C.

Eubanks, C.F. and Ishii, K., AI methods for life-cycle serviceability design of mechanical systems. *Artificial Intelligence in Engineering*, 1993, 8(2), 127–140.

Fabrycky, W.J. and Blanchard, B.S., *Life Cycle Cost and Economic Analysis*, 1991, Prentice Hall, Englewood Cliffs, NJ.

Fava, J.A., Denison, R., Jones, B., Curran, M.A., Vigon, B.W., Selke, S. and Barnum, J. *A Technical Framework for Life-Cycle Assessment*, 1991, SETAC Society of Environmental Toxicology and Chemistry, Washington, D.C.

Fiksel, J., *Design for the Environment: Creating Eco-Efficient Products and Processes*, 1996, McGraw Hill, New York.

Frosch, R.A., Manufactured products. In *Industrial Ecology, U.S.-Japan Perspectives*, D.J. Richards and A.B. Fullerton (eds.), pp. 28–36, 1994, National Academy Press, Washington, D.C.

Gale, R.J.P. and Stokoe, P.K., Environmental cost accounting and business strategy. In *Handbook of Environmentally Conscious Manufacturing*, C.N. Madu (ed.), pp. 119–137, 2001, Kluwer Academic Publishers, Norwell, MA.

Gatenby, D.A. and Foo, G., Design for X: Key to competitive, profitable markets. *AT&T Technical Journal*, 1990, 69(3), 2–13.

Gershenson, J. and Ishii, K., Life-cycle serviceability design. In *Concurrent Engineering: Automation, Tools and Techniques*, A. Kusiak (ed.), pp. 363–384, 1993, John Wiley & Sons, New York.

Giudice, F., La Rosa, G. and Risitano, A., Product recovery-cycles design: Extension of useful life. In *Feature Based Product Life-Cycle Modelling*, R. Soenen and G. Olling (eds.), pp. 165–185, 2003, Kluwer Academic Publishers, Dordrecht.

Graedel, T.E. and Allenby, B.R., *Design for Environment*, 1998, Prentice Hall, Upper Saddle River, NJ.

Guinée, J.B., Udo de Haes, H.A. and Huppes, G., Quantitative life cycle assessment of products: Goal definition and inventory. *Journal of Cleaner Production*, 1993, 1(1), 3–13.

Gungor, A. and Gupta, S.M., Issues in environmentally conscious manufacturing and product recovery: A survey. *Computers and Industrial Engineering*, 1999, 36, 811–853.

Gutowski, T., Murphy, C., Allen, D., Bauer, D., Bras, B., Piwonka, T., Sheng, P., Sutherland, J., Thurston, D. and Wolff, E. Environmentally benign manufacturing: Observations from Japan, Europe and the United States. *Journal of Cleaner Production*, 2005, 13, 1–17.

Hanssen, O.J., Preventive environmental strategies for product systems. *Journal of Cleaner Production*, 1995, 3(4), 181–187.

Harjula, T., Rapoza, B., Knight, W.A. and Boothroyd, G. Design for disassembly and the environment. *Annals of the CIRP*, 1996, 45(1), 109–114.

Heiskanen, E., The institutional logic of life cycle thinking. *Journal of Cleaner Production*, 2002, 10(5), 427–437.

Herrmann, J.W., Cooper, J., Gupta, S.K., Hayes, C.C., Ishii, K., Kazmer, D., Sandborn, P.A. and Wood, W.H. New directions in design for manufacturing, in ASME Design Engineering Technical Conference, 2004, DETC2004-57770.

Huang, G.Q., *Design for X: Concurrent Engineering Imperatives*, 1996, Chapman & Hall, London.

Hundal, M.S., Introduction to design for the environment and life cycle engineering. In *Mechanical Life Cycle Handbook*, M.S. Hundal (ed.), pp. 1–26, 2002, Marcel Dekker, New York.

International Electrotechnical Commission, Guide 109—IEC:2003(E), Environmental aspects—inclusion in electromechanical product standards, 2003.

International Standards Organization, ISO 14040:1997(E), Environmental management—life cycle assessment—principles and framework, 1997.

International Standards Organization, ISO 14020:2000(E), Environmental labels and declarations—general principles, 2000.

International Standards Organization, ISO/TR 14062:2002(E), Environmental management—integrating environmental aspects into product design and development, 2002.

Ishii, K., Eubanks, C.F. and Di Marco, P., Design for product retirement and material life cycle. *Materials and Design*, 1994, 15(4), 225–233.

Ishii, K., Eubanks, C.F. and Marks, M., Evaluation methodology for post manufacturing issues in life-cycle design. *Concurrent Engineering: Research and Applications*, 1993, 1(1), 61–68.

Ishii, K., Life-cycle engineering design. *Journal of Mechanical Design*, 1995, 117, 42–47.

Jackson, P., Wallace, D. and Kegg, R. An analytical method for integrating environmental and traditional design considerations. *Annals of the CIRP*, 1997, 46(1), 355–360.

Jelinski, L.W., Graedel, T.E., Laudise, R.A., McCall, D.W. and Patel, C.K.N. Industrial ecology: Concepts and approaches. *Proceedings of National Academy of Sciences*, 1992, 89(3), 793–797.

Jovane, F., Alting, L., Armillotta, A., Eversheim, W., Feldmann, K., Seliger, G. and Roth, N. A key issue in product life cycle: Disassembly. *Annals of the CIRP*, 1993, 42(2), 651–658.

Keoleian, G.A. and Menerey, D., Life cycle design guidance manual, 1993, EPA/600/R-92/226, US Environmental Protection Agency, Office of Research and Development, Cincinnati, OH.

Klement, M.A., Design for maintainability. In *Concurrent Engineering: Automation, Tools and Techniques*, A. Kusiak (ed.), pp. 385–400, 1993, John Wiley & Sons, New York.

Krishnan, V. and Ulrich, K.T., Product development decisions: A review of the literature. *Management Science*, 2001, 47(1), 1–21.

Kriwet, A., Zussman, E. and Seliger, G., Systematic integration of design for recycling into product design. *International Journal of Production Economics*, 1995, 38, 15–22.

Kumar, S. and Fullenkamp, J., Analysis of European Union environmental directives and producer responsibility requirements. *International Journal of Services and Standards*, 2005, 1(3), 379–398.

Kumaran, D.S., Ong, S.K., Tan, R.B.H. and Nee, A.Y.C. Environmental life cycle cost analysis of products. *Environmental Management and Health*, 2001, 12(3), 260–276.

Kusiak, A. and Lee, G., Design of parts and manufacturing systems for reliability and maintainability. *International Journal of Advanced Manufacturing Technology*, 1997, 13, 67–76.

Kusiak, A. and Wang, J., Decomposition in concurrent design. In *Concurrent Engineering: Automation, Tools and Techniques*, A. Kusiak (ed.), pp. 481–507, 1993, John Wiley & Sons: New York.

Lambert, A. and Gupta, S., *Disassembly Modeling for Assembly, Maintenance, Reuse, and Recycling*, 2005, CRC Press: Boca Raton, FL.

Li, W., Zhang, C., Wang, H.P.B. and Awoniyi, S.A. Design for disassembly analysis for environmentally conscious design and manufacturing, in ASME International Mechanical Engineering Congress and Exposition, 1995, pp. 969–976.

Lund, R.T., Remanufacturing. *Technology Review*, 1984, 87(2), 18–29.

Luttropp, C. and Karlsson, R., The conflict of contradictory environmental targets, in EcoDesign 2001: 2nd International Symposium on Environmentally Conscious Design and Inverse Manufacturing, 2001, pp. 43–48.

Mackenzie, D., *Green Design: Design for the Environment*, 1991, Laurence King, London.

Madge, P., Design, ecology, technology: A historiographical review. *Journal of Design History*, 1993, 6(3), 149–166.

Maffin, D., Engineering design models: Context, theory and practice. *Journal of Engineering Design*, 1998, 9(4), 315–327.

Makino, A., Barkan, P. and Pfaff, R., Design for serviceability, in *ASME Winter Annual Meeting*, 1989, pp. 117–120.

Manzini, E. and Vezzoli, C., *Lo Sviluppo di Prodotti Sostenibili*, 1998, Maggioli Editore, Rimini.

Molina, A., Sánchez, J.M. and Kusiak, A., *Handbook of Life Cycle Engineering: Concepts, Models and Technologies*, 1998, Kluwer Academic, Dordrecht.

Navin-Chandra, D., Design for environmentability, in *ASME Conference on Design Theory and Methodology*, 1991, DE-31, pp. 119–125.

Navin-Chandra, D., The recovery problem in product design. *Journal of Engineering Design*, 1994, 5(1), 67–87.

O'Riordan, T., *Environmentalism*, 1976, Pion, London.

OTA, Green products by design: Choices for a cleaner environment, 1992, Report OTA-E-541, Office of the Technology Assessment, Congress of the United States: Washington.

Overby, C., Product design for recyclability and life extension, in *American Society of Engineering Education Annual Conference*, 1979, pp. 181–196.

Overby, C., Design for the entire life-cycle: A new paradigm? in *American Society of Engineering Education Annual Conference*, 1990, pp. 552–563.

Pahl, G. and Beitz, W., *Engineering Design: A Systematic Approach*, 1996, Springer, London.

Papanek, V., *Design for the Real World: Human Ecology and Social Change*, 1971, Pantheon Books, New York.

Pnueli, Y. and Zussman, E., Evaluating the end-of-life value of a product and improving it by redesign. *International Journal of Production Research*, 1997, 35(4), 921–942.

Prasad, B., *Concurrent Engineering Fundamentals*, vol. 1, 1996, Prentice Hall, Englewood Cliffs, NJ.

Prudhomme, G., Zwolinski, P. and Brissaud, D., Integrating into the design process the needs of those involved in the product life-cycle. *Journal of Engineering Design*, 2003, 14(3), 333–353.

Rebitzer, G., Ekvall, T., Frischknecht, R., Hunkeler, D., Norris, G., Rydberg, T., Schmidt, W.-P., Suh, S., Weidema, B.P. and Pennington, D.W. Life cycle assessment—Part 1: Framework, goal and scope definition, inventory analysis, and applications. *Environment International*, 2004, 30(5), 701–720.

Ries, G., Winkler, R. and Zust, R., Barriers for a successful integration of environmental aspects in product design, in EcoDesign'99: 1st International Symposium on Environmentally Conscious Design and Inverse Manufacturing, 1999, pp. 527–532.

Robertson, J., *The Sane Alternative : A Choice of Futures*, 1980, River Basin Publishing, St. Paul, MN.

Rose, C.M., Masui, K. and Ishii, K., How product characteristics determine end-of-life strategies, in IEEE International Symposium on Electronics and the Environment, 1998, pp. 322–327.

Rounds, K.S. and Cooper, J.S., Development of product design requirements using taxonomies of environmental issues. *Research in Engineering Design*, 2002, 13, 94–108.

Sánchez, J.M., The concept of product design life cycle. In *Handbook of Life Cycle Engineering: Concepts, Models and Technologies*, A. Molina, J.M. Sánchez and A. Kusiak (eds.), pp. 399–412, 1998, Kluwer Academic, Dordrecht.

Schumacher, E.F., *Small is Beautiful: A Study of Economics as if People Mattered*, 1973, Blond & Briggs, London.

Shu, L. and Flowers, W., A structured approach to design for remanufacture, in ASME Winter Annual Meeting, 1993, pp. 13–19.

Simon, M., Design for dismantling. *Professional Engineering*, 1991, 4, 20–22.

Sodhi, R., Sonnenberg, M. and Das, S., Evaluating the unfastening effort in design for disassembly and serviceability. *Journal of Engineering Design*, 2004, 15(1), 69–90.

Srinivasan, H., Shyamsundar, N. and Gadh, R., A framework for virtual disassembly analysis. *Journal of Intelligent Manufacturing*, 1997, 8, 277–295.

Sun, J., Han, B., Ekwaro-Osire, S. and Zhang, H.-C. Design for environment: Methodologies, tools, and implementation. *Journal of Integrated Design and Process Science*, 2003, 7(1), 59–75.

Tipnis, V.A., Evolving issues in product life cycle design: Design for sustainability. In *Handbook of Life Cycle Engineering: Concepts, Models and Technologies*, A. Molina, J.M. Sánchez and A. Kusiak (eds.), pp. 413–459, 1998, Kluwer Academic, Dordrecht.

Todd, J.A. and Curran, M.A, Streamlined life-cycle assessment: A final report from the SETAC North America streamlined LCA workgroup, 1999, SETAC Society of Environmental Toxicology and Chemistry, Pensacola, FL.

Ufford, D.A., Leveraging commonalities between DFE and DFM/A, in IEEE International Symposium on Electronics and the Environment, 1996, pp. 197–200.

Ullman, D.G., *The Mechanical Design Process*, 2003, McGraw-Hill, New York.

Ulrich, K.T. and Eppinger, S.D., *Product Design and Development*, 2000, McGraw-Hill, New York.

van Nes, N., Cramer, J. and Stevels, A., A practical approach to the ecological lifetime optimization of electronic products, in EcoDesign'99: 1st International Symposium on Environmentally Conscious Design and Inverse Manufacturing, 1999, pp. 108–111.

Veroutis, A.D. and Fava, J.A., Framework for the development of metrics for design for environment assessment of products, in IEEE International Symposium on Electronics and the Environment, 1996, pp. 13–18.

Vigon, B.W., Tolle, D.A., Cornaby, B.W., Latham, H.C., Harrison, C.L., Boguski, T.L., Hunt, R.G. and Sellers, J.D. Life-cycle assessment: Inventory guidelines and principles, 1993, EPA/600/R-92/245, US Environmental Protection Agency, Office of Research and Development, Cincinnati, OH.

Vujosevic, R., Raskar, R., Yetukuri, N.V., Jothishankar, M.C. and Juang, S.-H. Simulation, animation, and analysis of design disassembly for maintainability analysis. *International Journal of Production Research*, 1995, 33(11), 2999–3022.

Wang, B., *Integrated Product, Process and Enterprise Design*, 1997, Chapman & Hall, London.

Wanyama, W., Ertas, A., Zhang, H.-C. and Ekwaro-Osire, S. Life-cycle engineering: Issues, tools and research. *International Journal of Computer Integrated Manufacturing*, 2003, 16(4–5), 307–316.

Wernick, I.K., Herman, R., Govind, S. and Ausubel, J.H. Materialization and dematerialization: Measures and trends. In *Technological Trajectories and the Human Environment*, J.H. Ausubel and H.D. Langford (eds.), pp. 135–156, 1997, National Academy Press, Washington, D.C.

Whitmer, C.I., Olson, W.W. and Sutherland, J.W., Methodology for including environmental concerns in concurrent engineering, in 1st World Conference on Integrated Design and Process Technology, 1995, pp. 8–13.

Woodward, D.G., Life cycle costing: Theory, information acquisition and application. *International Journal of Project Management*, 1997, 15(6), 335–344.

Yazdani, B. and Holmes, C., Four models of design definition: Sequential, design centered, concurrent and dynamic. *Journal of Engineering Design*, 1999, 10(1), 25–37.

Young, P., Byrne, G. and Cotterell, M., Manufacturing and the environment. *International Journal of Advanced Manufacturing Technology*, 1997, 13(7), 488–493.

Zhang, H.C., Kuo, T.C., Lu, H. and Huang, S.H. Environmentally conscious design and manufacturing: A state of the art survey. *Journal of Manufacturing Systems*, 1997, 16(5), 352–371.

Zust, R. and Caduff, G., Life-cycle modeling as an instrument for life-cycle engineering. *Annals of the CIRP*, 1997, 46(1), 351–354.

3

Product Life Cycle Monitoring via Embedded Sensors

Srikanth Vadde, Sagar V. Kamarthi,
Surendra M. Gupta, and Ibrahim Zeid

CONTENTS

3.1   Introduction

Reliability methods determine failure characteristics of the components using accelerated life tests. These tests often fail to simulate the myriad working conditions the products may be subjected to during their usage. This drawback frequently renders the reliability estimates provided by product manufacturers less accurate. Consider a business copier that breaks down after 700 copies, although its manufacturer-estimated mean time between failures is 4000 copies. Such a failure can make customers lose confidence in the manufacturer's estimate of the expected life of the copier. In contrast, some components of the copier may remain in good working condition even beyond their expected life. When products fail, accurate diagnosis of the failures can be a difficult and time-consuming process. This problem can shoot up maintenance costs and prolong repair time. Existing end-of-life (EOL) processing techniques use little or no prior knowledge about the condition and the remaining life of products [1–7]. In the case of product disassembly, these techniques often fail to categorize components according to their condition and remaining life. If sensors are embedded into products, they can serve two purposes: bring customers comfort during product lifetime and enable cost-effective EOL processing by considering product condition and the remaining life of the product.

The latest developments in sensor technology are opening up new avenues for addressing EOL processing issues. For example, radio frequency identification (RFID) tags are availed to monitor the real-time inventory levels of products and components in warehouses. Toward the management of waste, RFID tags are used to identify and collect recyclable products from waste disposal sites [8]. Sensor and RFID tags embedded in products can dramatically change many of the existing techniques in EOL processing, including disassembly, reuse, and recycling. Sensors, if embedded in products, can (a) provide features and functionalities that make product usage more convenient and comfortable to the consumer, (b) facilitate on-line product condition monitoring, (c) enable the detection or the prediction of product failures, and (d) estimate the remaining life of the components. The detection or the prediction of imminent product failures alerts the consumer to take necessary precautions to avert unexpected failures. Information related to product condition can be advantageously used for planning EOL activities. Recent trends in miniaturization of sensors are driving the current consumer market to embed increasing number of sensors in the products. The current sensor market is leaning toward the development of communication features in products to facilitate data transmission over a network [9]. For example, recently the automobile industry tried to bring forth regulations that mandate the use of pressure sensors in tires. This chapter provides an overview of product condition monitoring for EOL management and presents a framework for effective product life cycle management.

3.2   Related Research

Limited numbers of research papers are found in the published literature on after-sale product condition monitoring. Scheidt and Zong first proposed the idea of collecting data from products during product usage [10], and Klausner et al. [11–13], Karlsson [14,15], and Petriu et al. subsequently advocated this idea [16]. All these authors proposed installing devices with memory to record monitoring data generated during the product usage. Klausner et al. [12,13] presented an information system for product recovery that combined sensor data with demand and price information in second-hand markets. Cheng et al. [17] reported that they had developed a generic embedded device that could be installed in various kinds of equipment such as manufacturing machinery, portal servers, and automatic guided vehicles. This device consisted of real-time operating system to retrieve, collect, and manage equipment data. Mazhar et al. [18] presented a two-stage integrated approach to assess the reliability of components for reuse in refurbished products. To enable the assessment of reliability of components in EOL products, several researchers are engaged in designing life cycle data recording devices such as electronic log, life cycle data acquisition units, lifetime observer, life cycle unit, and watchdog [19–26]. Liu et al. [27] presented an Internet server controller–based intelligent maintenance system for information appliance products such as consumer appliances. They discussed the methods to develop products and manufacturing systems using Internet-based technologies to change maintenance practices from reactive to preventive. Bruce and Hathcock [28] introduced object models for the maintenance and the monitoring of high-availability network appliances. Tsai and Wu [29] developed a mobile agent–based multiagent architecture to meet the challenges of home network services to connect appliances such as dish washers and air conditioners.

3.3 Product Life Cycle Monitoring

A feasible study is conduced to identify whether embedding sensors promotes EOL processing. This study also aimed at assessing the impact of sensor-provided information on product EOL processing.

3.3.1 Product Life Cycle Monitoring Framework

To achieve effective product life cycle monitoring, we present a framework that interconnects sensor-embedded products (SEPs), a remote monitoring center (RMC), maintenance centers, disassembly centers, recycling centers, disposal centers, and remanufacturing centers (see Figure 3.1). This framework integrates several functions: (a) recording data from the point products are manufactured till the products are disposed off or recycled and (b) providing the maintenance, disassembly, recycling, disposal, and remanufacturing centers access to the recorded data and the product-specific information. Figure 3.1 depicts the functions and the characteristics of each entity in the framework.

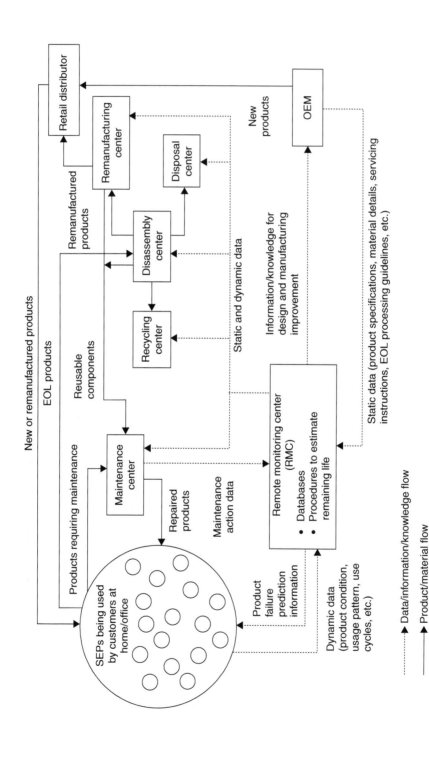

FIGURE 3.1
Framework for effective life cycle management of products.

3.3.1.1 Sensor-Embedded Products

SEPs contain sensors that are implanted during the production of SEPs. These sensors monitor the critical components of SEPs. By facilitating data collection during product usage, these embedded sensors enable the prediction of product or component failures and estimation of the remaining life of components as the products reach their EOL.

Sensors (see Figure 3.2) that are embedded in products contain [11,30] (a) a sensing element to register environmental parameters (such as temperature, vibration); (b) a microprocessor to extract features from the signals to perform local data processing; (c) memory (flash memory up to 512 KB) with limited capacity to store sensor data; (d) data transceiver; (e) unique identification code (UIC) that can be read using RFID tag readers; and (f) onboard power supply from the product power source, a separate battery, or an energy-harvesting mechanism (energy-scavenging sensors). The sensors can either be wired or wireless depending upon the product characteristics.

The following factors must be considered while designing and fabricating SEPs [31]: (a) the sensors being embedded in a product should not affect the product's portability, usability, esthetic appearance, or design; (b) the reliability of sensors is important since they have to be operational for the entire life of a product without the need for supervision; and (c) the additional cost and power consumption ensuing from the embedding sensors should be justified.

If a sensor breaks down during the product usage, the UIC should enable instant identification of the faulty sensor. The sensors should comply with the IEEE 1451 series of standards [32], which are a set of standards that provide guidelines to the manufacturers of sensors to manufacture universally compatible sensors. Compliance with IEEE standards would make replacement of faulty sensors easy.

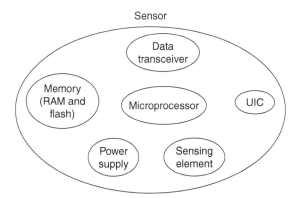

FIGURE 3.2
Components of a sensor.

3.3.1.2   Remote Monitoring Center

RMC is the nerve center of the product life cycle monitoring framework. This center documents the static data or information about products that is compiled by the original equipment manufacturer (OEM) in the product manufacturing stage [30] and the dynamic data or information that is chronicled by the sensors in SEPs during the use phase of the SEPs. Static data or information consists of the bill of materials, component suppliers, configuration options, servicing instructions, and EOL guidelines such as disassembly sequence [11]. Dynamic data or information generated during the use of a product consists of sensor data, patterns of usage, number of use cycles, runtime in each use cycle, and environmental conditions. Dynamic data also includes service history on inspections and replaced and repaired parts [11,12]. Much of the static and dynamic data or information is archived at the RMC in a database against a product's UIC. Sensors can use onboard radio frequency transmitter, telephone line, local area network (LAN), or the Internet to transmit the registered data or information to the RMC. Dynamic data can be transmitted to the RMC through a communication network, either continuously or intermittently, on an hourly, daily, weekly, or monthly basis, depending on the product-usage pattern. In intermittent transmission mode, sensors can send the entire data generated during every cycle of the product's operation, statistics of the measured data during the cycle [16], or an approximate signal form representing the data in each cycle [33].

The RMC contains procedures to estimate the remaining life and condition of the product or the component [34] and to predict the product or component failures in the immediate future. The RMC stores product condition data and remaining life information against the product's UIC. When product failure is either detected or predicted, the RMC alerts the user of the product so that the user can take precautionary measures to avert unexpected failures. The RMC facilitates better life cycle management of products by providing the archived data or information to the maintenance, disassembly, recycling, disposal, and remanufacturing centers. Wireless communication technologies, satellite links, and the Internet act as enablers for the RMC to receive and transmit data or information from geographically remote products.

3.3.1.3   Maintenance Center

Customers bring their products to the maintenance center when the products are predicted to fail or when the products are malfunctioning. For example, when sensors predict that a product is going to fail within a certain time period, the RMC sends an electronic message to the owner of the product and to the maintenance center notifying the imminent failure [35,36]. Before the customer sends the product to the maintenance center, the maintenance personnel can access the static and dynamic data of the product from the RMC to analyze its current condition, diagnose the reasons for failure, and prepare a maintenance plan. This prediagnosis, before the product arrives at the maintenance center, can save maintenance time and cost. Depending on the cause of failure, service personnel can repair the defective components or replace these components with either remanufactured or new components. Servicing information such as the reason(s) for a product's arrival at the maintenance center, problem diagnosed by maintenance personnel, action taken to rectify the problem, and new or refurbished components placed to fix the problem is communicated to the RMC for record keeping.

3.3.1.4 Disassembly Center

Products once sold to customers take different routes before they arrive at the disassembly center, thus making it difficult to determine the exact usage conditions, current state, and remaining life of products. Lack of such information on product life history is a major barrier that renders the current EOL practices inefficient and ineffective. Table 3.1 lists out the issues encountered at the disassembly center and the recommended solution.

TABLE 3.1

Issues and Recommended Solution

Issue	Recommended Solution
When components are disassembled, some of them may turn out to be nonfunctional. In most cases the components of products as given in the bill of materials are accurate. However in some cases it may not be true: components could have been upgraded, replaced, or eliminated. In the later case, it is not possible to find the actual listing of components prior to product disassembly. Although disassembly sequence algorithms are robust to handle component upgrades and replacements, they cannot reduce the uncertainty in disassembly time and cost estimation. With emphasis on design for disassembly and environmental regulations such as waste electrical and electronics equipment (WEEE) in Europe, it has become mandatory to tag the disassembly instructions with the product. To this end, manufacturers are developing the disassembly sequences of products at the design stage based on the bill of materials. Such an approach to disassembly sequence generation fails to account for upgraded, replaced, or missing components. When unexpected component changes are encountered, the predetermined disassembly sequence is rendered obsolete, requiring a revision of the sequence. Frequent revisions of disassembly sequences, although not time-consuming, certainly affect the balance of the disassembly line.	The knowledge available prior to disassembly regarding the condition and the remaining life of components can be used for selective component disassembly to minimize disassembly costs. In the cases in which the product components are upgraded, replaced, or lost, RFID tags affixed on each component can provide the actual listing of components before disassembly. Knowledge about the product's actual constituent components eliminates uncertainty in determining disassembly sequences.

(continued)

TABLE 3.1 (Continued)

Issues and Recommended Solution

Issue	Recommended Solution
The objective of a disassemble-to-order (DTO) system is to determine the optimum number of EOL products to be disassembled to satisfy the demand for components and materials. This determination is aimed at the maximization of total profits and the minimization of product recovery, holding, and disposal costs [37]. Lack of information on the condition and the quality of components can lead to disassembly of products that may contain inferior quality components, substituted components, or upgraded, replaced, or missing components. This can increase disassembly costs and back orders.	RFID tags attached to the components can report the information about the existence of the components. The sensors attached to the components can estimate the quality of these components. RFIDs and sensors can be extremely useful in determining the precise quantity of EOL products that have to be disassembled to satisfy the demand. Only the EOL products that contain good components can be disassembled and thus the quality of components can be maintained above the demanded level. If the demand exceeds the available quantity of components, then additional EOL products are procured to satisfy the demand. This approach can increase the total profit, minimize disassembly, holding, and disposal costs, and minimize back orders.
The layout of a disassembly line is designed on the basis of the disassembly sequence of products [7]. A change in the disassembly sequence directly affects the line balance. The balance of a disassembly line heavily depends upon the processing time at each workstation. Because of the uncertainty in the condition of the components being disassembled, the time to recover the components may vary significantly. This variance can throw the line balance out of gear.	Data provided by sensors embedded in products can be used to estimate the condition of components, which in turn can be used to determine the approximate time to disassemble each component. This determination can minimize the variability in disassembly operation times. To cope with the frequently changing disassembly sequence of products, a disassembly job shop can be used instead of the disassembly line. Using a disassembly job shop can increase the flexibility of the production system and minimize the adverse effects of events such as an early leaving workpiece, self-skipping workpiece, and skipping workpiece, which frequently occur in a disassembly line [7].

3.3.1.5 Recycling Center

Details about the composition of materials (static data) used in the product or the component are accessed from the RMC to determine the appropriate recycling processes.

3.3.1.6 Remanufacturing Center

Remanufactured products can be produced with components that have similar remaining life. Information about remaining life is obtained from the RMC. Remanufactured components are either sent to the maintenance center for use as spares or sold to retailers.

3.3.1.7 Disposal Center

The recovered components from disassembly center that are either physically damaged or have negligible remaining life are dispatched to disposal centers. At the disposal centers, these components are disposed off on the basis of the static data or information obtained from the RMC.

3.3.1.8 Sensor Data Mining

Data mining techniques can be run on the data or the information stored in the RMC to extract information or knowledge such as the patterns of product usage, components causing frequent failures, environmental working conditions of the product, and any information that can provide valuable feedback to the OEM. The OEM can incorporate this extracted information or knowledge to improve the design and the manufacturing of the product.

3.3.2 Conceived Benefits from Product Monitoring Framework

Sensors can have a significant positive impact on the life cycle management of products. Marketing advantage can be gained by extracting patterns of consumer use from the data stored in the RMC. Embedded sensors can enhance reliability, maintainability, serviceability, and recyclability and promote design improvements in each subsequent generation of the products. Table 3.2 lists out the advantages of SEPs over normal products.

3.4 Simulation Results

The product life cycle monitoring framework was implemented using 10,000 desktop computers with sensors embedded to monitor their hard drive as a simulation model. To evaluate the effectiveness of embedding sensors in computers, the performance measures in the two scenarios—with embedded

TABLE 3.2

Advantages of SEPs over Normal Products

Factor	SEPs	Normal Products
Percentage of products accurately predicted to fail	High	Low
Time required to diagnose and repair products	Short	Long
Maintenance cost	Low to medium	Medium to high
Warranty period on remanufactured products	Medium to long	Short
Longevity of products	Beyond expected life in some cases	May reach expected life
Percentage of surprise product failures	Low	High
Life cycle cost	Low	Medium

sensors (sensor embedded computers [SECs]) and without embedded sensors (normal computers)—were compared. The following performance measures were used to evaluate the two simulation models: average life cycle cost, average maintenance cost, average disassembly cost, and average downtime of a computer. The simulation results indicate the following:

- The average downtime and the average repair costs of a computer are lower in the case of SECs because sensors can pinpoint the exact reasons for computer failures and thus reduce the repair time and the downtime.
- The disassembly cost is lower in the case of SECs because (i) only those computers whose components have substantial remaining life are disassembled and (ii) no additional cost is required to estimate the reusability of recovered components because the remaining life is already known before disassembly.
- The average life cycle cost of a computer is lower in the case of SECs because of low repair and disassembly costs.

These results support our contention that embedded sensors provide significant advantage during the usage period and the EOL of computers.

3.5   Conclusions

Embedded sensors that register the working conditions of product components can facilitate online health monitoring of product components, detect or predict product or component failures, and estimate the remaining life of products or components. A product life cycle monitoring framework that interlinked SEPs, RMC, maintenance centers, disassembly centers, recycling centers, disposal centers, and remanufacturing centers was investigated for improved product life cycle management. The RMC archives the sensor-registered data and maintenance-activity data and warns the product user about imminent product failures. The archived data or information at the RMC can assist maintenance personnel in diagnosing product failures and enables reliable estimation of the remaining life of products and their components. The remaining life information can be used to develop better disassembly plans that selectively recover components with substantial life and make refurbished products using recovered components with substantial life. Simulation experiments on the product life cycle monitoring framework indicate that embedded sensors reduce repair time, repair costs, disassembly cost, and average life cycle cost. These factors contribute to improved customer satisfaction and to environmentally effective EOL product management.

Acknowledgment

The authors would like to acknowledge the National Science Foundation for supporting this research under the DMI-0330286 grant.

References

1. B. O'Shea, S. S. Grewal, and H. Kaebernick, State of the art literature survey on disassembly planning, *Concurrent Engineering*, 6, 4, 345–357, 1998.
2. A. Gungor and S. M. Gupta, Issues in environmentally conscious manufacturing and product recovery: A survey, *Computers and Industrial Engineering*, 36, 4, 811–853, 1999.
3. Y. Tang, M. C. Zhou, E. Zussman, and R. Caudill, Disassembly modelling, planning, and application: A review, *Proceedings of IEEE International Conference on Robotics and Automation*, 3, 2197–2202, 2000.
4. D. H. Lee, J. G. Kang, and P. Xirouchakis, Disassembly planning and scheduling: Review and further research, *Proceedings of Institute of Mechanical Engineers*, 215, 5, 695–710, 2001.
5. J. Dong and G. Arndt, A review of current research on disassembly sequence generation and computer aided design for disassembly, *Proceedings of Institute of Mechanical Engineers*, 217, 299–312, 2003.
6. A. J. D. Lambert, Disassembly sequencing: A survey, *International Journal of Production Research*, 41, 16, 3721–3759, 2003.
7. A. J. D. Lambert and S. M. Gupta, *Disassembly Modeling for Assembly, Maintenance, Reuse, and Recycling*, CRC Press, Boca Raton, FL, 2004.

8. V. M. Thomas, Product self-management: Evolution in recycling and reuse, *Environmental Sciences and Technology*, 37, 23, 5297–5302.

9. M. Rasche, Sensors: Bringing the world of tomorrow ... today, *Appliance*, 37–38, December 2003.

10. L. G. Scheidt and S. Zong, An Approach to achieve reusability of electronic modules, *Proceedings of the IEEE International Symposium on Electronics and the Environment*, 331–336, 1994.

11. M. Klausner, W. M. Grimm, C. Hendrickson, and A. Horvath, Sensor-based data recording of use conditions for product takeback, *Proceedings of IEEE International Symposium on Electronics and the Environment*, 138–143, 1998.

12. M. Klausner, W. M. Grimm, and C. Hendrickson, Reuse of electric motors in consumer products, *Journal of Industrial Ecology*, 2, 2, 89–102, 1998.

13. M. Klausner and W. M. Grimm, Integrating product takeback and technical service, *Proceedings of the 1999 IEEE International Symposium on Electronics and Environment*, 48–53, 1999.

14. B. Karlsson, A distributed data processing system for industrial recycling", *Proceedings of IEEE Instrumentation and Measurement Technology Conference*, vol. 1, 197–200, 1997.

15. B. Karlsson, Fuzzy handling of uncertainty in industrial recycling, *Proceedings of IEEE Instrumentation and Measurement Technology Conference*, 1, 832–836, 1998.

16. E. M. Petriu, N. D. Georganas, D. C. Petriu, D. Makrakis, and V. Z. Groza, Sensor-based information appliances, *IEEE Instrumentation and Measurement Magazine*, 3, 4, 31–35, 2000.

17. F. T. Cheng, G. W. Huang, and C. H. Chen, A generic embedded device for retrieving and transmitting information of various customized applications, *Proceedings of the 2004 IEEE International Conference on Robotics and Automation*, New Orleans, LA, 978–983, 2004.

18. M. I. Mazhar, S. Kara, and H. Kaebernick, Reusability assessment of components in consumer products—a statistical and condition monitoring data analysis strategy, *Australian LCA Conference*, Sydney, February 2005.

19. G. Seliger, A. Buchholz, and W. Grudzein, Multiple usage phases by component adaptation, *Proceedings of the 9th CIRP International Seminar on Life Cycle Engineering*, Erlangen, Germany, FAPS, 47–54, 2002.

20. M. Simon, G. Bee, P. R. Moore, J. Pu, and C. Xie, Life cycle acquisition methods and devices (online), available at http://www.mrg.dmu.ac.uk/whitebox/Mech_2000_Lcda.pdf, 2000.

21. P. R. Moore, J. Pu, C. Xie, M. Simon, and G. Bee, Life cycle data acquisition unit—design, implementation, economics and environmental benefits, *Proceedings of the IEEE International Symposium on Electronics and the Environment*, San Francisco, CA, 284–289, 2000.

22. M. Simon, J. Pu, and P. R. Moore, The Whitebox—capturing and using product life cycle data, *Proceedings of the 5th CIRP Seminar on Life Cycle Design*, Stockholm, 161–170, 1998.

23. J. Wallaschek, S. Wedman, and W. Wickod, Lifetime observer: An application of mechatronics in vehicle technology, *International Journal of Vehicle Design*, 28, 1–3, 121–130, 2002.

24. P. Weber, Towards easy-to-communicate and transparent LCA data: Application of life cycle design structure matrix, *Proceedings of the 7th CIRP International Seminar on Life Cycle Engineering*, 52–59, 2000.

25. J. Ni, J. Lee, and D. Djurdjanovic, Watchdog—information technology for proactive product maintenance and its implications to ecological product re-use, *Proceedings of the Symposium on Ecological Manufacturing*, Berlin, Germany, 101–110, 2003.

26. D. Djurdjanovic, J. Yan, H. Qiu, J. Lee, and J. Ni, Web—enabled remote spindle monitoring and prognostics, *Proceedings of the 2nd International CIRP Conference of Re-configurable Manufacturing Systems*, Ann Arbor, MI, August 21–24, 2003.

27. C. Liu, X. F. Zha, Y. Miao, and J. Lee, Internet server controller based intelligent maintenance system for information appliance products, *International Journal of Knowledge-Based and Intelligent Engineering Systems*, 9, 137–148, 2005.

28. J. W. Bruce and L. A. Hathcock, Maintenance and monitoring object models for high availability network appliances, *IEEE Transactions on Consumer Electronics*, 50, 2, 472–477, 2004.

29. C. F. Tsai and H. C. Wu, MASSIHN: A multi-agent architecture for intelligent home network service, *IEEE Transactions on Consumer Electronics*, 48, 3, 505–514, 2002.

30. M. Simon, G. Bee, P. Moore, J.-S. Pu, and C. Xie, Modelling of the life cycle of products with data acquisition features, *Computers in Industry*, 45, 111–122, 2001.

31. A. Schmidt and K. V. Laerhoven, How to build smart appliances? *IEEE Personal Communications*, 8, 4, 66–71, 2001.

32. K. Lee, IEEE 1451: A standard in support of smart transducer networking, *Proceedings of the 17th IEEE Instrumentation and Measurement Technology Conference*, 2, 525–528, 2000.

33. A. Deligiannakis, Y. Kotidis, and N. Roussopoulos, Compressing historical information in sensor networks, *Proceedings of the 2004 ACM SIGMOD International Conference on Management of Data*, 527–538, 2004.

34. A. D. Bucchianico, T. Figarella, G. Hulsken, M. H. Jansen, and H. P. Wynn, A multiscale approach to functional signature analysis for product end-of-life management, *Quality and Reliability Engineering International*, 20, 457–467, 2004.

35. *Life Cycle Management for Design and Maintenance.* LOGTECH Monograph Series, www.logtech.unc.edu, Version: A, 2002.

36. S. Gross, A. Parlikad, D. McFarlane, and E. Fleisch, *The Role of the Auto-ID Enabled Product Information in a Product's Usage: A Maintenance Example*, White Paper, Auto-ID Center, University of St. Gallen, Institute of Technology and Management, Switzerland, October 2003.

37. E. Kongar and S. M. Gupta, A multi-criteria decision making approach for disassembly-to-order system, *Journal of Electronics Manufacturing*, 11, 2, 171–183, 2002.

4

Quantitative Decision-Making Techniques for Reverse/Closed-Loop Supply Chain Design

Kishore K. Pochampally, Satish Nukala, and Surendra M. Gupta

CONTENTS

4.1 Introduction

Traditionally, a supply chain consists of all the stages involved, directly or indirectly, in fulfilling a customer desire [12]. The supply chain not only includes manufacturers and suppliers, but also transporters, warehouses, retailers, and customers. In today's highly competitive business environment, the success of any business depends to a large extent on the efficiency of the supply chain. Competition has moved beyond rivalry between companies to battle between supply chains. Managers in many industries now realize that the decisions made in each of the following three phases have a strong impact on the overall profitability and success of the respective businesses:

1. *Supply chain design.* During this phase, a company decides on its future supply chain structure. Decisions are made about the location, capacity of production, warehouse facilities, the products to be manufactured and stored at various locations, the modes of transportation to be made available, and the type of information system to be utilized. Supply chain design decisions are typically made for the long term (years) and are very expensive to alter on short notice. Consequently, during decision-making companies must take into

account the uncertainty factor in anticipated market conditions over the next few years.

2. *Supply chain planning.* During this phase, companies make decisions about subcontracting of manufacturing, inventory policies to be followed, and timing and size of marketing promotions. Decisions made during this phase cover the time frame from three months to a year. Companies must consider the uncertainty in demand, exchange rates, and competition over this time frame. In this phase, equipped with a shorter time frame and better forecasts than in the design phase, companies try to incorporate any flexibility built into the supply chain in the design phase and exploit it to optimize performance.

3. *Supply chain operation.* During this phase, companies make decisions about individual customer orders. The goal of supply chain operations is to handle incoming customer orders in the best possible manner. Decisions include allocating inventory or production to individual orders, setting a date for an order to be filled, generating pick lists at a warehouse, setting delivery schedules of trucks, and placing replenishment orders. Since operational decisions are made in the short term (minutes, hours, or days), there is less uncertainty about demand information.

Today, many supply chains are no longer limited to fulfilling a customer desire because the growing desire of customers to acquire the latest technology and the rapid technological development of new products have led to a new environmental problem: waste. Products that are discarded after their useful lives or discarded prematurely are called waste (we will use the term "used products" in place of "waste" throughout this chapter). Reprocessing, that is, processing of used products, is essential for the following:

i. *Saving natural resources.* Using reprocessed goods to make products, we conserve land and reduce the need to drill for oil and dig for minerals.

ii. *Saving energy.* Making products from reprocessed goods instead of virgin materials usually takes less energy.

iii. *Saving clean air and water.* Making products from reprocessed goods instead of virgin materials creates less air pollution and water pollution.

iv. *Saving landfill space.* Using reprocessed goods to make a product prevents these used products from going into landfills.

v. *Saving money.* Making products from reprocessed goods reduces cost of manufacture.

In addition to the above mentioned motivators, enforcement of environmental regulations by local governments also drives companies to engage in reprocessing.

Reprocessing of used products involves a series of activities (collection, disassembly, recycling, remanufacturing, disposal, etc.). These activities are performed by multiple parties and are collectively known as reverse supply chain. Today, many companies involved in a traditional supply chain (also known as a forward supply chain) also incorporate reprocessing of used products. This combined practice of forward and reverse supply chains is called closed-loop supply chain (see Figure 4.1 for a generic closed-loop supply chain). The past decade witnessed an explosive growth of reverse and closed-loop supply chains, both in scope and scale. However, decision-making in each of the three phases (design, planning, and operation) of a reverse or a closed-loop supply chain is a very difficult task because of the following primary challenges:

- Uncertainty in supply rate of used products
- Unknown condition of used products
- Imperfect correlation between supply (SU) of used products and demand for reprocessed goods

This chapter illustrates how numerous quantitative techniques available in the literature can be employed to address the aforementioned primary challenges and solve a variety of decision-making problems identified in the design phase of reverse and closed-loop supply chains. This chapter also shows that different techniques can be employed to resolve some of the problems faced in different situations. Table 4.1 lists out the quantitative techniques employed and the decision-making problems identified in this chapter.

The remainder of this chapter is organized as follows:

In Section 4.2, we present a brief review of the literature that addresses decision-making problems in the design phase of reverse and closed-loop supply chains. In Sections 4.3–4.14, we introduce 15 quantitative techniques and propose the employment of each technique appropriate for addressing decision-making problems faced by reverse and closed-loop supply chain designers. Finally, in Section 4.15, we state some conclusions.

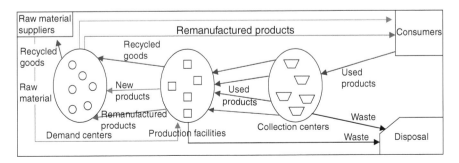

FIGURE 4.1
Generic closed-loop supply chain.

TABLE 4.1

Quantitative Techniques and Decision-Making Problems

Quantitative Techniques	Decision-Making Problems
1. Analytic hierarchy process	A. Selection of economical used products
2. Eigenvector method	B. Evaluation of collection centers
3. Fuzzy logic	C. Evaluation of recovery facilities
4. Bayesian updating	D. Optimal transportation of goods
5. Quality function deployment	E. Evaluation of marketing strategy
6. Method of total preferences	F. Evaluation of futurity of used products
7. Technique for order preference by similarity to ideal solution	G. Selection of secondhand markets
	H. Selection of new products
8. Borda's choice rule	I. Evaluation of production facilities
9. Neural networks	
10. Linear physical programming	
11. Goal programming	
12. Linear integer programming	
13. Cost–benefit function	
14. Analytic network process	
15. Extent analysis method	

4.2 Literature Review

Reverse and closed-loop supply chain design is a relatively new area of research and only a few quantitative models and case studies have been reported in the literature. In this section, we present a brief survey of those models and studies (see Ref. [14] for a detailed survey of many of the case studies):

- Louwers et al. [29] consider the design of a reverse supply chain for carpet waste in Europe. They propose a continuous location model in which all costs are considered volume-dependent. By solving the nonlinear model we can determine the appropriate locations and capacities for the regional recovery facilities, taking into consideration transportation, investment, and processing costs.

- Ravi et al. [33] propose a balanced scorecard and an analytic network process (ANP)–based approach to evaluate the alternative reverse logistics operations for used computers. This holistic approach links financial and nonfinancial, tangible and intangible, and internal and external factors for the selection of an alternative.

- Gupta and Veerakamolmal [16] propose a bidirectional supply chain optimization model for a reverse supply chain that demonstrates the management of demand and supply for the remanufacturing process. They present a system's decision-making approach to determine the aggregate number of a type of product to be disassembled to fulfill the demand in each period of the planning horizon and yet have an environmentally benign waste management policy.

- Beamon and Fernandes [4] propose a multiperiod mixed integer linear programming (MILP) model to study a closed-loop supply chain in which manufacturers produce new products and remanufacture used products. Their model determines which warehouses and collection centers should be opened and which warehouses should have sorting capabilities.

- Barros et al. [3] address the design of a supply chain for recycling sand from processing construction waste in the Netherlands. The four-level sand recycling network includes (i) companies engaged in crushing, yielding sieved sand from construction waste; (ii) regional depots specifying the pollution level and storing cleaned and half-cleaned sand; (iii) treatment facilities for cleaning and storing polluted sand; and (iv) infrastructure projects permitting sand reuse. The locations of the sand sources are known, and their supply volumes are estimated on the basis of historical data. The optimal number, capacities, and locations of depots and cleaning facilities are to be determined. The authors propose a multilevel capacitated facility location model for this problem, which is formulated as an MILP model, and solve the problem through iterative rounding of LP-relaxations strengthened by valid inequalities. Listes and Dekker [28] propose a stochastic programming-based approach for the sand-recycling network to account for the uncertainties. The stochastic model seeks to find a solution that is approximately balanced between some alternative scenarios identified by field experts.

- Savaskan et al. [39] study the problem of choosing the appropriate reverse channel structure for collecting the used products from customers. The researchers compare three decentralized closed-loop supply chain models, the manufacturer collecting the used products, the retailer collecting the used products, and a third party collecting the used products, with respect to the wholesale price, product return rate, and the total supply chain profits.

- Ammons et al. [2] address carpet recycling in the United States. A logistics network that includes the collection of used carpets from carpet dealerships and the separation of nylon and other reusable materials while landfilling the remainder is investigated. The authors assume that the delivery sites for the recovered materials are known, but the optimal number and location of both collection sites and processing plants for alternative configurations are to be determined. In addition, the amount of carpet collected from each site is to be determined. Facility capacity constraints is the main restriction in view of the vast volume that is landfilled. The authors propose a multilevel capacitated mixed integer LP model to address this problem.

- Biehl et al. [6] simulate a reverse logistics network for carpet recycling to manage highly variable return flows. They use experimental design technique to study the effect of system design factors and environmental factors on the operational performance of the reverse logistics network. From their study, the authors conclude that even with the design of an efficient reverse logistics network and the use of sophisticated recycling technologies, return flows cannot meet demand for nearly a decade. They also discuss possible managerial options to address this problem, which include legal responses to require return flows and utilization of market incentives for carpet recycling.

- Hu et al. [18] present a cost-minimization model for a multitime-step, multitype hazardous wastes reverse logistics system. The authors formulate a discrete-time analytical model that minimizes total hazardous waste reverse logistics costs subject to constraints, including business-operating strategies and government regulations. The authors consider waste collection, storage, processing, and distribution in their model. They found that use of the proposed methodology and the operational strategies can reduce the total reverse logistics costs by up to 49%.

- Lieckens and Vandaele [26] combine queuing models with the traditional reverse logistics location model and formulate MILP models to determine which facilities to open while minimizing the total cost of investment, transportation, disposal, and procurement. By combining the queuing models, they take care of some dynamic aspects such as lead time, inventory positions, and the high degree of uncertainty associated with reverse logistics networks . With these extensions, the problem is defined as a mixed integer nonlinear programming model. This model is presented for a single product, single-level network, and several case examples are solved using genetic algorithms based on the technique of differential evolution.

- Alshamrani et al. [1] study the reverse logistics network for blood distribution with the American Red Cross, where containers in which blood is delivered from a central processing unit to customers in one time period are available for return to the central processing unit the following period. Containers that are not picked up in the period following their delivery incur a penalty cost. This leads to a dynamic logistics planning problem, where, in each period, the vehicle dispatcher needs to design a multistop vehicle route while determining the number of containers to be picked up at each stop. A heuristic procedure is developed to solve the route design–pickup strategy problem.

- Vlachos et al. [46] address the capacity planning issues in remanufacturing facilities in reverse supply chains through a simulation model based on the principles of system dynamics methodology. In addition to economic issues, they consider environmental issues such as take back obligations and the green image effect on customer demand. The simulation model serves as an experimental tool that helps in evaluating long-term capacity planning policies using total supply chain profit as a measure of effectiveness.

- Lu and Bostel [30] study the facility location problem in a remanu-facturing network that has a strong interaction between the forward and reverse flows of products. Remanufactured products are intro-duced as new products into the forward flow. The authors assume that the demand is deterministic and the facilities are of three dif-ferent types: producers, remanufacturing centers, and intermediate centers. The problem is modeled as a 0–1 mixed integer program-ming problem that is solved using an algorithm based on Lagrang-ian heuristics.

- Spengler et al. [41] study the recycling networks for industrial by-products in the German steel industry. Recycling facilities with vari-able capacity levels and corresponding fixed and variable processing costs can be installed at a set of potential locations. Thus, we need to determine which recycling processes or process chains to install at different locations and their capacity levels. The authors propose a multilevel warehouse location model with piecewise linear costs, which is used for optimizing several scenarios.

- Thierry et al. [43] propose a conceptual model for a closed-loop sup-ply chain that addresses the situation of a manufacturing company collecting used products for recovery, in addition to producing and distributing new products. The recovered products are sold under the same conditions as the new products to satisfy a given market demand. The distribution network encompasses three levels, namely plants, warehouses, and markets. No fixed costs are considered in this model because all facilities are externally fixed. The objective is to determine the cost-optimal flow of goods in the network under the given capacity constraints. The problem is formulated as an LP model, which can be solved for optimality.

- Berger and Debaillie [5] address a situation similar to that of Thierry et al. [43] and propose a conceptual model for extending an exist-ing forward supply chain with disassembly centers to enable recov-ery of used products. The model is illustrated through a fictitious case of a computer manufacturer. Although the facilities in the for-ward supply chain are fixed, the number, locations, and capacities of the disassembly centers are to be determined. In a variant of this model, the recovery network is extended to another level by separat-ing inspection and disassembly or repair centers. After inspection, rejected items are disposed off, while recoverable items are sent to the disassembly centers. To this end, the authors propose a multi-level capacitated mixed integer LP model.

- Lim et al. [27] propose a mixed integer programming model that takes into account multiperiod planning horizons with uncertainties for a product with modular design and multiproduct configurations. In addition to maximizing overall profit, this model also considers minimizing environmental impacts by minimizing energy consumption.

- Reimer et al. [34] model the economics of electronics recycling from the perspective of recyclers, generators, and material processors individually. They propose a nonlinear mixed integer programming model for optimizing processing decisions in electronics recycling operations.

- Jayaraman et al. [21] analyze the logistics network of an electronic equipment remanufacturing company in the United States. The company's activities include collection of used products from customers, remanufacturing, and distribution of remanufactured products. In this network, the optimal number and locations of remanufacturing facilities and the number of used products collected are to be determined, while considering investment, transportation, processing, and storage costs. The authors propose a multiproduct capacitated warehouse location–mixed integer LP model, which is solved to optimality, for different supply and demand scenarios.

- Krikke et al. [24] report a case study on the implementation of the remanufacturing process at a copier manufacturing company in the Netherlands. The reverse supply chain is subdivided into three main stages, namely (i) disassembly of returned products to a fixed level; (ii) preparation, which encompasses the inspection and replacement of critical components; and (iii) reassembly of the remaining carcass together with repaired and new components into a remanufactured machine. While the supplying processes and disassembly are fixed, optimal locations and flow of goods are to be determined for both the preparation and the reassembly operations. Based on a mixed integer LP model, the optimal solution, minimizing operational costs, is compared with a number of preselected managerial solutions.

- Kroon and Vrijens [25] consider the design of a logistics system for reusable transportation packaging. More specifically, a closed-loop deposit-based system is considered for collapsible plastic containers that can be rented as secondary packaging material. The actors involved in the system include a central agency owning a pool of reusable containers; a logistics service provider responsible for storing, delivering, and collecting the empty containers; senders and recipients of full containers; and carriers transporting full containers from the sender to the recipient. In addition to determining the number of containers required for running the system and an appropriate fee per shipment, the probable location of the depots for empty containers poses a major question. Balancing the number of containers at the depots is also a requirement. The problem is formulated as a mixed integer LP model that is closely related to a classical uncapacitated warehouse location model.

- Salema et al. [38] point out that the majority of quantitative models existing in the area of reverse supply chain design are case-specific and hence lack generality. They propose a generic reverse supply chain model that incorporates multiproduct management, capacity limits, and uncertainty in product demands and returns. They also propose a mixed integer formulation that is solved using standard branch and bound (B&B) techniques.

4.3 Analytic Hierarchy Process and Eigen Vector Method

The methodology of analytic hierarchy process (AHP) is briefly described in Section 4.3.1 and is employed in Section 4.3.2 for selecting potential recovery facilities.

4.3.1 Methodology

The AHP is a tool, supported by simple mathematics, which enables decision makers to explicitly weigh tangible and intangible criteria against each other for the purpose of evaluating different alternatives. This process has been formalized by Saaty [35] and is used across a wide variety of problem areas (e.g., siting landfills [40], evaluating employee performance [36], and selecting a doctoral program [42]).

In a large number of cases (e.g., Ref. [32]), the tangible and intangible criteria (for evaluation) are considered independent of each other, that is, those criteria do not depend upon subcriteria and so on. The AHP in such cases is conducted in two steps: (1) weighing independent criteria using pairwise judgments and (2) computing the relative ranks of alternatives using pairwise judgments with respect to each independent criterion.

1. *Computation of relative weights of criteria.* AHP enables a person to make pairwise judgments of importance, between independent criteria, with respect to the scale shown in Table 4.2. The resulting matrix of comparative importance values is used to weigh the independent criteria by employing mathematical techniques such as eigenvalue or geometric mean.

2. *Computation of the relative ranks.* Pairwise judgments of importance using the scale shown in Table 4.2 are also computed for the alternatives. These judgments are obtained with respect to each independent criterion considered in step 1. The resulting matrix of comparative importance values is used to rank the alternatives by employing mathematical techniques such as eigenvalue or geometric mean.

TABLE 4.2

Scale for Pairwise Judgments

Comparative Importance	Definition
1	Equally important
3	Moderately more important
5	Strongly important
7	Very strongly more important
9	Extremely more important
2, 4, 6, 8	Intermediate judgment values

TABLE 4.3

Random Index Value for Each n Value

n	1	2	3	4	5	6	7	8	9	10
R	0	0	0.58	0.90	1.12	1.24	0.32	1.41	1.45	1.49

If eigenvalue is the employed mathematical technique for weighing the independent criteria or ranking the alternatives, the respective procedure is called eigenvector method.

The degrees of consistency of pairwise judgments in steps 1 and 2 are measured using an index called the consistency ratio (CR). Perfect consistency implies a value of zero for CR. However, perfect consistency cannot be demanded because we are often biased and inconsistent in our subjective judgments. Therefore, it is considered acceptable if CR $\leq$ 0.1. If CR $>$ 0.1, the pairwise judgments must be revised before the weights of the criteria and the ranks of the alternatives are computed. CR is computed using the formula:

$$CR = \frac{(\lambda_{max} - n)}{(n - 1)(R)} \tag{4.1}$$

where λ_{max} is the principal eigenvalue of the matrix of comparative importance values; n is the number of rows (or columns) in the matrix; and R is the random index for each n value that is greater than or equal to 1. Table 4.3 shows various R values for n values ranging from 1 to 10.

The AHP is illustrated in the form of a hierarchy of three levels, where the first level contains the primary objective, the second level contains the independent criteria, and the last level contains the alternatives. An important feature of the AHP is that the tangible and the intangible criteria in the second level must be chosen in such a way that they can somehow help the decision maker in comparing the different alternatives.

4.3.2 Evaluation of Recovery Facilities

In this section, we employ the AHP to identify the potential facilities in a set of candidate recovery facilities operating in a region where a reverse supply chain is to be designed.

The first level in the hierarchy contains our primary objective, that is, identifying potential facilities in a set of candidate recovery facilities. The last level in the hierarchy contains the candidate recovery facilities. The middle level contains criteria that must somehow be useful in comparing the candidate recovery facilities. For example, fixed cost and average skill level of the employees are criteria that can be used to compare the candidate facilities. Although the criteria to be considered in a reverse supply chain seem similar to those considered in a forward supply chain, there are three special factors in a reverse supply chain that need to be incorporated in the AHP without disturbing the hierarchy levels. These special factors are as follows:

- *Average quality of used products.* Unlike in a forward supply chain, components of used products of even the same type in a recovery facility are likely to be of varied quality (worn-out, low performing, etc). Although the average quality of reprocessed goods (QO) is a criterion for comparing two or more candidate facilities, using QO as an independent criterion for comparison is not justified, because QO depends on the average quality of incoming products (QI). So, the idea is to take the difference between QO and QI as a criterion in the hierarchy.

- *Average supply of used products.* The only driver to design a forward supply chain is the demand for new products and so if there is low demand for new products, there is practically no forward supply chain. However, this is not the case in some reverse supply chains where even a low supply of used products (SU) would trigger the administration of a reverse supply chain due to the possible drivers discussed in 1. In supply-driven cases like these, it is unfair to judge a recovery facility without considering SU in the hierarchy. Although throughput (TP) is a criterion that can be used to compare two or more candidate recovery facilities, using TP as an independent criterion is not justified, because TP depends on SU. In other words, a low SU might lead to a low TP and a high SU might lead to a high TP. Hence, the idea is to consider TP or SU as a criterion in the hierarchy. Thus, we compensate for the effect of a low TP by dividing TP with a possibly low SU to avoid underestimating the facility under consideration. Similarly, we dampen the effect of a high TP by dividing TP with a possibly high SU to avoid overestimating the facility under consideration.

- *Average disassembly time of used products.* In contrast to a forward supply chain, components of incoming goods (used products) in a recovery facility are likely to be deformed, broken, or different in number even for the same type of products. Hence, incoming products of the same type might have different average disassembly times (DTs). In a forward supply chain, assembly times are predetermined and equal for products of the same type. Since TP of a recovery facility depends upon DT, it is unfair not to consider DT in the hierarchy. Since a high DT might lead to a low TP and a low DT might lead to a high TP, the idea is to consider (TP) × (DT) as a criterion in the hierarchy. Thus, we compensate for the effect of a low TP by multiplying TP with a possibly high DT to avoid underestimating the facility under consideration. Similarly, we dampen the effect of a high TP by multiplying TP with a possibly low DT to avoid overestimating the facility under consideration.

In our approach, we consider customer service (CSE) as an intangible criterion. CSE provides an idea about how well a recovery facility is utilizing the incentives provided by the government, to what extent it is meeting the environmental regulations, what kind of incentives it is giving the collection centers that are supplying the used products, and what kind of incentives it is giving the customers who are buying the reprocessed goods. We are using the term "customer service" here because, in our opinion, any beneficiary is a customer, be it the government, the collection center, or the actual customer buying the reprocessed goods. In addition to the aforementioned criteria, we also consider the fixed cost of the facility (CO) in the hierarchy. Figure 4.2 illustrates the three-level hierarchy, in our approach, to implement the AHP for evaluation of recovery facilities.

Table 4.4 shows the comparative importance values given to the criteria in the second level of hierarchy in this example. It also gives the normalized (i.e., sum = 1) row eigenvalues (eigenvector method) for the comparative importance value matrix. These values represent the relative weights assigned by the decision maker to the independent criteria.

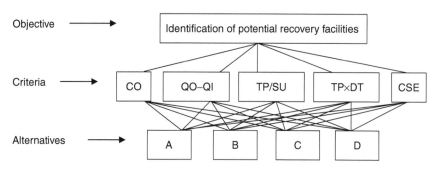

FIGURE 4.2
Three-level hierarchy for AHP.

TABLE 4.4

Comparative Importance Values at Second Hierarchy Level

Criteria	CO	QO – QI	TP/SU	TP × DT	CSE	Normalized Eigenvalues
CO	1	1/5	3	1	1/5	0.104
QO – QI	5	1	7	3	5	0.491
TP/SU	1/3	1/7	1	1/2	1/3	0.055
TP × DT	1	1/3	2	1	1/3	0.110
CSE	5	1/5	3	3	1	0.240

TABLE 4.5

Comparative Importance Values of Recovery Facilities with Respect to CO

CO Facilities	A	B	C	D	Normalized Eigenvalues
A	1	3	6	2	0.460
B	1/3	1	7	3	0.310
C	1/6	1/7	1	1/4	0.050
D	1/2	1/3	4	1	0.180

TABLE 4.6

Comparative Importance Values of Recovery Facilities with Respect to QO – QI

QO – QI Facilities	A	B	C	D	Normalized Eigenvalues
A	1	1	7	4	0.380
B	1	1	7	7	0.445
C	1/7	1/7	1	1/5	0.050
D	1/4	1/7	5	1	0.125

Tables 4.5 through 4.9 show comparative importance values of the alternatives (recovery facilities A, B, C, and D) with respect to the criteria CO, QO – QI, (TP) or (SU), (TP) × (DT), and CSE, respectively. They also show the normalized row eigenvalues representing the respective rankings. Note that each of the matrices in Tables 4.4 through 4.9 has a CR whose value is less than or equal to 0.1. For example, for the matrix in Table 4.7, $n = 4$, $R = 0.90$ (see Table 4.3), and λ_{max} is calculated as 4.15. Hence, from Equation 4.1, CR of that matrix is $(4.15 – 4)/(4 – 1) \times (0.90) = 0.06$, which is less than 0.1.

Table 4.10 shows the aggregate matrix of rankings of recovery facilities with respect to each criterion in the second level of hierarchy. This matrix is

TABLE 4.7

Comparative Importance Values of Recovery Facilities with Respect to (TP)/(SU)

TP/SU Facilities	A	B	C	D	Normalized Eigenvalues
A	1	1/7	1/3	1/2	0.072
B	7	1	2	7	0.574
C	3	1/2	1	1	0.212
D	2	1/7	1	1	0.142

TABLE 4.8

Comparative Importance Values of Recovery Facilities with Respect to (TP) × (DT)

TP × DT Facilities	A	B	C	D	Normalized Eigenvalues
A	1	1/5	1/2	1/2	0.091
B	5	1	3	7	0.595
C	2	1/3	1	1	0.171
D	2	1/7	1	1	0.143

TABLE 4.9

Comparative Importance Values of Recovery Facilities with Respect to CSE

CSE Facilities	A	B	C	D	Normalized Eigenvalues
A	1	1/6	1/3	1/7	0.053
B	6	1	5	1/3	0.298
C	3	1/5	1	1/6	0.101
D	7	3	6	1	0.548

TABLE 4.10

Aggregate of Rankings of Recovery Facilities with Respect to Each Criterion

	A	B	C	D
CO	0.460	0.310	0.050	0.180
QO − QI	0.380	0.445	0.050	0.125
TP/SU	0.072	0.574	0.212	0.142
TP × DT	0.091	0.595	0.171	0.143
CSE	0.053	0.298	0.101	0.548

nothing but the aggregate of the normalized eigenvectors (columns of normalized row eigenvalues) obtained in Tables 4.5 through 4.9.

Multiplying the matrix in Table 4.10 with the normalized eigenvector obtained in Table 4.4, we get the following normalized aggregate ranks for the facilities: $\text{rank}_A = 0.26$, $\text{rank}_B = 0.42$, $\text{rank}_C = 0.09$, and $\text{rank}_D = 0.23$. If the decision maker wishes to choose only those recovery facilities that have normalized aggregate ranks of at least 25%, as the potential recovery facilities, he will choose recovery facilities A and B.

4.4 Fuzzy Logic

This section is dedicated solely to the concepts of fuzzy logic [48]. Decision-making problems are addressed in later sections, where fuzzy logic is used in combination with other quantitative decision-making techniques.

Commonly used expressions such as "not very clear," "probably so," or "very likely," are imprecise. This imprecision or vagueness of human decision making is called "fuzziness" in the scientific literature. With different decision-making problems of varied intensity, the results can be misleading if fuzziness is not taken into account. However, since Zadeh [48] first proposed fuzzy logic, an increasing number of studies have dealt with imprecision (fuzziness) in problems by applying fuzzy logic. We will utilize the concepts of fuzzy logic in approaching many decision-making problems in this chapter.

4.4.1 Linguistic Values and Fuzzy Sets

When dealing with imprecision, decision makers may be provided with information characterized by vague language such as high risk, low profit, and good customer service. By using linguistic values such as "high," "low," "good," "medium," and "cheap," people usually attempt to describe factors with uncertain or imprecise values. For example, the weight of an object may be a factor with an uncertain or imprecise value and so its linguistic value can be "very low," "low," "medium," "high," "very high," etc. Fuzzy logic is primarily concerned with quantifying the vagueness in human thoughts and perceptions. The transition from vagueness to quantification is performed by applying fuzzy logic as depicted in Figure 4.3.

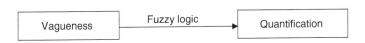

FIGURE 4.3
Application of fuzzy logic.

To quantify vagueness, Zadeh proposed a membership function that associates with each quantified linguistic value a grade of membership belonging to the interval [0, 1]. A fuzzy set is defined as

$$\forall x \in X \quad \mu_A(x) \in [0,1]$$

where $\mu_A(x)$ is the degree of membership, ranging from 0 to 1, of a quantity x of the linguistic value, A, over the universe of quantified linguistic values, X. This X is essentially a set of real numbers. The more x fits A, the larger the degree of membership of x. A quantity with a degree of membership equal to 1 reflects a complete fitness between the quantity and the vague description (linguistic value). However, if the degree of membership of a quantity is 0, then that quantity does not belong to the vague description.

The membership function can be viewed as an expert's opinion. We use the term "expert" because an expert usually possesses some required knowledge about relative problems, which a layperson may not have. For example, when a financial manager is asked what a "high annual interest rate" is, the possibility of 20% being "high annual interest rate" would be higher than that of 3%, 5%, or 9%. Thus, the membership function here can be explained as the possibility of an interest rate being considered as "high." A reasonable mapping from interest rate to its degree of membership about the fuzzy set high annual interest rate is depicted in Figure 4.4.

4.4.2 Triangular Fuzzy Numbers

A triangular fuzzy number (TFN) [45] is a fuzzy set with three parameters, each representing a quantity of a linguistic value associated with a degree of membership of either 0 or 1. The TFN is graphically depicted in Figure 4.5. The parameters a, b, and c denote the smallest possible quantity, the most promising quantity, and the largest possible quantity, respectively, which describe the linguistic value.

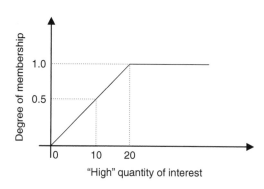

FIGURE 4.4
Mapping of quantified "high" interest values to their degrees of membership.

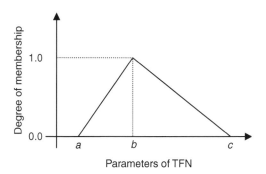

FIGURE 4.5
Triangular fuzzy number.

Each TFN, P, has linear representations on its left- and right-hand side, such that its membership function can be defined as

$$\mu_p = 0 \qquad x < a \tag{4.2}$$

$$= \frac{(x - a)}{(b - a)} \quad a \leq x \leq b \tag{4.3}$$

$$= \frac{(c - x)}{(c - b)} \quad b \leq x \leq c \tag{4.4}$$

$$= 0 \qquad x \geq c \tag{4.5}$$

For each quantity x increasing from a to b, its corresponding degree of membership linearly increases from 0 to 1. While x increases from b to c, its corresponding degree of membership linearly decreases from 1 to 0. The membership function is a mapping from any given x to its corresponding degree of membership.

The TFN is mathematically easy to implement, and more importantly, it represents the rational basis for quantifying the vague knowledge in most decision-making problems. The basic operations on TFNs are as follows [8,45]:

For example, $P_1 = (a, b, c)$ and $P_2 = (d, e, f)$.

$$P_1 + P_2 = (a + d, b + e, c + f) \quad \text{addition} \tag{4.6}$$

$$P_1 - P_2 = (a - f, b - e, c - d) \quad \text{subtraction} \tag{4.7}$$

$$P_1 \times P_2 = (a \times d, b \times e, c \times f) \quad \text{where } a \geq 0 \quad \text{and} \quad d \geq 0 \text{ multiplication} \tag{4.8}$$

$$\frac{P_1}{P_2} = \left(\frac{a}{f}, \frac{b}{e}, \frac{c}{d}\right) \quad \text{where } a \geq 0 \quad \text{and} \quad d > 0 \text{ division} \tag{4.9}$$

4.4.3 Defuzzification

Defuzzification is a technique to convert a fuzzy number into a crisp real number. There are several methods for this conversion. For example, the center-of-area method [47] converts a TFN, $P = (a, b, c)$ into a crisp real number Q, where

$$Q = \frac{(c - a) + (b - a)}{3} + a \tag{4.10}$$

Defuzzification might become necessary in two situations: (i) when comparison between two or more fuzzy numbers is difficult to perform and (ii) when a fuzzy number, to be operated on, has negative parameters (in other words, we ensure that we get only a TFN after performing an arithmetic operation on one or more TFNs, e.g., squaring TFN $(-1, 0, 1)$ using Equation 4.8 will lead to $(1, 0, 1)$, which is not a TFN, and so, we defuzzify $(-1, 0, 1)$ before squaring it).

4.5 Bayesian Updating

The concepts of Bayesian updating [17] are briefly described in Section 4.5.1 and are employed (along with fuzzy logic) in Section 4.5.2 for evaluation of futurity of used products.

4.5.1 Concepts

Bayesian updating is an uncertainty modeling technique. This technique assumes that it is possible for an expert in a domain to guess a probability to every hypothesis in that domain and that this probability can be updated in the light of evidence for or against the hypothesis.

Suppose the probability of a hypothesis H is $P(H)$. Then, the formula for the odds of that hypothesis, $O(H)$, is

$$O(H) = \frac{P(H)}{1 - P(H)} \tag{4.11}$$

A hypothesis that is absolutely certain, that is, has a probability of 1, has infinite odds. In practice, limits are often set on odds so that, for example, if $O(H) > 1000$, then H is true, and if $O(H) < 0.01$, then H is false.

4.5.1.1 *Updating Probabilities with Supporting Evidence*

The standard formula for updating the odds of hypothesis H, given that evidence E is observed, is

$$O(H|E) = (A) \cdot O(H) \tag{4.12}$$

where $O(H|E)$ is the odds of H, given the presence of evidence E, and A is the affirms weight of E. The definition of A is

$$A = \frac{P(E|H)}{P(E|\sim H)} \tag{4.13}$$

where $P(E|H)$ is the probability of E, given that H is true, and $P(E|\sim H)$ is the probability of E, not given that H is true.

4.5.1.2 Updating Probabilities with Opposing Evidence

Bayesian updating assumes that the absence of supporting evidence is equivalent to the presence of opposing evidence. The standard formula for updating the odds of a hypothesis H, given that the evidence E is absent, is

$$O(H|\sim E) = (D) \cdot O(H) \tag{4.14}$$

where $O(H|\sim E)$ is the odds of H, given the absence of evidence E, and D is the denies weight of E. The definition of D is

$$D = \frac{P(\sim E|H)}{P(\sim E|\sim H)} = \frac{1 - P(E|H)}{1 - P(E|\sim H)} \tag{4.15}$$

If a given piece of evidence E has an affirms weight A, which is greater than 1, then its denies weight must be less than 1 and vice versa. Also, if $A > 1$ and $D < 1$, then the presence of evidence E is supportive of hypothesis H. Similarly, if $A < 1$ and $D > 1$, then the absence of E is supportive of H. For example, while controlling a power station boiler, a rule—"IF (temperature is high) and NOT (water level is low) THEN (pressure is high)" can also be written as "IF (temperature is high—AFFIRMS A_1, DENIES D_1) AND (water level is low—AFFIRMS A_2, DENIES D_2) THEN (pressure is high)." Here,

$$A_1 = \frac{P(\text{temperature is high}|\text{pressure is high})}{P(\text{temperature is high}|\sim\text{pressure is high})} \qquad D_1 = \frac{P(\sim\text{temperature is high}|\text{pressure is high})}{P(\sim\text{temperature is high}|\sim\text{pressure is high})}$$

$$A_2 = \frac{P(\text{water level is low}|\text{pressure is high})}{P(\text{water level is low}|\sim\text{pressure is high})} \qquad D_2 = \frac{P(\sim\text{temperature is high}|\text{pressure is high})}{P(\sim\text{temperature is high}|\sim\text{pressure is high})}$$

4.5.1.3 Dealing with Uncertain Evidence

Sometimes, an evidence is neither definitely present nor definitely absent. For example, if we are diagnosing a TV set that is not functioning properly, it is not definite if the cause of the problem is a malfunctioning picture tube or not. In such a case, depending on the value of the probability of the evidence $P(E)$, the affirms and denies weights are modified using the following formulae:

$$A' = [2 \cdot (A - 1) \cdot P(E)] + 2 - A \tag{4.16}$$

$$D' = [2 \cdot (1 - D) \cdot P(E)] + D \tag{4.17}$$

When $P(E)$ is greater than 0.5, the affirms weight is used to calculate $O(H|E)$; when $P(E)$ is less than 0.5, the denies weight is used.

4.5.1.4 Combining Evidence

If n statistically independent pieces of evidence are found that support or oppose a hypothesis H, then the updating equations are

$$O(H|E_1 \& E_2 \& E_3 \cdots E_n) = (A_1) \cdot (A_2) \cdot (A_3) \cdots (A_n) \cdot O(H) \qquad (4.18)$$

and

$$O(H|{\sim}E_1 \& {\sim}E_2 \& {\sim}E_3 \cdots {\sim}E_n) = (D_1) \cdot (D_2) \cdot (D_3) \cdots (D_n) \cdot O(H) \qquad (4.19)$$

A_i and D_i are given by Equations 4.20 and 4.21, respectively.

$$A_i = \frac{P(E_i|H)}{P(E_i|{\sim}H)} \qquad (4.20)$$

$$D_i = \frac{P({\sim}E_i|H)}{P({\sim}E_i|{\sim}H)} \qquad (4.21)$$

4.5.2 Evaluation of Futurity of Used Products

Companies interested in collecting used products are driven by the idea of recoverable value through reprocessing. However, these companies seldom know when those products were bought and why they were discarded. Also, the remaining life periods are not indicated for these products.. Hence, these products often undergo partial or complete disassembly for subsequent reprocessing. For some used products, it might make more sense to make necessary repairs to the products and sell them in the secondhand markets than to disassemble them for subsequent reprocessing. In such cases, using a numerical example, we employ Bayesian updating and fuzzy logic to decide if it is sensible to repair a used product of interest for subsequent sale in a secondhand market (we assume here that the used product of interest functions improperly; it is obviously sensible to sell a properly functioning used product in a secondhand market).

Consider the used product, R, in Figure 4.6 (S1 and S2 are subassemblies and C1–C5 are components). Assuming that this product is not functioning properly, we shall decide whether it is sensible to repair it for subsequent sale in a secondhand market. Table 4.11 shows the probability values we use to implement Bayesian updating.

Section 4.5.2.1 presents the procedure to use fuzzy logic in Bayesian updating, Section 4.5.2.2 provides a list of the rules used in Bayesian updating, and Section 4.5.2.3 presents the implementation of Bayesian updating.

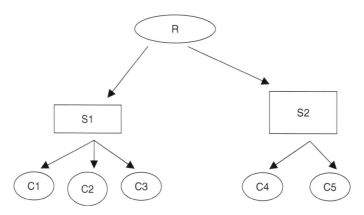

FIGURE 4.6
Used product.

TABLE 4.11

Probability Values Used in Bayesian Updating

H	E	P(H)	O(H)	P(E|H)	P(E|~H)	A	D
S1 needs repair	Product needs repair	0.60	1.50	1.00	0.60	1.67	0.00
S2 needs repair	Product needs repair	0.70	2.33	1.00	0.40	2.50	0.00
C1 needs repair	S1 needs repair	0.45	0.82	1.00	0.45	2.22	0.00
C2 needs repair	S1 needs repair	0.55	1.22	1.00	0.30	3.33	0.00
C3 needs repair	S1 needs repair	0.30	0.43	1.00	0.55	1.82	0.00
C4 needs repair	S2 needs repair	0.32	0.47	1.00	0.70	1.43	0.00
C5 needs repair	S2 needs repair	0.10	0.11	1.00	0.80	1.25	0.00
Sensible to repair product	C1 needs repair	0.60	1.50	**0.70**	**0.20**	3.50	0.38
Sensible to repair product	C2 needs repair	0.60	1.50	**0.60**	**0.30**	2.00	0.57
Sensible to repair product	C3 needs repair	0.60	1.50	**0.45**	**0.60**	0.75	1.38
Sensible to repair product	C4 needs repair	0.60	1.50	**0.10**	**0.75**	0.13	3.60
Sensible to repair product	C5 needs repair	0.60	1.50	**0.85**	**0.40**	2.13	0.25

4.5.2.1 Usage of Fuzzy Logic

Since it is difficult for an expert to guess the probabilities shown in bold (unlike the rest) in Table 4.11, we calculate them using fuzzy logic as follows:

I. Ask the expert to assign a linguistic rating to $P(E|H)$, for each component with respect to each of the following factors (see Table 4.12):

   a. Is it economical to repair or replace the component? (More economical implies higher rating.)

   b. If disposed, will the component be harmful to the environment? (More harmful implies higher rating.)

   c. What is the remaining life period of the component? (Longer life implies higher rating.)

   d. Is the raw material used to make the component depleting fast? (Faster depletion implies higher rating.)

   e. Is it difficult to repair the component? (More difficult implies lower rating.)

II. Use the data in Table 4.13 to convert the linguistic ratings into TFNs.

III. Calculate the average fuzzy $P(E|H)$ value for each component.

IV. Defuzzify the average $P(E|H)$ for each component.

TABLE 4.12

Linguistic $P(E|H)$ Ratings

	a	b	c	d	e
C1	H	H	M	M	H
C2	VH	H	VL	M	L
C3	L	L	VL	VH	M
C4	H	H	H	H	VL
C5	M	H	H	H	M

Note: VH = very high, H = high, M = medium, L = low, VL = very low.

TABLE 4.13

Conversion Table for Linguistic $P(E|H)$ Ratings in Bayesian Updating

Linguistic Rating	Triangular Fuzzy Number
VH	(0.7, 0.9, 1.0)
H	(0.5, 0.7, 0.9)
M	(0.3, 0.5, 0.7)
L	(0.1, 0.3, 0.5)
VL	(0.0, 0.1, 0.3)

Note: H = high, VH = very high, M = medium, L = low, VL = very low.

Apply steps I–IV to calculate appropriate $P(E|\sim H)$ values for each component. To save ourselves from the tedious calculations, we assume that the values shown in bold in Table 4.11 are the defuzzified average probabilities obtained after performing steps I–IV. For clarity, however, we present a numerical example to show the calculation procedure.

Suppose we wish to calculate the $P(E|H)$ value (numerical) for a component in a used product, we will implement four steps:

I. The expert linguistically rates the component with respect to the factors *a*, *b*, *c*, *d*, and *e*, as very high, high, medium, low, and very low, respectively.

II. Using Table 4.13, we convert the linguistic ratings into TFNs.

III. The average fuzzy $P(E|H)$ is equal to $((0.7 + 0.5 + 0.3 + 0.3 + 0.1)/5,$ $(0.9 + 0.7 + 0.5 + 0.5 + 0.3)/5, (1.0 + 0.9 + 0.7 + 0.7 + 0.5)/5)$, i.e., $(0.38, 0.58, 0.76)$.

IV. Defuzzifying the average fuzzy $P(E|H)$ using Equation 4.10, we get $(0.76 - 0.38) + (0.58 - 0.38)/3 + 0.38 = 0.57$.

4.5.2.2 Rules Used in Bayesian Updating

Rule 1: IF product needs repair (AFFIRMS: 1.67; DENIES: 0.00) THEN S1 needs repair.

Rule 2: IF product needs repair (AFFIRMS: 2.50; DENIES: 0.00) THEN S2 needs repair.

Rule 3: IF S1 needs repair (AFFIRMS: 2.22; DENIES: 0.00) THEN C1 needs repair.

Rule 4: IF S1 needs repair (AFFIRMS: 3.33; DENIES: 0.00) THEN C2 needs repair.

Rule 5: IF S1 needs repair (AFFIRMS: 1.82; DENIES: 0.00) THEN C3 needs repair.

Rule 6: IF S2 needs repair (AFFIRMS: 1.43; DENIES: 0.00) THEN C4 needs repair.

Rule 7: IF S2 needs repair (AFFIRMS: 1.25; DENIES: 0.00) THEN C5 needs repair.

Rule 8: IF C1 needs repair (AFFIRMS: 3.50; DENIES 0.38) AND C2 needs repair (AFFIRMS: 2.00; DENIES 0.57) AND C3 needs repair (AFFIRMS: 0.75; DENIES 1.38) AND C4 needs repair (AFFIRMS: 0.13; DENIES 3.60) AND C5 needs repair (AFFIRMS: 2.13; DENIES: 0.25) THEN it is sensible to repair the product.

4.5.2.3 Bayesian Updating

Refer to Table 4.11 while reading this section.

Rule 1: $H = $ S1 needs repair; $O(H) = 1.50$; $E = $ product needs repair; $A = 1.67$; $O(H|E) = O(H) \cdot (A) = 2.51$.

Rule 2: H = S2 needs repair; $O(H)$ = 2.33; E = product needs repair; A = 2.50; $O(H|E) = O(H) \cdot (A)$ = 5.83.

Rule 3: H = C1 needs repair; $O(H)$ = 0.82; E = S1 needs repair; $O(E)$ = 2.51; $P(E)$ = 0.72; A = 2.22; A' = $[2(A - 1) \times P(E)] + 2 - A$ = 1.54; $O(H|E) = O(H) \cdot (A')$ = (0.82) $\cdot$ (1.54) = 1.26.

Rule 4: H = C2 needs repair; $O(H)$ = 1.22; E = S1 needs repair; $O(E)$ = 2.51; $P(E)$ = 0.72; A = 3.33; A' = $[2(A - 1) \times P(E)] + 2 - A$ = 2.03; $O(H|E) = O(H) \cdot (A')$ = (1.22) $\cdot$ (2.03) = 2.48.

Rule 5: H = C3 needs repair; $O(H)$ = 0.43; E = S1 needs repair; $O(E)$ = 2.51; $P(E)$ = 0.72; A = 1.82; A' = $[2(A - 1) \times P(E)] + 2 - A$ = 1.36; $O(H|E) = O(H) \cdot (A')$ = (0.43) $\cdot$ (1.36) = 0.58.

Rule 6: H = C4 needs repair; $O(H)$ = 0.47; E = S2 needs repair; $O(E)$ = 5.83; $P(E)$ = 0.85; A = 1.43; A' = $[2(A - 1) \times P(E)] + 2 - A$ = 1.30; $O(H|E) = O(H) \cdot (A')$ = (0.47) $\cdot$ (1.30) = 0.61.

Rule 7: H = C5 needs repair; $O(H)$ = 0.11; E = S2 needs repair; $O(E)$ = 5.83; $P(E)$ = 0.85; A = 1.25; A'= $[2(A - 1) \times P(E)] + 2 - A$ = 1.18; $O(H|E) = O(H) \cdot (A')$ = (0.11) $\cdot$ (1.18) = 0.13.

Rule 8: H = sensible to repair product; $O(H)$ = 1.50;

$E1$ = C1 needs repair; $O(E1)$ = 1.26; $P(E1)$ = 0.56; $A1$ = 3.50; $A1'$ = $[2 \cdot (A1 - 1) \cdot P(E1)] + 2 - A1$ = 1.30;

$E2$ = C2 needs repair; $O(E2)$ = 2.48; $P(E2)$ = 0.71; $A2$ = 2.00; $A2'$ = $[2 \cdot (A2-1) \cdot P(E2)] + 2 - A2$ = 1.42;

$E3$ = C3 needs repair; $O(E3)$ = 0.58; $P(E3)$ = 0.37; $D3$ = 1.38; $D3'$ = $[2 \cdot (1 - D3) \cdot P(E3)] + D3$ = 1.09;

$E4$ = C4 needs repair; $O(E4)$ = 0.61; $P(E4)$ = 0.38; $D4$ = 3.60; $D4'$ = $[2 \cdot (1 - D4) \cdot P(E4)] + D4$ = 1.88;

$E5$ = C5 needs repair; $O(E5)$ = 0.13; $P(E5)$ = 0.12; $D5$ = 0.25; $D5'$ = $[2 \cdot (1 - D5) \cdot P(E5)] + D5$ = 0.43;

$O(H|E1 \,\&\, E2 \,\&\, E3 \,\&\, E4 \,\&\, E5) = O(H) \cdot (A1') \cdot (A2') \cdot (D3') \cdot (D4') \cdot (D5')$ = (1.50) $\cdot$ (1.30) $\cdot$ (1.42) $\cdot$ (1.09) $\cdot$ (1.88) $\cdot$ (0.43) = 2.44;

$P(H|E1 \,\&\, E2 \,\&\, E3 \,\&\, E4 \,\&\, E5)$ = (2.44)/(3.44) = 0.71.

That is, P (sensible to repair the product) = 0.71.

If the cutoff value as decided by the decision maker is, say, 0.55, he will send the used product for repair and for subsequent sale in a secondhand market.

4.6   Quality Function Deployment and Method of Total Preferences

The methodology of quality function deployment (QFD) [13] is briefly described in Section 4.6.1, the method of total preferences [13] is presented in Section 4.6.2, and both are employed (along with fuzzy logic) in Section 4.6.3 for selecting potential secondhand markets.

4.6.1 Quality Function Deployment

Erol and Ferrell Jr. [13] define "performance aspects" as the features that the decision maker wishes to consider in the selection process and "enablers" as the characteristics possessed by the alternatives, which can be used to satisfy the performance aspects.

The absolute technical importance ratings (ATIRs), which measure how effectively each enabler can satisfy all of the performance aspects, are computed by

$$\text{ATIR}_j = \sum_{i=1}^{I} d_i R_{ij} \quad \forall j = 1, ..., J \tag{4.22}$$

where d_i is the importance value of performance aspect i relative to the other performance aspects and R_{ij} is the relationship score for performance aspect i and enabler j. Since there is an ATIR for each enabler j, for the comparison of all enablers, it is normalized to form the relative technical importance rating (RTIR$_j$) as follows:

$$\text{RTIR}_j = \frac{\text{ATIR}_j}{\sum_{j=1}^{J} \text{ATIR}_j} \quad \forall j = 1, ..., J \tag{4.23}$$

4.6.2 Method of Total Preferences

RTIRs (see Section 4.6.1) are used to develop a single measure that reflects the rating of each alternative as follows:

$$\text{TUP}_n = \sum_{j=1}^{J} \text{RTIR}_j \text{WA}_{nj} \quad \forall n \tag{4.24}$$

where TUP$_n$ is the total user preference for alternative n, and WA$_{nj}$ is the degree to which alternative n can deliver enabler j.

For comparing all alternatives, TUP of each alternative is then normalized as follows:

$$\text{NTUP}_n = \frac{\text{TUP}_n}{\sum_{n-1}^{N} \text{TUP}_n} \quad \forall n \tag{4.25}$$

where NTUP$_n$ is the normalized total preference for alternative n and N is the total number of alternatives.

The alternative with the highest NTUP is considered the alternative with the highest potential.

4.6.3 Selection of Potential Secondhand Markets

This section is organized as follows: Section 4.6.3.1 presents the performance aspects and enablers that help us to select the most potential market to sell a used product from a set of candidate secondhand markets (by a secondhand market we mean, a store where secondhand products are sold along with new products). Section 4.6.3.2 presents the implementation of QFD, fuzzy logic, and the method of total preferences using a numerical example.

4.6.3.1 Performance Aspects and Enablers

In our approach, we consider the following performance aspects of the secondhand markets:

   a. *Before-sale-performance* (BSP). Reflects the ability to attract new customers to the secondhand market
   b. *While-sale-performance* (WSP). Reflects the ability to motivate the customers to buy secondhand products while the customers are in the secondhand market
   c. *After-sale-performance* (ASP). Reflects the ability to attract returning customers to the secondhand market

We consider the following enablers :

- Good advertisement (AD)
- High difference of prices (DP) between new and secondhand products
- Greenness of the sale (GS)
- Incentives (warranty, service, etc.) (IC)
- Low average price of products (LP)
- Good location of sale of secondhand products (placement in front of the new products) (LS)
- Proper maintenance of secondhand products (as in the case of new products) (MN)
- Discounts to returning customers (RC)
- Good return or exchange policy (RP)
- Reputation of the store (RS)
- Variety of secondhand products on the shelves (VS)

The enablers for BSP are LP, IC, RS, and AD, for WSP are DP, GS, IC, MN, AD, VS, RP, and LS, and for ASP are RC and AD.

4.6.3.2 Application of Quality Function Deployment, Fuzzy Logic, and Method of Total Preferences

An important feature of the QFD process is that human input is used to determine both the importance value of each performance aspect and the relationship scores that identify the degrees to which the enablers satisfy the performance aspect. Since the human input is vague in nature, we use fuzzy logic to convert linguistic importance values and linguistic relationship scores into TFNs (linguistic values are given by an expert). These TFNs are then defuzzified for further usage in the QFD process.

We present this approach using a numerical example. We compare two secondhand markets and choose the market that has more potential than the other. Suppose there are five experts (Ms) for assigning linguistic values to R_{ij} and d_i data. Table 4.14 shows the linguistic scale for R_{ij} as well as for d_i.

Tables 4.15 through 4.17 show the linguistic relationship scores (R_{ij}) as given by the five experts to the enablers of BSP, WSP, and ASP, respectively.

Table 4.18 lists out the linguistic importance values (d_i) of BSP, WSP, and ASP, as given by the five experts.

The ATIRs and RTIRs of the enablers are then calculated using Equations 4.22 and 4.23, respectively. It must be noted that when there are multiple experts, their TFNs are averaged before defuzzifying. For example, consider AD, which is an enabler for the three performance aspects BSP, WSP, and ASP. The linguistic relationship scores, as given by the five experts, for AD and BSP (see Table 4.15), for AD and WSP (see Table 4.16), and for AD and ASP (see Table 4.17) are converted into TFNs using Table 4.14. These TFNs are then averaged as follows: The average relationship score for AD and BSP, as calculated using Equations 4.6 and 4.9, is $= ((7.5 + 7.5 + 7.5 + 7.5 + 7.5)/5, (10 + 10 + 10 + 10 + 10)/5, (10 + 10 + 10 + 10 + 10)/5) = (7.5, 10, 10)$. Similarly, the average

TABLE 4.14

Conversion Table for Linguistic R_{ij} Data

Linguistic R_{ij} or d_i	TFN
VS	(7.5. 10, 10)
S	(5, 7.5, 10)
M	(2.5, 5, 7.5)
W	(0, 2.5, 5)
VW	(0, 0, 2.5)

Note: VS = very strong, S = strong, M = medium, W = weak, VW = very weak.

TABLE 4.15

Linguistic Relationship Scores of BSP and Its Enablers

	M1	M2	M3	M4	M5
LP	VS	VS	M	W	S
IC	S	S	M	VS	S
RS	M	S	VS	W	M
AD	VS	VS	VS	VS	VS

Note: VS = very strong, S = strong, M = medium, W = weak.

TABLE 4.16

Linguistic Relationship Scores of WSP and Its Enablers

	M1	M2	M3	M4	M5
DP	VS	VS	M	W	S
GS	M	M	VS	VS	S
IC	S	S	M	VS	S
MN	M	S	VS	W	M
AD	S	S	VS	M	W
VS	VS	VS	VS	VS	VS
RP	S	VS	M	VS	S
LS	M	S	VS	W	M

Note: VS = very strong, S = strong, M = medium, W = weak.

TABLE 4.17

Linguistic Relationship Scores of ASP and Its Enablers

	M1	M2	M3	M4	M5
RC	VS	VS	M	W	S
AD	VW	S	M	S	VS

Note: VS = very strong, S = strong, M = medium, W = weak, VW = very weak.

TABLE 4.18

Linguistic Importance Values of Performance Aspects

	M1	M2	M3	M4	M5
BSP	VS	VS	VS	S	S
WSP	S	S	M	S	VS
ASP	VS	M	S	S	VS

Note: VS = very strong, S = strong, M = medium, W = weak.

relationship score for AD and WSP and for AD and ASP are calculated as (4, 6, 8.5) and (4, 6, 8), respectively. Defuzzified average relationship scores for AD and BSP, for AD and WSP, and for AD and ASP, as calculated using Equation 4.10, are 9.17, 6.33, and 6, respectively. The linguistic importance values (d_is) of BSP, WSP, and ASP, as given by the five experts (see Table 4.18), are converted into TFNs using Table 4.14. The average importance values for BSP, WSP, and ASP are then calculated as (6.5, 9, 10) (defuzzified value = 8.5), (5, 7.5, 9.5) (defuzzified value = 7.33), and (5.5, 8, 9.5) (defuzzified value = 7.67), respectively. ATIR of AD is then calculated using Equation 4.22 as (8.5) × (9.17) + (7.33) × (6.33) + (7.67) × (6) = 212.79. The ATIRs of all the enablers are shown in Table 4.19. RTIR of each enabler is then calculated using Equation 4.23. For example, RTIR of AD is the ratio of its ATIR to the sum of the ATIRs of all the enablers, that is, 170.36/715.14 = 0.24. RTIRs too are shown in Table 4.19.

Table 4.20 shows the scale for converting linguistic WA_{nj} value given by the five experts (for arithmetic simplicity, we assume here that there is a consensus among the experts). The WA_{nj} (linguistic) values and the corresponding defuzzified TFNs (calculated using Equation 4.10) for each secondhand market are shown in Table 4.21.

Using Equation 4.24, we then calculate the TUPs for the two secondhand markets. For example, TUP for the secondhand market-1 is calculated using the $RTIR_j$ values (see Table 4.19) and the defuzzified WA_{1j} values (see Table 4.21) as follows:

$(0.24) \times (7.33) + (0.07) \times (1) + (0.06) \times (3) + (0.16) \times (7.33) + (0.08) \times (7.33) + (0.05) \times (9) + (0.05) \times (5) + (0.07) \times (1) + (0.06) \times (7.33) + (0.06) \times (1) + (0.09) \times (3) = 5.35$

TABLE 4.19

ATIRs and RTIRs of Enablers

Enabler	ATIR	RTIR
AD	170.36	0.24
DP	48.89	0.07
GS	46.44	0.06
IC	116.11	0.16
LP	56.67	0.08
LS	36.67	0.05
MN	36.67	0.05
RC	51.11	0.07
RP	42.50	0.06
RS	42.50	0.06
VS	67.22	0.10
Sum	715.14	1.00

TABLE 4.20

Conversion Table for Linguistic WA_{nj} Data

Linguistic WA_{nj}	TFN
VG	(7, 10, 10)
G	(5, 7, 10)
F	(2, 5, 8)
P	(1, 3, 5)
VP	(0, 0, 3)

Note: VG = very good, G = good, F = fair, P = poor, VP = very poor.

TABLE 4.21

Linguistic and Corresponding Defuzzified WA Values of
Secondhand Markets

Enabler	Market-1		Market-2	
AD	G	7.33	G	7.33
DP	VP	1	VG	9
GS	P	3	VG	9
IC	G	7.33	VG	9
LP	G	7.33	G	7.33
LS	VG	9	VP	1
MN	F	5	P	3
RC	VP	1	F	5
RP	G	7.33	G	7.33
RS	VP	1	VP	1
VS	P	3	G	7.33

Note: VG = very good, G = good, F = fair, P = poor, VP = very poor.

Similarly, TUP for the secondhand market-2 is calculated as 6.74.

Finally, using Equation 4.25, we calculate NTUPs for the two secondhand markets. For example, NTUP for the secondhand market-1 is calculated as follows:

$$\frac{5.35}{5.35 + 6.74} = 0.44$$

Similarly, NTUP for the secondhand market-2 is calculated as 0.56.

It is obvious that secondhand market-2 has more potential than secondhand market-1.

4.7 Technique for Order Preference by Similarity to Ideal Solution

The methodology of the TOPSIS (technique for order preference by similarity to ideal solution) is briefly described in Section 4.7.1. This methodology is applied (along with fuzzy logic) in Section 4.7.2 for evaluating production facilities (in a region where a closed-loop supply chain is to be designed) and in Section 4.7.3 for evaluating the marketing strategy of a reverse or a closed-loop supply chain.

4.7.1 Methodology

The basic concept of the TOPSIS [19] is that the rating of the alternative selected as the best from a set of different alternatives should have the shortest distance from the ideal solution and the farthest distance from the negative-ideal solution in a geometrical (i.e., Euclidean) sense.

The TOPSIS evaluates the following decision matrix, which refers to m alternatives that are evaluated in terms of n criteria [44]:

				Criteria		
	C_1	C_2	C_3	....	C_n
Alternatives	w_1	w_2	w_3	....	w_n
A_1	z_{11}	z_{12}	z_{13}	....	Z_{1n}
A_2	z_{21}	z_{22}	z_{23}	....	Z_{2n}
A_3	z_{31}	z_{32}	z_{33}	....	Z_{3n}
.	.	.	.		.
.	.	.	.		.
.	.	.	.		.
A_m	z_{m1}	z_{m2}	z_{m3}	....	z_{mn}

where

A_i = ith alternative

C_j = jth criterion, w_j is the weight (importance value) assigned to the jth criterion

z_{ij} = Rating (for example, on a scale of 1–10, the higher the rating, the better it is) of the ith alternative in terms of the jth criterion.

The following steps are performed:

Step 1: Construct the normalized decision matrix. This step converts the various dimensional measures of performance into nondimensional attributes. An element r_{ij} of the normalized decision matrix R is calculated as follows:

$$r_{ij} = \frac{z_{ij}}{\sqrt{\sum_{i=1}^{m} z_{ij}^2}} \tag{4.26}$$

Step 2: Construct the weighted normalized decision matrix. A set of weights $W = (w_1, w_2, ..., w_n)$ (such that $\sum w_j = 1$), specified by the decision maker, is used in conjunction with the normalized decision matrix R to determine the weighted normalized matrix V defined by $V = (v_{ij}) = (r_{ij}w_j)$.

Step 3: Determine the ideal and the negative-ideal solutions. The ideal (A^*) and the negative-ideal (A^-) solutions are defined as follows:

$$A^* = \left\{ \max_i v_{ij} \quad \text{for } i = 1, 2, 3, ..., m \right\}$$
$$= \{p_1, p_2, p_3, ..., p_n\} \tag{4.27}$$

$$A^- = \left\{ \min_i v_{ij} \quad \text{for } i = 1, 2, 3, ..., m \right\}$$
$$= \{q_1, q_2, q_3, ..., q_n\} \tag{4.28}$$

With respect to each criterion, the decision maker desires to choose the alternative with the maximum rating (this choice varies with the way the decision maker rates the alternatives). Obviously, A^* indicates the most preferable (ideal) solution and A^- indicates the least preferable (negative-ideal) solution.

Step 4: *Calculate the separation distances.* In this step, the concept of the n-dimensional Euclidean distance is used to measure the separation distances of the rating of each alternative from the ideal solution and the negative-ideal solution. The corresponding formulae are

$$S_{i^*} = \sqrt{\sum (v_{ij} - p_j)^2} \quad \text{for } i = 1, 2, 3, \ldots, m \qquad (4.29)$$

where S_{i^*} is the separation (in the Euclidean sense) of the rating of alternative i from the ideal solution, and

$$S_{i-} = \sqrt{\sum (v_{ij} - q_j)^2} \quad \text{for } i = 1, 2, 3, \ldots, m \qquad (4.30)$$

where S_i- is the separation (in the Euclidean sense) of the rating of alternative i from the negative-ideal solution.

Step 5: *Calculate the relative coefficient.* The relative closeness coefficient for alternative A_i with respect to the ideal solution A^* is defined as follows:

$$C_{i^*} = \frac{S_{i-}}{S_{i^*} + S_{i-}} \qquad (4.31)$$

Step 6: *Rank the preference order.* The best alternative can now be decided according to preference order of C_{i^*}. It is the one with the rating that has the shortest distance to the ideal solution. The processing of the alternatives in the previous steps reveals that if an alternative has the rating with the shortest distance to the ideal solution, then that rating is guaranteed to have the longest distance to the negative-ideal solution. Specifically, the higher the C_{i^*}, the better the alternative.

4.7.2   Evaluation of Production Facilities

In this section, we employ TOPSIS to evaluate production facilities in a region where a closed-loop supply chain is to be designed in terms of both environmental consciousness (mainly associated with the forward supply chain) and potentiality (mainly associated with the reverse supply chain). It must be noted here that since linguistic expressions such as "low," "medium," and "very good" are regarded as the natural representation of weights (importance values) assigned to the criteria for evaluation of production facilities (the same holds true for ratings, viz., performance measures of candidate production facilities with respect to criteria for evaluation), we apply fuzzy logic in our TOPSIS approach.

We frame the problem of evaluation of production facilities in a closed-loop supply chain as a four-level hierarchy. The first level in the hierarchy contains our objective, that is, evaluation of the efficiency of each production facility in the region where the closed-loop supply chain is to be designed, the second

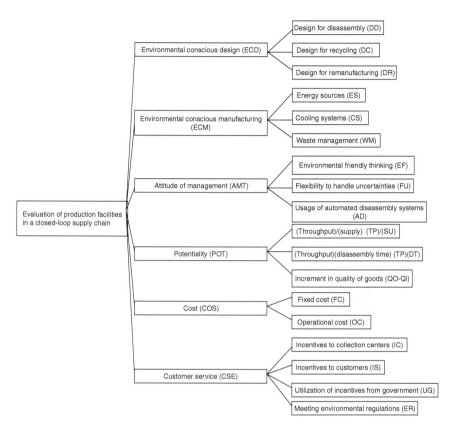

FIGURE 4.7
Levels of hierarchy.

level contains the main criteria for evaluation, the third level contains the sub-criteria under each main criterion, and the fourth (last) level contains all the production facilities of interest. Figure 4.7 illustrates these hierarchy levels.

Sections 4.7.2.1 gives brief descriptions of the criteria for evaluation of production facilities in a closed-loop supply chain. Section 4.7.2.2 presents a numerical example to demonstrate the implementation of TOPSIS (along with fuzzy logic).

4.7.2.1 Criteria for Evaluation of Production Facilities

The following are brief descriptions of the main criteria at the second level in the hierarchy (for implementation of the TOPSIS) and their corresponding subcriteria at the third level.

4.7.2.1.1 Environmentally Conscious Design

Environmentally conscious design (ECD) involves the designing of products on the basis of certain environmental considerations. The following subcriteria fall under ECD.

Design for disassembly (DD). Disassembly is used both in recycling and remanufacturing to increase the recovery rate by enabling selective separating of parts and materials. Thus, DD initiatives lead to the correct identification of design specifications to minimize the complexity of the structure of the product by minimizing the number of parts, increasing the use of common materials, and choosing the fastener and joint types, which are easily removable.

Design for recycling (DC). DC suggests making better choices for material selection such that the processes of material selection and material recovery become more efficient. Some important characteristics of DC are long product life with the minimized use of raw materials (source reduction), more adaptable materials for multiproduct applications, and fewer components within a given material in an engineered system.

Design for remanufacturing (DR). DR suggests the use of reusable parts and packaging in the design of new products, for source reduction.

4.7.2.1.2 *Environmentally Conscious Manufacturing (ECM)*

In addition to environment-friendly product designs, manufacturing issues must also be addressed to arrive at a complete concept of environmentally conscious production. These issues (we place them at the third level in the hierarchy) include the following:

- Selecting low pollution energy sources for manufacturing (ES)
- Designing cooling systems in such a way that the coolant can be reused and the heat collected by the system can be utilized as an energy source (CS)
- Monitoring waste generation as a result of manufacturing (WM)

4.7.2.1.3 *Attitude of Management (AMT)*

The attitude of the decision makers (managers) in the facility is crucial when it comes to implementing the aforementioned practices (ECD and ECM). All the managers in the facility must have the following credentials (we place them at the third level in the hierarchy):

- Environment-friendly thoughts (EF)
- Flexibility to handle uncertainties in the supply and quality of used products (FU)
- Readiness to the usage of automated disassembly systems (AD) to avoid high lead time, expensive labor use, and possible human exposure to hazardous by-products

4.7.2.1.4 *Potentiality (POT)*

We also evaluate a facility in terms of its potentiality to efficiently reprocess the incoming used products. The following factors (we place them at the third level in the hierarchy) serve as potentiality measures (these factors are explained in detail in Section 4.3.2).

- TP/SU
- TP × DT
- QO – QI

4.7.2.1.5 Cost (COS)

The costs incurred by a production facility can be divided into the following types (we place them at the third level in the hierarchy):

- Fixed cost (FC), which is the sum of space cost, machinery cost, personnel cost, etc.
- Operational cost (OC), which is the sum of rent, salaries, maintenance cost, etc.

4.7.2.1.6 Customer Service

Customer service (CSE) provides an idea about how well a production facility is

- giving incentives to the collection centers supplying used products (IC);
- giving incentives to the customers buying reprocessed goods (IS);
- utilizing incentives provided by the government (UG); and
- meeting environmental regulations laid down by the government (ER)

(Note that the term "customer service" is used here because, in our opinion, any beneficiary is a customer, be it the government, the collection center, or the actual customer buying reprocessed goods.)

4.7.2.2 Numerical Example

In our example, three experts assign linguistic (high, medium, etc.) weights (importance values) to the main criteria (second level in the hierarchy) and to the subcriteria (third level in the hierarchy). These linguistic weights are quantified using TFNs in Table 4.22.

Table 4.23 shows the linguistic weights assigned by the three experts to the main criteria. These weights are quantified using Table 4.22 and then averaged to form another TFN called the average weight. For example, average weight of the main criterion, ECD, is $(H + H + M)/3$, which is $((0.5 + 0.5 + 0.3)/3, (0.7 + 0.7 + 0.5)/3, (0.9 + 0.9 + 0.7)/3) = (0.43, 0.63, 0.83)$ (see Table 4.23).

The sum of the average weights of all the main criteria is calculated using Equation 4.6 as (2.29, 3.45, 4.59). The ratio of the average weight of each main criterion to the sum of the average weights of all the main criteria gives the

TABLE 4.22

Linguistic Weight Conversion Table for Criteria and Subcriteria

Linguistic Weight	Triangular Fuzzy Number
VH	(0.7, 0.9, 1.0)
H	(0.5, 0.7, 0.9)
M	(0.3, 0.5, 0.7)
L	(0.1, 0.3, 0.5)
VL	(0.0, 0.1, 0.3)

Note: VH = very high, H = high, M = medium, L = low, VL = very low.

TABLE 4.23

Relative Weights of Main Criteria

Criterion	Expert E1	Expert E2	Expert E3	Average Weight	Relative Weight
ECD	H	H	M	(0.43, 0.63, 0.83)	(0.09, 0.18, 0.36)
ECM	VH	H	VH	(0.63, 0.83, 0.97)	(0.14, 0.24, 0.42)
AMT	L	L	VL	(0.07, 0.23, 0.43)	(0.02, 0.07, 0.19)
POT	H	H	H	(0.50, 0.70, 0.90)	(0.11, 0.20, 0.39)
COS	M	H	H	(0.43, 0.63, 0.83)	(0.09, 0.18, 0.36)
CSE	M	L	M	(0.23, 0.43, 0.63)	(0.05, 0.12, 0.28)

Note: VH = very high, H = high, M = medium, L = low, VL = very low.

TABLE 4.24

Relative Weights of Subcriteria of ECD

Subcriterion	E1	E2	E3	Average Weight	Relative Weight
DD	H	H	M	(0.43, 0.63, 0.83)	(0.21, 0.42, 0.89)
DC	L	L	VL	(0.07, 0.23, 0.43)	(0.03, 0.15, 0.46)
DR	M	H	H	(0.43, 0.63, 0.83)	(0.21, 0.42, 0.89)

Note: H = high, M = medium, L = low, VL = very low.

corresponding relative weight. For example, relative weight of ECD is ((0.43, 0.63,0.83)/(2.29,3.45,4.59)), which is simplified using Equation 4.9 as (0.09, 0.18, 0.36) (see Table 4.23).

Similarly, the linguistic weights, average weights, and relative weights of subcriteria of ECD, ECM, AMT, POT, COS, and CSE are calculated and presented in Tables 4.24 through 4.29, respectively.

Since the weights considered in the TOPSIS must sum up to unity, we multiply the weight of each subcriterion at the third level in the hierarchy with the weight of its corresponding main criterion at the second level in the hierarchy. The weights of the criteria at the third level in the hierarchy, which are ready for use in the TOPSIS, are shown in Table 4.30. Table 4.31 shows the linguistic ratings (performance measures) and their corresponding TFNs,

TABLE 4.25

Relative Weights of Subcriteria of ECM

Subcriterion	E1	E2	E3	Average Weight	Relative Weight
ES	H	H	H	(0.50, 0.70, 0.90)	(0.21, 0.40, 0.78)
CS	M	H	H	(0.43, 0.63, 0.83)	(0.18, 0.36, 0.72)
WM	M	L	M	(0.23, 0.43, 0.63)	(0.10, 0.24, 0.54)

Note: H = high, M = medium, L = low.

TABLE 4.26

Relative Weight of Subcriteria of AMT

Subcriterion	E1	E2	E3	Average Weight	Relative Weight
EF	L	L	VL	(0.07, 0.23, 0.43)	(0.04, 0.18, 0.59)
FU	M	H	H	(0.43, 0.63, 0.83)	(0.23, 0.49, 1.14)
AD	M	L	M	(0.23, 0.43, 0.63)	(0.12, 0.33, 0.86)

Note: H = high, M = medium, L = low, VL = very low.

TABLE 4.27

Relative Weight of Subcriteria of POT

Subcriterion	E1	E2	E3	Average Weight	Relative Weight
TP/SU	H	H	M	(0.43, 0.63, 0.83)	(0.21, 0.42, 0.89)
TP × DT	L	L	VL	(0.07, 0.23, 0.43)	(0.03, 0.15, 0.46)
QO − QI	M	H	H	(0.43, 0.63, 0.83)	(0.21, 0.42, 0.89)

Note: H = high, M = medium, L = low, VL = very low.

TABLE 4.28

Relative Weight of Subcriteria of COS

Subcriterion	E1	E2	E3	Average Weight	Relative Weight
FC	H	H	M	(0.43, 0.63, 0.83)	(0.34, 0.73, 1.66)
OC	L	L	VL	(0.07, 0.23, 0.43)	(0.06, 0.27, 0.86)

Note: H = high, M = medium, L = low, VL = very low.

TABLE 4.29

Relative Weights of Subcriteria of CSE

Subcriterion	E1	E2	E3	Average Weight	Relative Weight
IC	M	H	H	(0.43, 0.63, 0.83)	(0.17, 0.35, 0.78)
IS	L	L	VL	(0.07, 0.23, 0.43)	(0.03, 0.13, 0.40)
UG	H	H	H	(0.50, 0.70, 0.90)	(0.19, 0.39, 0.84)
ER	L	L	VL	(0.07, 0.23, 0.43)	(0.03, 0.13, 0.40)

Note: H = high, M = medium, L = low, VL = very low.

TABLE 4.30

Weights of Subcriteria for the TOPSIS

Subcriterion	Weight for the TOPSIS (Relative Weight of Subcriterion × Relative Weight of Corresponding Main Criterion)
DD	(0.02, 0.08, 0.32)
DC	(0.00, 0.03, 0.17)
DR	(0.02, 0.08, 0.32)
ES	(0.03, 0.10, 0.33)
CS	(0.03, 0.09, 0.30)
WM	(0.01, 0.06, 0.23)
EF	(0.00, 0.01, 0.11)
FU	(0.00, 0.03, 0.22)
AD	(0.00, 0.02, 0.16)
TP/SU	(0.02, 0.08, 0.35)
TP × DT	(0.00, 0.03, 0.18)
QO − QI	(0.02, 0.08, 0.35)
FC	(0.03, 0.13, 0.60)
OC	(0.00, 0.05, 0.31)
IC	(0.01, 0.04, 0.22)
IS	(0.00, 0.02, 0.11)
UG	(0.01, 0.05, 0.24)
ER	(0.00, 0.02, 0.11)

TABLE 4.31

Linguistic Rating Conversion Table for Production Facilities

Linguistic Rating	Triangular Fuzzy Number
VG	(7, 10, 10)
G	(5, 7, 10)
F	(2, 5, 8)
P	(1, 3, 5)
VP	(0, 0, 3)

Note: VG = very good, G = good, F = fair, P = poor, VP = very poor.

which are used to evaluate the production facilities with respect to each subcriterion, except (TP/SU), (TP × DT), (QO − QI), FC, and OC. The reason for this exception is the easy availability of historical crisp (nonfuzzy) measures of these subcriteria (viz., [TP/SU], [TP × DT], [QO − QI], FC, and OC), for each production facility. Table 4.32 shows the crisp measures of these subcriteria for each production facility.

Now, we are ready to perform the six steps in the TOPSIS.

Step 1: *Construct the normalized decision matrix.* Table 4.33 shows our decision matrix for the TOPSIS approach. The elements of this table

TABLE 4.32

Crisp Measures of Subcriteria for Evaluation

	Production Facilities			
Subcriterion	A	B	C	D
TP/SU	0.9	0.7	0.9	0.5
TP × DT	25	30	15	40
QO – QI	0.6	0.7	0.3	0.5
FC ($)	100,000	150,000	70,000	200,000
OC ($)	500	300	450	400

TABLE 4.33

Decision Matrix for TOPSIS

	Production Facilities			
Subcriterion	A	B	C	D
DD	(7, 10, 10)	(2, 5, 8)	(1, 3, 5)	(5, 7, 10)
DC	(5, 7, 10)	(0, 0, 3)	(5, 7, 10)	(5, 7, 10)
DR	(2, 5, 8)	(7, 10, 10)	(2, 5, 8)	(5, 7, 10)
ES	(1, 3, 5)	(5, 7, 10)	(1, 3, 5)	(1, 3, 5)
CS	(0, 0, 3)	(5, 7, 10)	(1, 3, 5)	(0, 0, 3)
WM	(1, 3, 5)	(0, 0, 3)	(2, 5, 8)	(2, 5, 8)
EF	(1, 3, 5)	(5, 7, 10)	(5, 7, 10)	(7, 10, 10)
FU	(5, 7, 10)	(7, 10, 10)	(2, 5, 8)	(7, 10, 10)
AD	(0, 0, 3)	(7, 10, 10)	(1, 3, 5)	(0, 0, 3)
TP/SU	(0.9, 0.9, 0.9)	(0.7, 0.7, 0.7)	(0.9, 0.9, 0.9)	(0.5, 0.5, 0.5)
TP × DT	(25, 25, 25)	(30, 30, 30)	(15, 15, 15)	(40, 40, 40)
QO – QI	(0.6, 0.6, 0.6)	(0.7, 0.7, 0.7)	(0.3, 0.3, 0.3)	(0.5, 0.5, 0.5)
FC	(10, 10, 10)	(15, 15, 15)	(7, 7, 7)	(20, 20, 20)
OC	(5, 5, 5)	(3, 3, 3)	(4.5, 4.5, 4.5)	(4, 4, 4)
IC	(0, 0, 3)	(0, 0, 3)	(5, 7, 10)	(1, 3, 5)
IS	(1, 3, 5)	(2, 5, 8)	(7, 10, 10)	(5, 7, 10)
UG	(2, 5, 8)	(0, 0, 3)	(1, 3, 5)	(0, 0, 3)
ER	(1, 3, 5)	(5, 7, 10)	(5, 7, 10)	(2, 5, 8)

are the fuzzy ratings of the production facilities with respect to each subcriterion, as given by the three experts (for arithmetic simplicity, we assume here that the experts give a consensus rating). For example, the rating of facility C with respect to subcriterion, DD, is (1, 3, 5) (see Table 4.33) because the experts unanimously rate it as "poor" with respect to DD (the TFN for the linguistic rating "poor," is (1, 3, 5); see Table 4.31). For consistency in the TOPSIS, we convert crisp measures of subcriteria, (TP/SU), (TP × DT), (QO – QI), FC, and OC

TABLE 4.34

Normalized Decision Matrix

Subcriterion	Production Facilities			
	A	**B**	**C**	**D**
DD	(0.41, 0.74, 1.13)	(0.12, 0.37, 0.90)	(0.06, 0.22, 0.56)	(0.29, 0.52, 1.13)
DC	(0.28, 0.58, 1.15)	(0, 0, 0.35)	(0.28, 0.58, 1.15)	(0.28, 0.58, 1.15)
DR	(0.11, 0.35, 0.88)	(0.39, 0.71, 1.1)	(0.11, 0.35, 0.88)	(0.28, 0.50, 1.1)
ES	(0.11, 0.58, 2.89)	(0, 0, 1.73)	(0.11, 0.58, 2.89)	(0.11, 0.58, 2.89)
CS	(0, 0, 0.59)	(0.42, 0.92, 1.96)	(0.08, 0.39, 0.98)	(0, 0, 0.59)
WM	(0.08, 0.39, 1.67)	(0, 0, 1)	(0.16, 0.65, 2.67)	(0.16, 0.65, 2.67)
EF	(0.06, 0.21, 0.5)	(0.28, 0.49, 1)	(0.28, 0.49, 1)	(0.39, 0.7, 1)
FU	(0.26, 0.42, 0.89)	(0.37, 0.6, 0.89)	(0.1, 0.3, 0.71)	(0.37, 0.6, 0.89)
AD	(0, 0, 0.42)	(0.59, 0.96, 1.41)	(0.08, 0.29, 0.71)	(0, 0, 0.42)
TP/SU	(0.59, 0.59, 0.59)	(0.46, 0.46, 0.46)	(0.59, 0.59, 0.59)	(0.33, 0.33, 0.33)
TP × DT	(0.43, 0.43, 0.43)	(0.52, 0.52, 0.52)	(0.26, 0.26, 0.26)	(0.69, 0.69, 0.69)
QO – QI	(0.55, 0.55, 0.55)	(0.64, 0.64, 0.64)	(0.28, 0.28, 0.28)	(0.46, 0.46, 0.46)
FC	(0.36, 0.36, 0.36)	(0.54, 0.54, 0.54)	(0.25, 0.25, 0.25)	(0.72, 0.72, 0.72)
OC	(0.6, 0.6, 0.6)	(0.36, 0.36, 0.36)	(0.54, 0.54, 0.54)	(0.48, 0.48, 0.48)
IC	(0, 0, 0.59)	(0, 0, 0.59)	(0.42, 0.92, 1.96)	(0.08, 0.39, 0.98)
IS	(0.06, 0.22, 0.56)	(0.12, 0.37, 0.90)	(0.41, 0.74, 1.13)	(0.29, 0.52, 1.13)
UG	(0.19, 0.86, 3.58)	(0, 0, 1.34)	(0.1, 0.51, 2.24)	(0, 0, 1.34)
ER	(0.06, 0.26, 0.67)	(0.29, 0.61, 1.35)	(0.29, 0.61, 1.35)	(0.12, 0.44, 1.08)

into TFNs, each of whose parameters are equal. For example, crisp measure "10" of subcriterion, FC, for facility A is converted into the TFN (10, 10, 10) (see Table 4.33). Table 4.34 shows the normalized decision matrix formed by applying Equation 4.26 on each element of Table 4.33. For example, the normalized fuzzy rating of facility C, with respect to subcriterion DD (see Table 4.34), is calculated using Equation 4.26 as follows:

$$r_{13} = \frac{(1, 3, 5)}{\sqrt{(7, 10, 10)^2 + (2, 5, 8)^2 + (1, 3, 5)^2 + (5, 7, 10)^2}} = (0.06, 0.22, 0.56)$$

Note that Equations 4.6, 4.8, and 4.9 are used to perform the basic operations in the calculation of r_{13}.

Step 2: Construct the weighted normalized decision matrix. Table 4.35 shows the weighted normalized decision matrix. This is constructed using the weights of the subcriteria listed in Table 4.30 and the normalized decision matrix in Table 4.34. For example, the weighted normalized fuzzy rating of facility C, with respect to subcriterion DD, that is, (0, 0.02, 0.18) (see Table 4.35), is calculated by multiplying the weight of DD, that is, (0.02, 0.08, 0.32) (see Table 4.30), with the normalized

TABLE 4.35

Weighted Normalized Decision Matrix

Subcriterion	Production Facilities			
	A	B	C	D
DD	(0.01, 0.06, 0.36)	(0, 0.03, 0.29)	(0, 0.02, 0.18)	(0.01, 0.04, 0.36)
DC	(0, 0.02, 0.20)	(0, 0, 0.06)	(0, 0.02, 0.20)	(0, 0.02, 0.20)
DR	(0, 0.03, 0.28)	(0.01, 0.06, 0.35)	(0, 0.03, 0.28)	(0.01, 0.04, 0.35)
ES	(0, 0.06, 0.95)	(0, 0, 0.57)	(0, 0.06, 0.95)	(0, 0.06, 0.95)
CS	(0, 0, 0.18)	(0.01, 0.08, 0.59)	(0, 0.03, 0.29)	(0, 0, 0.18)
WM	(0, 0.02, 0.38)	(0, 0, 0.23)	(0, 0.04, 0.61)	(0, 0.04, 0.61)
EF	(0, 0, 0.06)	(0, 0, 0.11)	(0, 0, 0.11)	(0, 0.01, 0.11)
FU	(0, 0.01, 0.20)	(0, 0.02, 0.20)	(0, 0.01, 0.16)	(0, 0.02, 0.20)
AD	(0, 0, 0.07)	(0, 0.02, 0.23)	(0, 0.01, 0.11)	(0, 0, 0.07)
TP/SU	(0.01, 0.05, 0.20)	(0.01, 0.04, 0.16)	(0.01, 0.05, 0.20)	(0.01, 0.03, 0.11)
TP × DT	(0, 0.01, 0.08)	(0, 0.02, 0.09)	(0, 0.01, 0.05)	(0, 0.02, 0.12)
QO − QI	(0.01, 0.04, 0.19)	(0.01, 0.05, 0.22)	(0.01, 0.02, 0.10)	(0.01, 0.04, 0.16)
FC	(0.01, 0.05, 0.22)	(0.02, 0.07, 0.32)	(0.01, 0.03, 0.15)	(0.02, 0.09, 0.43)
OC	(0, 0.03, 0.18)	(0, 0.02, 0.11)	(0, 0.03, 0.17)	(0, 0.02, 0.15)
IC	(0, 0, 0.13)	(0, 0, 0.13)	(0, 0.04, 0.43)	(0, 0.02, 0.22)
IS	(0, 0, 0.06)	(0, 0.01, 0.10)	(0, 0.01, 0.12)	(0, 0.01, 0.12)
UG	(0, 0.04, 0.86)	(0, 0, 0.32)	(0, 0.03, 0.54)	(0, 0, 0.32)
ER	(0, 0.01, 0.07)	(0, 0.01, 0.15)	(0, 0.01, 0.15)	(0, 0.01, 0.12)

fuzzy rating of facility C, with respect to DD, that is, (0.06, 0.22, 0.56), (see Table 4.34). Equation 4.8 is used for the multiplication.

Step 3: *Determine the ideal and the negative-ideal solution.* Each row in the decision matrix shown in Table 4.35 has a maximum rating and a minimum rating. They are the ideal and the negative-ideal solutions, respectively, for the corresponding subcriterion. For arithmetic simplicity, we assume here that the rating with the highest "most promising quantity" (second parameter in the TFN) is the maximum and the rating with the lowest "most promising quantity" is the minimum. For example (see Table 4.35), with respect to subcriterion DD, the maximum rating is (0.01, 0.06, 0.36) and the minimum rating is (0, 0.02, 0.18), because in the row for that subcriterion, the TFN with the highest second parameter is (0.01, 0.06, 0.36), and the TFN with the lowest second parameter is (0, 0.02, 0.18).

Step 4: *Calculate the separation distances.* The separation distances (see Table 4.36) for each production facility are calculated using Equations 4.29 and 4.30. For example, the positive separation distance for facility C (see Table 4.36) is calculated using Equation 4.29, which contains the weighted normalized fuzzy ratings of C (see Table 4.35) and the ideal solution (obtained in step 3) for each subcriterion.

TABLE 4.36

Separation Measures of Facilities

Production Facility	Positive Distance S^*	Negative Distance S^-
A	0.239	0.288
B	0.320	0.208
C	0.200	0.290
D	0.305	0.233

TABLE 4.37

Relative Closeness Coefficients of Production Facilities

Production Facility	Relative Closeness Coefficient
A	0.547
B	0.394
C	0.592
D	0.434

It is important to note that since we obtain some TFNs with negative "smallest possible quantities" and negative "most promising quantities" in this step, we defuzzify those TFNs using Equation 4.10 before squaring them in the process of calculating separation distances.

Step 5: Calculate the relative closeness to the ideal solution. Using Equation 4.31, we calculate the relative closeness coefficient for each facility in the supply chain (see Table 4.37). For example, relative closeness coefficient (i.e., 0.592) for facility C (see Table 4.37) is the ratio of the facility's negative separation distance (i.e., 0.29) to the sum (i.e., 0.29 + 0.2 = 0.49) of its negative and positive separation distances (see Table 4.36).

Step 6: Rank the preference order. Since the relative closeness coefficients of facilities A and C (0.547 and 0.592, respectively) are much higher than those of facilities B and D (0.394 and 0.434, respectively) (see Table 4.37), it is obvious that facilities A and C are much better than facilities B and D. If the cutoff value of the relative closeness coefficient decided by the decision maker is, say, 0.45, the decision maker will identify facilities A and C as the efficient ones in the region where the closed-loop supply chain is to be designed.

4.7.3   Evaluation of Marketing Strategy

The success of a reverse or a closed-loop supply chain depends heavily on the level of public participation (in the chain), which in turn, depends on the marketing strategy of that chain. Hence, evaluating the marketing strategy of a reverse or a closed-loop supply chain is equivalent to evaluating the efficiency of the strategy in driving the public to participate in the chain. Studies are conducted in numerous cities around the world to assess the level of participation of the public in the respective reverse or closed-loop supply chains. The officials of each chain painstakingly approach many houses in the city, questioning the residents about public satisfaction with the chain and asking for suggestions to improve the chain. While environmental consciousness and profitability drive governments and companies to implement these chains and evaluate these chains' marketing strategies, it has been found that the factors behind the public participation in these chains are numerous and often conflicting (e.g., the more regularly a reverse supply chain offers to collect used products from consumers, the more taxes the consumers will have to pay; high regularity of collection and low tax levied on the consumers are conflicting drivers here). The literature does not identify all the important drivers of public participation, and it does not have a systematic method to evaluate the marketing strategy of a reverse or a closed-loop supply chain with respect to those drivers.

In Section 4.7.3.1, we identify a list of drivers for the public to participate in a reverse or a closed-loop supply chain, and in Section 4.7.3.2, using a numerical example, we propose an approach (using TOPSIS and fuzzy logic) to evaluate the marketing strategy of a reverse or a closed-loop supply chain with respect to the drivers identified in Section 4.7.3.1.

4.7.3.1 Drivers of Public Participation

The following is a fairly exhaustive list of self-explanatory drivers for the public to participate in a reverse or a closed-loop supply chain:

- Knowledge of drivers of implementation of the reverse or closed-loop supply chain (KD)
- Awareness of the reverse or closed-loop supply chain being implemented (AR)
- Simplicity of the reverse or closed-loop supply chain (SR)
- Convenience for the disposal of used products at collection centers (CD)
- Incentives for the disposal of used products (ID)
- Effectiveness of collection methods (EC)
- Information supplied about used products being collected (IU)
- Regularity of collection of used products (RC)
- Design of special methods for abusers of the reverse or closed-loop supply chain (AB)
- Good locations of centers where reprocessed goods are sold (LR)
- Incentives to buyers of reprocessed goods (IB)
- Cooperation of the program organizers with the local government (CL)

4.7.3.2 Approach

Suppose there are three representatives of a community to weigh the drivers of public participation, depending on the driver that greatly motivates the public to participate, the driver that is not so important for the public, and so on. Since the representatives find it difficult to assign numerical weights, they give linguistic weights like very high, low, medium etc. Table 4.38 illustrates the linguistic weights. Using fuzzy logic, these linguistic weights are converted into TFNs (Table 4.39 shows one of the many ways for such a conversion) and then averaged to form another TFN called the average weight. For example, the average weight of the driver, ID (see Tables 4.38 and 4.39), is (Low + High + Low)/3, which is ((0.1 + 0.5 + 0.1)/3, (0.3 + 0.7 + 0.3)/3, (0.5 + 0.9 + 0.5)/3) = (0.23, 0.43, 0.63) (see Table 4.40).

TABLE 4.38

Linguistic Weights of Drivers of Public Participation

Driver	Representative1	Representative2	Representative3
KD	L	M	H
AR	M	H	VH
SR	L	L	VH
CD	VH	VH	M
ID	L	H	L
EC	H	M	VH
IU	L	L	H
RC	M	L	L
AB	M	L	M
LR	VH	H	H
IB	H	H	M
CL	M	H	L

Note: VH = very high, H = high, M = medium, L = low, VL = very low.

TABLE 4.39

Conversion Table for Weights of Drivers

Linguistic Weight	TFN
VH	(0.7, 0.9, 1.0)
H	(0.5, 0.7, 0.9)
M	(0.3, 0.5, 0.7)
L	(0.1, 0.3, 0.5)
VL	(0.0, 0.1, 0.3)

Note: VH = very high, H = high, M = medium, L = low, VL = very low.

TABLE 4.40

Average Weights of Drivers
of Public Participation

Driver	Average Weight
KD	(0.3, 0.5, 0.7)
AR	(0.5, 0.7, 0.87)
SR	(0.3, 0.5, 0.67)
CD	(0.57, 0.77, 0.9)
ID	(0.23, 0.43, 0.63)
EC	(0.5, 0.7, 0.87)
IU	(0.23, 0.43, 0.63)
RC	(0.17, 0.37, 0.57)
AB	(0.17, 0.43, 0.63)
LR	(0.57, 0.77, 0.93)
IB	(0.43, 0.63, 0.83)
CL	(0.3, 0.5, 0.7)
Sum	(4.27, 6.73, 8.93)

TABLE 4.41

Normalized Weights of Drivers
of Public Participation

Driver	Normalized Weight
KD	(0.03, 0.07, 0.16)
AR	(0.06, 0.10, 0.20)
SR	(0.03, 0.07, 0.16)
CD	(0.06, 0.11, 0.21)
ID	(0.03, 0.06, 0.15)
EC	(0.07, 0.10, 0.20)
IU	(0.03, 0.06, 0.15)
RC	(0.02, 0.05, 0.13)
AB	(0.02, 0.06, 0.15)
LR	(0.06, 0.11, 0.22)
IB	(0.05, 0.09, 0.20)
CL	(0.03, 0.07, 0.16)

TABLE 4.42

Conversion for Ratings of Marketing Strategies

Linguistic Rating	TFN
VG	(7, 10, 10)
G	(5, 7, 10)
F	(2, 5, 8)
P	(0, 3, 5)
VP	(0, 0, 3)

Note: VG = very good, G = good, F = fair, P = poor,
VP = very poor.

The sum of the average weights of all the drivers is calculated using Equation 4.6 as (4.27, 6.33, 8.93). The ratio of the average weight of each driver to the sum of the average weights of all the drivers gives the corresponding normalized weight. For example, normalized weight of ID is ((0.23, 0.43, 0.63)/(4.27, 6.33, 8.93)), which is simplified using Equation 4.9 as (0.03, 0.06, 0.15) (see Table 4.41).

Suppose we evaluate marketing strategies of two different reverse supply chains. After the weights (normalized) of the drivers of public participation are ready, the two marketing strategies are linguistically rated by the three representatives with respect to each driver. Table 4.42 is used for conversion of linguistic ratings into TFNs. For arithmetic simplicity, we assume that the representatives arrive at a consensus about the rating of each marketing strategy with respect to each driver. Now, we arrive at the decision matrix shown

TABLE 4.43

Decision Matrix

Driver	S1	S2
KD	(7, 10, 10)	(0, 0, 3)
AR	(2, 5, 8)	(7, 10, 10)
SR	(2, 5, 8)	(2, 5, 8)
CD	(0, 0, 3)	(7, 10, 10)
ID	(7, 10, 10)	(5, 7, 10)
EC	(5, 7, 10)	(2, 5, 8)
IU	(0, 3, 5)	(2, 5, 8)
RC	(0, 0, 3)	(7, 10, 10)
AB	(2, 5, 8)	(5, 7, 10)
LR	(5, 7, 10)	(5, 7, 10)
IB	(0, 3, 5)	(7, 10, 10)
CL	(0, 0, 3)	(5, 7, 10)

TABLE 4.44

Normalized Decision Matrix

Driver	S1	S2
KD	(0.67, 1.00, 1.43)	(0.00, 0.00, 0.43)
AR	(0.16, 0.45, 1.10)	(0.55, 0.89, 1.37)
SR	(0.18, 0.71, 2.83)	(0.18, 0.71, 2.83)
CD	(0.00, 0.00, 0.43)	(0.67, 1.00, 1.43)
ID	(0.50, 0.82, 1.16)	(0.35, 0.57, 1.16)
EC	(0.39, 0.81, 1.86)	(0.16, 0.58, 1.49)
IU	(0.00, 0.51, 2.50)	(0.21, 0.86, 4.00)
RC	(0.00, 0.00, 0.43)	(0.67, 1.00, 1.43)
AB	(0.16, 0.58, 1.49)	(0.39, 0.81, 1.86)
LR	(0.35, 0.71, 1.41)	(0.35, 0.71, 1.41)
IB	(0.00, 0.29, 0.71)	(0.63, 0.96, 1.43)
CL	(0.00, 0.00, 0.60)	(0.48, 1.00, 2.00)

in Table 4.43 (S1 and S2 are the marketing strategies). For example, the rating (TFN) of marketing strategy S2, with respect to driver ID, is (5, 7, 10) (see Table 4.43), because the representatives unanimously rate it as good with respect to ID (the TFN for the linguistic rating good is (5, 7, 10) [see Table 4.42]).

Now, we are ready to perform the six steps in the TOPSIS.

Step 1: *Construct the normalized decision matrix.* Table 4.44 shows the normalized decision matrix formed by applying Equation 4.26 on each element of Table 4.43. For example, the normalized fuzzy rating of

TABLE 4.45

Weighted Normalized Decision Matrix

Driver	S1	S2
KD	(0.02, 0.07, 0.23)	(0.00, 0.00, 0.07)
AR	(0.01, 0.05, 0.22)	(0.03, 0.09, 0.28)
SR	(0.01, 0.05, 0.44)	(0.01, 0.05, 0.44)
CD	(0.00, 0.00, 0.09)	(0.04, 0.11, 0.30)
ID	(0.01, 0.05, 0.17)	(0.01, 0.04, 0.17)
EC	(0.03, 0.08, 0.38)	(0.01, 0.06, 0.30)
IU	(0.00, 0.03, 0.37)	(0.01, 0.06, 0.59)
RC	(0.00, 0.00, 0.06)	(0.01, 0.05, 0.19)
AB	(0.00, 0.04, 0.22)	(0.01, 0.05, 0.28)
LR	(0.02, 0.08, 0.31)	(0.02, 0.08, 0.31)
IB	(0.00, 0.03, 0.14)	(0.03, 0.09, 0.28)
CL	(0.00, 0.00, 0.10)	(0.02, 0.07, 0.33)

strategy, S2, with respect to driver, ID (see Table 4.44), is calculated using Equation 4.26 as follows:

$$r_{52} = \frac{(5, 7, 10)}{\sqrt{(7, 10, 10)^2 + (5, 7, 10)^2}} = (0.35, 0.57, 1.16)$$

Note that Equations 4.6, 4.8, and 4.9 are used to perform the basic operations in the calculation of r_{52}.

Step 2: Construct the weighted normalized decision matrix. Table 4.45 shows the weighted normalized decision matrix. This is constructed using the normalized weights of the drivers listed in Table 4.43 and the normalized decision matrix in Table 4.44. For example, the weighted normalized fuzzy rating of strategy, S2, with respect to driver, ID, that is, (0.01, 0.04, 0.17) (see Table 4.45), is calculated by multiplying the normalized weight of ID, that is, (0.03, 0.07, 0.15) (see Table 4.43) with the normalized fuzzy rating of S2 with respect to ID, that is, (0.35, 0.57, 1.16) (see Table 4.44). Equation 4.8 is used for the multiplication.

Step 3: Determine the ideal and the negative-ideal solution. Each row in the decision matrix shown in Table 4.45 has a maximum rating and a minimum rating. They are the ideal and the negative-ideal solutions, respectively, for the corresponding driver. For arithmetic simplicity, we assume here that the rating with the highest "most promising quantity" (second parameter in the TFN) is the maximum and the rating with the lowest "most promising quantity" is the minimum. For example, with respect to driver, ID, the maximum rating is (0.01, 0.05, 0.17) and the minimum rating is (0.01, 0.04, 0.17). (see Table 4.45), because in the row for that driver, the TFN with the highest second

TABLE 4.46

Separation Distances of Marketing Strategies

Marketing Strategy	Positive Distance S*	Negative Distance S⁻
S1	0.215	0.097
S2	0.096	0.215

TABLE 4.47

Relative Closeness Coefficients of Marketing Strategies

Marketing Strategy	Relative Closeness Coefficient
S1	0.311
S2	0.692

parameter is (0.01, 0.05, 0.14) and the TFN with the lowest second parameter is (0.01, 0.04, 0.14).

Step 4: Calculate the separation distances. Using Equations 4.29 and 4.30 we calculate the separation distances for each marketing strategy (see Table 4.46). For example, the positive separation distance for strategy, S2 (see Table 4.46) is calculated using Equation 4.29 that contains the weighted normalized fuzzy ratings of S2 (see Table 4.45) and the ideal solution (obtained in step 3) for each driver of public participation.

Step 5: Calculate the relative closeness coefficient. Using Equation 4.31, we calculate the relative closeness coefficient for each marketing strategy (see Table 4.47). For example, relative closeness coefficient (i.e., 0.692) for strategy, S2 (see Table 4.47) is the ratio of the strategy's negative separation distance (i.e., 0.215) to the sum (i.e., 0.215 + 0.097 = 0.312) of its negative and positive separation distances (see Table 4.46).

Step 6: Rank the preference order. Since the relative closeness coefficients of strategy S2 (i.e., 0.692) is much higher than strategy S1 (i.e., 0.308) (see Table 4.47), it is obvious that S2 is much better than S1.

4.8   Borda's Choice Rule

The rule is briefly described in Section 4.8.1 and is applied (along with eigenvector method and TOPSIS) in Section 4.8.2 for evaluating collection centers (in a region where a reverse or a closed-loop supply chain is to be designed) and in Section 4.8.3 for evaluating recovery facilities (in a region where a reverse supply chain is to be designed).

4.8.1 Methodology

Borda proposed a method [20] in which marks of $m - 1, m - 2, ..., 1, 0$ are assigned to the best, second-best, ..., worst alternatives, for each decision maker. A larger mark corresponds to more preference. Borda score (maximized consensus mark) for each alternative is then determined as the sum of the individual marks for that alternative. The alternative with the highest Borda score is declared the winner, that is, different decision makers unanimously choose the alternative that obtains the largest Borda score as the most preferred alternative.

4.8.2 Evaluation of Collection Centers

The designing of a prospective reverse or closed-loop supply chain must involve selection of collection centers with sufficient success potentials for the efficient operation of that chain. These success potentials depend heavily on the participation (in the reverse or closed-loop supply chain) of three important categories of decision makers who have multiple, conflicting, and incommensurate goals. Therefore, the potentials must be evaluated on the basis of the maximized consensus among the following three categories:

i. *Consumers*, whose primary concern is convenience
ii. *Local government officials*, whose primary concern is environmental consciousness
iii. *Supply chain company executives*, whose primary concern is profit

We consider the following sets of criteria for the three categories of decision makers for evaluating the candidate collection centers:

Criteria of consumers:
- Incentives from collection center (IC) (higher incentives imply higher motivation to participate)
- Proximity to the residential area (PH) (higher proximity implies higher motivation to participate)
- Proximity to roads (PR) (higher proximity implies higher motivation to participate)
- Simplicity of the collection process (SP) (simpler process implies higher motivation to participate)
- Employment opportunity (EO) (the more the better)
- Salary (SA) (the higher the better)

Criteria of local government officials:

- Proximity to residential area (PH) (higher proximity implies greater collection and hence lower disposal)
- Proximity to roads (PR) (higher proximity implies greater collection and hence lower disposal)

Criteria of supply chain company executives:

- Per capita income of people in the residential area (PI) (the higher it is, the more the number of "resourceful" used products and the less the people will care about the incentives from the collection center)
- Space cost (SC) (the lower the better)
- Labor cost (LC) (the lower the better)
- Utilization of incentives from local government (UI) (the higher the better)
- Proximity to residential area (PH) (higher proximity implies greater collection and hence greater profit)
- Proximity to roads (PR) (higher proximity implies greater collection and hence greater profit)
- Incentives from local government (IG) (higher incentives from local government imply higher incentives to consumers)

The approach to select collection centers of sufficient success potentials is implemented in two phases. In the first phase, using the eigenvector method (see Section 4.3.1), we assign weights to the criteria identified for each category of decision makers and then employ the TOPSIS (see Section 4.7.1) to find the success potential of each candidate collection center, as evaluated by that category. In the second phase, we use Borda's choice rule that, for each candidate collection center, combines individual success potentials into a group success potential or maximized consensus ranking.

We present our approach using a numerical example. Three collection centers, C1, C2, and C3, are considered for evaluation.

4.8.2.1 Phase I of the Approach (Individual Decision Making)

Tables 4.48 through 4.50 show the pairwise comparison matrices as formed by the consumers, the local government officials, and the supply chain company executives, respectively.

Tables 4.51 through 4.53 show the relative weights given by the consumers, the local government officials, and the supply chain company executives, respectively, to the respective criteria for evaluation. These sets of weights, calculated using the eigenvector method, are the elements of the normalized eigenvectors of pairwise comparison matrices shown in Tables 4.48 through 4.50, respectively. For example, the relative weights of the criteria, PH (proximity to residential area) and PR (proximity to roads) shown in Table 4.52

TABLE 4.48

Pairwise Comparison Matrix Formed by Consumers

Criteria	IC	PH	PR	SP	EO	SA
IC	1	1	1	2	1/2	1
PH	1	1	2	1	1	1/2
PR	1	1/2	1	1	1/7	1/2
SP	1/2	1	1	1	1	1
EO	2	1	7	1	1	1
SA	1	2	2	1	1	1

TABLE 4.49

Pairwise Comparison Matrix Formed by Local Government Officials

Criterion	PH	PR
PH	1	2
PR	1/2	1

TABLE 4.50

Pairwise Comparison Matrix Formed by Supply Chain Company Executives

Criteria	PI	SC	LC	UI	PH	PR	IG
PI	1	2	4	6	8	9	4
SC	1/2	1	2	6	8	9	1
LC	1/4	1/2	1	1	1	1	1
UI	1/6	1/6	1	1	1	1	1
PH	1/8	1/8	1	1	1	1	1
PR	1/9	1/9	1	1	1	1	1
IG	1/4	1	1	1	1	1	1

TABLE 4.51

Relative Weights Given by Consumers to Criteria for Evaluation

Criteria for Evaluation	Relative Weights
IC	0.1621
PH	0.1515
PR	0.0960
SP	0.1434
EO	0.2533
SA	0.1938

TABLE 4.52

Relative Weights Given by Local Government Officials to Criteria for Evaluation

Criteria for Evaluation	Relative Weights
PH	0.6667
PR	0.3333

TABLE 4.53

Relative Weights Given by Supply Chain Company Executives to Criteria for Evaluation

Criteria for Evaluation	Relative Weights
PI	0.3876
SC	0.2599
LC	0.0781
UI	0.0634
PH	0.0598
PR	0.0585
IG	0.0927

TABLE 4.54

Decision Matrix Formed by Consumers

Collection Center	IC	PH	PR	SP	EO	SA
C1	8	1	6	2	2	3
C2	2	1	7	3	2	3
C3	3	2	4	1	1	5

are the elements of the normalized eigenvector of the pairwise comparison matrix shown in Table 4.49.

The decision matrices formed by the consumers, the local government officials, and the supply chain company executives, for implementation of the TOPSIS for each category of decision makers, are shown in Tables 4.54 through 4.56, respectively. The elements of these matrices are the ranks (ranging from 1 to 10) assigned to the collection centers with respect to each criterion for evaluation. A lower rank implies higher success potential (with respect to that criterion).

Now, we are ready to perform the six steps in the TOPSIS for each category. The following steps show the implementation of the TOPSIS for the consumers to evaluate C1, C2, and C3.

TABLE 4.55

Decision Matrix Formed by Local Government Officials

Collection Center	PH	PR
C1	2	1
C2	2	1
C3	3	2

TABLE 4.56

Decision Matrix Formed by Supply Chain Company Executives

Collection Center	PI	SC	LC	UI	PH	PR	IG
C1	1	3	2	1	3	4	3
C2	2	4	7	1	2	2	2
C3	10	9	8	7	5	1	6

TABLE 4.57

Normalized Decision Matrix Formed by Consumers

Collection Center	IC	PH	PR	SP	EO	SA
C1	0.9117	0.4082	0.5970	0.5345	0.6667	0.4575
C2	0.2279	0.4082	0.6965	0.8018	0.6667	0.4575
C3	0.3419	0.8165	0.3980	0.2673	0.3333	0.7625

Step 1: *Construct the normalized decision matrix.* Table 4.57 shows the normalized decision matrix formed by applying Equation 4.26 on each element of Table 4.54 (decision matrix formed by the consumers). For example, the normalized rank of collection center, C2, with respect to criterion, PH (see Tables 4.54 and 4.57), is calculated using Equation 4.26 as follows:

$$r_{22} = \frac{1}{\sqrt{1^2 + 1^2 + 2^2}} = 0.4082$$

Step 2: *Construct the weighted normalized decision matrix.* Table 4.58 shows the weighted normalized decision matrix for the consumers. This is constructed using the relative weights of the criteria listed in Table 4.51 and the normalized decision matrix in Table 4.57. For example, the weighted normalized rank of collection center, C2, with respect to criterion, PH, that is, 0.0619 (see Table 4.58), is calculated by multiplying the relative weight of PH, that is, 0.1515 (see Table 4.51)

TABLE 4.58

Weighted Normalized Decision matrix Formed by Consumers

Collection Center	IC	PH	PR	SP	EO	SA
C1	0.1478	0.0619	0.0573	0.0767	0.1689	0.0887
C2	0.0370	0.0619	0.0669	0.1150	0.1689	0.0887
C3	0.0554	0.1237	0.0382	0.0383	0.0844	0.1478

TABLE 4.59

Separation Distances Calculated by Consumers

Collection Center	S*	S⁻
C1	0.1458	0.0942
C2	0.1176	0.1400
C3	0.0875	0.1495

with the normalized rank of C2 with respect to PH, that is, 0.4082 (see Table 4.57).

Step 3: Determine the ideal and the negative-ideal solution. Each column in the weighted normalized decision matrix shown in Table 4.58 has a minimum rank and a maximum rank. They are the ideal and the negative-ideal solutions, respectively, for the corresponding criterion. For example (see Table 4.58), with respect to criterion, PH, the ideal solution (minimum rank) is 0.0619 and the negative-ideal solution (maximum rank) is 0.1237.

Step 4: Calculate the separation distances. The separation distances (see Table 4.59) for each collection center are calculated using Equations 4.19 and 4.30. For example, the positive separation distance for collection center, C2 (see Table 4.59), is calculated using Equation 4.29 that contains the weighted normalized ranks of C2 (see Table 4.58) and the ideal solutions (obtained in step 3) for the criteria.

Step 5: Calculate the relative closeness coefficient. Using Equation 4.31, we calculate the relative closeness coefficient for each collection center (see Table 4.60). For example, relative closeness coefficient (i.e., 0.5435) for collection center C2 (see Table 4.60) is the ratio of C2's negative separation distance (i.e., 0.1400) to the sum (i.e., 0.1400 + 0.1176 = 0.2576) of its negative and positive separation distances (see Table 4.59).

Step 6: Form the preference order. Since the best alternative is the one with the highest relative closeness coefficient, the preference order for the collection centers is C3, C2, and C1 (that means, C3 is the best collection center, as evaluated by the consumers).

TABLE 4.60

Relative Closeness Coefficients Calculated by Consumers

Collection Center	C*
C1	0.3926
C2	0.5435
C3	0.6308

TABLE 4.61

Relative Closeness Coefficients Calculated by Local Government Officials and Supply Chain Company Executives

Collection Center	Local Govt Officials	Supply Chain Company Executives
C1	1.000	0.902
C2	1.000	0.851
C3	0.739	0.091

TABLE 4.62

Marks and Borda Scores of Collection Centers

Collection Center	Consumers	Local Government Officials	Supply Chain Company Executives	Borda Score
C1	0	2	1	3
C2	1	2	2	5
C3	2	1	0	3

The TOPSIS is implemented for the local government officials and the supply chain company executives as well. The relative closeness coefficients of the collection centers, as calculated by those two categories, are shown in Table 4.61.

4.8.2.2 Phase II of the Approach (Group Decision Making)

Table 4.62 shows the marks of the collection centers, provided by using Borda's choice rule, for the consumers, the local government officials, and the supply chain company executives. Borda scores (group success potentials) calculated for C1, C2, and C3 (viz., 3, 5, and 3, respectively) are also shown in the table. For example, Borda score for C2 (i.e., 5) is calculated by summing the marks of C2 for the consumers, the local government officials, and the supply chain company executives (i.e., $1 + 2 + 2$). Since C2 has the highest Borda score, it is the best of the lot.

4.8.3 Evaluation of Recovery Facilities

We consider the following sets of criteria for the three categories of decision makers for evaluating the candidate recovery facilities:

Criteria of consumers (whose primary concern is convenience):

- Proximity to surface water (PS) (lower proximity implies more suitability, i.e., less hazardous)
- Proximity to residential area (PH) (lower proximity implies more suitability, i.e., less hazardous)
- Employment opportunity (EO) (the more the better)
- Salary (the higher the better) (SA)

Criteria of local government officials (whose primary concern is environmental consciousness):

- Proximity to surface water (PS) (lower proximity implies more suitability, i.e., less hazardous)
- Proximity to residential area (PH) (lower proximity implies more suitability, i.e., less hazardous)

Criteria of supply chain company executives (whose primary concern is profit):

- Space cost (SC) (the lower the better)
- Labor cost (LC) (the lower the better)
- Proximity to roads (PR) (higher proximity implies easier transportation)
- QO − QI (the higher the better)
- TP/SU (the higher the better)
- TP × DT (the higher the better)
- Utilization of incentives from local government (UI) (the higher the better)
- Pollution control (PC) (the higher the better)

The approach to select efficient recovery facilities is implemented in two phases (like in the evaluation of collection centers; see Section 4.8.2.1). In the first phase, using the eigenvector method (see Section 4.3.1), we give weights to the criteria identified for each category of decision makers, and then employ the TOPSIS (see Section 4.7.1) to find the success potential of each candidate recovery facility, as evaluated by that category. In the second phase, we use Borda's choice rule that combines individual success potentials into a group success potential or maximized consensus ranking for each candidate recovery facility.

We present our approach using a numerical example. Three recovery facilities, R1, R2, and R3 are considered for evaluation.

4.8.3.1 Phase I of the Approach (Individual Decision Making)

Tables 4.63 through 4.65 show the pairwise comparison matrices as formed by the consumers, the local government officials, and the supply chain company executives, respectively.

Tables 4.66 through 4.68 show the relative weights given by the consumers, the local government officials, and the supply chain company executives respectively, to the respective criteria for evaluation. These sets of weights, calculated using the eigenvector method, are the elements of the normalized eigenvectors of pairwise comparison matrices shown in Tables 4.63 through 4.65, respectively. For example, the relative weights of the criteria, PH (proximity to residential area) and PR (proximity to roads) shown in Table 4.67

TABLE 4.63

Pairwise Comparison Matrix Formed by Consumers

Criterion	PS	PH	EO	SA
PS	1	1/2	1/2	1/3
PH	2	1	1	1
EO	2	1	1	1
SA	3	1	1	1

TABLE 4.64

Pairwise Comparison Matrix Formed by Local Government Officials

Criterion	PS	PH
PH	1	1/3
PR	3	1

TABLE 4.65

Pairwise Comparison Matrix Formed by Supply Chain Company Executives

Criterion	SC	LC	PR	QO – QI	TP/SU	TP × DT	UI	PC
SC	1	1	1	1	1/2	6	1	1
LC	1	1	1	1	1	5	2	1
PR	1	1	1	1/9	1/7	1	1	1/3
AQ	1	1	1	1	2	2	1	1/4
TP/SU	2	1	7	1	1	5	1	1
TP × DT	1/6	1/5	1	1/2	0.2	1	1/9	1
UI	1	1/2	1	1	1	9	1	1
PC	1	1	3	4	1	1	1	1

TABLE 4.66

Relative Weights Given by Consumers to Criteria for Evaluation

Criteria for Evaluation	Relative Weights
PS	0.1277
PH	0.2804
EO	0.2804
SA	0.3116

TABLE 4.67

Relative Weights Given by Local Government Officials to Criteria for Evaluation

Criteria for Evaluation	Relative Weights
PS	0.25
PH	0.75

TABLE 4.68

Relative Weights Given by Supply Chain Company Executives to Criteria for Evaluation

Criteria for Evaluation	Relative Weights
SC	0.1233
LC	0.1437
PR	0.0717
QO – QI	0.1198
TP/SU	0.1905
TP × DT	0.0491
UI	0.1356
PC	0.1663

are the elements of the normalized eigenvector of the pairwise comparison matrix shown in Table 4.64.

The decision matrices formed by the consumers, the local government officials, and the supply chain company executives, for implementation of the TOPSIS for each category are shown in Tables 4.69 through 4.71, respectively. The elements of these matrices are the ranks (ranging from 1 to 10) assigned to the recovery facilities with respect to each criterion for evaluation. A lower rank implies higher success potential (with respect to that criterion).

Now, we are ready to perform the six steps in the TOPSIS for each category of decision makers. The following steps show the implementation of the TOPSIS for the consumers to evaluate R1, R2, and R3.

TABLE 4.69

Decision Matrix Formed by Consumers

Recovery Facility	PS	PH	EO	SA
R1	2	2	4	1
R2	4	1	7	3
R3	5	3	3	1

TABLE 4.70

Decision Matrix Formed by Local Government Officials

Recovery Facility	PS	PH
R1	4	2
R2	3	1
R3	1	5

TABLE 4.71

Decision Matrix Formed by Supply Chain Company Executives

Recovery Facility	SC	LC	PR	QO − QI	TP/SU	TP × DT	UI	PC
R1	3	2	1	3	4	1	1	2
R2	4	7	1	2	2	2	3	4
R3	9	8	7	5	1	6	5	6

TABLE 4.72

Normalized Decision Matrix Formed by Consumers

Recovery Facility	PS	PH	EO	SA
R1	0.2981	0.5345	0.4650	0.3015
R2	0.5963	0.2673	0.8137	0.9045
R3	0.7454	0.8018	0.3487	0.3015

Step 1: *Construct the normalized decision matrix.* Table 4.72 shows the normalized decision matrix formed by applying Equation 4.26 on each element of Table 4.69 (decision matrix formed by the consumers). For example, the normalized rank of recovery facility, R2, with respect to criterion, PH (see Tables 4.69 and 4.72), is calculated using Equation 4.26 as follows:

$$r_{22} = \frac{1}{\sqrt{2^2 + 1^2 + 3^2}} = 0.2673$$

TABLE 4.73

Weighted Normalized Decision Matrix Formed by Consumers

Recovery Facility	PS	PH	EO	SA
R1	0.0381	0.1499	0.1304	0.0940
R2	0.0761	0.0749	0.2282	0.2819
R3	0.0952	0.2248	0.0978	0.0940

TABLE 4.74

Separation Distances Calculated by Consumers

Recovery Facility	S*	S⁻
R1	0.0817	0.2318
R2	0.2318	0.1511
R3	0.1604	0.2287

Step 2: *Construct the weighted normalized decision matrix.* Table 4.73 shows the weighted normalized decision matrix for the consumers. This is constructed using the relative weights of the criteria listed in Table 4.66 and the normalized decision matrix in Table 4.72. For example, the weighted normalized rank of recovery facility, R2, with respect to criterion, PH, that is, 0.0749 (see Table 4.73), is calculated by multiplying the relative weight of PH, that is, 0.2804 (see Table 4.66) with the normalized rank of R2 with respect to PH, that is, 0.2673 (see Table 4.72).

Step 3: *Determine the ideal and the negative-ideal solution.* Each column in the weighted normalized decision matrix shown in Table 4.73 has a minimum rank and a maximum rank. They are the ideal and the negative-ideal solutions, respectively, for the corresponding criterion. For example (see Table 4.73), with respect to criterion, PH, the ideal solution (minimum rank) is 0.0749, and the negative-ideal solution (maximum rank) is 0.2248.

Step 4: *Calculate the separation distances.* The separation distances (see Table 4.74) for each recovery facility are calculated using Equations 4.29 and 4.30. For example, the positive separation distance for recovery facility, R2 (see Table 4.74), is calculated using Equation 4.29 that contains the weighted normalized ranks of R2 (see Table 4.73) and the ideal solutions (obtained in step 3) for the criteria.

Step 5: *Calculate the relative closeness coefficient.* Using Equation 4.31, we calculate the relative closeness coefficient for each recovery facility (see Table 4.75). For example, relative closeness coefficient (i.e., 0.3945) for recovery facility R2 (see Table 4.75) is the ratio of R2's

TABLE 4.75

Relative Closeness Coefficients Calculated by Consumers

Recovery Facility	C*
R1	0.7394
R2	0.3945
R3	0.5878

TABLE 4.76

Relative Closeness Coefficients Calculated by Local Government Officials and Supply Chain Company Executives

Recovery Facility	Local Government Officials	Supply Chain Company Executives
R1	0.6715	0.5952
R2	0.8487	0.5962
R3	0.2941	0.4057

TABLE 4.77

Marks and Borda Scores of Recovery Facilities

Recovery Facility	Consumers	Local Government Officials	Supply Chain Company Executives	Borda Score
R1	2	1	2	5
R2	0	2	2	4
R3	1	0	1	2

negative separation distance (i.e., 0.1511) to the sum (i.e., 0.1511 + 0.2318 = 0.3829) of its negative and positive separation distances (see Table 4.74).

Step 6: Form the preference order. Since the best alternative is the one with the highest relative closeness coefficient, the preference order for the recovery facilities is R1, R3, and R2 (that means, R1 is the best recovery facility, as evaluated by the consumers).

The TOPSIS is implemented for the local government officials and the supply chain company executives as well. The relative closeness coefficients of the recovery facilities, as calculated by those two categories of decision makers, are shown in Table 4.76.

4.8.3.2 Phase II of the Approach (Group Decision Making)

Table 4.77 shows the marks of the recovery facilities given using Borda's choice rule for the consumers, the local government officials, and the supply chain company executives. Borda scores (group success potentials) calculated for R1, R2, and R3 (viz., 5, 4, and 2, respectively) are also shown in the table. For example, Borda score for R2 (i.e., 4) is calculated by summing the marks of R2 for the consumers, the local government officials, and the supply chain company executives (i.e., $0 + 2 + 2$). Since R1 has the highest Borda score, it is the best of the lot.

4.9   Neural Networks

A neural network [7] is an information processing paradigm that is inspired by the way biological nervous systems, such as the brain, process information. The novel structure of the information processing system is the key element of this paradigm. This paradigm is composed of a large number of highly interconnected processing elements (neurons) working in unison to solve specific problems. Neural networks, like people, learn by example. A neural network is configured for a specific application, such as pattern recognition or data classification, through a learning process. Learning in biological systems involves adjustments to the synaptic connections that exist between the neurons. This is true of neural networks as well.

Neural networks, with their remarkable ability to derive meaning from complicated or imprecise data, can be used to extract patterns and detect trends that are too complex to be noticed by either humans or other computer techniques. A trained neural network can be thought of as an expert in the category of information it has been given to analyze. This expert can then be used to provide projections in new situations of interest and answer "what if" questions.

Neural networks (along with fuzzy logic, TOPSIS, and Borda's choice rule) are applied in Section 4.9.1 for evaluating collection centers, and in Section 4.9.2 for evaluating recovery facilities.

4.9.1   Evaluation of Collection Centers

In this section, we evaluate the success potential of collection centers of interest, which are being considered for inclusion in a reverse or a closed-loop supply chain using the available linguistic data of collection centers that already exist in the reverse or closed-loop supply chain.

Fuzzy Logic (see Section 4.4) can be employed to convert linguistic ratings (viz., performance measures) of existing collection centers with respect to criteria for evaluation and their (collection centers') overall ratings, into numerical ratings. However, this is not necessarily true in the case of weights of the criteria, because an expert may find it very difficult to assign linguistic weights in the first place. To this end, our approach to evaluate the success potentials of the collection centers of interest is carried out in three phases as follows: In phase I, we use fuzzy ratings of existing collection centers to construct a neural network that gives weights of criteria identified for each category of decision makers discussed in Section 4.8.2. Then, in phase II, using the weights obtained in phase I, we use fuzzy logic and the TOPSIS (see Section 4.7.1) to obtain the overall ratings of the collection centers of interest, as calculated by each category. Finally, in phase III, we employ Borda's choice rule (see Section 4.8.1) to calculate the maximized consensus (among the categories considered) rating, that is, success potentials, of the collection centers of interest.

4.9.1.1 Phase I of the Approach (Derivation of Weights of Criteria)

Suppose we have the linguistic ratings of 10 existing collection centers, as given by an expert in each category of decision makers described in Section 4.8.2. Using fuzzy logic, these linguistic ratings are converted into TFNs (fuzzy ratings). Table 4.78 shows not only one of the many ways for conversion of linguistic ratings into TFNs, but also the defuzzified ratings of the corresponding TFNs. Tables 4.79 through 4.81 show the defuzzified

TABLE 4.78

Conversion Table for Ratings

Linguistic Rating	TFN	Defuzzified Rating
VG	(7, 10, 10)	9
G	(5, 7, 10)	7.3
F	(2, 5, 8)	5
P	(1, 3, 5)	3
VP	(0, 0, 3)	1

Note: VG = very good, G = good, F = fair, P = poor, VP = very poor.

TABLE 4.79

Consumer Ratings of Existing Collection Centers

Collection Center	IC	PH	PR	SP	EO	SA	Overall
C1	1	3	5	3	5	9	5
C2	9	1	3	5	7.3	9	7.3
C3	3	1	3	1	9	1	3
C4	3	9	1	7.3	1	7.3	5
C5	5	1	3	5	1	3	7.3
C6	9	3	7.3	3	5	7.3	3
C7	5	7.3	9	1	7.3	9	1
C8	1	5	1	5	3	1	9
C9	1	5	5	9	9	5	5
C10	5	9	5	3	9	3	1

TABLE 4.80

Local Government Officials Ratings of Existing Collection Centers

Collection Center	PH	PR	Overall
C1	1	3	5
C2	9	1	7.3
C3	3	1	3
C4	3	9	5
C5	5	1	7.3
C6	9	3	3
C7	5	7.3	1
C8	1	5	9
C9	1	5	5
C10	5	9	1

TABLE 4.81

Supply Chain Company Executives' Ratings of Existing Collection Centers

Collection Center	PI	SC	LC	UI	PH	PR	IG	Overall
C1	1	3	1	3	5	1	3	5
C2	9	1	3	7.3	3	5	7.3	7.3
C3	3	1	7.3	9	1	7.3	9	3
C4	3	9	5	1	5	3	1	5
C5	5	1	5	5	9	9	5	7.3
C6	9	3	9	5	3	9	3	3
C7	5	7.3	3	1	7.3	9	1	1
C8	1	5	1	3	1	3	5	9
C9	1	5	3	5	9	9	1	5
C10	5	9	1	3	5	7.3	7.3	1

overall rating of each existing collection center as well as its (collection center's) defuzzified rating with respect to each criterion as evaluated by the consumers, the local government officials, and the supply chain company executives, respectively.

A neural network is constructed and trained for each category using the defuzzified ratings of the existing collection centers with respect to criteria as input sets and their (collection centers') defuzzified overall ratings as corresponding outputs. In our example, we have 10 input–output pairs for each neural network because of 10 existing collection centers. Also, we consider three layers in each network, with five nodes in the hidden layer. The number of nodes in the output layer is 1 (for overall rating), and the number of nodes in the input layer is the number of criteria considered by the corresponding category of decision makers. For example, Figure 4.8 shows the neural network constructed and trained for the category of consumers.

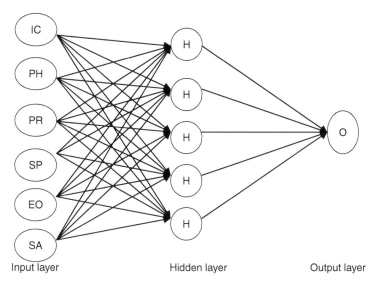

FIGURE 4.8
Neural network for consumers (*vis-à-vis* collection centers).

TABLE 4.82

Weights of Criteria of Consumers

Criterion	IC	PH	PR	SP	EO	SA
Impact	0.01	0.13	0.06	0.18	0.19	0.43

After each neural network is trained, Equation 4.32 [7] is used to calculate the weights of criteria considered by the corresponding category of decision makers. Here, absolute value of W_v is the weight of the vth input node upon the output node, n_V the number of input nodes, n_H the number of hidden nodes, I_{ij} the connection weight (not to be mistaken for weight of a criterion) from the ith input node to the jth hidden node, and O_j the connection weight from the jth hidden node to the output node.

$$|W_v| = \frac{\sum_j^{n_H}\left(I_{vj}\Big/\sum_i^{n_V}|I_{ij}|\right)O_j}{\sum_i^{n_V}\left(\sum_j^{n_H}\left(\left(I_{vj}\Big/\sum_i^{n_V}|I_{ij}|\right)O_j\right)\right)} \tag{4.32}$$

Tables 4.82 through 4.84 show the weights of the criteria considered by the consumers, the local government officials, and the supply chain company executives, respectively.

TABLE 4.83

Impacts of Criteria of Local Government Officials

Criterion	PH	PR
Impact	0.33	0.67

TABLE 4.84

Impacts of Criteria of Supply Chain Company Executives

Criterion	PI	SC	LC	UI	PH	PR	IG
Impact	0.24	0.09	0	0.18	0.25	0.1	0.13

TABLE 4.85

Decision Matrix Formed by Consumers

Collection Center	IC	PH	PR	SP	EO	SA
C11	3	9	1	7.33	1	7.33
C12	5	1	3	5	1	3
C13	9	3	7.33	3	5	7.33

TABLE 4.86

Decision Matrix Formed for Local Government Officials

Collection Center	PH	PR
C11	3	9
C12	5	1
C13	9	3

4.9.1.2 Phase II of the Approach (Individual Decision Making)

Suppose there are three collection centers, C11, C12, and C13. We use the weights obtained in phase I and employ fuzzy logic and the TOPSIS to calculate the overall ratings of the three collection centers.

The decision matrices formed by the consumers, the local government officials, and the supply chain company executives (with defuzzified ratings for C11, C12, and C13) in this example are shown in Tables 4.85 through 4.87, respectively (we use Table 4.78 here too to convert linguistic ratings given by each category into TFNs).

Now, we are ready to perform the six steps in the TOPSIS for each category of decision makers. The following steps show the implementation of the TOPSIS for the consumers to evaluate C11, C12, and C13.

Step 1: *Construct the normalized decision matrix.* Table 4.88 shows the normalized decision matrix formed by applying Equation 4.26 on each

TABLE 4.87

Decision Matrix Formed by Supply Chain Company Executives

Collection Center	PI	SC	LC	UI	PH	PR	IG
C11	5	1	5	5	9	9	5
C12	9	3	9	5	3	9	3
C13	5	7.33	3	1	7.33	9	1

TABLE 4.88

Normalized Decision Matrix Formed by Consumers

Collection Center	IC	PH	PR	SP	EO	SA
C11	0.278	0.943	0.125	0.783	0.192	0.679
C12	0.466	0.105	0.376	0.534	0.192	0.278
C13	0.839	0.314	0.918	0.320	0.962	0.679

TABLE 4.89

Weighted Normalized Decision Matrix Formed by Consumers

Collection Center	IC	PH	PR	SP	EO	SA
C11	0.004	0.122	0.007	0.140	0.036	0.294
C12	0.006	0.014	0.022	0.095	0.036	0.120
C13	0.011	0.041	0.053	0.057	0.181	0.294

element of Table 4.85 (decision matrix formed by the consumers). For example, the normalized rating of collection center, C12, with respect to criterion, PH (see Tables 4.85 and 4.88), is calculated using Equation 4.26 as follows:

$$r_{22} = \frac{1}{\sqrt{9^2 + 1^2 + 3^2}} = 0.105$$

Step 2: Construct the weighted normalized decision matrix. Table 4.89 shows the weighted normalized decision matrix for the consumers. This is constructed using the weights of the criteria listed in Table 4.82 and the normalized decision matrix in Table 4.88. For example, the weighted normalized rank of collection center, C12, with respect to criterion, PH, that is, 0.014 (see Table 4.89), is calculated by multiplying the impact of PH, that is, 0.129 (see Table 4.82) with the normalized rating of C12 with respect to PH, that is, 0.105 (see Table 4.88).

Step 3: Determine the ideal and the negative-ideal solution. Each column in the weighted normalized decision matrix shown in Table 4.89 has a maximum rating and a minimum rating. They are the ideal and the negative-ideal solutions, respectively, for the corresponding criterion. For example (see Table 4.89), with respect to criterion, PH, the ideal solution (maximum rating) is 0.122, and the negative-ideal solution (minimum rating) is 0.014.

Step 4: Calculate the separation distances. The separation distances (see Table 4.90) for each collection center are calculated using Equations 4.29 and 4.30. For example, the positive separation distance for collection center, C12 (see Table 4.90), is calculated using Equation 4.29 that contains the weighted normalized ratings of C12 (see Table 4.89) and the ideal solutions (obtained in step 3) for the criteria.

Step 5: Calculate the relative closeness coefficient. Using Equation 4.31, we calculate the relative closeness coefficient for each collection center (see Table 4.91). For example, relative closeness coefficient (i.e., 0.137) for collection center C12 (see Table 4.91) is the ratio of C12's negative separation distance (i.e., 0.041) to the sum (i.e., 0.041 + 0.257 = 0.298) of its negative and positive separation distances (see Table 4.90).

Step 6: Rank the preference order. Since the best alternative is the one with the highest relative closeness coefficient, the preference order for the collection centers is C13, C11, and C12 (that means, C13 is the best collection center, as evaluated by the consumers).

The TOPSIS is implemented for the local government officials and the supply chain company executives as well. The relative closeness coefficients of the collection centers, as calculated by those two categories, are shown in Table 4.92.

TABLE 4.90

Separation Distances Calculated by Consumers

Collection Center	S^*	S^-
C11	0.152	0.221
C12	0.257	0.041
C13	0.116	0.233

TABLE 4.91

Relative Closeness Coefficients Calculated by Consumers

Collection Center	C^*
C11	0.592
C12	0.137
C13	0.668

TABLE 4.92

Relative Closeness Coefficients Calculated by Local Government Officials and Supply Chain Company Executives

Collection Center	Local Government Officials	Supply Chain Company Executives
C11	0.754	0.619
C12	0.096	0.502
C13	0.354	0.415

TABLE 4.93

Marks and Borda Scores of Collection Centers

Collection Center	Consumers	Local Govt. Officials	Supply Chain Company Executives	Borda Score
C11	1	2	2	5
C12	0	0	1	1
C13	2	1	0	3

4.9.1.3 Phase III of the Approach (Group Decision Making)

Table 4.93 shows the marks of the collection centers given using Borda's choice rule (see Section 4.8.1) for the consumers, the local government officials, and the supply chain company executives. Borda scores (maximized success potentials) calculated for C11, C12, and C13 (viz., 5, 1, and 3, respectively) are also shown in the table. For example, Borda score for C12 (i.e., 1) is calculated by summing the marks of C12 for the consumers, the local government officials, and the supply chain company executives (i.e., $0 + 0 + 1$). Since C11 has the highest Borda score, it is the best of the lot.

4.9.2 Evaluation of Recovery Facilities

In this section, we evaluate the success potentials of recovery facilities of interest, which are considered for inclusion in a reverse supply chain using the available linguistic data of recovery facilities that already exist in the reverse supply chain.

While fuzzy logic (see Section 4.4) can be employed to convert linguistic ratings (viz., performance measures) of existing recovery facilities with respect to criteria for evaluation and their (recovery facilities) overall ratings into numerical ratings, it is not necessarily true in the case of weights of the criteria, because it is very difficult in the first place for an expert to give linguistic weights. To this end, our approach to evaluate the success potentials of the recovery facilities of interest is carried out in three phases: in phase I, we use fuzzy ratings of existing recovery facilities to construct a neural network that gives weights of criteria identified for each category of decision makers discussed in Section 4.8.3. Then, in phase II, using the weights obtained in phase I, we use fuzzy logic and the TOPSIS (see Section 4.7.1) to obtain the overall ratings of the recovery facilities of interest, as calculated by each category. Finally, in phase III, we employ Borda's choice rule (see Section 4.8.1) to calculate the maximized consensus (among the categories considered) rating, that is, success potential, of the recovery facilities of interest.

4.9.2.1 Phase I of the Approach (Derivation of Weights of Criteria)

Suppose we have the linguistic ratings of 10 existing recovery facilities, as given by an expert in each category of decision makers described in

Section 4.8.3. Using fuzzy logic, these linguistic ratings are converted into TFNs (fuzzy ratings). Table 4.94 shows not only one of the many ways for conversion of linguistic ratings into TFNs but also the defuzzified ratings of the corresponding TFNs. Tables 4.95 through 4.97 show the defuzzified overall rating of each existing recovery facility as well as its (recovery facility's) defuzzified rating with respect to each criterion, as evaluated by the consumers, the local government officials, and the supply chain company executives, respectively.

A neural network is constructed and trained for each category of decision makers using the defuzzified ratings of the existing recovery facilities with respect to criteria as input sets and their (recovery facilities) defuzzified overall ratings as corresponding outputs. In our example, there are 10 input–output pairs for each neural network because there are 10 existing recovery facilities. Also, we consider three layers in each network, with five nodes in the hidden layer. The number of nodes in the output layer is one (for overall rating), and the number of nodes in the input layer is the number of criteria

TABLE 4.94

Conversion Table for Ratings

Linguistic Rating	TFN	Defuzzified Rating
VG	(7, 10, 10)	9
G	(5, 7, 10)	7.3
F	(2, 5, 8)	5
P	(1, 3, 5)	3
VP	(0, 0, 3)	1

Note: VG = very good, G = good, F = fair, P = poor, VP = very poor.

TABLE 4.95

Consumer Ratings of Existing Recovery Facilities

Recovery Facility	PS	PH	EO	SA	Overall
R1	9	3	7.33	3	3
R2	5	7.33	9	1	1
R3	1	5	1	5	9
R4	1	5	5	9	5
R5	5	9	5	3	7.33
R6	1	3	5	3	5
R7	9	1	3	5	1
R8	3	1	3	1	9
R9	3	9	1	7.33	5
R10	5	1	3	5	7.33

TABLE 4.96

Local Government Officials' Ratings of Existing Recovery Facilities

Recovery Facility	PS	PH	Overall
R1	9	3	5
R2	5	7.33	3
R3	1	5	9
R4	1	5	7.33
R5	5	9	1
R6	1	3	1
R7	9	1	7.33
R8	3	1	5
R9	3	9	5
R10	5	1	3

TABLE 4.97

Supply Chain Company Executives' Ratings of Existing Recovery Facilities

Recovery Facility	SC	LC	PR	QO − QI	TP/SU	TP × DT	UI	PC	Overall
R1	9	3	9	5	3	9	3	3	5
R2	5	7.33	3	1	7.33	9	1	1	3
R3	1	5	1	3	1	3	5	9	1
R4	1	5	3	5	9	9	1	5	7.33
R5	5	9	1	3	5	7.33	7.33	1	9
R6	9	3	9	5	3	9	3	3	9
R7	5	7.33	3	1	7.33	9	1	1	1
R8	1	5	1	3	1	3	5	9	3
R9	1	5	3	5	9	9	1	5	5
R10	5	9	1	3	5	7.33	7.33	1	7.33

considered by the corresponding category. For example, Figure 4.9 shows the neural network constructed and trained for the category of consumers.

After each neural network is trained, Equation 4.32 is used to calculate the weights of criteria considered by the corresponding category.

Tables 4.98 through 4.100 show the weights of the criteria considered by the consumers, the local government officials, and the supply chain company executives, respectively.

4.9.2.2 Phase II of the Approach (Individual Decision-Making)

Suppose that there are three recovery facilities, R11, R12, and R13. We use the weights obtained in phase I and employ fuzzy logic and the TOPSIS to calculate the overall ratings of the three recovery facilities.

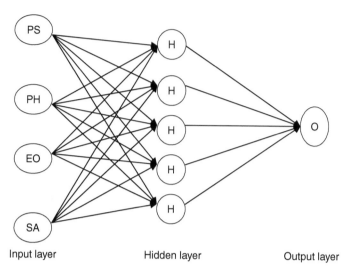

FIGURE 4.9
Neural network for consumers (*vis-à-vis* recovery facilities).

TABLE 4.98

Weights of Criteria of Consumers

Criterion	PS	PH	EO	SA
Impact	0.17	0.18	0.27	0.38

TABLE 4.99

Weights of Criteria of Local Government Officials

Criterion	PS	PH
Impact	0.61	0.39

TABLE 4.100

Weights of Criteria of Supply Chain Company Executives

Criterion	SC	LC	PR	QO − QI	TP/SU	TP × DT	UI	PC
Impact	0.06	0.02	0.10	0.19	0.11	0.39	0.10	0.02

The decision matrices formed by the consumers, the local government officials, and the supply chain company executives (with defuzzified ratings for R11, R12, and R13) in this example are shown in Tables 4.101 through 4.103, respectively (we also use Table 4.94 here to convert linguistic ratings given by each category into TFNs).

TABLE 4.101

Decision Matrix Formed by Consumers

Recovery Facility	PS	PH	EO	SA
R11	1	5	1	5
R12	1	5	5	9
R13	5	9	5	3

TABLE 4.102

Decision Matrix Formed for Local Government Officials

Recovery Facility	PS	PH
R11	3	1
R12	3	9
R13	5	1

TABLE 4.103

Decision Matrix Formed by Supply Chain Company Executives

Recovery Facility	SC	LC	PR	QO − QI	TP/SU	TP × DT	UI	PC
R11	5	9	1	3	5	7.33	7.33	1
R12	9	3	9	5	3	9	3	3
R13	5	7.33	3	1	7.33	9	1	1

Now, we are ready to perform the six steps in the TOPSIS for each category of decision makers. The following steps show the implementation of the TOPSIS for the consumers to evaluate R11, R12, and R13.

Step 1: *Construct the normalized decision matrix.* Table 4.104 shows the normalized decision matrix formed by applying Equation 4.26 on each element of Table 4.101 (decision matrix formed by the consumers). For example, the normalized rating of recovery facility, R12, with respect to criterion, PH (see Tables 4.104 and 4.101), is calculated using Equation 4.26 as follows:

$$r_{22} = \frac{5}{\sqrt{5^2 + 5^2 + 9^2}} = 0.4369$$

Step 2: *Construct the weighted normalized decision matrix.* Table 4.105 shows the weighted normalized decision matrix for the consumers. This is constructed using the weights of the criteria listed in Table 4.98 and the normalized decision matrix in Table 4.104. For example, the weighted normalized rank of recovery facility, R12, with respect to criterion, PH, that is, 0.0793 (see Table 4.105), is calculated by multiplying the

TABLE 4.104

Normalized Decision Matrix Formed by Consumers

Recovery Facility	PS	PH	EO	SA
R11	0.1923	0.4369	0.1400	0.4663
R12	0.1925	0.4369	0.7001	0.8393
R13	0.9623	0.7863	0.7001	0.2798

TABLE 4.105

Weighted Normalized Decision Matrix Formed by Consumers

Recovery Facility	PS	PH	EO	SA
R11	0.0330	0.0793	0.0371	0.1780
R12	0.0330	0.0793	0.1857	0.3203
R13	0.1650	0.1428	0.1857	0.1068

TABLE 4.106

Separation Distances Calculated by Consumers

Recovery Facility	S*	S⁻
R11	0.2526	0.0712
R12	0.1464	0.2602
R13	0.2136	0.2086

impact of PH, that is, 0.18 (see Table 4.98) with the normalized rating of R12 with respect to PH, that is, 0.4369 (see Table 4.104).

Step 3: Determine the ideal and the negative-ideal solution. Each column in the weighted normalized decision matrix shown in Table 4.105 has a maximum rating and a minimum rating. They are the ideal and the negative-ideal solutions, respectively, for the corresponding criterion. For example (see Table 4.105), with respect to criterion, PH, the ideal solution (maximum rating) is 0.1428 and the negative-ideal solution (minimum rating) is 0.0793.

Step 4: Calculate the separation distances. The separation distances (see Table 4.106) for each recovery facility are calculated using Equations 4.19 and 4.30. For example, the positive separation distance for recovery facility, R12 (see Table 4.106), is calculated using Equation 4.29 that contains the weighted normalized ratings of R12 (see Table 4.105) and the ideal solutions (obtained in step 3) for the criteria.

Step 5: Calculate the relative closeness coefficient. Using Equation 4.31, we calculate the relative closeness coefficient for each recovery facility

TABLE 4.107

Relative Closeness Coefficients Calculated by Consumers

Recovery Facility	C*
R11	0.2199
R12	0.6398
R13	0.4942

TABLE 4.108

Relative Closeness Coefficients Calculated by Local Government Officials and Supply Chain Company Executives

Recovery Facility	Local Government Officials	Supply Chain Company Executives
R11	0	0.458016
R12	0.646968	0.695261
R13	0.588066	0.29753

(see Table 4.107). For example, relative closeness coefficient (i.e., 0.6398) for recovery facility R12 (see Table 4.107) is the ratio of R12's negative separation distance (i.e., 0.2602) to the sum (i.e., 0.2602 + 0.1464 = 0.4066) of its negative and positive separation distances (see Table 4.106).

Step 6: Rank the preference order. Since the best alternative is the one with the highest relative closeness coefficient, the preference order for the recovery facilities is R12, R13, and R11 (that means, R12 is the best recovery facility, as evaluated by the consumers).

The TOPSIS is implemented for the local government officials and the supply chain company executives as well. The relative closeness coefficients of the recovery facilities, as calculated by those two categories, are shown in Table 4.108.

4.9.2.3   Phase III of the Approach (Group Decision-Making)

Table 4.109 shows the marks of the recovery facilities given using Borda's choice rule (see Section 4.8.1) for the consumers, the local government officials, and the supply chain company executives. Borda scores (maximized success potentials) calculated for R11, R12, and R13 (viz., 1, 6, and 2, respectively) are also shown in the table. For example, Borda score for R12 (i.e., 6) is calculated by summing the marks of R12 for the consumers, the local government officials, and the supply chain company executives (i.e., 2 + 2 + 2). Since R12 has the highest Borda score, it is the best of the lot.

TABLE 4.109

Marks and Borda Scores of Recovery Facilities

Recovery Facility	Consumers	Local Government	Supply Chain Company	Borda Score
R11	0	0	1	1
R12	2	2	2	6
R13	1	1	0	2

4.10   Linear Physical Programming

In the linear physical programming (LPP) method [31], four distinct classes (1S, 2S, 3S, and 4S) are used to allow the decision maker to express his preferences for the value of each criterion (for decision making) in a more detailed, quantitative, and qualitative way than when using a weight-based method like AHP. These classes are defined as follows: smaller-is-better (1S), larger-is-better (2S), value-is-better (3S), and range-is-better (4S). Figure 4.10 depicts these classes.

The value of the pth criterion, g_p, for evaluating the alternative of interest, is categorized according to the preference ranges shown on the horizontal axis. Consider, for example, the case of Class-1S. The preference ranges are

Ideal range	$g_p \leq t^+_{p1}$
Desirable range	$t^+_{p1} \leq g_p \leq t^+_{p2}$
Tolerable range	$t^+_{p2} \leq g_p \leq t^+_{p3}$
Undesirable range	$t^+_{p3} \leq g_p \leq t^+_{p4}$
Highly undesirable range	$t^+_{p4} \leq g_p \leq t^+_{p5}$
Unacceptable range	$g_p \geq t^+_{p5}$

The quantities t^+_{p1} through t^+_{p5} represent the physically meaningful values that quantify the preferences associated with the pth generic criterion. Consider, for example, the cost criterion for class 1S. The decision maker could specify a preference vector by identifying t^+_{p1} through t^+_{p5} in dollars as $10, $20, $30, $40, $50. Thus, an alternative having a cost of $15 would lie in the desirable range, an alternative with a cost of $45 would lie in the highly undesirable range, and so on. We can accomplish this specification for a non-numerical criterion also, such as color, by (i) specifying a numerical preference structure and (ii) quantitatively assigning each alternative a specific criterion value within a preference range (e.g., desirable, tolerable).

The class function, Z_p, on the vertical axis in Figure 4.10, is used to map the criterion value, g_p, into a real, positive, and dimensionless parameter (Z_p is, in fact, a piecewise linear function of g_p). Such a mapping ensures that

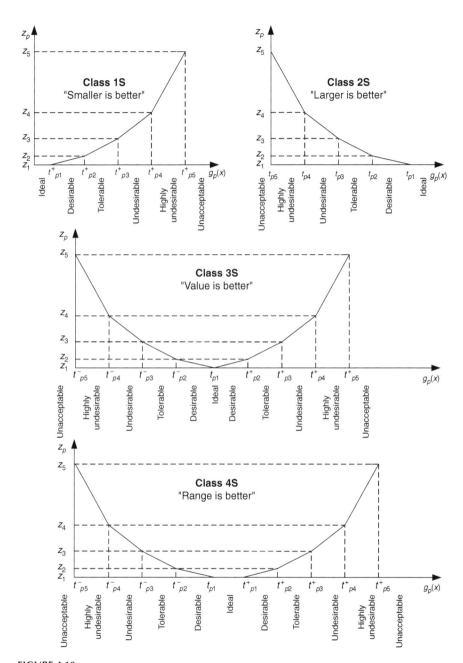

FIGURE 4.10
Soft-class functions for physical programming.

different criteria values, with different physical meanings, are mapped to a common scale. Consider class 1S again. If the value of a criterion, g_p, is in the ideal range, then the value of the class function is small (zero), while if the value of the criterion is greater than t_{p5}^+, that is, in the unacceptable range, then the value of the class function is very high. Class functions have several important properties such as (i) they are nonnegative, continuous, piecewise linear, and convex and (ii) the value of the class function, Z_p, at a given range intersection (say, desirable, tolerable) is the same for all class types.

Basically, ranking of the alternatives is performed in four steps, as follows:

Step 1: Identify criteria for evaluating each of the alternatives.

Step 2: Specify preferences for each criterion, based on one of the four classes (see Figure 4.10).

Step 3: Calculate incremental weights: Based on the preference structures for the different criteria, the LPP weight algorithm determines incremental weights, Δw_{pr}^+ and Δw_{pr}^- (used in step 4) that represent the incremental slopes of the class functions, Z_p. Here, r denotes the range intersection.

Step 4: Calculate total score for each alternative: The formula for the total score, J, of the alternative of interest is constructed as a weighted sum of deviations over all ranges ($r = 2$ to 5) and criteria ($p = 1$ to P), as follows:

$$J = \sum_{p=1}^{P} \sum_{r=2}^{5} (\Delta w_{pr}^- d_{pr}^- + \Delta w_{pr}^+ d_{pr}^+) \qquad (4.33)$$

where

P = Total number of criteria (each belonging to one of the four classes in Figure 4.10)

Δw_{pr}^+ and Δw_{pr}^- = Incremental weights for the pth criterion

d_{pr}^+ and d_{pr}^- = Deviations of the pth criterion value of the alternative of interest from the corresponding target values

An alternative with a lower total score is more desirable than one with a higher total score.

The most significant advantage of using LPP is that no weights need to be specified for the criteria for evaluation. The decision maker only needs to specify a preference structure for each criterion, which has more physical meaning than a physically meaningless weight arbitrarily assigned to the criterion.

Note that there are no decision variables in the aforementioned ranking procedure. LPP can be used in a problem consisting of decision variables too, by minimizing J in Equation 4.33, and subjecting (if necessary) each criterion,

g_p, to a constraint that falls into either one of the four classes (also called *soft* classes) in Figure 4.10 or one of the following four *hard* classes:

Class 1H	Must be smaller, that is, $g_p \leq t_{p,max}$
Class 2H	Must be larger, that is, $g_p \geq t_{p,min}$
Class 3H	Must be equal, that is, $g_p = t_{p,val}$
Class-4H	Must be in range, that is, $t_{p,min} \leq g_p \leq t_{p,max}$

The proposed LPP model can be used to identify simultaneously the most economically used product to reprocess in a closed-loop supply chain, the efficient production facilities that remanufacture used products and produce new products, and the right mix and quantity of goods to be transported across the supply chain. A numerical example will illustrate the methodology.

We consider the following scenario in our model: Suppose that the manufacturer has incorporated a remanufacturing process for used products into the original production system so that products can be manufactured directly from raw materials or remanufactured from used products. The final demand for the product is met either with new or remanufactured products.

We present the nomenclature for the model in Section 4.10.1, the model formulation in Section 4.10.2, and a numerical example in Section 4.10.3.

4.10.1 Nomenclature

A_{iuv}	Decision variable representing number of used products of type i transported from collection center u to remanufacturing facility v
B_{ivw}	Decision variable representing number of used products of type i transported from production facility v to demand center w
b_i	Probability of breakage of product i
TA_{uv}	Cost to transport one unit from collection center u to remanufacturing facility v
TB_{vw}	Cost to transport one unit from remanufacturing facility v to demand center w
CC_u	Cost per product retrieved at collection center u
CNP_v	Cost to produce one unit of new product at production facility v
CR_v	Cost to remanufacture at production facility v
C_{di}	Disposal cost of product i
DI_i	Disposal cost index of component y in product x (0 = lowest, 10 = highest)
DT_i	Disassembly time for product i
DC	Disassembly cost/unit time
i	Product type
MINTPS	Minimum throughput per supply
N_{ivw}	Decision variable representing number of new product type i transported from production facility v to demand center w

Nd_{iw}	Net demand for product type i (remanufactured or new) at demand center w
PRC_i	Percent of recyclable contents by weight in product i
RCYR_i	Total recycling revenue of product i
RSR_i	Total resale revenue of product i
RCRI_i	Recycling revenue index of component y in product x
S_{1v}	Storage capacity of remanufacturing facility v for used products
S_{2v}	Storage capacity of remanufacturing facility v for remanufactured and new products
S_u	Storage capacity of collection center u
SP_i	Selling price of a unit of new product of type i
SU_{iu}	Supply of used product i at collection center U
SF_v	Supply of used products at production facility v, different from SU_i, these are products that are fit for remanufacturing, after accounting for recycled and disposed products + new products
TP_v	Throughput (considering only remanufactured products) of production facility v
U	Collection center
V	Remanufacturing facility
W	Demand center
W_i	Weight of product i
x_1	Space occupied by one unit of used product (square units per product)
x_2	Space occupied by one unit of remanufactured or new product (square units per product)
Y_v	Decision variable signifying selection of production facility V (1 if selected, 0 if not)
Z_{iu}	Decision variable representing number of units of product type i picked for remanufacturing at collection center u ($\text{SU}_{iu} - Z_{iu}$ = recycled or disposed)
δ_v	Factor that accounts for unassignable causes of variations at production facility v

4.10.2   Model Formulation

4.10.2.1   Costs: Class 1S (Smaller the Better)

1. Collection/retrieval cost.

$$\sum_u \sum_i \text{CC}_u\, \text{SU}_{iu}$$

2. Processing cost = disassembly cost of used products + remanufacturing cost of used products + production cost of new products in the forward supply chain.

$$\left(DC \sum_i \sum_u \sum_v \text{DT}_i\, A_{iuv} \right) + \sum_i \sum_u \sum_v \text{CR}_v B_{ivw} + \sum_i \sum_u \sum_v \text{CNP}_v N_{ivw}$$

3. Transportation costs = cost of transporting used products from collection centers to production facility + cost of transporting remanufactured and new products from production facilities to demand centers.

$$TA_{uv} \sum_i \sum_u \sum_v A_{iuv} + TB_{vw} \sum_i \sum_v \sum_w (B_{ivw} + N_{ivw})$$

4. Disposal cost = cost of disposing of products that cannot be remanufactured (broken products) or recycled.

$$\sum_i \sum_u \{(SU_{iu} - Z_{iu}) \, DI_i \, W_i \, (1 - PRC_i)\}$$

4.10.2.2 Revenues: Class 2S (Larger the Better)

1. Reuse revenue

$$\sum_i \sum_u \{Z_{iu} \, RSR_i\}$$

2. Recycle revenue

$$\sum_i \sum_u \{(SU_{iu} - Z_{iu}) \, RCYI_i \, W_i \, PRC_i\}$$

3. New product sale revenue

$$SP_i {}^* N_{ivw}$$

4.10.2.3 System Constraints

1. The number of used products sent to all production facilities from a collection center u must be equal to the number of used products picked for remanufacturing at that collection center.

$$\sum_v A_{iuv} = Z_{iu}$$

2. Demand at each center w must be met with either new or remanufactured goods.

$$\sum_v (B_{ivw} + N_{ivw}) = Nd_{iw} \quad \forall w$$

3. Number of remanufactured products transported from a production facility v to a demand center w = (number of used products fit for remanufacturing, transported from collection center u to that production facility)*δ_v v, that is, no loss of products in the supply

chain due to reasons other than common cause variations, beyond control. δ_v accounts for the unassignable causes of variation at the production facility v.

$$\sum_w B_{ivw} = \sum_u A_{iuv} * \delta_v \quad \forall v$$

4. Total number of used products of type i picked for remanufacturing at u must be at most equal to the total number of used products fit for remanufacturing.

$$Z_{iu} \leq SU_{iu}(1 - b_i)$$

5. Total number of used products of all types collected at all collection centers must be at least equal to the net demand.

$$\sum_i \sum_u SU_{iu} \geq \sum_i \sum_w Nd_{iw}$$

6. Number of remanufactured products must be at most equal to the net demand; this is to avoid excess remanufacturing.

$$\sum_i \sum_u Z_{iu} \leq \sum_i \sum_w Nd_{iw}$$

7. Space constraints for used products at production facility v.

$$x_1 \sum_i \sum_u A_{iuv} \leq S_{1v} \cdot Y_v$$

8. Space constraint for new and remanufactured products at production facility v, assuming new and remanufactured products occupy the same space.

$$\sum_i \sum_w x_2(B_{ivw} + N_{ivw}) \leq S_{2v} * Y_v$$

9. Space constraint for used products at collection center.

$$x_1 \sum_i \sum_v a_{uv} \leq S_u$$

10. Production facility's potentiality constraints, valid only for remanufactured products:

$$\left(\frac{TP_v}{SF_v}\right) Y_v \geq MINTPS$$

Nonnegativity constraints:

$$A_{iuv}, B_{ivw}, N_{ivw}, Z_{iu} \geq 0 \quad \forall u, v, w$$

$$Y_v \in [0,1] \quad \forall v, 0 \text{ if facility } v \text{ not selected, 1 if selected}$$

4.10.3 Numerical Example

We consider a closed-loop supply chain with three collection centers, two production facilities to choose from, two demand centers to be served, and three brands of similar products. We consider the collection cost, transportation cost, disposal cost, recycling revenue, reuse revenue, and revenue from the sale of new product criteria in our numerical example.

The example data we take to implement the LPP model are

$CCu = 0.01$; $SU_{11} = 20$; $SU_{12} = 25$; $SU_{13} = 15$; $SU_{21} = 25$; $SU_{22} = 18$; $SU_{23} = 15$; $SU_{31} = 17$; $SU_{32} = 9$; $SU_{33} = 15$; $TA_{11} = 0.001$; $TA_{12} = 0.009$; $TA_{21} = 0.01$; $TA_{22} = 0.002$; $TA_{31} = 0.004$; $TA_{32} = 0.003$; $TB_{11} = 0.004$; $TB_{12} = 0.003$; $TB_{21} = 0.009$; $TB_{22} = 0.005$; $DI_1 = 4$; $DI_2 = 6$; $DI_3 = 5$; $W_1 = 0.8$; $W_2 = 1.0$; $W_3 = 0.9$; $PRC_1 = 0.65$; $PRC_2 = 0.6$; $PRC_3 = 0.75$; $Cd_1 = 0.02$; $Cd_2 = 0.05$; $Cd_3 = 0.03$; $RSR_1 = 80$; $RSR_2 = 80$; $RSR_3 = 65$; $RCYR_1 = 5$; $RCYR_2 = 7$; $RCYR_3 = 10$; $RCRI_1 = 7$; $RCRI_2 = 4$; $RCRI_3 = 6$; $SP_1 = 100$; $SP_2 = 110$; $SP_3 = 95$; $Nd_{11} = 20$; $Nd_{12} = 15$; $Nd_{21} = 16$; $Nd_{22} = 22$; $Nd_{31} = 25$; $Nd_{32} = 20$; $\delta_1 = 0.85$; $\delta_2 = 0.75$; $b_1 = 0.2$; $b_2 = 0.4$; $b_3 = 0.3$; $X_1 = 0.7$; $S_{11} = 400$; $S_{12} = 400$; $S_1 = 150$; $S_2 = 150$; $S_3 = 150$; $X_2 = 0.7$; $S_{21} = 500$; $S_{22} = 500$; $MINTPS = 0.25$.

The target values for the criteria are given in Table 4.110 (target values are scaled by a factor of 10) and Table 4.111 shows the incremental weights obtained using the LPP weight algorithm.

TABLE 4.110

Target Values of Criteria

Criteria	$t_{p1}+$	$t_{p2}+$	$t_{p3}+$	$t_{p4}+$	$t_{p5}+$	$t_{p1}-$	$t_{p2}-$	$t_{p3}-$	$t_{p4}-$	$t_{p5}-$
g_1	1	3	5	7	9	–	–	–	–	–
g_2	5	10	13	17	20	–	–	–	–	–
g_3	2	2.5	5	7	10	–	–	–	–	–
g_4	2	5	7	9	13	–	–	–	–	–
G_5	–	–	–	–	–	10	15	20	25	30
G_6	–	–	–	–	–	10	15	17	19	22
G_7	–	–	–	–	–	15	17	20	25	35

TABLE 4.111

Output of LPP Weight Algorithm

Criteria	$\Delta w_{p2}+$	$\Delta w_{p3}+$	$\Delta w_{p4}+$	$\Delta w_{p5}+$	$\Delta w_{p2}-$	$\Delta w_{p3}-$	$\Delta w_{p4}-$	$\Delta w_{p5}-$
g_1	0.05	0.17	0.748	3.291	–	–	–	–
g_2	0.02	0.12667	0.337	2.355467	–	–	–	–
g_3	0.2	0.024	1.344	4.285867	–	–	–	–
g_4	0.033	0.24667	1.288	2.822	–	–	–	–
G_5	–	–	–	–	0.02	0.024	0.052	0.11616
G_6	–	–	–	–	0.02	0.09	0.132	0.112933
G_7	–	–	–	–	0.05	0.03	0.035	0.023

Solving the LPP model using LINGO, we get the following optimal solution:

$A_{211} = 10$, $A_{212} = 5$, $A_{222} = 11$, $A_{231} = 9$, $A_{311} = 3$, $A_{321} = 6$, $B_{211} = 16$, $B_{222} = 12$, $B_{312} = 8$, $N_{111} = 20$, $N_{112} = 15$, $N_{212} = 10$, $N_{321} = 25$, $N_{322} = 12$, $Z_{21} = 15$, $Z_{22} = 11$, $Z_{23} = 9$, $Z_{31} = 3$, $Z_{32} = 6$, $Z_{33} = 6$, $Y_1 = 1$, $Y_2 = 2$.

It is obvious from the preceding solution that both the production facilities were chosen for the network design.

4.11 Goal Programming

Goal programming (GP), generally applied to linear problems, deals with the achievement of specific targets or goals. In 1961, Charnes and Cooper first reported this technique [11]. The basic purpose of GP is to simultaneously satisfy several goals relevant to the decision-making situation. To this end, several criteria are to be considered in the problem situation at hand. A target value is determined for each criterion. Next, the deviation variables are introduced, which may be positive or negative (represented by ρ_k and η_k, respectively). The negative deviation variable, η_k, represents the under-achievement of the kth goal. Similarly, the positive deviation variable, ρ_k, represents the overachievement of the kth goal. Finally, the desire to over-achieve (minimize η_k) or underachieve (minimize ρ_k), or satisfy the target value exactly (minimize $\rho_k + \eta_k$) is articulated for each criterion.

The following steps are used to solve the GP model:

Step 1: Read all the relevant data, set the first goal as the current goal.

Step 2: Obtain a LP solution with the current goal as the objective function.

Step 3: If the current goal is the last goal, set it equal to the LP objective function value found in step 2, and stop. Otherwise, go to step 4.

Step 4: If the current goal is just achieved or overachieved, set it equal to its aspiration level and add this equation to the constraint set, and go to step 5. Else, if the value of the current goal is underachieved, set the aspiration level of the current goal to the LP objective function value found in step 2, go to step 5.

Step 5: Set the next goal as the current goal, go to step 2.

The proposed GP model can be used to identify simultaneously the most economically used product to reprocess in a closed-loop supply chain, the efficient production facilities that remanufacture used products or produce new products, and the right mix and quantity of goods to be transported across the supply chain. A numerical example will illustrate the methodology.

We consider the following scenario in our model: Suppose the manufacturer has incorporated a remanufacturing process for used products into the original production system, so that products can be manufactured directly from raw materials or remanufactured from used products. The final demand for the product is met either with new or remanufactured products.

We consider three goals in our GP model:

1. Maximize the total profit in the closed-loop supply chain (TP)
2. Maximize the revenue from recycling (RR)
3. Minimize the number of disposed items (NDIS)

The first two goals involve minimizing the negative deviation from the respective target values, while the third goal, which has an environmentally benign character rather than financial character, involves minimizing the positive deviation from the target value.

The model formulation is presented in Section 4.11.1 and a numerical example is presented in Section 4.11.2. We use the nomenclature given in Section 4.10.1.

4.11.1 Model Formulation

4.11.1.1 Revenues

1. Reuse revenue (all product types together, taken one unit at a time)

$$\sum_i \sum_u \{Z_{iu} RSR_i\}$$

2. Recycle revenue (all product types together, taken one unit at a time)

$$\sum_i \sum_u \left[(SU_{iu} - Z_{iu}) RCYI_i W_i PRC_i \right] C_{rf}$$

3. New product sale revenue

$$SP_i * N_{ivw}$$

4.11.1.2 Costs

1. Collection/retrieval cost.

$$\sum_u \sum_i CC_u SU_{iu}$$

2. Processing cost = disassembly cost of used products + remanufacturing cost of used products + new products production cost in the forward supply chain.

$$\left(DC \sum_i \sum_u \sum_v DT_i A_{iuv} \right) + \sum_i \sum_u \sum_v CR_v B_{ivw} + \sum_i \sum_u \sum_v CNP_v N_{ivw}$$

3. Inventory cost: assuming inventory carrying cost at collection center (used products) is 20% of collection cost (CC_u); inventory carrying cost at production facility (remanufactured or new products) is 25% of remanufacturing or new product production cost, whichever is applicable.

$$\sum_i \sum_u \sum_v \left(\frac{CC_u}{5}\right) \cdot A_{iuv} + \left(\sum_i \sum_v \sum_w \left\{\left(\frac{CR_v}{4}\right) B_{ivw} + \left(\frac{CNP_v}{4}\right) * N_{ivw}\right\}\right)$$

4. Transportation costs: used products from collection centers to production facility + remanufactured and new products from production facilities to demand centers.

$$TA_{uv} \sum_i \sum_u \sum_v A_{iuv} + TB_{vw} \sum_i \sum_v \sum_w (B_{ivw} + N_{ivw})$$

5. Disposal cost: products that cannot be remanufactured or recycled (all product types together, taken one unit at a time).

$$\sum_i \sum_u \{(SU_{iu} - Z_{iu})DI_i \, W_i (1 - PRC_i)\} C_{df}$$

4.11.1.3   System Constraints

1. The number of used products sent to all production facilities from a collection center u must be equal to the number of used products picked for remanufacturing at that collection center.

$$\sum_v A_{iuv} = Z_{iu}$$

2. Demand at each center w must be met with new or remanufactured goods.

$$\sum_v (B_{ivw} + N_{ivw}) = Nd_{iw} \quad \forall w$$

3. Number of remanufactured products transported from a production facility v to a demand center w = (number of used products fit for remanufacturing, transported from collection center u to that production facility)*δ_v v, that is, no loss of products in the supply chain due to reasons other than common cause variations beyond control. δ_v is a factor that accounts for the unassignable causes of variation at the production facility v.

$$\sum_w B_{ivw} = \sum_u A_{iuv} * \delta_v \quad \forall v$$

4. Total number of used products of type i picked for remanufacturing at u must be at most equal to the total number of used products fit for remanufacturing.

$$Z_{iu} \leq SU_{iu}(1 - b_i)$$

5. Total number of used products of all types collected at all collection centers must be at least equal to the net demand.

$$\sum_i \sum_u SU_{iu} \geq \sum_i \sum_w Nd_{iw}$$

6. Number of remanufactured products must be at most equal to the net demand; this is to avoid excess remanufacturing.

$$\sum_i \sum_u Z_{iu} \leq \sum_i \sum_w Nd_{iw}$$

7. Space constraints for used products at production facility v.

$$x_1 \sum_i \sum_u A_{iuv} \leq S_{1v} \cdot Y_v$$

8. Space constraint for new and remanufactured products at production facility v, assuming new and remanufactured products occupy the same space.

$$\sum_i \sum_w x_2 (B_{ivw} + N_{ivw}) \leq S_{2v} * Y_v$$

9. Space constraint for used products at collection center.

$$x_1 \sum_i \sum_v a_{uv} \leq S_u$$

10. Production facility's potentiality constraints, valid only for remanufactured products:

$$\left(\frac{TP_v}{SF_v} \right) Y_v \geq MINTPS$$

4.11.1.4 Nonnegativity Constraints

$$A_{iuv}, B_{ivw}, N_{ivw}, Z_{iu} \geq 0 \quad \forall u, v, w$$

$$Y_v \in [0, 1] \quad \forall v, 0 \text{ if facility } v \text{ not selected, 1 if selected}$$

4.11.2 Numerical Example

We consider a closed-loop supply chain with three collection centers, two production facilities to choose from, two demand centers to be served, and three brands of similar products.

The example data we take to implement the GP model are

$CCu = 0.01; SU_{11} = 50; SU_{12} = 45; SU_{13} = 25; SU_{21} = 35; SU_{22} = 38; SU_{23} = 22; SU_{31} = 30; SU_{32} = 35; SU_{33} = 28; DC = 0.05; DT_1 = 10; DT_2 = 12; DT_3 = 9; CR_1 = 13; CR_2 = 10; CNP_1 = 60; CNP_2 = 45; TA_{11} = 0.01; TA_{12} = 0.09; TA_{21} = 0.5; TA_{22} = 0.1; TA_{31} = 0.02; TA_{32} = 0.04; TB_{11} = 0.04; TB_{12} = 0.03; TB_{21} = 0.09; TB_{22} = 0.05; DI_1 = 4; DI_2 = 6; DI_3 = 5; W_1 = 0.8; W_2 = 1.0; W_3 = 0.9; PRC_1 = 0.5; PRC_2 = 0.6; PRC_3 = 0.75; Cd_1 = 0.2; Cd_2 = 0.5; Cd_3 = 0.3; RSR_1 = 30; RSR_2 = 40; RSR_3 = 45; RCYR_1 = 1.5; RCYR_2 = 2; RCYR_3 = 2.5; RCRI_1 = 7; RCRI_2 = 4; RCRI_3 = 5; SP_1 = 65; SP_2 = 55; SP_3 = 60; Nd_{11} = 20; Nd_{12} = 15; Nd_{21} = 16; Nd_{22} = 22; Nd_{31} = 25; Nd_{32} = 20; \delta_1 = 0.4; \delta_2 = 0.6; b_1 = 0.2; b_2 = 0.4; b_3 = 0.3; X_1 = 0.7; S_{11} = 400; S_{12} = 400; S_1 = 150; S_2 = 150; S_3 = 150; X_2 = 0.7; S_{21} = 500; S_{22} = 500; MINTPS = 0.25.$

Upon solving the GP model using LINGO, we get the following optimal solution:

$TP = 3945 \text{ (target} = 2500); RR = 951 \text{ (target} = 750); NDIS = 74 \text{ (target} = 50); Z_{12} = 35; Z_{21} = 2; Z_{22} = 23; Z_{23} = 13; Z_{31} = 20; Z_{32} = 12; Z_{33} = 12; N_{111} = 3; N_{112} = 5; N_{211} = 8; N_{222} = 1; N_{211} = 3; N_{312} = 14; A_{121} = 15; A_{122} = 20; A_{212} = 2; A_{221} = 5; A_{222} = 18; A_{231} = 5; A_{232} = 8; A_{311} = 11; A_{312} = 9; A_{321} = 12; A_{331} = 3; B_{111} = 8; B_{112} = 4; B_{211} = 5; B_{212} = 3; B_{311} = 15; B_{312} = 6; B_{121} = 9; B_{122} = 6; B_{221} = 3; B_{222} = 18; B_{321} = 7; Y_1 = 1; Y_2 = 2.$

It is obvious from the preceding solution that both the production facilities were chosen for the network design.

4.12   Linear Integer Programming

Optimization problems concerning yes/no decisions can often be modeled as 0–1 linear integer programming problems. The general integer programming problem is to find the minimum of a function over the set of integer vectors that satisfy a given collection of constraints. Detailed introduction to linear integer programming is beyond the scope of this chapter and can be found in hundreds of optimization textbooks [15].

In this section, we formulate a linear integer programming problem to select the most economically used product to process in a reverse or a closed-loop supply chain. The following notations are used to formulate the model:

C_{df}	Disposal cost factor dollar/unit weight
C_{dx}	Disposal cost of product x
C_r	Reprocessing cost per unit time
C_{rf}	Recycling revenue factor dollar/unit weight

C_{rpx} Reprocessing cost of total product x
C_{rx} Retrieval cost of product
CC_x Collection cost of product x
D_{xy} Disposal cost index of component y in product x (0 = lowest, 10 = highest)
E_{xk} Subassembly k in product x
m_{xy} Probability of missing component y in product x
M_x Number of subassemblies in product x
N_{xy} Multiplicity of component y in product x
p_{xy} Probability of breakage of component y in product x
P_{xy} Component y in product x
PRC_{xy} Percent of recyclable contents by weight in component y of product x
R_{rcx} Total recycling revenue of product x
R_{rsx} Total resale revenue of product x
R_{rsxy} Resale value of component y in product x
RC_{xy} Recycling revenue index of component y in product x
$Root_x$ Root node of product x
$T(E_{xk})$ Time to disassemble subassembly k in product x
$T(Root_x)$ Time to disassemble $Root_x$
W_{xy} Weight of component y in product x
x Product type
X_{xy} Decision variable signifying selection of component y to be retrieved from product x for reuse (X_{xy} = 1 for reuse, 0 for recycle)
y Component type
Z Overall profit

Objective function:

$$\text{Maximize } Z = R_{rsx} + R_{rcx} - C_{rpx} - C_{dx} - CC_x \tag{4.34}$$

where

$$R_{rsk} = \sum_{j \exists Pxy \in (Root_x)} \{R_{rsxy} \cdot N_{xy} \cdot (1 - p_{xy} - m_{xy}) \cdot X_{xy}\} \tag{4.35}$$

$$R_{rcx} = \sum_{j \exists Pxy \in (Root_x)} \{PRC_{xy} \cdot W_{xy} \cdot R_{cxy}(N_{xy}(1 - m_{xy}) - N_{xy}(1 - p_{xy} - m_{xy})X_{xy})\} \tag{4.36}$$

$$C_{prx} = \left[T(Root_x) + \sum_{k=1}^{Mx} T(E_{xk}) \right] \cdot C_r \tag{4.37}$$

$$C_{dx} = \sum_{j \exists Pxy \in (Root_x)} \{D_{xy} \cdot W_{xy} \cdot (1 - PRC_{xy}) \cdot (N_{xy}(1 - m_{xy}) - N_{xy}(1 - p_{xy} - m_{xy}) \cdot X_{xy})\} \cdot C_{df} \tag{4.38}$$

subject to

$$X_{xy} = 0 \text{ or } 1 \quad \text{for all } x \text{ and } y$$

The above mentioned formulation assesses the feasible combination of sets of components that should be retrieved from each used product and compares the combination's overall profit against other products.

We consider two used products whose product structures are shown in Figures 4.11 and 4.12, respectively. The data required to implement the mathematical model for both the products are shown in Tables 4.112 and 4.113, respectively. We also assume the following :
$C_{r1} = 25$, $C_{r2} = 40$, $C_{rf} = 0.5$, $C_r = 0.8/\text{min}$, $C_{df} = 0.25/\text{lb}$, $T(\text{Root}_1) = 6$ min, $T(\text{Root}_2) = 4$ min, $T(E_{11}) = 3$ min, $T(E_{12}) = 5$ min, $T(E_{21}) = 2$ min, $T(E_{22}) = 4$ min.

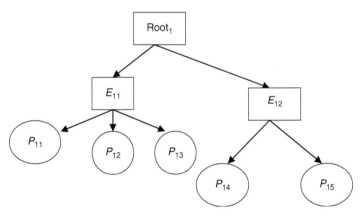

FIGURE 4.11
Structure of product 1 (linear integer programming).

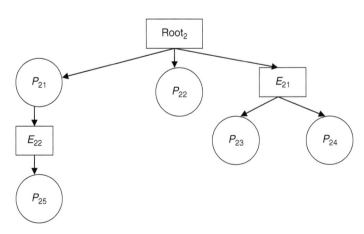

FIGURE 4.12
Structure of product 2 (linear integer programming).

TABLE 4.112

Data of Product 1

Part	R_{rs1y}	N_{1y}	W_{1y}	R_{c1y}	$PRC_{1y}(\%)$	D_{1y}	p_{1y}	m_{1y}
P_{11}	10	4	3	7	75	3	0.02	0.0
P_{12}	2	2	2.5	9	60	2	0.05	0.0
P_{13}	3.75	1	5	5	50	6	0.0	0.5
P_{14}	5	5	7	3	70	1	0.1	0.3
P_{15}	3	6	4	6	40	7	0.4	0.02

TABLE 4.113

Data of Product 2

Part	R_{rs2y}	N_{2y}	W_{2y}	R_{c2y}	$PRC_{2y}(\%)$	D_{2y}	p_{2y}	m_{2y}
P_{21}	2.5	2	4	5	40	2	0.0	0.05
P_{22}	5	3	1	6	35	6	0.0	0.1
P_{23}	3	1	3	2	70	8	0.1	0.0
P_{24}	0.5	2	4.5	8	80	5	0.15	0.2
P_{25}	2	2	5	7	25	7	0.1	0.25

Solving the model with the above data using LINGO, we get total profit for product 1 as $46.16 and for product 2 as $68.5. Hence, the decision maker should choose product 2 in this case.

4.13 Cost–Benefit Function

Economic analyses are of two kinds [22,23]: single criterion analysis and multicriteria analysis. They are divided into two subgroups: deterministic and nondeterministic. Single criterion and deterministic analysis contains discounted cash flow techniques (present worth, annual worth, etc.). Single criterion and nondeterministic analysis contains sensitivity analysis, decision tree, Monte Carlo simulation, etc. Multicriteria deterministic analysis contains AHP, GP, dynamic programming, etc. Multicriteria nondeterministic analysis contains fuzzy set theory, expert systems, and game theoretical models. In this section, we use fuzzy logic to perform multicriteria nondeterministic economic analysis and select the most economical new product to produce in a closed-loop supply chain.

The cost–benefit function–based technique provides for a more understandable approach of economic analysis than the techniques involving rate of return, present worth, and future worth. The cost–benefit function, in our context, can be defined as the ratio of the equivalent value of benefits associated with the object of interest to the equivalent value of costs associated with the same object. The equivalent value can be present worth, annual worth, and future worth. In this phase, the object of our interest is the product to be produced in a closed-loop supply chain. The cost–benefit function (F) is formulated as

$$F = \frac{B}{C} \tag{4.39}$$

where B represents the equivalent value of the benefits (revenues) and C represents the equivalent value of the costs. An F value greater than 1.0 indicates that the object is economically advantageous. It should be noted that due to uncertainties in supply, quality, and disassembly times of used products, decision makers must rely on experts' knowledge to obtain fuzzy data for calculating B, C, and hence, F values.

Our cost–benefit function consists of equivalent values of the following terms: new product sale revenue (revenue from selling new products), reuse revenue (revenue from direct sale or usage in remanufacturing of usable components of used products), recycle revenue (revenue from selling material obtained from recycling of unusable components of used products), new product production cost (cost to produce new products), collection cost (cost to collect used products from consumers), reprocessing cost (cost to remanufacture or recycle used products), disposal cost (cost to dispose off the material left over after remanufacturing or recycling of used products), loss-of-sale cost (cost due to loss of sale, which might occur every now and then, due to lack of SU), and investment cost (capital required for the production facility and its machinery).

We make the following assumptions while formulating the cost–benefit function:

1. The product of interest in the reverse supply chain will be completely disassembled.
2. All usable components of the product of interest in the reverse supply chain will be reused (for direct sale or in remanufacturing), and all the remaining ones will be recycled or disposed off.

The nomenclature is presented in Section 4.13.1, the formulation in Section 4.13.2, the methodology in Section 4.13.3, and a numerical example in Section 4.13.4.

4.13.1 Nomenclature

b_{ij} Probability of bad quality (broken, worn-out, low performing, etc.) of component j in product i

C_i Cost to produce one product i in the forward supply chain ($)

CC_i Total collection cost of product i per period ($)

CD Cost of reprocessing per unit time (dollar/unit time)

CF	Recycling revenue factor (dollar/unit weight)
CR_i	Total recycle revenue of product i per period ($)
CO_i	Cost to collect one product i ($)
DC_i	Total disposal cost of product i per period ($)
D_i	Demand for product i in the forward supply chain (number of products)
DI_{ij}	Disposal cost index of component j in product i (index scale 0 = lowest, 10 = highest)
DF	Disposal cost factor (dollar/unit weight)
E_{ik}	Subassembly k in product i
FCB_i	Fuzzy cost–benefit function for product i
i	Product type
IC_i	Investment cost of product i ($)
j	Component type
LC_i	Loss-of-sale cost of product i ($)
M_i	Total number of subassemblies in product i
m_{ij}	Probability of missing component j in product i
MC_i	Total production cost of product i per period ($)
N_{ij}	Multiplicity of component j in product i
RCP_{ij}	Percentage of recyclable contents by weight in component j of product i
RC_i	Total reprocessing cost of product i per period ($)
RI_{ij}	Recycling revenue index of component j in product i (index scale 0 = lowest, 10 = highest)
$Root_i$	Root node (e.g., outer casing) of product i
RV_{ij}	Resale value of component j in product i ($)
SP_i	Selling price of product i in the forward supply chain ($)
SR_i	Total new product sale revenue of product i per period ($)
SU_i	Supply of product i per period in the reverse supply chain (number of products)
$T(Root_i)$	Time to disassemble $Root_i$ (time units)
$T(E_{ik})$	Time to disassemble subassembly k in product i (time units)
UR_i	Total reuse revenue of product i per period ($)
W_{ij}	Weight of component j in product i (lb)
ΔBZ	Incremental total revenues (between the challenger and the defender)
ΔCZ	Incremental total costs (between the challenger and the defender)

4.13.2 Formulation of Cost–Benefit Function

The cost–benefit function (FCB) of product i of interest consists of equivalent values (EV) of nine terms (viz., total new product sale revenue per period, SR_i; total reuse revenue per period, UR_i; total recycle revenue per period, CR_i; total new product production cost per period, MC_i; total collection cost per period, CC_i; total reprocessing cost per period, RC_i; total disposal cost per period, DC_i; loss-of-sale cost, LC_i; and investment cost, IC_i) as follows:

$$FCB_i = \frac{EV \text{ of } (SR_i + UR_i + CR_i)}{EV \text{ of } (MC_i + CC_i + RC_i + DC_i + LC_i + IC_i)} \tag{4.40}$$

The following subsections explain how the abovementioned nine terms are calculated.

4.13.2.1 Total New Product Sale Revenue per Period

SR of product i per period is influenced by the demand for new products per period (D_i) and the selling price of each new product (SP_i). This revenue equation can be written as

$$SR_i = D_i \cdot SP_i \tag{4.41}$$

Often, in practice, objective data is available to express D_i and SP_i as crisp real numbers. Hence, SR_i is also a crisp real number.

4.13.2.2 Total Reuse Revenue per Period

UR of product i is influenced by the fuzzy supply of the product per period (SU_i) and the following data of component of each type j in the product: the resale value (RV_{ij}), the number of components (N_{ij}), the fuzzy probability of missing (m_{ij}), and the fuzzy probability of bad quality (broken, worn-out, low performing, etc.) (b_{ij}). This revenue equation can be written as

$$UR_i = \sum_j SU_i \cdot RV_{ij} \cdot N_{ij} \cdot (1 - b_{ij} - m_{ij}) \tag{4.42}$$

Since SU_i, b_{ij}, and m_{ij} are expressed as fuzzy numbers, the resulting UR_i is also a fuzzy number.

4.13.2.3 Total Recycle Revenue per Period

CR of product i is calculated by multiplying the component recycling revenue factors by the number of components recycled per period as follows:

$$CR_i = \sum_j \left[\begin{array}{l} SU_i \cdot RI_{ij} \cdot W_{ij} \cdot RCP_{ij} \cdot \\ \{N_{ij}(1 - m_{ij}) - N_{ij} \cdot (1 - b_{ij} - m_{ij})\} \end{array} \right] \cdot CF \tag{4.43}$$

Note that each component has a percentage of recyclable contents (RCP_{ij}). RI_{ij} is the recycling revenue index (varying in value from 1 to 10) representing the degree of benefit generated by the recycling of component of type j (the higher the value of the index, the more profitable it is to recycle the component), W_{ij} the weight of the component of type j, and CF the recycling revenue factor. Since SU_i, b_{ij}, and m_{ij} are expressed as fuzzy numbers, the resulting CR_i is also a fuzzy number.

4.13.2.4 Total New Product Production Cost per Period

MC of product i is calculated by multiplying the demand for new products per period (D_i) by the cost to produce one new product (C_i) as follows:

$$MC_i = D_i \cdot C_i \qquad (4.44)$$

Often, in practice, objective data is available to express D_i and C_i as crisp real numbers. Hence, MC_i is also a crisp real number.

4.13.2.5 Total Collection Cost per Period

CC of product i is calculated by multiplying the supply of the product per period (SU_i) by the cost of collecting one used product from consumers (CO_i)

$$CC_i = SU_i \cdot CO_i \qquad (4.45)$$

Since SU_i is expressed as a fuzzy number, the resulting CC_i is also a fuzzy number.

4.13.2.6 Total Reprocessing Cost per Period

RC of product i can be calculated from the supply of the product per period (SU_i), DT of the root node (e.g., outer casing) of the product [$T(Root_i)$], the DT of each subassembly in the product [$T(E_{ik})$] and the reprocessing cost per unit time (CD) as follows:

$$RC_i = SU_i \cdot \left[T(Root_i) + \sum_{k=1}^{M_i} T(E_{ik}) \right] \cdot CD \qquad (4.46)$$

Depending upon the type (vague or objective) of data available of the DTs, RC_i is a fuzzy or crisp real number.

4.13.2.7 Total Disposal Cost per Period

DC of product i is calculated by multiplying the component disposal cost by the number of component units disposed per period as follows:

$$DC_i = \sum_j \begin{bmatrix} SU_i \cdot DI_{ij} \cdot W_{ij} \cdot (1 - RCP_{ij}) \cdot \\ \{N_{ij}(1 - m_{ij}) - N_{ij} \cdot (1 - b_{ij} - m_{ij})\} \end{bmatrix} \cdot DF \qquad (4.47)$$

Note that DI_{ij} is the disposal cost index (varying in value from 1 to 10) representing the degree of nuisance created by the disposal of component of type j

(the higher the value of the index, the more nuisance the component creates and hence it costs more to dispose it off), W_{ij} the weight of the component of type j, and DF the disposal cost factor. Since SU_i, b_{ij}, and m_{ij} are expressed as fuzzy numbers, the resulting CR_i is also a fuzzy number.

4.13.2.8   Loss-of-Sale Cost per Period

LC of product i represents the cost of not meeting the demand of the product in time. This occurs because of the unpredictable supply of end-of-life products, as consumers do not discard them in a predictable manner. LC is difficult to predict and experts have to guess the LC. LC_i is expressed as a fuzzy number because of the involvement of the guess factor.

4.13.2.9   Investment Cost

IC of product i is the fixed cost of the production facility and the machinery required to produce product i. Whether IC_i is a fuzzy or crisp real number depends on the type (vague or objective) of data available of the product and the location of the existing or future production facility.

4.13.3   Methodology

We use the following steps to select the most economical product to produce in a closed-loop supply chain from a set of candidate products:

Step 1: Eliminate every candidate product whose FCB less than 1.0.

Step 2: Assign the candidate product that has the lowest IC as the defender and the product with the next lowest *IC* as the challenger.

Step 3: Calculate the ratio of the EV of incremental total revenue ΔBZ (between the challenger and the defender) to the EV of incremental total cost ΔCZ (between the challenger and the defender). If the ratio is less than 1.0, eliminate the challenger. Otherwise, eliminate the defender.

Step 4: Repeat steps 2 and 3 until only one product (the most economical one in the set) is left.

4.13.4   Numerical Example

We take three different products (product 1, product 2, and product 3) whose structures are shown in Figures 4.13 through 4.15, respectively. We assume that the supplies of all these products are perpetual. Hence, we take capitalized worth (CW) as the EV. Therefore, FCB is nothing but the ratio of CW of total revenues to CW of total costs. The data necessary to calculate FCB of product 1, product 2, and product 3 are given in Tables 4.114 through Table 4.116,

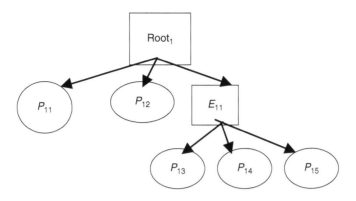

FIGURE 4.13
Structure of product 1 (cost–benefit function).

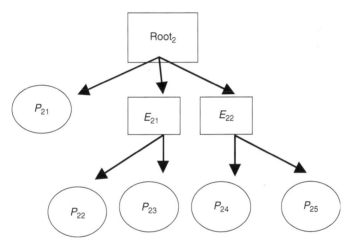

FIGURE 4.14
Structure of product 2 (cost–benefit function).

respectively. Also, $T(Root_1) = 2$ min; $T(Root_2) = 1.5$ min; $T(Root_3) = 1.5$ min; $T(E_{11}) = 9$ min; $T(E_{21}) = 7$ min; $T(E_{22}) = 8$ min; $T(E_{31}) = 7$ min; $T(E_{32}) = 8$ min; $SU_1 = (200, 230, 250)$ products per year; $SU_2 = (210, 220, 230)$ products per year; $SU_3 = (600, 650, 700)$ products per year; $CO_1 = \$20$; $CO_2 = \$21$; $CO_3 = \$18$; $IC_1 = \$20{,}000$; $IC_2 = \$25{,}000$; $IC_3 = \$30{,}000$; $D_1 = 900$ products per year; $D_2 = 850$ products per year; $D_3 = 1000$ products per year; $SP_1 = \$70$; $SP_2 = \$28$; $SP_3 = \$58$; $C_1 = \$25$; $C_2 = \$30$; $C_3 = \$28$; $LC_1 = \$(300, 500, 700)$ per year; $LC_2 = \$(100, 400, 500)$ per year; $LC_3 = \$(900, 1000, 1100)$ per year; $CF = 0.2$ \$/lb; $DF = 0.1$ \$/lb; $CD = 0.55$ \$/min.

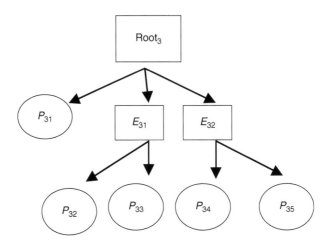

FIGURE 4.15
Structure of product 3.

TABLE 4.114

Data of Product 1

Component	RV_{1j} ($)	N_{1j}	W_{1j} (lb)	RI_{1j}	RCP_{1j} (%)	DI_{1j}	b_{1j}	m_{1j}
P_{11}	7.0	3	4.5	5	65	6	(0.1, 0.1, 0.2)	(0.3, 0.4, 0.4)
P_{12}	8.0	4	6.5	5	50	4	(0.5, 0.6, 0.7)	(0.1, 0.2, 0.2)
P_{13}	9.0	2	7.0	3	75	4	(0.2, 0.3, 0.4)	(0.3, 0.4, 0.4)
P_{14}	6.9	1	2.7	9	35	5	(0.2, 0.2, 0.3)	(0.1, 0.1, 0.2)
P_{15}	8.4	5	7.5	6	70	1	(0.1, 0.1, 0.2)	(0.3, 0.4, 0.5)

TABLE 4.115

Data of Product 2

Component	RV_{2j} ($)	N_{2j}	W_{2j} (lb)	RI_{2j}	RCP_{2j} (%)	DI_{2j}	b_{2j}	m_{2j}
P_{21}	1.0	1	3.9	2	40	3	(0.1, 0.1, 0.2)	(0.0, 0.1, 0.1)
P_{22}	1.5	3	1.5	4	20	1	(0.1, 0.2, 0.2)	(0.0, 0.0, 0.0)
P_{23}	1.2	7	4.1	1	70	2	(0.2, 0.3, 0.4)	(0.2, 0.2, 0.3)
P_{24}	2.5	4	3.2	5	90	4	(0.3, 0.4, 0.5)	(0.1, 0.1, 0.2)
P_{25}	3.1	3	2.0	2	50	2	(0.3, 0.4, 0.4)	(0.1, 0.1, 0.2)

Calculating revenues and benefits for each product, we get FCB_1 = (2.13, 2.45, 2.88), FCB_2 = (0.77, 0.84, 0.91), and FCB_3 = (1.76, 2.11, 2.68). Defuzzifying these numbers using Equation 4.10, we get FCB_1 = 2.48, FCB_2 = 0.94, and FCB_3 = 2.17. Since FCB_2 is less than 1.0, we eliminate it from further analysis.

Now, since IC_1 is less than IC_3, we consider product 1 the defender and product 3 the challenger. Calculating the defuzzified ratio of CW of ΔBZ to

TABLE 4.116

Data of Product 3

Component	$RV_{3j}($\$$)$	N_{3j}	$W_{3j}($lb$)$	RI_{3j}	$RCP_{3j}(\%)$	DI_{3j}	b_{3j}	m_{3j}
P_{31}	9.0	2	4.0	9	30	4	$(0.2, 0.3, 0.4)$	$(0.1, 0.2, 0.3)$
P_{32}	8.0	5	5.0	7	60	3	$(0.1, 0.2, 0.2)$	$(0.1, 0.2, 0.2)$
P_{33}	9.0	3	2.0	8	70	1	$(0.3, 0.4, 0.4)$	$(0.1, 0.2, 0.2)$
P_{34}	7.0	2	6.0	9	25	3	$(0.2, 0.3, 0.3)$	$(0.3, 0.3, 0.4)$
P_{35}	7.0	1	5.2	6	50	2	$(0.3, 0.3, 0.4)$	$(0.1, 0.1, 0.2)$

CW of ΔCZ, we get 1.86, which is greater than 1.0. Hence, we eliminate the defender, that is, product 1. Therefore, the remaining product, that is, product 3 is the most economical product among the three products.

4.14 Analytic Network Process and Extent Analysis Method

The methodology of ANP (37) and extent analysis (9,10) is briefly described in Section 4.14.1 and is employed in Section 4.14.2 for selecting potential recovery facilities in a closed-loop supply chain network.

4.14.1 Methodology

The ANP is a multiattribute decision-making tool based on reasoning, knowledge, and experience of experts in the field, supported by simple mathematics that enables the decision maker to weigh tangible and intangible criteria against each other for resolving conflict or setting priorities. AHP forms the starting point for ANP. AHP assumes independence among the criteria and subcriteria considered in the decision making, but real-life situations warrant against such assumption. ANP allows for dependence within a set of criteria (inner dependence) as well as between sets of criteria (outer dependence). Therefore, ANP goes beyond AHP. ANP allows for a more complex relationship among decision levels and attributes as it does not require a strict hierarchical structure, whereas AHP assumes unidirectional hierarchical relationships among the decision levels.

4.14.1.1 Extent Analysis Method to Derive Weight Vectors

In the first step, TFNs are used for pairwise comparisons. Then, the synthetic extent value S_i of the pairwise comparison is introduced by using extent analysis method, and the weight vectors with respect to each element under a certain criterion are calculated by applying the principle of the comparison of fuzzy numbers. The details of the methodology are presented in the following steps:

Let $X = \{x_1, x_2, ..., x_n\}$ be an object set and $U = \{u_1, u_2, ..., u_m\}$ be a goal set. According to the method of extent analysis, each object is taken and an extent analysis for each goal, g_i, is performed. Therefore, m extent analysis values for each object can be obtained with the following signs:

$M^1_{gi}, M^2_{gi}, ..., M^m_{gi}, i = 1, 2, ..., n$, where all the M^j_{gi} ($j = 1, 2, ..., m$) are TFNs.

Step 1: The value of fuzzy synthetic extent with respect to the ith object is defined as

$$S_i = \sum_{j=1}^{m} M^j_{gi} \otimes \left[\sum_{i=1}^{n} \sum_{j=1}^{m} M^j_{gi} \right]^{-1} \qquad (4.48)$$

To obtain $\sum_{j=1}^{m} M^j_{gi}$, perform the fuzzy addition operation of m extent analysis values for a particular matrix such that

$$\sum_{j=1}^{m} M^j_{gi} = \left(\sum_{j=1}^{m} l_j, \sum_{j=1}^{m} m_j, \sum_{j=1}^{m} u_j \right) \qquad (4.49)$$

To obtain $\{\sum_{i=1}^{n} \sum_{j=1}^{m} M^j_{gi}\}^{-1}$, perform the fuzzy addition operation of M^j_{gi} ($j = 1, 2, ..., m$) values such that

$$\sum_{i=1}^{n} \sum_{j=1}^{m} M^j_{gi} = \left(\sum_{i=1}^{n} l_i, \sum_{i=1}^{n} m_i, \sum_{i=1}^{n} u_i \right) \qquad (4.50)$$

and then compute the inverse of the vector in Equation 4.50.

Step 2: The degree of possibility of $M_2 = (l_2, m_2, u_2) \geq M_1 = (l_1, m_1, u_1)$ is expressed as

$V(M_2 \geq M_1) = \mathrm{hgt}(M_1 \geq M_2)$

$$= \left\{ 1, \text{ if } m_2 \geq m_1; 0, \quad \text{if } l_1 \geq u_2; \frac{(l_1 - u_2)}{((m_2 - u_2) - (m_1 - l_1))} \right\} \qquad (4.51)$$

Both $V(M_2 \geq M_1)$ and $V(M_1 \geq M_2)$ are required to compare M_1 and M_2.

Step 3: The degree of possibility for a convex fuzzy number to be greater than k convex fuzzy numbers M_i ($i = 1, 2, ..., k$) can be defined as

$V(M \geq M_1, M_2, ..., M_k) = V[(M \geq M_1) \text{ and } (M \geq M_2) \cdots (M \geq M_k)]$
$$= \min V(M_i \geq M_i), i = 1, 2, ..., k \qquad (4.52)$$

Let $d'(A_i) = \min V(S_i \geq S_k)$, for $k = 1, 2, ..., n; k \neq i$. Then the weight vector is given by

$$W' = (d'(A_1), d'(A_2), ..., d'(A_n))^{\mathrm{T}} \qquad (4.53)$$

Step 4: The weight vector obtained in step 3 is normalized to get the normalized weights.

4.14.1.2 Steps Involved in the ANP Methodology

Step 1: *Model development and problem formulation.* In this step, the decision problem is structured into its constituent components. The relevant criteria, the subcriteria, and the alternatives are chosen and structured in the form of a control hierarchy, as shown in Figure 4.7.

Step 2: *Pairwise comparisons.* In this step, the decision maker is asked to carry out a series of pairwise comparisons, where two main criteria are simultaneously compared with respect to the goal (evaluation of production facilities), two subcriteria are simultaneously compared with respect to their main criteria, pairwise comparisons are performed to address the interdependencies among the subcriteria, and the pairwise comparisons for the relative impact of the alternatives on the subcriteria, in influencing the main criteria, are performed. The decision maker's linguistic values are then converted into TFNs (Table 4.117 shows one of the many ways for such a conversion). The weight vectors are deducted from the pairwise comparisons by applying the principle of extent analysis reported earlier.

Step 3: *Super matrix formulation.* The super matrix enables resolution of interdependencies that exist among the subcriteria. It is a partitioned matrix, where each submatrix is composed of a set of relationships between and within the levels, as represented by the decision maker's model. In our problem, there are six super matrices for each of the six main criteria. The super matrix M is made to converge to obtain a long-term stable set of weights. For convergence, M must be made column stochastic, which is done by rising M to the power of 2^{k+1}, where k is an arbitrarily large number. Convergence is reached at $k = 59$, i.e., by rising M to the power of 2^{60}.

TABLE 4.117

Linguistic Weight Conversion Table for Criteria and Subcriteria

Linguistic Weight	TFN
VH	(0.7, 0.9, 1.0)
H	(0.5, 0.7, 0.9)
M	(0.3, 0.5, 0.7)
L	(0.1, 0.3, 0.5)
VL	(0.0, 0.1, 0.3)

Note: VH = very high, H = high, M = medium, L = low, VL = very low.

Step 4: *Selection of the best alternative.* The selection of the best alternative depends on the desirability index. The desirability index, D_i for alternative i is defined as

$$D_i = \sum_{J-1}^{J} \sum_{k=1}^{K_j} P_j A_{kj}^D A_{kj}^I S_{ikj} \qquad (4.54)$$

where

   P_j = Relative importance weight of main criteria j

   A^D_{kj} = Relative importance weight for subcriteria k of main criteria j for the dependency (D) relationships among subcriteria (pairwise comparisons among subcriteria)

   A^I_{kj} = Stabilized relative importance weight (determined by the super matrix) for subcriteria k of main criteria j for interdependency (I) relationships among subcriteria

   S_{ikj} = Relative impact of alternative i on subcriteria k of main criteria j

4.14.2   Evaluation of Production Facilities

In this section, we employ the ANP and extent analysis methodology to identify potential production facilities in a set of candidate production facilities operating in a region where a closed-loop supply chain network is to be designed.

Tables 4.117 and 4.118 show the linguistic weight conversion of TFNs for the main criteria and subcriteria and the production facilities, respectively.

Table 4.119 illustrates the comparative linguistic weights (H = high, M = medium, L = low) assigned to the main criteria in the second level of the hierarchy in this example. Using fuzzy logic, these linguistic weights are converted into TFNs (Table 4.117 shows one of the many ways for such a conversion) and then averaged to form another TFN called the average weight. For example, the average weight of criteria, ECD with respect to ECM

TABLE 4.118

Linguistic Weight Conversion Table for Production Facilities

Linguistic Weight	TFN
VG	(7, 10, 10)
G	(5, 7, 10)
F	(2, 5, 8)
P	(1, 3, 5)
VP	(1, 1, 3)

Note: VG = very good, G = good, F = fair, P = poor, VP = very poor.

TABLE 4.119

Linguistic Weights of Main Criteria

Criteria	ECD	ECM	AMT	POT	COST	CSE
ECD	(1,1,1)	(H,H,M)	(M,L,M)	(L,VL,M)	(L,VL,L)	(M,L,L)
ECM	1/(H,H,M)	(1,1,1)	(VH,H,H)	(H,M,H)	(M,H,M)	(H,VH,M)
AMT	1/(M,L,M)	1/(VH,H,H)	(1,1,1)	(H,M,H)	(M,L,M)	(M,H,L)
POT	1/(L,VL,M)	1/(H,M,H)	1/(H,M,H)	(1,1,1)	(H,M,H)	(M,L,H)
COST	1/(L,VL,L)	1/(M,H,M)	1/(M,L,M)	1/(H,M,H)	(1,1,1)	(H,M,H)
CSE	1/(M,L,L)	1/(H,VH,M)	1/(M,H,L)	1/(M,L,H)	1/(H,M,H)	(1,1,1)

Note: VH = very high, H = high, M = medium, L = low, VL = very low.

TABLE 4.120

Average Weights of Main Criteria

Criteria	ECD	ECM	AMT	POT	COST	CSE
ECD	(1,1,1)	(0.43,0.63,0.83)	(0.23,0.43,0.63)	(0.14,0.3,0.5)	(0.07,0.23,0.43)	(0.17,0.37,0.57)
ECM	(1.20,1.58,2.32)	(1,1,1)	(0.57,0.77,0.63)	(0.43,0.63,0.83)	(0.37,0.57,0.77)	(0.5,0.7,0.57)
AMT	(1.58,2.32,4.34)	(1.58,1.29,1.75)	(1,1,1)	(0.43,0.63,0.83)	(0.23,0.43,0.63)	(0.3,0.5,0.7)
POT	(2,3.33,7.14)	(1.2,1.58,2.32)	(1.2,1.58,2.32)	(1,1,1)	(0.43,0.63,0.83)	(0.3,0.5,0.7)
COST	2.32,4.34,4.28)	(1.29,0.07,2.7)	(1.58,2.32,4.34)	(1.2,1.58,2.32)	(1,1,1)	(0.43,0.63,0.83)
CSE	(1.75,2.7,5.88)	(1.75,1.42,2)	(1.42,2,3.33)	(1.42,2,3.33)	(1.2,1.58,2.32)	(1,1,1)
SUM	(9.87,15.29, 34.98)	(7.27,6.01, 10.61)	(6.02,8.11,12.2)	(4.63,6.14,8.81)	(3.3, 4.44,5.98)	(2.7,3.7,4.37)

is (H + H + M)/3, which is ((0.5 + 0.5 + 0.3)/3, (0.7 + 0.7 + 0.5)/3, (0.9 + 0.9 + 0.7)/3) = (0.43, 0.63, 0.83) (see Table 4.120).

The steps of fuzzy extent analysis are applied to the average weights to get the normalized weight vectors of main criteria. For example, consider the fuzzy synthetic extent value of main criteria ECD shown in Table 4.121. Applying Equations 4.49 and 4.50 to the average weights in Table 4.120, we get the fuzzy synthetic extent value of main criteria ECD = (0.052, 0.12, 0.181). By applying Equations 4.51 through 4.53 to the fuzzy synthetic extent values, the weights vectors are obtained for the main criteria, which are then normalized to get the normalized weight vectors shown in Table 4.122.

Table 4.123 shows normalized weight vectors of each subcriteria, with respect to its main criteria, obtained after carrying out pairwise comparisons among them and applying the steps of extent analysis method described earlier. For brevity's sake we show only one such matrix. For pairwise comparison, the decision maker is asked: "What is the relative impact of subcriteria *a* on main criteria *X* compared to subcriteria *b*, in evaluating each available production facility?"

TABLE 4.121

Fuzzy Synthetic Extent Values
of Main Criteria

Criteria	Fuzzy Synthetic Extent Value
ECD	(0.052, 0.12, 0.181)
ECM	(0.06, 0.14, 0.273)
AMT	(0.07, 0.197, 0.423)
POT	(0.101, 0.227, 0.754)
COST	(0.111, 0.245, 0.528)
CSE	(0.026, 0.067, 0.117)

TABLE 4.122

Weights of Main Criteria (P_i)

Criteria	Weight
ECD	0.008351847
ECM	0.093393234
AMT	0.158907973
POT	0.22579517
COS	0.253366077
CSE	0.260185699

TABLE 4.123

Weights of Subcriteria with
Respect to Main Criteria (A_{kj}^p)

Subcriteria	Weight
DD	0.365181452
DC	0.332675525
DR	0.302143023
ES	0.05687219
CS	0.29054319
WM	0.65258462
EF	0.06678159
FU	0.33525506
AD	0.59796335
TP/SU	0.13614722
TP × DT	0.33686398
QO − QI	0.5269888
FC	0.12134932
OC	0.87865068
IC	0.09669799
UG	0.47671802
ER	0.426584

Pairwise comparisons are carried out to consider interdependencies among the subcriteria. For evaluating interdependencies, the decision maker is asked: "When considering DD with regard to evaluating the production facilities, what is the relative impact of DC compared to DR?" For brevity's sake, we are not showing these pairwise comparison results here.

Table 4.124 shows a portion of the converged super matrix.

Pairwise comparisons are also carried out among the four available production facilities, with respect to the subcriteria, to obtain the respective weights. In the present case, we have 17 subcriteria that lead to 17 pairwise comparison matrices. One such matrix is shown in Table 4.125.

Table 4.126 shows the desirability indices obtained from Equation 4.54.

The overall weighted index for each of the four production facilities is calculated by multiplying the desirability index of each alternative, for each criterion, by the weight of the criteria and summing up over all the criteria. Table 4.127 shows the overall weighted indices for the four production facilities.

Facility D's overall weighted index is the largest; hence, the decision maker would choose facility D.

TABLE 4.124

Converged Super Matrix (A_{kj}^l)

	DD	DC	DR
DD	0.121663195	0.121663195	0.121663195
DC	0.42786297	0.42786297	0.42786297
DR	0.450473835	0.450473835	0.450473835

TABLE 4.125

Comparative Importance Values of Alternatives with Respect to Subcriteria (S_{ikj})

Subcriteria	A	B	C	D
Design for disassembly	0.263028215	0.25637619	0.241718731	0.238876864
Design for recycling	0.263695249	0.250633282	0.241695093	0.243976375
Design for remanufacturing	0.263028215	0.25637619	0.241718731	0.238876864

TABLE 4.126

Desirability Indices (D_i)

Criteria	A	B	C	D
ECD	0.085020498	0.081960329	0.078041862	0.0778535
ECM	0.110365306	0.108803779	0.103249723	0.1035761
AMT	0.088893536	0.086194204	0.081657039	0.0841138
POT	0.092035754	0.092471275	0.088886589	0.1491573
Cost	2.07501E-31	1.84313E-31	2.01608E-31	1.954E-31
CSE	0.097026636	0.092598846	0.091298302	0.0904604

TABLE 4.127

Overall Weighted Indices for Production Facilities

Production Facility	Weighted Index
A	0.246529487
B	0.240799902
C	0.232416329
D	0.280254282

4.15 Conclusions

This chapter illustrated how numerous quantitative techniques in the literature could be employed to address a variety of decision-making problems we identified in the design phase of reverse and closed-loop supply chains. Specifically, we addressed the following decision-making problems: selection of economically used products, evaluation of collection centers, evaluation of recovery facilities, optimal transportation of goods, evaluation of marketing strategy, evaluation of futurity of used products, selection of secondhand markets, selection of new products, and evaluation of production facilities. The quantitative techniques that we used to address these problems are AHP, eigenvector method, fuzzy logic, Bayesian updating, QFD, method of total preferences, TOPSIS, Borda's choice rule, neural networks, LPP, GP, linear integer programming, cost–benefit function, ANP, and extent analysis method.

References

 1. Alshamrani, A., Mathur, K. and Ballou, R. H., Reverse logistics: simultaneous design of delivery routes and return strategies, *Computers and Operations Research*, 34, 595–619, 2007.
 2. Ammons, J. C., Realff, M. J. and Newton, D. J., Carpet recycling: determining the reverse production system design, *Polymer-Plastics Technology and Engineering*, 38(3), 547–567, 1999.
 3. Barros, A. I., Dekker, R. and Scholten, V., A two-level network for recycling sand: a case study, *European Journal of Operational Research*, 110, 199–214, 1998.
 4. Beamon, M. B. and Fernandes, C., Supply-chain network configuration for product recovery, *Production Planning and Control*, 15(3), 270–281, 2004.
 5. Berger, T. and Debaillie, B., Location of Disassembly Centers for Re-Use to Extend an Existing Distribution Network, Master's Thesis, University of Leuven, Belgium (in Dutch), 1997.
 6. Biehl, M., Prater, E. and Realff, M. J., Assessing performance and uncertainty in developing carpet reverse logistics systems, *Computers and Industrial Engineering*, 34, 443–463, 2007.
 7. Cha, Y. and Jung, M., Satisfaction assessment of multi-objective schedules using neural fuzzy methodology, *International Journal of Production Research*, 41(8), 1831–1849, 2003.
 8. Chan, F. T., Chan, H. K. and Chan, M. H., An integrated fuzzy decision support system for multi-criterion decision making problems, *Journal of Engineering Manufacture*, 217, 11–27, 2003.
 9. Chang, D. Y., *Extent Analysis and Synthetic Decision, Optimization Techniques and Applications*, Vol. 1, World Scientific, Singapore, 1992, p. 352.
10. Chang, D. Y., Applications of the extent analysis method on fuzzy AHP, *European Journal of Operations Research*, 95, 649–655, 1996.

11. Charnes, A. and Cooper, W. W., *Management Models and Industrial Applications of Linear Programming (Vol. 1)*, Wiley, New York, 1961.
12. Chopra, S. and Meindl, P., *Supply Chain Management*, 3rd edition, Prentice-Hall, New York, 2006.
13. Erol, I. and Ferrell, Jr. W. G., A methodology for selecting problems with multiple, conflicting objectives and both qualitative and quantitative data, *International Journal of Production Economics*, 86, 187–199, 2003.
14. Fleischmann, M., *Quantitative Models for Reverse Logistics: Lecture Notes in Economics and Mathematical Systems*, Springer, Germany, 2001.
15. Gass, S. I., *Linear Programming: Methods and Applications*, 5th edition, Dover Publications, New York, 2003.
16. Gupta, S. M. and Veerakamolmal, P., A bi-directional supply chain optimization model for reverse logistics, *IEEE International Symposium on Electronics and the Environment*, 2000, pp. 254–259.
17. Hopgood, A. A., *Knowledge-Based Systems for Engineers and Scientists*, CRC Press, Boca Raton, FL, 1993.
18. Hu, T., Sheu, J. and Huan, K., A reverse logistics cost minimization model for the treatment of hazardous wastes, *Transportation Research Part E*, 38, 457–473, 2002.
19. Hwang, C. L. and Yoon, K., *Multiple Attribute Decision Making Methods and Application*, Springer, Berlin, 1981.
20. Hwang, C. L., *Group Decision Making Under Multi-Criteria: Methods and Applications*, Springer, New York, 1987.
21. Jayaraman, V., Guide, Jr. V. D. R. and Srivastava, R., A closed-loop logistics model for remanufacturing, *Journal of the Operational Research Society*, 50(5), 497–508, 1999.
22. Kahraman, C., Tolga, E. and Ulukan, Z., Justification of manufacturing technologies using fuzzy benefit/cost ratio analysis, *International Journal of Production Economics*, 66, 45–52, 2000.
23. Kolli, S., Wilhelm, M. R. and Liles, D. H., A classification of scheme for traditional and non-traditional approaches to the economic justification of advanced automated manufacturing systems, *Economic and Financial Justification of Advanced Manufacturing Technologies*, Elsevier, Amsterdam, 1992, pp. 165–187.
24. Krikke, H. R., Van harten, A. and Schuur, P. C., Business case: reverse logistic network re-design for copiers, *OR Spectrum*, 21(3), 381–409, 1999.
25. Kroon, L. and Vrijens, G., Returnable containers: an example of reverse logistics, *International Journal of Physical Distribution and Logistics Management*, 25(2), 56–68, 1995.
26. Lieckens, K. and Vandaele, N., Reverse logistics network design with stochastic lead times, *Computers and Operations Research*, 34, 395–416, 2007.
27. Lim, G. H., Kasumastuti, R. D. and Piplani, R., Designing a reverse supply chain network for product refurbishment, Proceedings of the International Conference of Simulation and Modeling, 2005.
28. Listes, O. and Dekker, R., A stochastic approach to a case study for a product recovery network design, *European Journal of Operational Research*, 160, 268–287, 2005.
29. Louwers, D., Kip, B. J., Peters, E., Souren, F. and Flapper, S. D. P., A facility location allocation model for reusing carpet materials, *Computers and Industrial Engineering*, 36(4), 855–869, 1999.

30. Lu, Z. and Bostel, N., A facility location model for logistics system including reverse flows: the case of remanufacturing activities, *Computers and Operations Research*, 34, 299–323, 2007.
31. Messac, A., Gupta, S. M. and Akbulut, B., Linear physical programming: a new approach to multiple objective optimization, *Transactions on Operational Research*, 8, 39–59, 1996.
32. Prakash, G. P., Ganesh, L. S. and Rajendran, C., Criticality analysis of spare parts using the analytic hierarchy process, *International Journal of Production Economics*, 35(1-3), 293–297, 1994.
33. Ravi, V., Shankar, R. and Tiwari, M. K., Analyzing alternatives in reverse logistics for end-of-life computers: ANP and balanced scorecard approach, *Computers and Industrial Engineering*, 48(2), 327–356, 2005.
34. Reimer, B., Sodhi, M. S. and Knight, W. A., Optimizing electronics end-of-life disposal costs, Proceedings of IEEE Symposium on Electronics and the Environment, 2000, pp. 342–347.
35. Saaty, T. L., *The Analytic Hierarchy Process*, McGraw-Hill, New York, 1980.
36. Saaty, T. L., How to make a decision: the analytic hierarchy process, *European Journal of Operational Research*, 48(1), 9–26, 1990.
37. Saaty, T. L., *Decision Making with Dependence and Feedback: The Analytic Network Process*, RWS Publications, Pittsburgh, PA, 1996.
38. Salema, M. I. G., Barbosa-Povoa, A. P. and Novais, A. Q., An optimization model for the design of a capacitated multi-product reverse logistics network with uncertainty, *European Journal of Operational Research*, 179, 1063–1077, 2007.
39. Savaskan, R. C., Bhattacharya, S. and Van Wassenhove Luk, N., Closed-loop supply chain models with product remanufacturing, *Management Science*, 50(2), 239–252, 2004.
40. Siddiqui, M. Z., Everett, J. W. and Vieux, B. E., Landfill siting using geographic information systems: a demonstration, *Journal of Environmental Engineering*, 122(6), 515–523, 1996.
41. Spengler, T., Puchert, H., Penkuhn, T. and Rentz, O., Environmental integrated production and recycling management, *European Journal of Operations Research*, 97, 308–326, 1997.
42. Tadisina, S. K., Troutt, M. D. and Bhasin, V., Selecting a doctoral programme using the analytic hierarchy process: the importance of perspective, *Journal of the Operational Research Society*, 42(8), 631–638, 1991.
43. Thierry, M., Salomon, M., van Nunen, J. and van Wassenhove, L. N., Strategic issues in product recovery management. *California Management Reviews*, 37, 114–135, 1995.
44. Triantaphyllou, E. and Lin, C., Development and evaluation of five fuzzy multi-attribute decision-making methods, *International Journal of Approximate Reasoning*, 14, 281–310, 1996.
45. Tsaur, S., Chang, T. and Yen, C., The evaluation of airline service quality by fuzzy MCDM, *Tourism Management*, 23, 107–115, 2002.
46. Vlachos, D., Georgiadis, P. and Iakovou, E., A systems dynamic model for dynamic capacity planning of remanufacturing in closed-loop supply chains, *Computers and Operations Research*, 34, 367–394, 2007.
47. Wang, M. and Liang, G., Benefit/cost analysis using fuzzy concept, *The Engineering Economist*, 40(4), 359–376, 1995.
48. Zadeh, L. A., Fuzzy sets, *Information and Control*, 8, 338–353, 1965.

5

Proactive Yesterday, Responsive Today: Use of Information to Enhance Planning in Closed-Loop Supply Chains

Muhammad N. Jalil, Rob A. Zuidwijk, and Harold Krikke

CONTENTS

5.1 Introduction

In this chapter, we emphasize that especially in closed-loop supply chains (CLSCs), uncertainty needs to be managed by using information. We distinguish between two planning approaches, proactive and responsive, to indicate that information can be used in different ways. For example, data that capture the technical state of products in the installed base can be used to estimate the average lifetime of products and to forecast end-of-life

(EOL) returns. This data can also be used to respond to possible mismatches between product performance and customer expectations, which may result in product replacement. In both the cases, information is used in a particular way. We study enabling techniques and the use of information in CLSCs from three perspectives, namely customers, products, and processes. We revisit the CopyMagic case to illustrate some of the issues and provide some conclusions. First, we briefly introduce the notion of CLSCs.

5.1.1   Closed-Loop Supply Chains

CLSC management aims at making operations more sustainable by extracting more value from products while addressing environmental and societal concerns. A typical CLSC may be described as a supply chain that extends from the forward supply chain to incorporate the collection, selection, recovery, and reuse of the returned products. Figure 5.1 depicts the physical flows in CLSCs in a simplified way.

The typical forward supply chain generates product returns during manufacturing, distribution, use, and maintenance. The collection and selection processes require a product testing and sorting scheme, so that the products can be forwarded to the appropriate recovery option such as recycling or remanufacturing and can be reused. Recovering product returns for further use has been profitable for a number of companies (Stock et al., 2002). However, managing product returns poses several challenges. First, there are

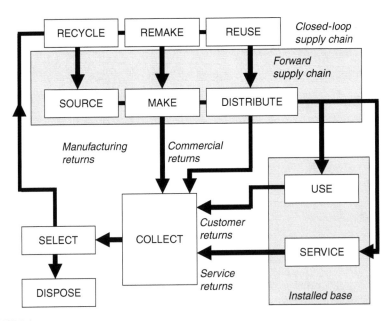

FIGURE 5.1
Typical closed-loop supply chain (CLSC). (Adapted from Van Nunen, J.A.E.E., and Zuidwijk, R.A., *California Management Review*, 46:2, 40–54, 2004).

additional sources of uncertainty. Demand and process uncertainties in forward supply chains are accompanied by uncertainties in the timing, volume, and quality of product returns and the outcomes of collection, selection, and recovery processes. Second, CLSCs encounter heterogeneous return flows consisting of many product types and vintages (Fleischmann, 2001).

In this chapter, we emphasize that the use of information in CLSCs enables the decision maker to deal with uncertainty and heterogeneity. The need for information management in CLSCs has been stressed by Thierry et al. (1995), where the uncertainty within a CLSC is reduced by fulfilling the CLSC's specific information requirements.

5.1.2   Uncertainty in CLSC

Uncertainty in forward supply chains has been categorized as demand uncertainty and supply uncertainty (Fisher et al., 1994; Lee et al., 1997). From a demand perspective, uncertainty is typically low for functional products and high for innovative items. Similarly, the supply uncertainty is lower for a stable process, which usually inhibits higher variability, than an evolving process (Lee, 2002). In their research, Lee et al. (1997) analyzed bullwhip phenomena, and the findings of these authors strongly indicate the impact of information distortion on the bullwhip effect in supply chains.

We discuss that uncertainty in CLSCs is even more pertinent and may require additional managerial levers. A first assessment reads as follows: First, end customers are suppliers of product returns and are typically a source of uncertainty, as their role is different from other supply chain partners. Second, returned products constitute a heterogeneous population, as these products not only show a large variety in product types, but each individual product also has its own usage history. As a result, individual products have varying configurations and quality levels. Third, operational processes such as disassembly are characterized by uncertain outcomes because of the aforementioned variety of returned products. These returned products serve as input for the operational processes, which themselves may involve destructive disassembly or separation steps.

Before discussing managerial levers to deal with uncertainty, we discuss the concept of uncertainty in detail. We focus on the causes of uncertainty.

"Uncertainty" is a term used in a number of fields, including philosophy, statistics, economics, finance, insurance, psychology, engineering, and science. In general, uncertainty is defined as a doubt or precariousness in the information at hand. However, most definitions of uncertainty ignore its contextual nature and do not signify the role of an uncertainty-processing model. Zimmermann (2000) states that uncertainty is a situational property and provides the following definition:

> Uncertainty implies that in a certain situation a person does not dispose about information which quantitatively and qualitatively is appropriate to describe, prescribe or predict deterministically and numerically a system, its behavior or other characteristics.

TABLE 5.1

Causes of Uncertainty Linked to Quality Aspects of Information and Examples in CLSC

Cause	Quality Aspect	Example in CLSC
Lack of information	Completeness	No information on configuration of returned computer systems
Abundance of information	Complexity	Complete usage history of returned machines without diagnostic capabilities to determine optimal recovery option
Conflicting evidence	Consistency	Sales and service data on products in the market that are to be returned do not match
Ambiguity	Ambiguity	Subjective or unclear statements on the quality of returned items
Measurements	Precision	Imprecise weight measurements of returned items

This definition clearly indicates the contextual nature of uncertainty and the close relationship of uncertainty with the system or the problem at hand. Uncertainty can be caused by various factors. We list out a number of these factors that are adapted from Zimmermann (2000) and link them to the quality aspects of information and examples in CLSC (see Table 5.1).

As stated earlier, uncertainties could be due to the lack of information about the decision problem at hand. However, it is also possible that uncertainty arises out of the lack of control over the decision situation, in which case we assume that it is inherently impossible to obtain information on some future events. Customer behavior and recovery processes are two important types of processes in CLSC that are difficult to control. In both these cases, opportunities exist to reduce uncertainty by creating incentives that influence behavior (for example, studies on the tire rethreading industry by Debo and Van Wassenhove [2005] and Yadav et al. [2003]) and by designing products in such a way that recovery processes such as disassembly or material extraction become less erratic. Krikke et al. (2003) cite an example on refrigerators, while Thierry's work (1997) includes an example on copiers. We elaborate on the latter example in the CopyMagic case (see case description in Section 5.3) in which many changes were introduced in the product design process. These changes were made by promoting more durable materials using a more modular setup and involving the suppliers more closely.

5.1.3   Uncertainty Management in CLSC

We discern two basic approaches in managing uncertainty. First, we may reduce uncertainty by enhancing information provision and gaining control. In this case, the value of information of control (or coordination) can be expressed in terms of the benefits resulting from better decision-making, for example, cost reductions or service enhancement. Second, we may mitigate the effects of uncertainty through decision-making that incorporates uncertainty. Decision theory describes how different decision strategies result in decisions that anticipate on, for example, a worst-case situation or when future scenarios can be weighed with probabilities, or average scenarios based on expected values. Another approach develops the so-called robust decisions that are satisfactory under all possible scenarios or provide recourse actions that "repair" the effects of *a posteriori* suboptimal decision. The works of Realff et al. (2000) and Listes and Dekker (2005) contain examples of the second approach. In this chapter, we focus on the first approach toward managing uncertainty. Both approaches may benefit from risk management in which the probabilities and the effects of possible future scenarios are used to express the vulnerability of the supply chain processes; see, for example, Repo (1989).

We should mention that CLSC management is not only challenged by additional sources of uncertainty but may also be supported by new levers to mitigate uncertainty. Zhao et al. (2006) state that the bullwhip effect can be "absorbed" by return flows in the case of considerable return rates. In production planning, dual sourcing (i.e., sourcing from new and remanufactured stocks) provides the opportunity to respond quickly to demand changes when remanufacturing lead times are relatively short (Teunter et al., 2004). The installed base, i.e. the population of products in the market, is a potential rich source of spare parts for future service requests, and may help reduce risk in parts supply in the end-of-life phase of the product (Spengler and Schröter, 2003).

5.1.4 Planning and Execution in CLSC

Planning and execution in CLSCs under uncertain conditions can be divided into the following temporal categories: strategic, tactical, and operational (Gupta and Maranas, 2003). Strategic decisions aim to identify the timing, extent, and allocation of long-term investment decisions such as decisions on the location of recovery facilities over a long time horizon. Operational decisions deal with immediate or short-term problems, for example, planning remanufacturing operations for the next two weeks while considering the available product returns and customer orders (Thierry, 1997). Tactical decisions, for example, network planning models in which tactical problems such as carry-over inventory or resource availability need to be considered fall in between these two categories. In each of these decision situations, the uncertainties encountered vary in nature and in effects on the system. Short-term uncertainty may include day-to-day variation in the quality of returns or remanufacturing yields. Medium-term uncertainty could be the price fluctuations in scrap markets and the demand for recovered products or parts. Long-term variations could be new legislation, new recovery technologies, consumer attitude toward "green" products, and economic lifetime of products. Planning and execution in CLSCs should account for and attempt to manage the uncertainty in each of these aforementioned temporal classifications.

In the planning phase, which involves the actual decision-making or execution (Fleischmann et al., 2002), different approaches toward managing uncertainty become apparent. Indeed, planning can be subdivided into categories that are labeled as "proactive" and "responsive." Proactive planning usually involves some time lag before the execution; therefore, we need to forecast the decision situation at execution using historical information. In contrast, responsive planning uses immediate feedback events from the system. For example, production plan updates that aim to incorporate new demand information releases are of a responsive nature. These two types of planning may be undertaken simultaneously. Such a hybrid planning approach could consist of a proactive planning phase in which a base remanufacturing plan is established. In response to new arrivals of information, for example, product returns and remanufacturing yields, the base plan can be updated. In Section 5.2, we consider the proactive and responsive planning approaches in more detail.

5.1.5   Information Management in CLSC

Over the years, the role of information management has shifted from that of support tool for general management to the development of solutions to provide supply chains with strategic advantage over their competitors. As noted by Glazer (1993), owing to the developments in information technology (IT), organizations in the supply chain are equipped with process, product, and customer information. However, very few supply chains have gained from this massive gathering of information, and organizations need to rethink how to use retrieved information from the supply chain in an effective way. As argued by Carr (2003), the key is to align the information management strategy with the overall organizational strategy.

In the maintenance industry, IT has been an enabler of new products and services. By creating remote sensing, diagnostics, and prognostic capabilities in products that are in the installed base, maintenance services can be delivered proactively. As maintenance service packages include replacements of consumables, parts, and possibly complete products, there is a strong link with the management of product returns. Linton et al. (2000) report how Nortel uses remanufacturing to establish product upgrades in a cost-efficient and timely manner. Moreover, managing the complete life cycle of products is one of the objectives of modern service organizations (Takata et al., 2004).

IT has enabled organizations to obtain more accurate information about their customers and to improve their responsiveness to customers. For CLSCs, customer relationship management (CRM) tools provide opportunities to retrieve information and to analyze customer behavior to assess the reasons for returns and forecast the timing, quantity, and quality of returns. Obviously, managing product returns is not an aim in itself but should be incorporated into tailoring processes, products, and services to meet customer requirements in a timely manner. Retrieval and analysis of point-of-sales data help organizations to respond to customer needs in a timely manner. In particular, the maintenance and the management of installed base information and the use of condition base maintenance technologies such as inbuilt quality sensors and radio frequency identification (RFID) technologies can be crucial to map customer needs and predict the rate as well as the probable quality levels of commercial and end-of-use returns. In the next section, we discuss various information-enabling mechanisms in greater detail.

5.2 Information as an Enabler in Proactive and Responsive Planning

As discussed earlier, efficient planning and execution of a CLSC requires information as an essential input at various stages of the supply chain. Various information processing methods turn historical demand and customer and product data into useful contextual information. The information obtained through retrieval mechanisms is used at various stages in CLSCs to enable coordination and decision-making.

We discuss these information processing methods or information-enabling mechanisms in the contexts of proactive planning and responsive planning. The distinction between proactive planning and responsive planning is that proactive planning aims at anticipating the decision situation, while responsive planning typically awaits the development of the decision situation and attempts to form an appropriate response. Proactive planning usually deploys information processing methods such as forecasting to extract relevant information from historical data on the future decision situation. For example, the proactive decision maker would aim to take a decision that incorporates all possible outcomes, possibly weighing future scenarios in terms of probability or impact. In contrast, responsive planning aims to incorporate all information available about the current decision situation, typically under time pressure, so that the decision made will be a function of the actual decision situation. In a way, the two approaches are of a conflicting nature. Indeed, proactive planning by its very nature allows for a time lag between information processing and decision-making. In the situation where such a time lag is unavoidable, proactive planning is the only option. Responsive planning allows no time lag between information processing and decision-making, hence, accepting time pressure. In the case of decision environment changing erratically to the extent that the decision situation cannot be anticipated in a meaningful way, response is the proper approach. In the cases in which the decision situation can be anticipated to a certain extent and (corrective) decisions can be made at a later stage, a hybrid approach may serve the purpose. Proactive planning attempts to maximize the amount of information that could be attained from any specific data, whereas in the responsive case, the focus is toward processing the particular information relevant to the current situation.

It is relevant to state that the timescale in the planning process and the enabling technologies do not necessarily dictate the method of preference. For example, a plan made an hour in advance of execution can be either of a pro-active or a responsive nature, depending on the amount of information arrivals in that hour. For the same reason, an optimal planning obtained from a mathematical decision support tool can be associated with both approaches. However, a planning system used in a volatile environment will typically provide robust solutions in the proactive mode and adaptive solutions in the responsive mode. Consider the case of Obermeyer, a skiwear manufac-turer with long replenishment lead times and seasonal and erratic demand. The company develops the production plans at the start of the planning horizon with stochastic methods and adjusts the production plans as more information becomes available throughout the planning horizon. The plans developed at the start of the planning horizon carefully distinguish between high variable demand and low variable demand. These plans attempt to allocate the production capacities with longer replenishment lead times to the demand with low variability and vice versa. In this way, the company can rather easily adjust the production plans of the high variability demand products since these products are manufactured at the production capacities with lower replenishment lead times (Fisher, 1997). From an IT perspective also, we conduct this discussion along similar lines. Both approaches do not necessarily restrict to either mathematical modeling tools such as an optimi-zation module or information extracting and interchanging systems based on RFID technologies, although one would associate the proactive planning approach with the first type and the responsive planning approach with the second type of technologies.

In many practical situations, organizations opt for a hybrid system that attempts to combine the proactive and responsive mechanisms. For example, some original equipment manufacturers (OEMs) predict their parts returns based on the historical reliability estimates and the sales figures while aware of the fact that forecasts produced by such methods could still deviate from true returns. Therefore, the company also opts for condition-based monitor-ing to monitor customer returns. This is typical of the automotive industry in which OEMs usually have extensive reliability and sales data for each model sold but still encourage their customers for regular condition-based checks.

We have considered two planning approaches that can be used, possibly in combination, depending on the decision situation at hand. In the next few sections, we explore various information-enabling methods that are used for planning and execution in CLSCs from three perspectives: customer, product, and process.

5.2.1 Three Perspectives for Information-Enabling Methods in CLSC Planning

Products, processes, and customers and their interactions are vital sources of information in CLSCs (Van Nunen and Zuidwijk, 2004). The customer perspective in CLSC relates to the activities that attempt to acquire, retain, and manage customer relationships. In addition to the customer behavior and demand information, there is a need to manage the product disposal attitudes of the customer. The product perspective in a CLSC differs significantly from the product perspective in a traditional forward chain. Product information in traditional forward chains does not reach beyond the point of sale. In general, service operations in forward chains are not considered part of the supply chain. However, owing to the nature of CLSCs, additional product information during the product's complete life cycle is a requisite for the management of product recovery and remanufacturing operations. From the process perspective, there have been significant changes in many processes to facilitate the product recovery and remanufacturing aspects. As noted in our case description of CopyMagic (Thierry, 1997), the design, manufacturing, and distribution processes changed significantly to accommodate the product recovery operations. Product design changed to more modular designs, and the usage of durable materials was promoted. The manufacturing processes changed to more flexible operations to accommodate the product remanufacturing cycle. Additionally, the testing and sorting stage was introduced for the returned products. Each of these steps required additional product information, which the MRP system was unable to provide.

5.2.2 Forecasting

Forecasting methods are probably the most recognized information-enabling method within (closed-loop) supply chains and are used for all kinds of planning. Forecasting in itself does not provide a plan, but forecasting is essential for almost any kind of supply chain planning because proactive planning requires some form of forecast as an input. By definition, forecasts are with errors, and we usually observe that reduction of the forecast error is vital for efficient planning. The usual representation of uncertainty in forecast values is in probabilistic terms.

Despite the maturity of the field, many recent developments have occurred in forecasting techniques, for example, in forecasting product returns (to be discussed below). Forecasting methods range from simple heuristics to rigorous mathematical and statistical methods. In the literature, often a distinction is made between *ad hoc* and "structural" forecast methods as a classification (Talluri and van Ryzin, 2005). Examples of *ad hoc* methods or heuristics-based methods are moving average and exponential smoothing, while structural forecasting methods include time series forecasting methods, Bayesian forecasting methods, Kalman filtering, and neural networks. These forecasting methods are used at various stages in CLSCs to predict demand and customer returns. The use of appropriate forecasting methods is a debatable subject. Appropriate process dynamics should be considered before selecting any forecasting method. To some extent, companies can influence product returns by choosing appropriate returns policies (Toktay et al., 2003). At the same time, appropriate prediction of the timing and the volume of product returns is required to manage return handling (Thierry, 1997).

The important characteristics that may impact product returns are past product sales, return cycle time, and return probability. Toktay et al. (2003) discuss the data-driven product returns forecasting methods by considering the product returns as a function of historical sales and return information. Lembke and Amato (2001) attempt to compare the performance of various methods to forecast product returns. The compared methods are weighted moving average, exponential smoothening, and failure rate–based forecasting. The simulation results show that forecasts can be improved by combining product failure rates with sales information.

In practice, the decision to use a specific method is often based on the ease of use and the availability of data. As discussed earlier, process and product dynamics and characteristics are important aspects to consider for choosing an appropriate forecasting method. For example, for a functional product where demand variation is low, it might be appropriate to use simple heuristics such as moving average, whereas for innovative products where demand is erratic and product life cycle is short, it might be suitable to use a tailored method that considers process dynamics thoroughly. Indeed, in this case, forecasting methods such as Bayesian forecasting may provide a more systematic and robust approach for the future demand forecasts (Talluri and van Ryzin, 2005).

5.2.3   Sensing

Sensing refers to the use of state-of-the-art sensing technologies to monitor products, customers, and processes. In CLSCs, RFID technologies are more frequently used to track the movement of returnable packaging materials. In this manner, sensing technologies help make investment decisions for appropriate capacity of returnable packaging materials. A chip-in-crate study (van Dalen et al., 2005) was performed at Heineken to observe the movement of the company's returnable packaging materials such as crates. This pilot study provided the product movement information for returned crates, which are distributed through various distribution channels in the network. Further analysis enabled the computation of appropriate quantities of packaging material requisites for various distribution channels throughout the year. This analysis also indicated that movement of packaging material systematically deviated from many common understandings such as the "first in first out" concept. The results of the chip-in-crate pilot study were also used to estimate the average cycle times for the return of packaging material through various channels. Moreover, sensing technologies also provided data that enhanced return forecasts in this case (van Dalen and van Nunen, 2004).

Other examples of such studies include the Product Lifecycle Manufacturing and Information Tracking Using Smart Embedded Systems (PROMISE) and environmental life cycle information management acquisition (ELIMA) projects. The ELIMA project was funded by the European Commission and intended to develop better ways to manage product life cycle (Muller et al., 2002). This project included the embedding of sensing technologies in various consumer products such as video game consoles and refrigerators to monitor the usage history of these products throughout the product life cycle. The data was processed in life cycle information management software to identify the remaining life of the components and devise look-ahead strategies to estimate the product's recycling or reuse value. The PROMISE project intends to design smart embedded technologies, communication technologies, data management technologies, and adaptive product management for beginning-of-life (BOL), middle-of-life (MOL), and EOL product management.

A series of researches have attempted to investigate the economics of sensing technology usage for commercial applications. Klausner et al. (1998) developed the electronic data logging system for electric motors, which enabled the prediction of the reusability of the motor components by combining the logged data with classification algorithm. The researchers further investigated the economics associated with the installation of such a logging system and established the recovery rate parameters that make the embedding of electronic data logging system feasible. Simon et al. (2001) performed the economic analysis of installing data acquisition technologies in washing machines. The work of these authors also shows that the reusability of components is higher for the components recovered during servicing operation than the EOL product recovery.

Researchers at Cambridge University analyzed various tracking technologies and indicated the problems of accuracy and the confusions caused by multiple standards of traditional barcode technologies. Efficient use of RFID technology would eliminate such problems to a greater extent (Parlikad and McFarlane, 2004; Parlikad et al., 2006). In addition to monitoring the product and its location information, sensing technologies could also efficiently monitor the product reliability state. Such sensing technologies already exist and are being used for many high-value products. For example, IBM uses sensing technologies to monitor the expensive critical components in the company's high-end business machines (Russell et al., 2001). Efficient use of sensing technologies can provide real-time information about product location and the current state of the product, which would aid the prediction of product returns and the capacity planning of remanufacturing operations.

5.2.4 Customer Relationship Management

In the competitive world of business, the importance of maintaining a close relationship with the customer is increasing. Organizations have adopted various customer-focused strategies to maintain and enhance the close relationship with their customers.

CRM and installed base management are examples of such strategies. Generally, CRM is defined by its four elements: know, target, sell, and service (Rygielski et al., 2002). The idea is to understand and manage the customer by learning and maintaining a close relationship. In essence, the CRM process entails relationship initiation, relationship maintenance, and relationship termination stages (Reinartz et al., 2004). From a technological perspective, the CRM technology consists of a data warehouse coupled with tailored data mining techniques to extract the information about various customer behaviors and trends. Case-based reasoning, cross-tabulation, neural networks, decision trees, rules-based systems, and genetic algorithms are some of the widely used data mining approaches in CRM (Rygielski et al., 2002). An integration of CRM software with ERP systems generates company-wide benefits in terms of better customer management as well as improved demand forecasts and subsequent planning.

In forward supply chains, the use of installed base information is limited and typically employed within the after-sales service departments of the company. The integration of installed base information with company-wide ERP is also limited and, consequently, it is hardly used to obtain return forecast information. This state of affairs exists because after-sales service organizations typically are not considered a part of the forward supply chain. In the context of CLSCs, the importance of exploiting the installed base becomes apparent when customer returns can be efficiently predicted and managed. As mentioned in Section 5.2.2, Lembke and Amato (2001) indicate that the return forecast can be highly improved if the data on product sales and product reliability are used in conjunction with product returns data. Moreover, the further integration of installed base information into the CRM database can help companies to improve the customer maintenance dramatically. Since installed base information is typically managed from maintenance or service perspective, it contains many aspects of information (such as machine service history) that are coherent with the information requirements to assess the EOL returns and the remanufacturability of the returned goods.

CRM technologies help organizations to maintain close relationships with their customers. At the same time, the information obtained through CRM regarding customer trends helps improve forecasting. The effective use of installed base information can also enable better prediction of product returns. IBM's service organization maintains complete information about the company's installed base. This installed base information helps the company to maintain and service its premium customers according to its contractual requirements. The company further uses the installed base information in demand forecasting and spare parts inventory planning. For service logistics planning, IBM forecasts demand information by combining the historical demand with the company's installed base information. This method enables the company to tackle the problem of slow moving demand, which is usually observed in service logistics. However, in such a situation the quality of installed base information becomes critical, since the demand is allocated according to the geographical spread of the installed base.

Many auto OEMs have traditionally combined the product sales and reliability data with various forecasting methods to obtain superior forecasts for their service requirements. Auto OEMs typically possess extensive reliability and sales data for each of the models sold, which enables them to predict the spare parts requirements for their service network; still, they encourage their customers for regular checks. This continued monitoring helps the auto OEMs to update their spare parts requirements and in turn enables them to provide their customers with better service and keep the spare parts inventories at a reduced level.

5.2.5 Product Data Management

New technologies have changed the way product design and development is managed today. Traditionally, product conceptualization and design was a tedious and time-consuming process in which a series of engineering design steps were accomplished by performing manual calculations and a subsequent manual engineering drawing process. The development process was also time-consuming with little possibility of prototype development and testing. Any design modification proved to be rather lengthy and expensive. Advanced computer aided design (CAD)/computer aided manufacturing (CAM) technologies have improved the design and development processes by reducing design time and providing means for collaborative design, simulation testing, and prototype testing. These technologies have also enabled more accurate prediction of product reliabilities and life cycles by providing numerous life cycle simulation environments. Because of these technologies, companies are better equipped to adapt to product design, innovate, and stay competitive. In CLSCs, such information can also be used to reduce the remanufacturing cycle uncertainty by incorporating product design for recovery and to provide reliable information to the remanufacturing organizations. As mentioned in Section 5.2.1.2, combining the product reliability and the product sales information can lead to better return forecasts. At the same time, companies are attempting to equip products with data logging system that can monitor the components' performance during the entire life cycle of the product. Klausner et al. (1998) designed the electronic data log system for electric motors and analyzed the product life cycle data for potential product recovery. Many xerox machines are equipped with a self-diagnostic system that helps the company during product sorting and remanufacturing processes by providing information about the current state of the product (Maslennikova and Foley, 2000).

To manage product recovery operations, the recovery chain typically requires additional product data to cope with the uncertainties encountered during the product recovery operations. The product data management operations such as bill of material (BOM) contain the details of the products and their components during the manufacturing cycle to incorporate additional information that supports product recovery aids to reduce the uncertainties during recovery operations (Thierry, 1997).

Meanwhile, owing to increased product warranty and after-sales service requirements, service and maintenance organizations typically maintain the detailed history of the service performed in their installed base to manage their service operations in a better way. Such information could also be incorporated in CLSCs to organize and plan the resources and the capacities in remanufacturing organizations.

The use of sales data also helps in return forecasts. However, point-of-sale systems are typically not designed to track the products after sales. Due to this loss of information associated with the sold product, the point-of-sale system has been identified as a major obstacle for the recovery of value from returned products (Thomas et al., 1999).

In many instances, owing to the strategic initiations of product recovery strategies in many companies, the OEM adopted the modular design of the product. This was typically done to reduce the product recovery time and lessen the uncertainty in the returned product quality. One notable observation in such cases was that the recovery initiatives forced the OEM to involve its suppliers in the design process for the suppliers' input on product recovery aspects. The suppliers were able to contribute effectively due to their knowledge of the operational aspects of the recovery process (Thierry, 1997).

In many instances, organizations employ condition-based monitoring or failure prediction technologies to observe the product state. Such technologies are quite common in aviation, where the consequences of failure are tremendous. This real-time monitoring helps organizations to improve maintenance and provides the necessary input for future research and development of the products (Lee and Ni, 2004).

5.2.6   Process Data Management

Recovery operations such as remanufacturing are characterized by uncertain outcomes. An important factor is yield, which is related to the quality of the input and the characteristics of the process itself. Operational data generated during the product manufacturing cycle not only aids the future production planning for the forward supply chain stage but also supports remanufacturing. For example, quality test plans are generated and implemented during manufacturing. Similar tests are employed in the later stages during returned product testing for acceptable and failed components. The test results generated in the manufacturing stages can also be used to identify certain failure trends in various parts of the product. This could lead to the design of intelligent test standards for product returns, which could save time and resources at the returned product testing stage. Monitoring the recovery processes can support a better estimate of remanufacturing yields, which could result in value creation as shown by Ferrer et al. (2004).

5.3 CopyMagic Case Revisited

The case study in this section is an extended version of the CopyMagic case from Thierry (1997), where we used material from several recent case studies in the copier industry.

CopyMagic is a large multinational copier manufacturer with operations in many European countries. The company manufactures a range of copiers for different market segments. The majority of copiers, especially the expensive models are leased out to the customers for a specific period. Figure 5.2 shows the scheme of manufacturing and distribution operations at CopyMagic.

Many changes were required at various stages of this manufacturing and distribution scheme to accommodate CopyMagic's product recovery initiatives. First, the role of the distributor was altered from distributor and after-sales service provider to accommodate the product take-back facility. Previously, the distributors used to dispose off the returned end-of-lease products. Owing to product recovery initiatives, the returned products were transported to the central recovery center, which tests and sorts the returned copiers according to their remanufacturability and reusability. The assembly operations were evolved to more flexible manufacturing operations to accommodate the remanufacturing of the returned products. The role of the supplier progressed to that of a partner in the design process for feedback regarding the remanufacturability of the component. Figure 5.3 depicts the processes added and the modifications made to the existing processes.

The changes at each step prompted the additional information requirements to coordinate the manufacturing and distribution stages due to the following reasons:

- Uncertain quantity of the returned products (supply uncertainty)
- Uncertain quality of the returned products (supply uncertainty)
- Inadequate accuracy of inspection or testing and its communication to other stages
- Inadequate production planning system to handle remanufacturing operations on a large scale

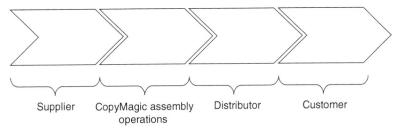

FIGURE 5.2
CopyMagic manufacturing and distribution operations.

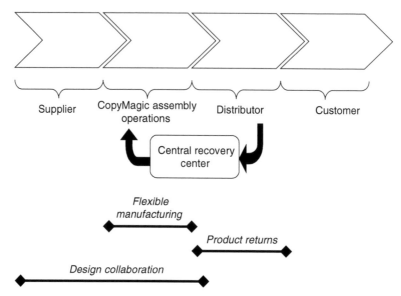

FIGURE 5.3
CopyMagic manufacturing and distribution operations.

The uncertain quantity of the returned products was addressed to a large extent by enhancing the information system and by promoting the leasing option. The central recovery center closely coordinates with the distribution, manufacturing, and remanufacturing channels through an ERP system. Through the leasing option, CopyMagic receives sufficient information about the installed base of its machines. Information about its installed base of leased products and their expiry dates enables CopyMagic to forecast the number of product returns of each type in each geographical area with high precision. To monitor the quality of the returned products, recently developed products contain a monitoring chip that records the service history of the product. Sensing and diagnostics capabilities in copiers are used to enable predictive maintenance, which may result in better planning of spare parts supply and maintenance management. Moreover, consumables such as toner cartridges can be replenished automatically through the use of sensor information. The service information used during service operations can also be used to enhance recovery operations and is therefore transferred to the central database of the company, enabling the linkage of the forecasted quantity of the product returns to their quality. CopyMagic also promoted the leasing pricing option based on the actual usage of the machines, thereby creating incentives for customers to provide their usage history of the copiers, such as the number of copies made.

To address the inspection and production planning issues, two product data management approaches were considered, namely the overhaul method and the dual BOM. The overhaul method extends the existing BOM with information about the components that are used in a specific product. The dual BOM uses two separate BOMs: the assembly BOM from the traditional MRP system and the disassembly BOM providing additional information for product recovery operations. The analysis shows, in each case, that by providing additional information for the product recovery operations, significant improvements are achieved in these operations. The results also suggest that, in general, the dual BOM approach performs better than the overhaul method for product recovery management. Besides the dual BOM, products also contain a safety information card with data on the hazardous materials in the product. Dismantling technicians use such information to identify and sort the hazardous material during dismantling operations. Many products also contain information about the remanufactured components within that product for better customer awareness and to comply with environmental legislations. CopyMagic aims to feed information retrieved during recovery operations back into the design and production phases to enable product and process redesign.

5.4 Concluding Remarks

The various information-enabling mechanisms discussed in Section 5.2 and exemplified in Section 5.3 use information in an integrated way to support CLSC planning. For example, return forecasting may use the combination of past product sales and reliability data to predict future returns. Similarly, customer information together with installed base information support customer response as well as material and capacity planning. Table 5.2 depicts the contribution of information-enabling methods toward proactive or responsive planning from product, customer, and process perspectives in CLSC.

The contribution of forecasting methods is typically limited to proactive planning. Sensing technologies can be conducive to the real-time monitoring of products and customers as well as facilitate precise monitoring of customer behavior for future planning. The CRM methods find their application both in proactive and responsive planning, since the information generated by these methods helps improve responsiveness and future planning. Product operational data enables improvements in process parameters for process planning and optimization as well as increase in process flexibility and responsiveness.

The information-enabling mechanisms aim to reduce uncertainty in two ways, which we associate with two planning approaches. First, these mechanisms may improve the reliability of forecasting methods by providing more data that can be fed into the forecasting models. Second, these mechanisms may provide information on events that enables appropriate response. We distinguished three perspectives that roughly define the domains of application, namely customers, products, and processes. In CLSCs, these three domains are strongly related, for example, customers are suppliers of product

TABLE 5.2

Role of Information-Enabling Methods in CLSC Planning

	Proactive	Responsive
Customer	Demand and return forecasts by using historical data	Monitor service and sales events to respond to demand and returns
	Customer acquisition and retention	Customer responsiveness, complaint handling
	Manage customer incentives	Real-time customer behavior monitoring and management
	Customer feedback for product design	Monitor product use to enhance customer services
Product	Product returns forecasts by using product reliability data	Monitor product condition to manage handling of product or parts returns and maintenance service operations
	Product installed base information for capacity and resource planning	
	Enhanced product life cycle information from sensor data	
	Usage history and recovery data for product and process redesign	
Process	Forecasts (process yields, market prices, etc.) for (re)manufacturing capacity planning and inventory planning	Customer responsiveness
	Process design and quality control design	Process and sourcing flexibility
		Service operations management

returns, the quality of returned products depends also on the use by customers, and the recovery process yields depend on the quality of product returns. Therefore, an integrated approach is required in CLSCs.

References

Carr, N.G., It doesn't matter. *Harvard Business Review*, May 2003.

Debo, L., and Van Wassenhove, L.N., Tire recovery: The RetreadCo case, in Flapper, Simme D.P., van Nunen, J.A.E.E., Van Wassenhove, Luk N. (eds.), *Managing Closed-Loop Supply Chains*, Springer: Berlin, 2005.

Ferrer, G., and Ketzenberg, M.E., Value of information in remanufacturing complex products, *IIE Transactions*, 2004, 36:3, 265–277.

Fisher, M.L., What is the right supply chain for your product? *Harvard Business Review*, March 1997.

Fisher, M.L., Hammond, J.H., Obermeyer, W.R., and Raman, A., Making supply meet demand in an uncertain world. *Harvard Business Review*, May 1994.

Fisher, M., and Raman, A., Reducing the cost of demand uncertainty through accurate response to early sales. *Operation Research*, 1996, 44:1, 87–99.

Fleischmann, B., Meyr, H., and Wagner, M., Advanced planning, in Stadtler, H., Kilger, C. (eds.), *Supply Chain Management and Advanced Planning*, second edition, Chapter 4, Springer: Berlin, 2002.

Fleischmann, M., *Quantitative Models for Reverse Logistics, Lecture Notes in Economics and Mathematical Systems*, Springer: Berlin, 2001, pp. 501.

Glazer, R., Measuring the value of information: The Information-Intensive Organization. *IBM Systems Journal*, 1993, 32: 1.

Gupta, A., and Maranas, C.D., Managing demand uncertainty in supply chain planning. *Computers and Chemical Engineering*, 2003, 27, 1219–1227.

Klausner, M., Grimm, W., and Hendrickson, C., Reuse of electric motors in consumer products. *Journal of Industrial Ecology*, 1998, 2:2, 89–102.

Krikke, H., Bloemhof, J., and Van Wassenhove, L.N., Design of a production and return network for refrigerators. *International Journal of Production Research*, 2003, 41:16, 36–89.

Lee, H.L., Aligning supply chain strategies with product uncertainties. *California Management Review*, 2002, 44:3, 105–119.

Lee, H.L., Padmanabhan, V., and Whang, S., Information distortion in a supply chain: The bullwhip effect. *Management Science*, 1997, 43:4, 546–558.

Lee, J., and Ni, J., Infotronics-based intelligent maintenance system and its impacts to closed-loop product life cycle systems, invited keynote paper. Proceedings of the Intelligent Maintenance Systems 2004 International Conference, Arles, France, 2004.

Lembke, R.S.T., and Amato, H., Replacement parts management: The value of information. *Journal of Business Logistics*, 2001, 22:2, 149–164.

Linton, J.D., and Johnson, D.A. A decision support system for the planning of remanufacturing at Nortel. *Interfaces*, 2000, 30:6, 17–31.

Listes O., and Dekker R., A stochastic approach to a case study for product recovery network design. *European Journal of Operational Research*, 2005, 160:1, 268–287.

Maslennikova, I., and Foley, D., Xerox's approach to sustainability. *Interfaces*, 2000, 30:3, 226–233.

Muller, K., Schneider A., Simon, M., Moore, P., Pu, J., Yang, K., Kopacek, B., and Keif, T., *Environmental Life Cycle Information Management Needs for Standardization*. CARE Innovation, Brussels, Belgium, November 2002.

Parlikad, A., and McFarlane, D., Investigating the role of product information in the end-of-life decision making. Proceedings of the 11th IFAC Symposium on Information Control Problems in Manufacturing, Brazil, April 2004.

Parlikad, A., McFarlane, D., Kulkarni, A., Ralph, D., and Wong, A., Quantifying the value of RFID in product recovery decisions. 7th Annual Conference of POMS, Boston 2006.

Realff, M.J., Ammons, J.C., and Newton, D. Srategic design of reverse production systems. *Computers and Chemical Engineering*, 2000, 24:2, 991–996.

Reinartz, W., Krafft, M., and Hoyer, W., The customer relationship management process: Its measurement and impact on performance. *Journal of Marketing Science*, 2004, XLI, 290–305.

Repo, A. J., The value of information: Approachesin economics, accounting, and management science. *Journal of the American Society for Information Science*, 1989, 40:2, 68–85.

Russell, S., Gonzalez, O. L. B., De Coutere, B., and Furniss, D., *High Availability without Clustering*. IBM Redbooks-International Technical Support Organization, March 2001.

Rygielski, C., Wang, J. C., and Yen, D. C., Data mining techniques for customer relation management. *Technology in Society*, 2002, 24, 482–502.

Simon, M., Bee, G., Moore, P., Pu, J.S., and Xie, C, Modeling of the life cycle of products with data acquisition features. *Computers in Industry*, 2001, 45, 111–122.

Spengler, T., and Schröter, M., Strategic management of spare parts in closed-loop supply chains—a system dynamics approach. *Interfaces*, 2003, 33, 7–17.

Stock, J., Speh, T., and Shear, H., Many happy (product) returns. *Harvard Business Review*, July 2002.

Takata, S., Kimura, F., van Houten, F.J.A.M., Westkämper, E., Shpitalni, M., Ceglarek, D., and Lee, J. Maintenance: Changing role in life cycle management. *CIRP Annals of Manufacturing Technology*, 2004, 53:2, 643–655.

Talluri, K. T. and van Ryzin, G. J., *The Theory and Practice of Revenue Management*, Springer: New York, 2005.

Teunter, R., van der Laan, E., and Vlachos, D., Inventory strategies for systems with fast remanufacturing. *Journal of the Operational Research Society*, 2004, 55:5, 475–484.

Thierry, M., An analysis of the impact of product recovery management on manufacturing companies. PhD dissertation, Erasmus University, Rotterdam, 1997.

Thierry, M., Salomon, M., van Nunen, J., and van Wassenhove, L., Strategic issues in product recovery management. *California Management Review*, 1995, 37:2, 114–136.

Thomas, V., Neckel, W., and Wagner, S., Information technology and product life-cycle management. Proceedings of the IEEE International Symposium on Electronics and Environment, 1999, pp. 54–57.

Toktay, B., van der Laan, E.A., and de Brito, M.P., Managing product returns: The role of forecasting. ERIM Report Series research in Management, 2003.

Yadav, P., Miller, D.M. and Schmidt, C.P. Drake, R., McGriff Treading Company Implements Service Contracts with Shared Savings. *Interfaces*, 2003, 33:6, 18–29.

van Dalen, J., and van Nunen, J.A.E.E., Predicting turnaround times of product carriers. Conference Proceedings Euroma, Fontainebleau, June 27–29, 2004.

van Dalen J., van Nunen, J.A.E.E, and Wilens, C., The chip-in-crate: The Heineken case. van Wassenhove, L., van Nunen, J.A.E.E., and Flapper, S. (eds.) *Managing Closed-Loop Supply Chains*, Springer: Berlin, 2005.

Van Nunen, J.A.E.E., and Zuidwijk, R.A., E-enabled closed-loop supply chain. *California Management Review*, 2004, 46:2, 40–54.

Zhou, L., and Disney, S.M. Bullwhip and inventory variance in a closed loop supply chain. *OR Spectrum*, 2006, 28:1, 127–149.

Zimmermann, H.-J., An application-oriented view of modeling uncertainty. *European Journal of Operation Research*, 2000, 122, 190–198.

6

Disassembly Line Balancing

Seamus M. McGovern and Surendra M. Gupta

CONTENTS

6.1 Introduction

Manufacturers are increasingly recycling and remanufacturing their postconsumed products because of new and more rigid environmental legislations, increased public awareness, and extended manufacturer responsibility. In addition, the economic attractiveness of reusing products, subassemblies, or parts instead of disposing of them, has further fueled this effort. Recycling is a process performed to retrieve the material content of used and nonfunctioning products. Remanufacturing, however, is an industrial process in which worn-out products are restored to make them almost new. Thus, remanufacturing provides the quality standards of new products with used parts.

Product recovery seeks to obtain materials and parts from old or outdated products through recycling and remanufacturing to minimize the amount of waste sent to landfills. Therefore, product recovery includes the reuse of parts and products. Several attributes of a product, for example, ease of disassembly, modularity, type and compatibility of materials used, material identification markings, and efficient cross-industrial reuse of common parts or materials enhance product recovery. Disassembly is the first crucial step in product recovery.

Disassembly is defined as the methodical extraction of valuable parts or subassemblies and materials from discarded products through a series of operations. After disassembly, the reusable parts or the subassemblies are cleaned, refurbished, tested, and directed to the part or the subassembly inventory for remanufacturing operations. The recyclable materials can be sold to raw-material suppliers, while the residuals are sent to landfills.

Recently, disassembly has found a significant place in the literature because of its role in product recovery. A disassembly system faces many unique challenges, for example, it has significant inventory problems because of the disparity between the demand for certain parts or subassemblies and their yield from disassembly. The flow process is also different. As opposed to the normal "convergent" flow in regular assembly environment, in disassembly the flow process is "divergent" (a single product is broken down into many subassemblies and parts). There is also a high degree of uncertainty in the structure and the quality of the returned products. The condition of the products received is usually unknown, and the reliability of the components is suspect. In addition, some parts of the product may cause pollution or pose hazards. These parts tend to have a higher possibility of being damaged and hence may require special handling. This requirement can also influence the use of the disassembly workstations. For example, an automobile slated for disassembly contains a variety of parts and material such as battery, airbags, fuel, and oil that are dangerous to remove or present a hazard to the environment. Various demand sources may also lead to complications in disassembly line balancing. Although the reusability of parts creates a demand for these parts, the demand and the availability of the reusable parts is significantly less predicable than in the assembly process. Most products contain parts that are installed (and must be removed) in different attitudes, from different areas of the main structure, or in different directions. Since any required directional change increases the setup time for the disassembly process, it is desirable to minimize the number of directional changes in the selected disassembly sequence. Finally, disassembly line balancing is critical in minimizing the use of valuable resources (such as time and money) invested in disassembly and in maximizing the level of automation of the disassembly process as well as the quality of the parts (or materials) recovered.

In this chapter, the disassembly line balancing problem (DLBP) is solved using exhaustive search and combinatorial optimization methodologies. Exhaustive search consistently provides the optimal solution, although the exponential time complexity of this search quickly reduces its practicality. Combinatorial optimization techniques are instrumental in obtaining optimal or near-optimal solutions to problems with intractably large solution spaces. Combinatorial optimization is an emerging field that combines techniques from applied mathematics, operations research, and computer science to solve optimization problems over discrete structures. These techniques include greedy algorithms, integer and linear programming, branch-and-bound, divide and conquer, dynamic programming, local optimization, simulated annealing (SA), genetic algorithms (GAs), and approximation algorithms. In this chapter, an exhaustive search algorithm is presented for obtaining the optimal solution to small instances of the DLBP. A GA is then presented. The GA considered involves a randomly generated initial population with crossover, mutation, and fitness competition performed over many generations. The DLBP is then solved using an ant colony optimization (ACO) metaheuristic. The ACO used is an ant system (AS) algorithm known as the ant-cycle model [7] that is enhanced for DLBP. A deterministic hybrid process consisting of a greedy sorting algorithm followed by a hill-climbing heuristic (adjacent element hill climbing, AEHC) is then applied. Finally, a new general-purpose heuristic algorithm is presented for obtaining near-optimal solutions to the problem. Influenced by the hunter–killer (H–K) search tactics of military helicopters, this heuristic easily lends itself to the DLBP. All of these combinatorial optimization techniques seek to provide a feasible disassembly sequence, minimize the number of workstations, minimize the total idle time, and minimize the variation in idle times between workstations while attempting to remove hazardous and high-demand product components as early as possible and to remove parts with similar part removal directions together. Examples are considered to illustrate the implementation of these methodologies. The conclusions drawn from the study include the consistent generation of optimal or near-optimal solutions, the ability to preserve precedence relationships, the superior speed of the methods, and the practicality of the methods because of the ease of implementation in solving DLBPs.

6.2 Literature Review

Many steps are involved in product recovery and remanufacturing. Disassembly is the first crucial step in product recovery. Disassembly is the methodical extraction of valuable parts or subassemblies and materials from postused products through a series of operations. After disassembly, the reusable parts or subassemblies are cleaned, refurbished, tested, and directed to the part or the subassembly inventory for remanufacturing operations. The recyclable materials can be sold to raw material suppliers, and the residuals are disposed off. Many papers have discussed the different aspects of product recovery. Brennan et al. [4] and Gupta and Taleb [20] investigated the problems associated with disassembly planning and scheduling. Torres et al. [49] reported a study for nondestructive automatic disassembly of personal computers (PCs). Gungor and Gupta [15,16,19] introduced the concept of disassembly line balancing and developed an algorithm for solving the DLBP in the presence of failures with the goal of assigning tasks to workstations in a way that probabilistically minimizes the cost of defective parts [18]. For a review of environment-conscious manufacturing and product recovery, see Gungor and Gupta [17], and for a comprehensive review of disassembly sequencing, see Lambert [27] and Lambert and Gupta [28].

Gutjahr and Nemhauser [21] were the first to describe a solution to the assembly line balancing problem with an algorithm that was developed to minimize the delay times at each workstation. The heuristic accounts for precedence relationships and seeks to find the shortest path through a network, with the resulting technique being similar to dynamic programming. Erel and Gokcen [11] developed a modified version of Gutjahr and Nemhauser's line balancing problem algorithm by allowing for mixed-model lines (assembly lines used to assemble different models of the same product) through multiple state times and then, during construction of the solution network, considering all state times before categorizing a given set of completed tasks at a workstation. Suresh et al. [48] were the first to present a GA to provide a near-optimal solution to the assembly line balancing problem that minimized idle time and minimized the probability of line stoppage, that is, minimized the probability that the workstation time exceeds cycle time. Hackman et al. [22] proposed a branch-and-bound heuristic for the simple assembly line balancing type I problem, while Ponnambalam et al. [45] provided quantitative evaluations of various assembly line balancing techniques.

Dorigo et al. [9] introduced ACO. Dorigo et al. [8] provided a summary and a review of ant algorithms. McMullen and Tarasewich [42] used ACO techniques to solve the assembly line balancing problem with parallel workstations, stochastic task durations, and mixed models. Bautista and Pereira [2] used ACO techniques to solve the assembly line balancing problem and compared these techniques to their hill climbing heuristics.

McGovern et al. [40] were the first to propose combinatorial optimization techniques for the DLBP. McGovern and Gupta [31] demonstrated an early version of the H–K heuristic.

6.3   Notation

The following notation is used in the remainder of the chapter:

$\leq_p$	Polynomial time reduction or polynomial transformation; read as: "can be converted to," "is easier than," "is a subset of," or "is a smaller problem than"
$<$	Partial ordering; that is, x precedes y is written $x < y$
$\langle 1, 2, ..., n \rangle$	Ordered n-tuple
$\{1, 2, ..., n\}$	Set (formal definition used here; i.e., listing of n distinct items)
$(1, 2, ..., n)$	List of n items
$\|X\|$	Cardinality of the set X
$\lceil x \rceil$	Ceiling function of x; assigns the smallest integer $\geq x$, for example, $\lceil 1.3 \rceil = 2$
$\forall$	"For all," "for every"
$\in$	"An element of," "is in"
$\exists$	"There exists," "there exists at least one," "for some";
O	"Big-O," $g(x)$ is $O(h(x))$ whenever $\exists\, a:\|g(x)\| \leq a\|h(x)\| \forall\, x \geq 0$
$!$	Factorial
$:$	"Such that"
$\Delta\psi_k$	kth element's delta skip measure; difference between problem size, n, and skip size, ψ_k (i.e., for $\Delta\psi = 10$ and $n = 80$, $\psi = 70$)
ψ_k	kth element's skip measure (i.e., for the solution's third element, visit every second possible task, if $\psi_3 = 2$)
α	Weight of existing pheromone (trail) in path selection
β	Weight of the edges in path selection
η_{pq}	Visibility value of edge (arc for DLBP) pq at time t
ρ	Variable such that $1 - \rho$ represents the pheromone evaporation rate
$\tau_{pq}(NC)$	Amount of trail on edge pq (arc for DLBP) during cycle NC
a	Multiprocessor scheduling problem task variable
A	Multiprocessor scheduling problem task set
b	Function variable in complexity theory
B	Multiprocessor scheduling problem deadline bound
BST_k	kth part in temporary best solution sequence during AEHC
c	Initial amount of pheromone on all of the paths at time $t = 0$
CT	Cycle time; maximum time available at each workstation
d_k	Demand; quantity of part k requested
D	Demand rating for a given solution sequence; also demand bound for the decision version of DLBP
D^*	Lower demand bound (optimal) for a given instance; also used to refer to the set of solutions optimal in D
D_{nom}	Upper demand bound (nominal) for a given instance;
DP	The set of demanded parts
F	Measure of balance for a given solution sequence; fitness measure in GA

F^*	Lower measure of balance bound (optimal) for a given instance; also used to refer to the set of solutions optimal in F
F_{nom}	Upper measure of balance bound (nominal) for a given instance
F_{tr}	Measure of balance of ant r's sequence at time t
FS	Feasible sequence binary value; FS = 1 if feasible, 0 otherwise
h_k	Binary value; 1 if part k is hazardous, else 0
H	Hazard rating for a solution; also hazard bound for the decision version of DLBP
H^*	Lower hazard bound (optimal) for a given instance; also used to refer to the set of solutions optimal in H
H_{nom}	Upper hazard bound (nominal) for a given instance
HP	Set of hazardous parts
i	Counter variable
I	Total idle time for a given solution sequence; also refers to an instance of a problem in complexity theory
I^*	Lower idle time bound (optimum) for a given instance
I_j	Total idle time of workstation j
I_{nom}	Upper idle time bound (nominal) for a given instance
ISS_k	Binary value; 1 if kth part is in solution sequence, else 0
j	Workstation count (1, ..., NWS)
k	part identification (1, ..., n)
$l(a)$	Multiprocessor scheduling problem task length
L	A given problem in complexity theory
L_r	ACO delta trail divisor value; set equal to F_{nr} for DLBP
m	Number of processors in the multiprocessor scheduling problem; also number of ants
n	Number of parts for removal
N	Number of chromosomes (population)
$\mathbf{N}$	Set of natural numbers; that is, {0, 1, 2, ...}
NC_{max}	Maximum number of cycles for ACO
NPW_j	Number of parts in workstation j
NWS	Number of workstations required for a given solution sequence
NWS*	Minimum possible number of workstations for n parts
NWS_{nom}	Maximum possible number of workstations for n parts
p	Countervariable; also edge/arc variable providing node/vertex identification
$p_{pq}^r(t)$	Probability of ant r taking an edge (arc for DLBP) pq at time t during cycle NC
P	Set of n part removal tasks
PRT	Set of part removal times
PRT_k	Part removal time required for kth part
PS_k	kth part in a solution sequence; that is, for solution $\langle 3, 1, 2 \rangle$, $PS_2 = 1$
PSG	Solution sequence after application of the greedy algorithm

PSS_k	kth part in solution sequence after sorting
PST	Solution sequence after AEHC
q	Counter variable; also edge/arc variable (node/vertex identification)
Q	Amount of pheromone added if a path is selected
r	Set of unique part removal directions; also ant count
r_k	Integer value corresponding to the kth part's removal direction
R	Direction rating for a given solution sequence; also direction bound for the decision version of DLBP
R^*	Lower direction bound (optimal) for a given instance; also used to refer to the set of solutions optimal in R
R_k	Binary value; 0 if part k can be removed in the same direction as part $k + 1$, else 1
R_m	Mutation rate
R_{nom}	Upper direction bound (nominal) for a given instance
R_x	Crossover rate
ST_j	Station time; total processing time requirement in workstation j
t	Time within a cycle; ranges from 0 to n
TMP_k	kth part in temporary sequence during AEHC
V	Maximum range for a workstation's idle time
x	General variable; also refers to a part removal direction
y	General variable; also refers to a part removal direction
z	General variable; also refers to a part removal direction
Z	Set of integers; that is $\{\ldots, -2, -1, 0, 1, 2, \ldots\}$
Z^+	Set of positive integers; that is $\{1, 2, \ldots\}$
Z_p	pth objective to minimize or maximize

6.4 Considerations Related to Disassembly Lines

In 2002, Gungor and Gupta [19] became the first authors to formally propose the disassembly line. The following is a summary of some of the considerations in a disassembly line setting based on these authors' observation that a disassembly line is fraught with a variety of disassembly-unique complications. The following is a reprinted listing and description of these considerations.

6.4.1 Product Considerations

The number of different products disassembled on the same line is an important characteristic of a disassembly line. The line may deal with only one type of product whose original configuration is the same for every product received, for example, only Pentium-based personal computers (PCs) with certain specifications. The line may also be used to disassemble products belonging to the same family, that is, products whose original configurations vary only slightly from each other. For example, we may have different models of PCs such as Pentium-based PCs and 486-based PCs on the same line. The line may receive several types of products, including partially disassembled products and subassemblies such as PCs, printers, digital cameras, motherboards, and monitors whose configurations are significantly or completely different from one another.

The changing characteristics of the products complicate the disassembly operations on the line. Intuitively, balancing a disassembly line used for the disassembly of several types of products can become very complex. Such a line may be balanced for a group of products, yet its status may become largely or completely unbalanced when a new type of product is received.

6.4.2   Line Considerations

Various disassembly line configurations may be possible. Some layouts are probably inspired by assembly lines layouts. Layouts such as serial, parallel, circular, U-shaped, cellular, and two-sided lines [47] may also find their way onto disassembly lines. However, new layouts may still be required to create more efficient disassembly lines.

One of the most important considerations in a disassembly line is the line speed. Disassembly lines can be either paced or unpaced. Paced lines may be used to regulate the flow of parts on disassembly lines. However, if there is too much variability in the task times (which might depend on the conditions of the products being disassembled or the variety of the products processed on the line), it might be preferable to have an unpaced line. In such lines, each station works at its own pace and advances the part to the next station whenever it completes the assigned tasks. The advantages of paced lines over unpaced lines include considerably less work in process, less space requirements, more predictable and higher throughput and, if properly handled, less chance of bottlenecks. To take advantage of the positive aspects of a paced line, its speed can be dynamically modified throughout the disassembly process to minimize the negative effects of variability (including variability in demand).

6.4.3   Part Considerations

6.4.3.1   Quality of Incoming Products

When a disassembly system receives the returned products, the condition of these products is usually not known; sometimes the products are in good shape and are relatively new, while at other times they are old and nonfunctioning items. Therefore, there is a high level of uncertainty in the quality of the products and their constituent parts. There is a trade-off between the level of uncertainty and the efficiency of the disassembly line. When the level of uncertainty increases, the efficiency measures worsen, especially if the disassembly line has not been designed to cope with the uncertainty factors.

A part is considered to be defective if the part has a different structure or different operational specifications from its original structure or specifications. A part can become defective after exposure to an accident or to a hostile operating environment. There are mainly two types of defects:

- *Physical defect.* A part can be physically defective, indicating that the geometric specifications (dimensions and shape) of the part are different from its original design. For example, the cathode ray tube (CRT) of a computer monitor can be broken or the cover of a PC can be dented.

- *Functional defect.* A part may not contain any physical defect; however, this part may not function as per its original design. For example, the central processing unit (CPU) of a PC may be perfect in its physical aspects but may not function because of an internal problem.

6.4.3.1.1   Removable Defective Parts (Type A Defect)

Some of the physically defective parts in the product can be disassembled, although they sustain some level of damage. We call this type of physical defect "type A defect." The precedence relationships must be satisfied during the assignment of the disassembly tasks to the disassembly workstations. When a part is physically defective and yet removable, it might take longer to disassemble that part. However, this disassembly does not affect the disassembly of other parts since the precedence relationship is not affected by the longer disassembly time.

6.4.3.1.2   Nonremovable Defective Parts (Type B Defect)

Sometimes a physically defective part in the product cannot be disassembled because it is either badly damaged (and thus gets stuck in its place) or its fixture is of the type that cannot be undone. This is a "type B defect" and has a tremendous impact on the efficiency of the disassembly line. For example, even if the nonremovable part does not have a demand, this part may still precede other demanded part(s). This case may have various levels of negative effects on the disassembly line depending on the number of demanded (and not necessarily all) parts that are preceded by the nonremovable part. A part with a type B defect may also result in what we call the disappearing workpieces phenomena (DWP) (see Section 6.4.4.5 for a discussion on DWP).

6.4.3.1.3   Parts with Functional Defects

Assume that part k does not have a physical defect (types A or B) but sustains a functional defect. If this fact were known in advance, then the disassembler may or may not have an incentive to disassemble this part (unless, of course, this part precedes other demanded parts). Therefore, in addition to concerns related to physical defects in parts, the possibility of functional defects in parts belonging to the incoming products should also be incorporated in operating and balancing a disassembly line.

6.4.3.2 *Quantity of Parts in Incoming Products*

Another complication is related to the quantity of parts in the incoming products. Because of upgrading (or downgrading) of the product during its use, the number of parts in the product may not match with the product's original configuration. When the product is received, the actual number of parts may be more or less than expected.

The following aspects should be considered:

- The demanded part of the product may be absent or its quantity may be less than expected. This may require a provision in the calculation to ensure that the demand for the missing parts is satisfied. Another approach would be to carry out a preliminary evaluation of the product to ensure that all the parts of interest exist in the product before it is pushed down on the disassembly line. This evaluation may be used to determine the route of the product on the disassembly line.

- The number of demanded parts may be more than expected. The most common example is that of a PC disassembly system. Assume that there is a demand for the memory modules of old PCs. Further assume that the PCs undergoing disassembly were originally configured to have one 32 megabyte (MB) memory module. However, if the owner of the PC upgraded the computer's memory to 64 MB using another 32 MB module, two 32 MB memory modules would be retrieved as a result of the disassembly of the PC, which is one more than expected. This situation would affect the memory demand constraints and the workstation time (since the disassembly of two memory modules would take longer than the removal of just one module).

6.4.4 Operational Considerations

6.4.4.1 *Variability of Disassembly Task Times*

Similar to the assembly line case, the disassembly task times may vary depending on several factors that are related to the condition of the product and the state of the disassembly workstation (or worker). Task times may be considered as deterministic, stochastic, or dynamic. Dynamic task times are possible because of learning effects, which allow systematic reduction in disassembly times. For example, in their case-based reasoning (CBR) approach, Zeid et al. [51] demonstrated that the disassembly of products can be assisted by the reuse of solutions that were applied to similar problems encountered in the past, thus reducing disassembly times.

6.4.4.2 *Early Leaving Workpieces*

If one or more (not all) tasks of a workpiece that has been assigned to the current workstation cannot be completed because of some defect, the workpiece might leave the workstation early. This phenomenon is termed as the early leaving workpiece (ELWP).

Due to ELWP, the workstation experiences an unscheduled idle time for the duration of the tasks that cause the workpiece to leave early. Note that the cost of unscheduled idle time is high because the disassembly cost of failed tasks has been incurred, although the demand for the parts associated with the failed tasks is unfulfilled. Assume that disassembly of the hard disk and the floppy drive of a PC is carried out at workstation 1. After the removal of the hard disk, if the holding screws of the floppy drive are stuck or damaged, we may not be able to remove the floppy. Therefore, the workpiece (the PC being disassembled) has to leave workstation 1 earlier than its scheduled time. When this problem occurs, workstation 1 remains idle for the duration of the normal disassembly time of floppy.

6.4.4.3 Self-Skipping Workpieces

If all tasks of a workpiece that have been assigned to the current workstation are disabled due to some defect in the tasks or precedence relationships, the workpiece leaves the workstation early without being worked upon. This phenomenon is known as self-skipping workpiece (SSWP).

Consider the disassembly of the PC presented as an example in Section 6.4.4.2. Assume that the disassembly of the hard disk must be carried out before the disassembly of the floppy drive. If the hard disk of the PC cannot be removed because its holding screws are damaged, neither of the tasks assigned to workstation 1 (disassembly of the hard disk and the floppy drive) can be performed. Therefore, the workpiece leaves the workstation without being worked upon; in other words, the workpiece self-skips workstation 1.

6.4.4.4 Skipping Workpieces

At workstation j, if one or more defective tasks of a workpiece directly or indirectly precede all the tasks of workstation $j + 1$ (i.e., the workstation immediately succeeding workstation j), the workpiece "skips" workstation $j + 1$ and moves on to workstation $j + 2$. This phenomenon is known as a skipping workpiece (SWP).

The number of workstations a workpiece skips is the strength of skipping. If a workpiece skips only one workstation, the workpiece is a 1-SWP; if two workstations are skipped, the workpiece is a 2-SWP, and so on.

In addition to unscheduled idle time, both SSWP and SWP experience added complexities in material handling (e.g., how to transfer the workpiece out of turn to the downstream workstation) and the status of the downstream workstation (e.g., it may be busy working on other workpieces and may require some sort of buffer allocation procedure to hold the skipped workpiece until the machine becomes available). Using the disassembly of a PC given before, assume the disassembly of the floppy drive precedes the removal of the random access memory (RAM) modules and the sound card of the PC scheduled to be carried out at workstation 2. If the floppy drive cannot be removed at workstation 1 because of its damaged holding screws, the workpiece leaves workstation 1 early and may skip workstation 2 because neither the RAM modules nor the sound card can be removed from the workpiece. In this case, workstation 2 remains idle for the duration of its originally scheduled tasks, namely the removal of the RAM modules and the sound card.

6.4.4.5 Disappearing Workpieces

If a defective task disables the completion of all the remaining tasks on a workpiece, the workpiece may simply be taken off the disassembly line before it reaches any downstream workstation. In other words, the workpiece effectively disappears, which is referred to as a disappearing workpiece (DWP).

DWP may result in starvation of subsequent workstations, leading to a higher overall idle time that is highly undesirable in a balanced line. DWP, in a way, is a special case of SWP in which the workpiece skips all succeeding workstations. The consequences of the DWP are similar to the SWP but to a greater extent.

For example, assume that the disassembly of the top cover of the PC is carried out at workstation 1 in addition to the disassembly of the hard disk and the floppy drive. If the top cover cannot be removed because of damage, the PC (workpiece) can be taken off the line since we cannot reach any of the parts inside the PC. In a way, the PC disappears, thus leading to idle times in the remaining workstations.

6.4.4.6 Revisiting Workpieces

A workpiece currently at workstation j may revisit a preceding workstation $(j - p)$, where $(j - p) \geq 1$ and $p \geq 1$ and integer, to perform task x if the completion of current task y enables one to work on task x, which was originally assigned to workstation $(j - p)$ and was, however, disabled due to the failure of another preceding task. These are termed revisiting workpieces (RWP).

An RWP results in overloading of one of the previous workstations. As a result, complications may occur in the material-handling system because of reverse flows, decoupling from the revisited workstation, or introduction of a buffer to hold the revisiting workpiece until the workstation becomes available. This problem would obviously have financial consequences as well.

To exemplify RWP, assume that at workstation 1 the hard disk and the floppy drive are disassembled, at workstation 2 the RAM modules and the sound card are disassembled, and at workstation 3 the power unit of a PC is disassembled. Also assume that the RAM modules can be removed after the removal of either the floppy drive or the power unit. Further assume that disassembly of the floppy drive precedes the removal of the sound card. If the floppy drive cannot be removed because of preexisting damage, the PC skips workstation 2 and goes to workstation 3 where the power unit is removed. Since the power unit has been taken out of the PC, the RAM modules can be removed. Thus, the PC (workpiece) is sent back to workstation 2 for disassembly of the RAM modules. Complicated RWPs that could make the material handling and flow control more difficult may also exit the line.

6.4.4.7 Exploding Workpieces

A workpiece may split into two or more workpieces (subassemblies) as it moves on the disassembly line because of the disassembly of certain parts that hold the workpiece together. Each of these subassemblies acts as an individual workpiece on the disassembly line. This phenomenon is known as exploding workpieces (EWP).

EWP complicates the flow mechanism of the disassembly line; however, EWP can be planned in advance since we know the removal of which part would result in the EWP. In a disassembly line, a complication occurs when the part that would normally result in EWP cannot be removed because of some defect.

6.4.5 Demand Considerations

Demand is one of the most crucial issues in disassembly line design and optimization since we want to maximize the use of the disassembly line while meeting the demand for parts in associated planning periods. In disassembly, the following demand scenarios are possible: demand for one part only (single part disassembly—a special case of partial disassembly); demand for multiple parts (partial disassembly); and demand for all parts (complete disassembly).

Parts with physical defects or with functional defects may influence the performance of the disassembly line. If part x is not demanded and directly or indirectly precedes a demanded part y, then part x must be disassembled before the removal of part y. (Note that when part a is precedent to part b, which is precedent to part c, then part a is said to be a direct precedent to part b and an indirect precedent to part c.) Removal of part x may require additional time because of the presence of a defect, which, for example, may extend the access time. This may cause the processing time to exceed the cycle time, which could result in the starvation of subsequent workstations and in the blockage of the previous workstations. Placing buffers at the workstations may be necessary to avoid starvation, blockage, and incomplete disassembly. If part x has a demand, then the various types of demand sources may lead to complications. These demand considerations affect the number of products to be disassembled and eventually the disassembly line balance. There are three types of demand.

The first type of demand is as follows: the demand source may accept part k "as is." This situation is possible, for example, when part k is demanded for its material content (in which case the defect in the part may not be important).

The second type of demand is as follows: the demand source may not accept parts with any type of defect. This situation occurs when a part is used "as is" (e.g., in remanufacturing or repairs of other products). Thus, if a demanded part has a defect, its disassembly does not satisfy the demand. Since the primary objective of a disassembly line is to meet the demand, the objective function and demand constraints must cope with this type of complication. If the part has second-type demand and does not directly or indirectly precede another demanded part, the disassembly of the part is redundant. This redundancy may have an effect on the efficiency of the disassembly line since the workstation responsible for removing the defective part would remain idle for some time.

In the third type of demand, the demand source may accept certain defective parts depending on the seriousness of the defect. This type of demand may be received from a refurbishing environment where the parts undergo several correction processes (such as cleaning and repair) before reuse. This type of demand introduces a further complication that requires some sort of tracing mechanism to identify the type of defect and the associated demand constraint.

6.4.6 Assignment Considerations

In addition to precedence relationships, several other restrictions limit the assignment of tasks to workstations. While similar restrictions are present in assembly lines, the following are some restrictions related specifically to disassembly.

Certain tasks must be grouped and assigned to a specific workstation for a reason. For example, the removed parts will be sorted and packaged together for shipment to the demand source. Thus, assigning the disassembly of these components (parts or joining elements) to the same workstation minimizes the distance that the components travel in the disassembly system. Moreover, the tasks requiring similar operating conditions (e.g., temperature and lighting) can be restricted to certain workstations as well.

The availability of special machining and tooling at certain workstations may necessitate the assignment of certain tasks to these workstations. For example, disassembly of hazardous parts may be assigned to highly specialized workstations to minimize the possibility of contamination in the rest of the system. Similarly, the skills of human workers can be a factor in the task assignment restrictions.

Tasks may be assigned to workstations such that the amount of repositioning of the workpieces on the disassembly line (e.g., the disassembly direction changes) is minimized. Similarly, tasks may have to be assigned to minimize the number of tool changes throughout the disassembly process.

6.4.7 Other Considerations

Additional uncertainty factors are associated with the reliability of the disassembly workstations. Some parts of the product may cause pollution or nuisance due to the nature of their contents (e.g., oil and gas), which may increase the possibility of breakdowns or workstation downtime. Furthermore, hazardous parts may require special handling, which can also influence the use of the workstations.

6.5 DLBP Model Considerations

The DLBP investigated in this chapter is associated with a paced disassembly line for a single model of product that undergoes complete disassembly. Part removal times are known and discrete; hence, these times do not posses any probabilistic component. A summary of model assumptions is listed at the end of this section.

The desired solution to a DLBP instance consists of an ordered sequence (*n*-tuple) of work elements (also referred to as tasks, components, or parts). For example, if a solution consisted of the 8-tuple ⟨5, 2, 8, 1, 4, 7, 6, 3⟩, then component 5 would be removed first, followed by component 2, then component 8, and so on.

While different authors use a variety of definitions for the term "balanced" in reference to assembly [10] and disassembly lines, we propose the following definition [30,40] that will be used consistently throughout this chapter:

> *Definition:* A disassembly line is optimally balanced when the fewest possible number of workstations is needed and the variation in idle times between all workstations is minimized while observing all constraints. This is mathematically described by

$$\text{Minimize NWS}$$

then

$$\text{Minimize } [\max (ST_x) - \min (ST_y)] \quad \forall\, x,\, y \in \{1, 2, ..., \text{NWS}\}$$

Except for some minor variation in the greedy or hill-climbing approach, all of the combinatorial optimization techniques described here use a similar methodology to address the multicriteria aspects of DLBP. Since measure of balance is the primary consideration in this chapter, additional objectives are only considered subsequently, that is, the methodologies first seek to select the best performing measure of balance solution as given by Formula 6.4; equal balance solutions are then evaluated for hazardous part removal positions as per Formula 6.13; equal balance and hazard measure solutions are evaluated for high-demand part removal positions as measured by Formula 6.21; and equal balance, hazard measure, and high-demand part removal position solutions are evaluated for the number of direction changes as measured by Formula 6.26. Based on lexicographic goal programming, this priority ranking approach was selected over a weighting scheme for several reasons. These include the simplicity of implementing lexicographic goal programming, ease in re-ranking the priorities, ease in expanding or reducing the number of priorities, because other weighting methods can be readily addressed at a later time, and primarily to enable unencumbered efficacy analysis of the solution-generating methodologies and instances.

Our model is based on the following assumptions:

- The line is paced.
- Part removal times are deterministic, constant, and discrete (or able to be converted to integer form).
- Each product undergoes complete disassembly (even if demands are zero).
- All products contain all parts with no additions, deletions, modifications, or physical defects.
- Each part is assigned to only one workstation.
- The sum of the part removal times of all the parts assigned to a workstation must not exceed CT.
- The precedence relationships among the parts must not be violated.

6.6 Mathematical Description of the DLBP Model

The problem investigated in this chapter seeks to fulfill five objectives:

1. Minimize the number of disassembly workstations and hence minimize the total idle time
2. Ensure that the idle times at each workstation are similar
3. Remove hazardous components early in the disassembly sequence
4. Remove high-demand components before low-demand components
5. Minimize the number of direction changes required for disassembly

A major constraint is the requirement to provide a feasible (i.e., precedence preserving) disassembly sequence for the product being investigated. The result is an integer, deterministic, n-dimensional, multiple-criteria decision-making problem with an exponential search space. Testing a given solution against the precedence constraints fulfills the major constraint of precedence preservation. Minimizing the sum of the workstation idle times, which will also minimize the total number of workstations, attains objective 1 and is described by

$$I = (\text{NWS} \cdot \text{CT}) - \sum_{k=1}^{n} \text{PRT}_k \tag{6.1}$$

or

$$I = \sum_{j=1}^{NWS}(CT - ST_j) \tag{6.2}$$

This objective is represented as

$$\text{Minimize } Z_1 = \sum_{j=1}^{NWS}(CT - ST_j) \tag{6.3}$$

Line balancing seeks to achieve perfect balance (all idle times equal to zero). When this balance is not achievable, either line efficiency (LE) or the smoothness index (SI) is used as a performance evaluation tool [10]. We use a measure of balance that combines the two and is easier to calculate. SI rewards similar idle times at each workstation, but at the expense of allowing for a large (suboptimal) number of workstations. This problem occurs because SI compares workstation elapsed times to the largest ST_j instead of the CT. LE rewards the minimum number of workstations but allows unlimited variance in idle times between workstations because no comparison is made between ST_js. This chapter makes use of the balancing method developed by McGovern and Gupta [30,40]. The McGovern–Gupta balancing method simultaneously minimizes the number of workstations while aggressively ensuring that idle times at each workstation are similar, though at the expense of the generation of a nonlinear objective function. This method is computed based on the minimum number of workstations required as well as the sum of the square of the idle times for each of the workstations. This method penalizes solutions in the cases in which the number of workstations may be minimized but one or more workstation has an exorbitant amount of idle time compared with the other workstations. The balancing method provides for leveling the workload between different workstations on the disassembly line. Therefore, a resulting minimum numerical performance value is the more desirable solution indicating both a minimum number of workstations and similar idle times across all workstations. The McGovern–Gupta measure of balance is represented as

$$F = \sum_{j=1}^{NWS}(CT - ST_j)^2 \tag{6.4}$$

with the DLBP balancing objective represented as

$$\text{Minimize } Z_2 = \sum_{j=1}^{NWS}(CT - ST_j)^2 \tag{6.5}$$

Perfect balance is indicated by

$$Z_2 = 0 \tag{6.6}$$

Note that mathematically, Formula 6.5 effectively makes Formula 6.3 redundant because Formula 6.5 concurrently minimizes the number of workstations.

Theorem 6.1: *Let* PRT_k *be the part removal time for the kth of n parts, where CT is the maximum amount of time available to complete all tasks assigned to each workstation. Then for the most efficient distribution of tasks, the minimum number of workstations, NWS* satisfies*

$$NWS^* \geq \left\lceil \frac{\sum\limits_{k=1}^{n} PRT_k}{CT} \right\rceil \qquad (6.7)$$

Proof: If the above inequality is not satisfied, then there must be at least one workstation completing tasks requiring more than *CT* of time, which is a contradiction.

Subsequent bounds are shown to be true in a similar fashion and are presented throughout the chapter without proof.

The upper bound for the number of workstations is given by

$$NWS_{nom} = n \qquad (6.8)$$

therefore

$$\left\lceil \frac{\sum\limits_{k=1}^{n} PRT_k}{CT} \right\rceil \leq NWS \leq n \qquad (6.9)$$

The lower bound on *F* is given by

$$F^* \geq \left(\frac{I}{NWS^*} \right)^2 \cdot NWS^* \qquad (6.10)$$

while the upper bound is described by

$$F_{nom} = \sum_{k=1}^{n} (CT - PRT_k)^2 \qquad (6.11)$$

therefore

$$\left(\frac{I}{NWS^*} \right)^2 \cdot NWS^* \leq F \leq \sum_{k=1}^{n} (CT - PRT_k)^2 \qquad (6.12)$$

A hazard measure was developed to quantify each solution sequence's performance, with a lower calculated value being more desirable [33]. This measure is based on binary variables that indicate whether a part contains hazardous material (the binary variable is equal to one if the part is hazardous, else zero) as well as the position of this material in the sequence. A given solution sequence hazard measure is defined as the sum of hazard binary flags multiplied by their position in the solution sequence, thereby rewarding the removal of hazardous parts early in the part removal sequence. This measure is represented as

$$H = \sum_{k=1}^{n} \left(k \cdot h_{PS_k} \right) \quad h_{PS_k} = \begin{cases} 1, & \text{hazardous} \\ 0, & \text{otherwise} \end{cases} \qquad (6.13)$$

with the DLBP hazardous part objective represented as

$$\text{Minimize } Z_3 = \sum_{k=1}^{n} \left(k \cdot h_{PS_k} \right) \tag{6.14}$$

The lower bound on the hazardous part measure is given by

$$H^* = \sum_{p=1}^{|HP|} p \tag{6.15}$$

where the set of hazardous parts is defined as

$$HP = \{k : h_k \neq 0 \ \forall \ k \in P\} \tag{6.16}$$

and its cardinality can be calculated with

$$|HP| = \sum_{k=1}^{n} h_k \tag{6.17}$$

For example, a product with three hazardous parts would give an H^* value of $1 + 2 + 3 = 6$. The upper bound on the hazardous part measure is given by

$$H_{\text{nom}} = \sum_{p=n-|HP|+1}^{n} p \tag{6.18}$$

or alternatively

$$H_{\text{nom}} = (n \cdot |HP|) - |HP| \tag{6.19}$$

For example, three hazardous parts in a product having a total of 20 would give an H_{nom} value of $18 + 19 + 20 = 57$ or equivalently, $H_{\text{nom}} = (20 \cdot 3) - 3 = 60 - 3 = 57$. Formulae 6.15, 6.18, and 6.19 are combined to give

$$\sum_{p=1}^{|HP|} p \leq H \leq \sum_{p=n-|HP|+1}^{n} p = (n \cdot |HP|) - |HP| \tag{6.20}$$

In addition, a demand measure was developed to quantify each solution sequence's performance, with a lower calculated value more desirable [33]. This measure is based on positive integer values that indicate the quantity required of this part after it is removed—or zero if it is not desired—and its position in the sequence. Any given solution sequence's demand measure is defined as the sum of the individual part's demand values multiplied by each part's position in the sequence, thereby rewarding the removal of high-demand parts early in the part removal sequence with a numerically low value for the corresponding demand measure. This measure is represented as

$$D = \sum_{k=1}^{n} \left(k \cdot d_{PS_k} \right) \quad d_{PS_k} \in N, \ \forall PS_k \tag{6.21}$$

with the DLBP demand part objective represented as

$$\text{Minimize } Z_4 = \sum_{k=1}^{n} \left(k \cdot d_{PS_k} \right) \tag{6.22}$$

The lower bound on the demand measure (D^*) is given by Formula 6.21, where

$$d_{PS_1} \geq d_{PS_2} \geq \cdots \geq d_{PS_n} \tag{6.23}$$

For example, three parts with demands of 4, 5, and 6, respectively, would give a best-case value of $(1 \cdot 6) + (2 \cdot 5) + (3 \cdot 4) = 28$. The upper bound on the demand measure (D_{nom}) is given by Formula 6.21, where

$$d_{PS_1} \leq d_{PS_2} \leq \cdots \leq d_{PS_n} \tag{6.24}$$

For example, three parts with demands of 4, 5, and 6, respectively, would give a worst-case value of $(1 \cdot 4) + (2 \cdot 5) + (3 \cdot 6) = 32$.

Finally, a direction measure was developed to quantify each solution sequence's performance, with a lower calculated value indicating minimal direction changes and a more desirable solution [35]. This measure is based on a count of the direction changes. Integer values represent each possible direction (typically $r = \{+x, -x, +y, -y, +z, -z\}$; in this case $|r| = 6$). These directions are expressed as

$$r_{PS_k} = \begin{cases} +1, & direction + x \\ -1, & direction - x \\ +2, & direction + y \\ -2, & direction - y \\ +3, & direction + z \\ -3, & direction - z \end{cases} \tag{6.25}$$

and are easily expanded to other or different directions in a similar manner. The direction measure is represented as

$$R = \sum_{k=1}^{n-1} R_k \qquad R_k = \begin{cases} 1, & r_{PS_k} \neq r_{PS_{k+1}} \\ 0, & \text{otherwise} \end{cases} \tag{6.26}$$

with the DLBP direction part objective represented as

$$\text{Minimize } Z_5 = \sum_{k=1}^{n-1} R_k \tag{6.27}$$

The lower bound on the direction measure is given by

$$R^* = |r| - 1 \tag{6.28}$$

For example, for a given product containing six parts that are installed or removed in directions $r_k = (-y, +x, -y, -y, +x, +x)$ the resulting best-case value would be $2 - 1 = 1$ (e.g., one possible R^* solution containing the optimal, single-change of product direction would be: $\langle -y, -y, -y, +x, +x, +x \rangle$). In the specific case, where the number of unique direction changes is one less than the total number of parts n, the upper bound on the direction measure would be given by

$$R_{nom} = |r| \quad \text{where } |r| = n - 1 \tag{6.29}$$

Otherwise, the measure varies depending on the number of parts having a given removal direction and the total number of removal directions. It is bound by

$$|r| \le R_{nom} \le n - 1 \quad \text{where } |r| < n - 1 \tag{6.30}$$

For example, six parts installed or removed in directions $r_k = (+x, +x, +x, -y, +x, +x)$ would give an R_{nom} a value of 2 as given by the lower bound of Formula 6.30 with a solution sequence of $\langle +x, +x, -y, +x, +x, +x \rangle$. Six parts installed or removed in directions $r_k = (-y, +x, -y, -y, +x, +x)$ would give an R_{nom} value of $6 - 1 = 5$ as given by the upper bound of Formula 6.30 with a solution sequence of $\langle -y, +x, -y, +x, -y, +x \rangle$, for example.

In the special case, where each part has a unique removal direction, the measures for R^* and R_{nom} are equal and given by

$$R^* = R_{nom} = n - 1 \quad \text{where } |r| = n \tag{6.31}$$

Note that the preceding optimal and nominal balance, hazard, demand, and direction formulae are dependent upon favorable precedence constraints that will enable generation of these optimal or nominal measures.

6.7 Computational Complexity of DLBP

Although similar in name to the assembly line balancing problem (a claimed NP-hard combinatorial optimization problem), the DLBP contains a variety of enhancements that increase complexity. The general assembly line balancing problem can be formulated as a scheduling problem, specifically as a flow shop [44]. In its most basic form, the objective of DLBP is to minimize the idle times found at each machine (workstation). Since finding the optimal solution requires investigating all permutations of the sequence of n part removal times, there are $n!$ possible solutions. The time complexity of searching this space is $O(n!)$, referred to as exponential time. Though a possible solution to an instance of this problem can be verified in polynomial time, the problem cannot always be optimally solved. Therefore, this problem is classified as nondeterministic polynomial (NP); if the problem is solved in polynomial time, it would be classified as polynomial (P). (Note that a great deal of the explanations that follow can be found in Garey and Johnson [12] and Tovey [50].)

NP-completeness provides strong evidence that the problem cannot be solved with a polynomial time algorithm (the theory of NP-completeness is applied only to decision problems). Many problems that are polynomial solvable differ only slightly from other problems that are NP-complete. While 3-satisfiability and 3-dimensional matching are NP-complete, the related 2-satisfiability and 2-dimensional matching problems can be solved in polynomial time (polynomial time algorithm design is detailed in Refs. 1 and 46). If a problem is NP-complete, some subproblem (created by additional restrictions being placed on the allowed instances) may be solvable in polynomial time. Certain encoding schemes and mathematical techniques can be used on some size-limited cases of NP-complete problems (an example is a dynamic programming approach to partition that results in a search space size equal to the size of the instance multiplied by the log of the sum of each number in the instance) enabling the solution by what is known as a "pseudopolynomial time algorithm." In the PARTITION case, this results in a polynomial time solution as long as extremely large input numbers are not found in the instance. Problems for which no pseudopolynomial time algorithm exists (assuming $P \neq NP$) are known as "NP-complete in the strong sense" (also, *unary* NP-complete—referring to allowance for a nonconcise or unary encoding scheme notation where a string of n 1s represents the number n—with an alternative being *binary* NP-complete). Typical limiting factors for NP-complete problems to have a pseudopolynomial time algorithm include the requirement that the problem be a number problem (versus a graph problem, logic problem, etc.) and be size constrained, typically in the number of instance elements (n in DLBP) and the size of the instance elements (PRT_k in DLBP). Formally, an algorithm that solves a problem L will be called a pseudopolynomial time algorithm for L if its time complexity function is bound above as a function of the instance I by a polynomial function of the two variables: Length[I] and Max[I]. A general NP-completeness result implies that the problem cannot be solved in polynomial time in all the chosen parameters.

Knapsack has been shown to be solvable by a pseudopolynomial time algorithm in a dynamic programming fashion similar to partition by Dantzig [6]. Multiprocessor scheduling and sequencing within intervals have both been shown to be NP-complete in the strong sense. The same is true for 3-partition, which is similar to partition but has m (instead of 2) disjoint sets, and each of the disjoint sets has exactly three elements. By considering subproblems created by placing restrictions on one or more of the natural problem parameters, useful information about what types of algorithms are possible for the general problem can be gleaned.

The decision version of DLBP is NP-complete in the strong sense. This can be shown through a proof by restriction to multiprocessor scheduling. Garey and Johnson [12] describe multiprocessor scheduling as follows:

> **Instance:** A finite set A of tasks, a length $l(a) \in Z^+$ for each $a \in A$, a number $m \in Z^+$ of processors, and a deadline $B \in Z^+$.

> **Question:** Is there a partition $A = A_1 \cup A_2 \cup \cdots \cup A_m$ of A into m disjoint sets, such that

$$\max\left\{\sum_{a\in A_i} l(a) : 1 \leq i \leq m\right\} \leq B?$$

In DLBP, NWS is equivalent to m in multiprocessor scheduling, P is equivalent to A, while CT is equivalent to B.

The following theorem provides the proof:

Theorem 6.2: *DLBP is NP-complete in the strong sense.*

Instance: A finite set P of tasks, partial order $<$ on P; task time $\mathrm{PRT}_k \in Z^+$, hazardous part binary value $h_k \in \{0, 1\}$, part demand $d_k \in N$, and part removal direction $r_k \in Z$ for each $k \in P$; workstation capacity $CT \in Z^+$; number NWS $\in Z^+$ of workstations; difference between largest and smallest idle time $V \in N$; hazard measure $H \in N$; demand measure $D \in N$; and direction change measure $R \in N$.

Question: Is there a partition of P into disjoint sets $P_A, P_B, \ldots, P_{NWS}$, such that the sum of the sizes of the tasks in each P_X is CT or less, the difference between largest and smallest idle times is V or less, the sum of the hazardous part binary values multiplied by their sequence position is H or less, the sum of the demanded part values multiplied by their sequence position is D or less, the sum of the number of part removal direction changes is R or less, and it obeys the precedence constraints?

Proof: DLBP $\in$ NP. Given an instance, it can be verified in polynomial time if the answer is yes by counting the number of disjoint sets and showing that they are NWS or less, summing the sizes of the tasks in each disjoint set and showing that they are CT or less, examining each of the disjoint sets and noting the largest and the smallest idle times, then subtracting the smallest from the largest and showing that this value is V or less, summing the hazardous part binary values multiplied by their sequence position and showing this summation is H or less, summing the demanded part values multiplied by their sequence position and showing that this summation is D or less, summing the number of changes in part removal direction and showing that this is R or less, and checking that each task has no predecessors listed after it in the sequence.

Multiprocessor scheduling $\leq_p$ DLBP. Restrict to multiprocessor scheduling by allowing only instances in which V = CT, $<$ is empty, and $h_x = h_y$, $d_x = d_y$, $r_x = r_y \ \forall \ x, y \in P$.

Therefore, the decision version of DLBP is NP-complete in the strong sense.

DLBP has also been shown to be NP-hard [36]. It is easy to see that the bin-packing problem is also similar to DLBP. The bin-packing problem is NP-hard in the strong sense (it includes 3-partition as a special case), which indicates that there is little hope of finding even a pseudopolynomial time optimization algorithm for this problem. To solve an exponential time problem, Papadimitriou and Steiglitz [43] propose that one of the following approaches be used:

Approximation: Application of an algorithm that quickly finds a suboptimal solution, which is within a certain (known) range or percentage of the optimal solution

Probabilistic: Application of an algorithm that provably yields good average runtime behavior for a distribution of the problem instances

Special cases: Application of an algorithm that is provably fast if the problem instances belong to a certain special case

Exponential: Strictly speaking, pseudopolynomial time algorithms are exponential; also, many effective search techniques (e.g., branch-and-bound, simplex) have exponential worst-case complexity

Local search: Recognized to be one of the most effective search techniques for hard combinatorial optimization problems, local (or neighborhood) search is the discrete analog of hill climbing

Heuristic: Application of an algorithm that works reasonably well in many cases but cannot be proven to be always fast or optimal

In this chapter, several promising heuristic techniques are demonstrated using some of the DLBP instances found in the current literature.

6.8 Graphical Representation

DLBP can be represented by a directed graph (digraph). The DLBP digraph is strongly connected when there are no precedence constraints and weakly connected otherwise. A modified digraph-based disassembly diagram is effective for displaying the additional and the relevant information in a disassembly problem's precedence diagram [34]. This modified diagram is based on the familiar format found in assembly line balancing problems (as depicted in Ref. 10; see Ref. 27 for other representations), which consists of a circular symbol (the *vertex*) with the task identification number at the center and the part removal time listed outside and above. This format is selectively enhanced to make it more applicable to disassembly problems. The part removal time is kept outside the vertex but in the upper right corner.

Working around in a clockwise manner, the additional considerations found in disassembly are listed using, as a default, the order of priority as given in this chapter that is, "H" (only if the part is labeled as hazardous, else blank), the part's demand, and the part's removal direction (Figure 6.1).

This representation provides a reference to all of a given problem's disassembly-unique quantitative data and concentrates this data in one location while still enabling all of the graphical relationship information that a traditional precedence diagram portrays. Directed edges (arcs) continue to be depicted as solid lines terminated with arrows. The arrows point (top to bottom, or left to right) toward subsequent disassembly tasks, that is, the arrows leave the predecessor parts. This format tends to be more angular, similar to an electronics schematic or a flowchart.

Another proposed modification is the use of broken lines to express OR relations by connecting them to each other. Both AND and OR relationships are depicted in Figure 6.2, that is, remove (1 AND 2) OR (2 AND 3) prior to 4

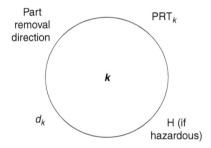

FIGURE 6.1
Proposed disassembly task/part vertex representation.

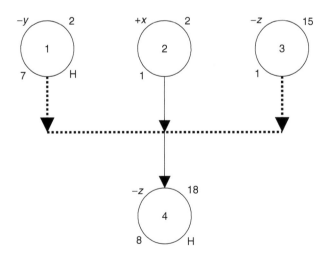

FIGURE 6.2
AND and OR example depicting the requirement to remove (1 AND 2) OR (2 AND 3) prior to 4.

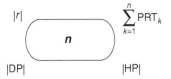

FIGURE 6.3
Disassembly precedence diagram *start* terminator box (optional).

(note that, e.g., part 1 takes 2 s to remove, is hazardous, has a demand of 7, and can only be removed in direction $-y$).

Although all parts should be included in the diagram, only parts with demands (as per Section 6.4.5, it is desirable to remove these parts and their predecessors) or parts that are predecessors to subsequently demanded parts are required to be listed when incomplete disassembly is possible. Parts without demands that do not precede other parts having demands can be deleted from the diagram to conserve space. However, all of any deleted part's additional information (PRT_k, h_k, and r_k) will no longer be available to the reader. Since, in this case, the diagram would be incomplete, the beginning and the end of the disassembly precedence diagram can optionally be depicted using flowchart-type terminator boxes, with the start box used to contain summary information about the product undergoing disassembly. The total number of parts is listed at the center and, working around the outside in a clockwise manner, the total part removal time for the entire product is listed in the upper right corner. These details are followed by the listing of the number of hazardous parts, the number of demanded parts, and the number of part removal directions; see Figure 6.3.

As in the case of the set of hazardous parts, the set of demanded parts is defined as

$$DP = \{k : d_k \neq 0 \; \forall \; k \in P\} \tag{6.32}$$

6.9 Case Study Instances

Since DLBP is a recent problem, there are very few instances that can be studied to find out about the performance of different heuristic solutions. The four instances that have been used in this chapter are the PC problem instance, the 10-part problem instance, the cell phone problem instance, and a variable-size, known-solution benchmark set of instances.

6.9.1 Personal Computer Problem Instance

Gungor and Gupta developed the PC problem instance on the basis of a study they conducted [19]. This practical and relevant example from the literature

TABLE 6.1

Knowledge Base of the PC Example

Task	Part Removal Description	Time	Hazardous	Demand	Direction
1	PC top cover	14	No	360	$-x$
2	Floppy drive	10	No	500	$+x$
3	Hard drive	12	No	620	$-x$
4	Back plane	18	No	480	$\pm x, \pm y$
5	PCI cards	23	No	540	$+y$
6	RAM modules (2)	16	No	750	$+z$
7	Power supply	20	Yes	295	$\pm x, \pm y$
8	Motherboard	36	No	720	$+z$

consists of the data for the disassembly of a PC as shown in Table 6.1. The objective is to completely disassemble a PC consisting of eight subassemblies on a paced disassembly line operating at a speed that allows 40 s for each workstation to perform its required disassembly tasks (see Figure 6.4; note the dotted lines as described in Section 6.8, which represent the OR relationship in the product; e.g., remove part (2 OR 3) prior to part 6).

6.9.2 Ten-Part Problem Instance

Kongar and Gupta provided the basis for the 10-part DLBP instance [25]. McGovern and Gupta then modified this instance from its original use in disassembly sequencing and augmented it with disassembly line-specific attributes [35]. Here the objective is to completely disassemble a notional product (see Figure 6.5) consisting of $n = 10$ components and several precedence relationships (e.g., parts 5 AND 6 need to be removed prior to part 7). The problem and its data were modified for use on a paced disassembly line operating at a speed that allows $CT = 40$ s for each workstation to perform its required disassembly tasks. This example consists of the data for the disassembly of a product as shown in Table 6.2.

6.9.3 Cell Phone Problem Instance

The increasing use of cellular phones and the rapid changes in the usage, technology, and features of these phones have prompted the entry of new models into the market on a regular basis. Unwanted cell phones typically end up in landfills and contain numerous hazardous parts, including mercury, cadmium, lead, gallium arsenide, and beryllium, any of which can pose a threat to the environment. McGovern and Gupta [35,37] selected a 2001 model year Samsung SCH-3500 cell phone for disassembly analysis. The result is an appropriate, real-world instance consisting of $n = 25$ components having several precedence relationships. The data set includes a paced

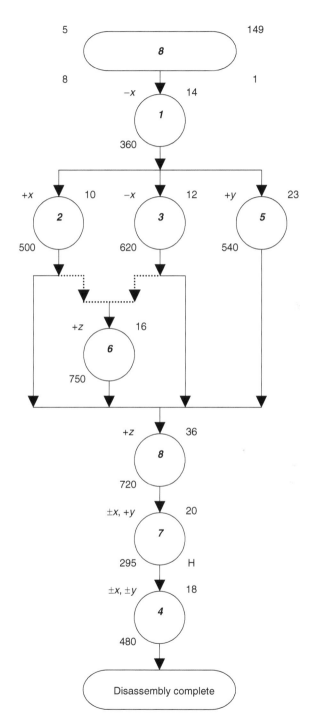

FIGURE 6.4
Personal computer precedence relationships.

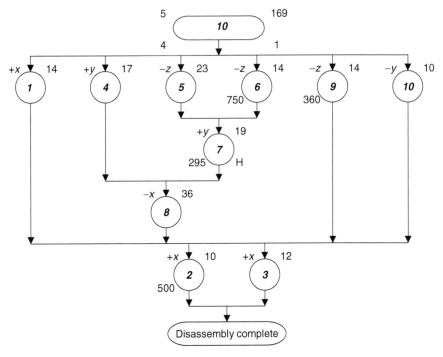

FIGURE 6.5
Ten-part product precedence relationships.

TABLE 6.2

Knowledge Base of the 10-Part Problem Instance

Task	Time	Hazardous	Demand	Direction
1	14	No	No	$+y$
2	10	No	500	$+x$
3	12	No	No	$+x$
4	17	No	No	$+y$
5	23	No	No	$-z$
6	14	No	750	$-z$
7	19	Yes	295	$+y$
8	36	No	No	$-x$
9	14	No	360	$-z$
10	10	No	No	$-y$

TABLE 6.3

Knowledge Base of the Cell Phone Problem Instance

Task	Part Removal Description	Time	Hazardous	Demand	Direction
1	Antenna	3	Yes	4	$+y$
2	Battery	2	Yes	7	$-y$
3	Antenna guide	3	No	1	$-z$
4	Bolt (type 1) a	10	No	1	$-z$
5	Bolt (type 1) b	10	No	1	$-z$
6	Bolt (type 2) 1	15	No	1	$-z$
7	Bolt (type 2) 2	15	No	1	$-z$
8	Bolt (type 2) 3	15	No	1	$-z$
9	Bolt (type 2) 4	15	No	1	$-z$
10	Clip	2	No	2	$+z$
11	Rubber seal	2	No	1	$+z$
12	Speaker	2	Yes	4	$+z$
13	White cable	2	No	1	$-z$
14	Red/blue cable	2	No	1	$+y$
15	Orange cable	2	No	1	$+x$
16	Metal top	2	No	1	$+y$
17	Front cover	2	No	2	$+z$
18	Back cover	3	No	2	$-z$
19	Circuit board	18	Yes	8	$-z$
20	Plastic screen	5	No	1	$+z$
21	Keyboard	1	No	4	$+z$
22	LCD	5	No	6	$+z$
23	Subkeyboard	15	Yes	7	$+z$
24	Internal IC	2	No	1	$+z$
25	Microphone	2	Yes	4	$+z$

disassembly line operating at a speed that allows CT = 18 s per workstation. Table 6.3 lists out the data collected on the SCH-3500. Demand was estimated on the basis of part value or recycling value; part removal times and precedence relationships (Figure 6.6) were determined experimentally. Part removal times were repeatedly collected until a consistent part removal performance was attained.

6.9.4 DLBP *A Priori* Optimal Solution Benchmark

Any DLBP solution methodology needs to be applied to a collection of test cases to demonstrate its performance as well as its limitations. In addition, new methodologies must first have their developed software thoroughly tested by undergoing a verification and validation (V&V) process. Verification consists of providing a wide range of inputs to a module of the software

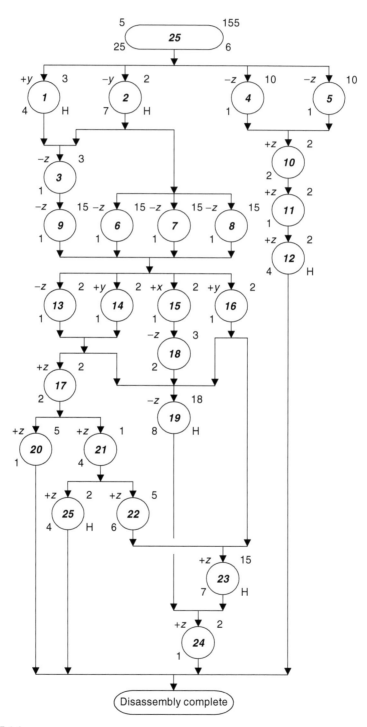

FIGURE 6.6
Cell phone precedence relationships.

to ensure proper operation of an individual software component, while validation determines whether or not the program as a whole provides a correct output for a given input, necessitating (in the case of DLBP and similar problems) varying-size data sets that have known optimal results.

Benchmark data sets are common for many NP-complete problems such as *Oliver30* and *RY48P* that are applied to the traveling salesman problem (TSP) and *Nugent15/20/30, Elshafei19*, and *Krarup30* that are applied to the quadratic assignment problem. Unfortunately, because of their size and their design, most of these existing data sets have no known optimal answer and new solutions are not compared to the optimal solution but to the best-known solution to date. In addition, since DLBP is a recently defined problem, no appropriate benchmark data sets exist. It was necessary to develop a set of instances for the DLBP to evaluate DLBP heuristics. We propose a known optimal solution benchmark line balance data set to determine efficacy (a method's effectiveness in finding good solutions).

This size-independent *a priori* benchmark data set was generated [35] on the basis of the following. In general, solutions to larger and larger instances cannot be verified as optimal (because of the time complexity in exhaustive search); therefore, it is proposed that instances be generated in such a way so as to always provide a known solution. This was done by using part removal times (PRT_k) consisting exclusively of prime numbers further selected to ensure that no combinations of these part removal times allowed for any equal summations (to reduce the number of possible optimal solutions). For example, part removal times 1, 3, 5, and 7 and CT = 16 would have minimum idle time solutions of not only one 1, one 3, one 5, and one 7 at each workstation, but various additional combinations of these as well since $1 + 7 = 3 + 5 = 1/2$ CT. In this example, workstations could contain optimal permutations from three different sets (e.g., {1, 1, 7, 7}, {3, 3, 5, 5}, and {1, 3, 5, 7}) rather than just one, dramatically increasing the number of possible optimal solutions. Subsequently, the chosen instances were made up of parts with removal times of 3, 5, 7, and 11 and CT = 26. As a result, the optimal balance for all subsequent instances would consist of a perfect balance of precedence-preserving combinations of 3, 5, 7, and 11 at each workstation with idle times of zero. This *a priori* data set is then constrained by

$$n = x \cdot |\text{PRT}| : x \in Z^+ \tag{6.33}$$

To further complicate the data (i.e., provide a large, feasible search space), only one part was listed as hazardous, and this was one of the parts that had the largest part removal time (the last one listed in the original data). In addition, one part (the last listed, second-largest part removal time component) was listed as demanded. This approach was followed so that only the hazardous and the demand sequencing would be demonstrated while providing a slight solution sequence disadvantage to any purely greedy methodology (since two parts with part removal times of 3 and 5 are needed along with the larger part removal time parts to reach F^*, assigning hazardous and demanded parts to those smaller part removal time parts may enable some methodologies to artificially obtain the initial F^* single workstation sequence). From each part removal time size, the first listed part was selected to have a removal direction differing from the other parts with the same part removal time. This approach was followed to demonstrate direction selection while requiring any solution-generating methodology to move these first parts of each part removal time size encountered to the end of the sequence (i.e., into the last workstation) to obtain the optimal direction value of $R^* = 1$ (i.e., if the solution technique being evaluated is able to successfully place the hazardous and demanded parts toward the front of the sequence). No precedence constraints were placed on the sequence, a deletion that further challenges any method's ability to attain an optimal solution. Known optimal results include $F^* = 0$, $H^* = 1$, $D^* = 2$, $R^* = 1$. While this chapter makes use of data with $|\mathrm{PRT}| = 4$ unique part removal times, in general, for any n parts consisting of this type of data, the following can be calculated:

$$\mathrm{NWS}^* = \frac{n}{|\mathrm{PRT}|} \tag{6.34}$$

$$\mathrm{NWS}_{\mathrm{nom}} = n \tag{6.35}$$

$$I^* = 0 \tag{6.36}$$

$$I_{\mathrm{nom}} = \frac{n \cdot \mathrm{CT} \cdot (|\mathrm{PRT}| - 1)}{|\mathrm{PRT}|} \tag{6.37}$$

$$F^* = 0 \tag{6.38}$$

with F_{nom} given by Formula 6.11. Hazard measures and values are given by

$$h_k = \begin{cases} 1, & k = n \\ 0, & \text{otherwise} \end{cases} \tag{6.39}$$

$$H^* = 1 \tag{6.40}$$

$$H_{\mathrm{nom}} = n \tag{6.41}$$

with demand measures and values given by

$$d_k = \begin{cases} 1, & k = \dfrac{n \cdot (|\mathrm{PRT}| - 1)}{|\mathrm{PRT}|} \\ 0, & \text{otherwise} \end{cases} \tag{6.42}$$

$$D^* = \begin{cases} 2, & H = 1 \\ 1, & \text{otherwise} \end{cases} \tag{6.43}$$

$$D_{\mathrm{nom}} = \begin{cases} n - 1, & H = n \\ n, & \text{otherwise} \end{cases} \tag{6.44}$$

and part removal direction measures and values given by

$$r_k = \begin{cases} 1, & k = 1, \dfrac{n}{|PRT|} + 1, \dfrac{2n}{|PRT|} + 1, \dots, \dfrac{(|PRT| - 1) \cdot n}{|PRT|} + 1 \\ 0, & \text{otherwise} \end{cases} \tag{6.45}$$

$$R^* = 1 \tag{6.46}$$

$$R_{\text{nom}} = \begin{cases} 0, & n = |PRT| \\ 2 \cdot |PRT| - 1, & n = 2 \cdot |PRT| \\ 2 \cdot |PRT|, & \text{otherwise} \end{cases} \tag{6.47}$$

Because $|PRT| = 4$ in this chapter, each part removal time is generated by

$$PRT_k = \begin{cases} 3, & 0 < k \le \dfrac{n}{4} \\ 5, & \dfrac{n}{4} < k \le \dfrac{n}{2} \\ 7, & \dfrac{n}{2} < k \le \dfrac{3n}{4} \\ 11, & \dfrac{3n}{4} < k \le n \end{cases} \tag{6.48}$$

Although the demand values as generated by Formula 6.42 are the preferred representation (because the resulting small numerical values make it easy to interpret demand efficacy as $D = k$), algorithms that allow incomplete disassembly may terminate after placing the single demanded part in the solution sequence. In this case, Formulae 6.42 through 6.44 may be modified to give

$$d_k = \begin{cases} 2, & k = \dfrac{n \cdot (|PRT| - 1)}{|PRT|} \\ 1, & \text{otherwise} \end{cases} \tag{6.49}$$

$$D^* = \begin{cases} 2 + \sum\limits_{p=1}^{n} p, & H = 1 \\ 1 + \sum\limits_{p=1}^{n} p, & \text{otherwise} \end{cases} \tag{6.50}$$

$$D_{\text{nom}} = \begin{cases} n - 1 + \sum\limits_{p=1}^{n} p, & H = n \\ n + \sum\limits_{p=1}^{n} p, & \text{otherwise} \end{cases} \tag{6.51}$$

TABLE 6.4

Number of Possible Entries in Each Element Position Resulting in Perfect Balance Using the DLBP *A Priori* Data with $n = 12$ and $|PRT| = 4$

K	1	2	3	4	5	6	7	8	9	10	11	12
Count	12	9	6	3	8	6	4	2	4	3	2	1

Note that a data set containing parts with equal PRTs and no precedence constraints will have more than one optimal solution. To properly gauge the performance of any solution-generating technique on the DLBP *a priori* data, the size of the optimal solution set needs to be quantified. From probability theory we know that, for example, with $n = 12$ and $|PRT| = 4$, the size of the set of optimally balanced solutions $|F^*|$ when using the DLBP *a priori* data could be calculated as $(12 \cdot 9 \cdot 6 \cdot 3) \cdot (8 \cdot 6 \cdot 4 \cdot 2) \cdot (4 \cdot 3 \cdot 2 \cdot 1) = 17{,}915{,}904$ from Table 6.4.

Grouping these counts by workstation and reversing their ordering makes it easier to recognize a pattern.

$$(1 \quad 2 \quad 3 \quad 4)$$
$$(2 \quad 4 \quad 6 \quad 8)$$
$$(3 \quad 6 \quad 9 \quad 12)$$

It can be seen that the first row can be generalized as $(1 \cdot 2 \cdot 3 \cdots \cdot |PRT|)$, the second as $(2 \cdot 4 \cdot 6 \cdots \cdot 2 \cdot |PRT|)$, and the third as $(3 \cdot 6 \cdot 9 \cdots \cdot 3 \cdot |PRT|)$. Expanding in this way, the number of optimally balanced solutions can be written as

$$|F^*| = (1 \cdot 2 \cdot 3 \cdots \cdot (1 \cdot |PRT|)) \cdot (2 \cdot 4 \cdot 6 \cdots \cdot (2|PRT|)) \cdots$$
$$\cdot \left(\frac{n}{|PRT|} \cdot \frac{2n}{|PRT|} \cdot \frac{3n}{|PRT|} \cdots n \right)$$

This can be written as

$$|F^*| = \prod_{x=1}^{|PRT|} x \cdot \prod_{x=1}^{|PRT|} 2x \cdot \prod_{x=1}^{|PRT|} 3x \cdots \prod_{x=1}^{|PRT|} \frac{n}{|PRT|} \cdot x$$

and finally as

$$|F^*| = \prod_{x=1}^{|PRT|} x^{\frac{n}{|PRT|}} \cdot \prod_{y=1}^{\frac{n}{|PRT|}} y$$

or

$$|F^*| = \prod_{x=1}^{|PRT|} \prod_{y=1}^{\frac{n}{|PRT|}} x^{\frac{n}{|PRT|}} \cdot y \tag{6.52}$$

Because $|PRT| = 4$ in this chapter, Formula 6.52 becomes

$$|F^*| = \prod_{x=1}^{4} \prod_{y=1}^{\frac{n}{4}} x^{\frac{n}{4}} \cdot y \tag{6.53}$$

In our example with $n = 12$ and $|\text{PRT}| = 4$, Formula 6.53 is solved as

$$|F^*| = \prod_{x=1}^{4} \prod_{y=1}^{\frac{12}{4}} x^{12/4} \cdot y = \prod_{x=1}^{4} \prod_{y=1}^{3} x^3 \cdot y = \prod_{x=1}^{4} x^3 \cdot (1 \cdot 2 \cdot 3)$$

or

$$|F^*| = 1 \cdot 1 \cdot 1 \cdot (1 \cdot 2 \cdot 3) \cdot 2 \cdot 2 \cdot 2 \cdot (1 \cdot 2 \cdot 3) \cdot 3 \cdot 3 \cdot 3 \cdot (1 \cdot 2 \cdot 3) \cdot 4 \cdot 4 \cdot 4 \cdot (1 \cdot 2 \cdot 3)$$

which, when rearranged, can be written as the more familiar

$$|F^*| = (12 \cdot 9 \cdot 6 \cdot 3) \cdot (8 \cdot 6 \cdot 4 \cdot 2) \cdot (4 \cdot 3 \cdot 2 \cdot 1) = 17{,}915{,}904$$

Even when all objectives are considered, there still exist multiple optimal solutions, again due to the use of a data set containing parts with equal PRTs and no precedence constraints. Using probability theory with the example having $n = 12$ and $|\text{PRT}| = 4$, it is known that the size of the set of solutions optimal in F, H, D, and R, $|F^* \cap H^* \cap D^* \cap R^*|$, when using the DLBP *a priori* data can be calculated as $(1 \cdot 1 \cdot 6 \cdot 3) \cdot (4 \cdot 3 \cdot 2 \cdot 1) \cdot (4 \cdot 3 \cdot 2 \cdot 1) = 10{,}368$ from Table 6.5.

Repeating the technique of grouping these counts by workstation and reversing their ordering again reveals a pattern:

$$(1 \quad 2 \quad 3 \quad 4)$$
$$(1 \quad 2 \quad 3 \quad 4)$$
$$(3 \quad 6 \quad 1 \quad 1)$$

The middle elements will always be the same as those given by Formula 6.52 but with two fewer sets (due to different first and last workstation elements). The first row is always $(1 \cdot 2 \cdot 3 \cdot \cdots \cdot |\text{PRT}|)$ since the directional elements in the *a priori* data should always be together at the end (the beginning in this case since we reversed the sequence for readability) of any optimal solution sequence. The last row (again, reversed) is always $((n/|\text{PRT}|) \cdot (2n/|\text{PRT}|) \cdot (3n/|\text{PRT}|) \cdot \cdots \cdot ((|\text{PRT}| - 2) \cdot n/|\text{PRT}|) \cdot 1 \cdot 1)$ since there is only one hazardous part (optimal element position $k = 1$) and only one demanded part

TABLE 6.5

Number of Possible Entries in Each Element Position Resulting in Optimal in F, H, D, and R Using *A Priori* Data with $n = 12$ and $|\text{PRT}| = 4$

K	1	2	3	4	5	6	7	8	9	10	11	12
Count	1	1	6	3	4	3	2	1	4	3	2	1

(optimal element position $k = 2$). Combining these components, the number of fully optimal solutions can be written as

$$|F^* \cap H^* \cap D^* \cap R^*| = (1 \cdot 2 \cdot 3 \cdot \cdots \cdot |PRT|) \cdot ((1 \cdot 2 \cdot 3 \cdot \cdots \cdot (1 \cdot |PRT|))$$

$$\cdot (2 \cdot 4 \cdot 6 \cdot \cdots \cdot (2 \cdot |PRT|)) \cdot \cdots \cdot \left[1 \cdot \left(\frac{n}{|PRT|} - 2\right) \cdot 2 \cdot \left(\frac{n}{|PRT|} - 2\right) \cdot 3 \cdot \left(\frac{n}{|PRT|} - 2\right) \cdot \cdots \right.$$

$$\left. \cdot |PRT| \cdot \left(\frac{n}{|PRT|} - 2\right)\right) \cdot \left(\left(\frac{n}{|PRT|}\right)\left(\frac{2n}{|PRT|}\right) \cdot \left(\frac{3n}{|PRT|}\right) \cdot \cdots \cdot \left(\frac{(|PRT| - 2) \cdot n}{|PRT|}\right) \cdot 1 \cdot \right] 1$$

By replacing the second term with a modified version of Formula 6.52 and simplifying the first and third terms, the preceding formula can be written as

$$|F^* \cap H^* \cap D^* \cap R^*| = \left(\prod_{y=1}^{|PRT|} y\right) \cdot \left(\prod_{y=1}^{|PRT|} \prod_{z=1}^{\frac{n}{|PRT|} - 2} y^{\frac{n}{|PRT|} - 2} \cdot z\right) \cdot \left(\prod_{x=1}^{|PRT| - 2} \frac{n}{|PRT|} \cdot x\right)$$

Expanding the second term gives

$$|F^* \cap H^* \cap D^* \cap R^*| = \left(\prod_{y=1}^{|PRT|} y\right) \cdot \left(\prod_{y=1}^{|PRT|} y^{\frac{n}{|PRT|} - 2} \cdot \left(\prod_{z=1}^{\frac{n}{|PRT|} - 2} z\right)\right) \cdot \left(\prod_{x=1}^{|PRT| - 2} \frac{n}{|PRT|} \cdot x\right)$$

Combining the first and second terms results in

$$|F^* \cap H^* \cap D^* \cap R^*| = \left(\prod_{y=1}^{|PRT|} y^{\frac{n}{|PRT|} - 1} \left(\prod_{z=1}^{\frac{n}{|PRT|} - 2} z\right)\right) \cdot \left(\prod_{x=1}^{|PRT| - 2} \frac{n}{|PRT|} \cdot x\right)$$

or

$$|F^* \cap H^* \cap D^* \cap R^*| = \prod_{x=1}^{|PRT| - 2} \frac{nx}{|PRT|} \cdot \prod_{y=1}^{|PRT|} \prod_{z=1}^{\frac{n}{|PRT|} - 2} y^{\frac{n}{|PRT|} - 1} \cdot z$$

with a constraint that

$$\frac{n}{|PRT|} > 2$$

Alternatively, this can be written as

$$|F^* \cap H^* \cap D^* \cap R^*| = \prod_{x=1}^{|PRT| - 2} \frac{nx}{|PRT|} \cdot \prod_{y=1}^{|PRT|} \prod_{z=1}^{a} y^{\frac{n}{|PRT|} - 1} \cdot z \qquad (6.54)$$

where

$$a = \begin{cases} \dfrac{n}{|PRT|} - 2, & \text{if } \dfrac{n}{|PRT|} > 2 \\ 1, & \text{otherwise} \end{cases}$$

Since $|\text{PRT}| = 4$ in this chapter, Formula 6.54 becomes

$$|F^* \cap H^* \cap D^* \cap R^*| = \prod_{x=1}^{2} \frac{nx}{4} \cdot \prod_{y=1}^{4} \prod_{z=1}^{\frac{n}{4}-2} y^{\frac{n}{4}-1} \cdot z \qquad (6.55)$$

In our example with $n = 12$ and $|\text{PRT}| = 4$, Formula 6.55 is solved as

$$|F^* \cap H^* \cap D^* \cap R^*| = \prod_{x=1}^{2} 3x \cdot \prod_{y=1}^{4} \prod_{z=1}^{1} y^2 \cdot z$$

or

$$|F^* \cap H^* \cap D^* \cap R^*| = \prod_{x=1}^{2} 3x \cdot \prod_{y=1}^{4} y^2$$

giving

$$|F^* \cap H^* \cap D^* \cap R^*| = ((3 \cdot 1) \cdot (3 \cdot 2)) \cdot ((1 \cdot 1) \cdot (2 \cdot 2) \cdot (3 \cdot 3) \cdot (4 \cdot 4))$$

which, when rearranged, can be written as the more familiar

$$|F^* \cap H^* \cap D^* \cap R^*| = (1 \cdot 1 \cdot 6 \cdot 3) \cdot (4 \cdot 3 \cdot 2 \cdot 1) \cdot (4 \cdot 3 \cdot 2 \cdot 1) = 10{,}368$$

Although the sizes of both DLBP *a priori* optimal solution sets are quite large in these examples, they are also significantly smaller than the search space of $n! = 479{,}001{,}600$. As shown in Table 6.6, the number of solutions that are optimal in balance alone goes from 100% of n at $n = 4$, to 22.9% at $n = 8$, and to less than 1% at $n = 16$; as n grows, this percentage gets closer and closer to 0%. The number of solutions optimal in all objectives goes from less than 8.3% of n at $n = 4$, to 0.12% at $n = 8$, dropping to effectively 0% at $n = 16$; again, as n grows, the percentage of optimal solutions gets closer and closer to zero.

TABLE 6.6

Comparison of Possible Solutions to Optimal Solutions for a Given n Using the DLBP *A Priori* Data

n	$n!$	Number Optimal in Balance	Number Optimal in All	Percentage Optimal in Balance (%)	Percentage Optimal in All (%)
4	24	24	2	100.00	8.33
8	40,320	9,216	48	22.86	0.12
12	479,001,600	17,915,904	10,368	3.74	0.00
16	2.09228E+13	1.10075E+11	7,077,888	0.53	0.00

6.10   The Combinatorial Optimization Searches

Five solution techniques (exhaustive search, GA, ACO, greedy or AEHC hybrid heuristic, and the H–K general-purpose heuristic) are considered and then the four heuristics are selected for analysis and evaluation using one of the instances from Section 6.9 (although the DLBP *a priori* benchmark data are also used with exhaustive search, this use is primarily to demonstrate exponential growth rather than to enable a thorough analysis of the chosen algorithm). Sections 6.10.5.4 and 6.10.5.5 demonstrate the graphical analysis tools developed by McGovern and Gupta [38].

For consistency in software architecture and data structures, we wrote all of the search algorithms; no off-the-shelf software was used. In addition, any sections of software code that could be used by multiple programs were reused. All computer programs were written in C++ and run on a 1.6 GHz PM x86-family workstation. After engineering, each program was first investigated on a variety of test cases for V&V purposes. All of the combinatorial optimization computer software was run at least three times to obtain an average of the computation time. Solution data for use in efficacy analysis were collected a minimum of five times when methodologies possessed a probabilistic component (ACO and GA) with the results averaged to avoid reporting unusually favorable or unusually poor results.

6.10.1   Exhaustive Search

6.10.1.1   Exhaustive Search Model Description

An exhaustive search algorithm was developed to confirm the exponential time growth of exhaustive search, to provide a time complexity benchmark for the subsequent algorithm studies, and to determine the optimal solutions for any data set up to a given size (based on a reasonable search time). The exhaustive algorithm was built around a recursive function (Figure 6.7) that generated all permutations of a sequence of n numbers (given the input $q = n$ when it is first called) with each number representing a disassembly task and the order of the numbers representing the disassembly sequence. The purpose of this exhaustive algorithm is to find every subset of the data set (every permutation) and from all those subsets, find the solution set that best satisfies the performance requirements by checking against all other permutations as described in Section 6.5.

6.10.1.2   Exhaustive Search Algorithm Numerical Results

Figure 6.8 provides the plot of runtime versus instance size using the DLBP *a priori* benchmark for the exhaustive algorithm and an n^3 curve for growth comparison as well as to demonstrate how the exhaustive search runtime rapidly increases with instance size. Using the DLBP *a priori* instances, $n = 8$ averaged 0.10 s, while $n = 12$ averaged 1476.44 s (just under 25 min). At this

```
Procedure GENERATE (q){
    IF ((q = = 0) ∧ (PRECEDENCE_PASS)){

        CALC_PERFORMANCE

        IF (FIRST_PERMUTATION_GENERATED){
            SET best_solution := new_solution
        }

        IF (BETTER_SOLUTION (new_solution, best_solution)){
            SET best_solution := new_solution
        }
    }

    ∀ k∈ P{
        EXCHANGE (k, q – 1)
        GENERATE (q – 1)
        EXCHANGE (k, q – 1)
    }
    RETURN
}

Procedure EXCHANGE (x, y){
        SET temp := PSₓ
        SET PSₓ := PSᵧ
        SET PSᵧ := temp
        RETURN
}
```

FIGURE 6.7
Recursive backtracking algorithm to generate all permutations of n numbers.

rate it could be conservatively estimated (assuming a 10-fold increase for each single increment of n; this rate approximates the observed growth but is actually slower than its true $n!$ growth) that $n = 13$ would take over 4 h, $n = 14$ would take almost 2 days, $n = 15$ would run over 2 weeks, $n = 16$ almost 6 months, $n = 17$ over 4.5 years, and so forth. Faster computers and programming tricks can speed this rate up but the growth in exhaustive search is exponential nonetheless. Using the DLBP *a priori* instances in the range of $8 \leq n \leq 80$, only $n \leq 12$ sized instances were found to be solved in a reasonable amount of time using the recursive exhaustive search algorithm. The algorithm's time complexity was noted to increase with instance size; in fact, it was seen to be in proportion to $1.199(n!)$ (using the actual growth rate gives an $n = 16$ runtime of 2.04 years).

6.10.2 Genetic Algorithm

6.10.2.1 The DLBP Genetic Algorithm and Model Description

A GA (a parallel neighborhood, stochastic-directed search technique) provides an environment in which solutions continuously crossbreed, mutate, and compete with one another until they evolve into an optimal or a

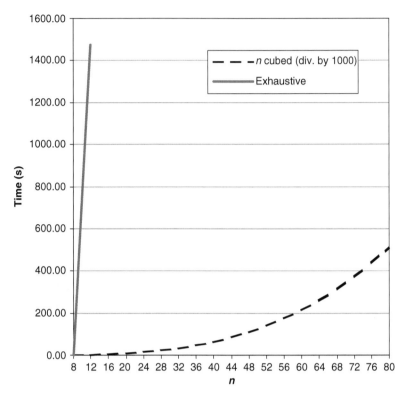

FIGURE 6.8
DLBP exhaustive search time complexity.

near-optimal solution. Owing to its structure and search method, a GA is often able to find a global solution, unlike many other heuristics that use hill climbing to find a best solution nearby, resulting only in a local optima. In addition, a GA does not need specific details about a problem, nor is the problem's structure relevant; a function can be linear, nonlinear, stochastic, combinatorial, or noisy.

GA has a solution structure, defined as a chromosome, which is made up of genes and generated by two parent chromosomes from the pool of solutions, each having its own measure of fitness. New solutions are generated from old using the techniques of crossover (sever parents genes and swap severed sections) and mutation (randomly vary genes within a chromosome). Typically, the main challenge with any GA implementation is to determine a chromosome representation that remains valid after each generation.

For DLBP, the chromosome (solution) consisted of a sequence of genes (parts). A population or a pool of size N was used. Only feasible disassembly sequences were considered as members of the population or as offspring.

The fitness (F) and the NWS were computed for each chromosome using the same method for solution performance determination as in the exhaustive search (H, D, and R calculations are not listed in the interest of space; see Section 6.5 for multicriteria treatment and Section 6.6 for the actual equations).

6.10.2.2 DLBP-Specific Genetic Algorithm Architecture

The GA for DLBP was constructed as follows [39] [32]. An initial, feasible population was randomly generated and the fitness of each chromosome in this generation was calculated. An even integer of $R_x \cdot N$ parents was randomly selected for crossover to produce $R_x \cdot N$ offspring (offspring make up $R_x \cdot N \cdot 100\%$ percent of each generation's population). An elegant crossover, the precedence preservative crossover (PPX) developed by Bierwirth et al. [3], was used to create the offspring. As shown in Figure 6.9, PPX first creates a mask (one for each child, every generation). The mask consists of random 1s and 2s, indicating which parent part information should be taken from. If, for example, the mask for child 1 reads

Parent 1:	1 3 2 6 5 8 7 4
Parent 2:	1 2 3 5 6 8 7 4
Mask:	2 2 1 2 1 1 1 2
Child:	

Parent 1:	1 3 2 6 5 8 7 4
Parent 2:	**1 2** 3 5 6 8 7 4
Mask:	2 2 1 2 1 1 1 2
Child:	1 2

Parent 1:	x **3** x 6 5 8 7 4
Parent 2:	x x 3 5 6 8 7 4
Mask:	1 2 1 1 1 2
Child:	1 2 3

Parent 1:	x x x 6 5 8 7 4
Parent 2:	x x x **5** 6 8 7 4
Mask:	2 1 1 1 2
Child:	1 2 3 5

Parent 1:	x x x **6** x **8** 7 4
Parent 2:	x x x x 6 8 7 4
Mask:	1 1 1 2
Child:	1 2 3 5 6 8 7

Parent 1:	x x x x x x x **4**
Parent 2:	x x x x x x x **4**
Mask:	2
Child:	1 2 3 5 6 8 7 4

FIGURE 6.9
PPX example.

221211, the first two parts (i.e., from left to right) in parent 2 would make up the first two genes of child 1 (and these parts would be stricken from the parts available to take from both parents 1 and 2); the first available (i.e., not stricken) part in parent 1 would make up gene three of child 1; the next available part in parent 2 would make up gene four of child 1; the last two parts in parent 1 would make up genes five and six of child 1. This technique is repeated using a new mask for child 2. After crossover, mutation is randomly conducted. Mutation was occasionally (based on the R_m value) performed by randomly selecting a single child and then exchanging two of its disassembly tasks while ensuring that precedence is preserved. The $R_x \cdot N$ least fit parents are removed by sorting the entire parent population from worst to best based on fitness. Since the GA saves the best parents from generation to generation and it is possible for duplicates of a solution to be formed using PPX, the solution set could contain multiple copies of the same answer resulting in the algorithm potentially becoming trapped in a local optima. This becomes more likely in a GA with solution constraints (such as precedence requirements) and small populations, both of which are seen in this chapter. To avoid this, DLBP GA was modified to treat duplicate solutions as if they had the worst fitness performance (highest numerical value), relegating them to replacement in the next generation. With this new ordering, the best unique $(1 - R_x) \cdot N$ parents were kept along with all of the $R_x \cdot N$ offspring to make up the next generation and then the process was repeated. To again avoid becoming trapped in local optima, DLBP GA—as with many combinatorial optimization techniques—was run not until a desired level of performance was reached but rather for as many generations as deemed acceptable by the user. Since DLBP GA always keeps the best solutions thus far from generation to generation, there is no risk of solution drift or bias, and the possibility of mutation allows for a diverse range of possible solution space visits over time.

6.10.2.3 DLBP-Specific Genetic Algorithm Qualitative Modifications

DLBP GA was modified from a general GA in several ways. Instead of the worst portion of the population being selected for crossover, in the DLBP GA the entire population was (randomly) considered for crossover. This better enables the selection of nearby solutions (i.e., solutions similar to the best solutions to date) common in many scheduling problems. Also, mutation was performed only on the children and not on the worst parents. This approach was followed to address the small population used in DLBP GA and to counter PPX's tendency to duplicate parents. Finally, duplicate children are sorted to make their deletion from the population likely since there is a tendency for the creation of duplicate solutions (due to PPX) and because of the small population saved from generation to generation.

6.10.2.4 DLBP-Specific Genetic Algorithm Quantitative Modifications

A small population was used (20 versus the more typical 10,000 to 100,000) to minimize data storage requirements and simplify analysis (except for the case study where the generations were varied) while a large number of generations were used (10,000 versus the more typical 10 to 1000) to compensate for this small population while not being so large as to take an excessive amount of processing time. This was also done to avoid solving all cases to optimality since it was desirable to determine the point at which the DLBP GA performance begins to break down and how that breakdown manifests itself. Lower than the recommended 90% [26], a 60% crossover was selected based on test and analysis. The 60% crossover provided better solutions and did so with one-third less processing time. Previous assembly line balancing literature that indicated best results have typically been found with crossover rates ranging from 0.5 to 0.7 and which also substantiated the selection of this lower crossover rate. A mutation was performed about 1% of the time. Although some texts recommend 0.01% mutation while applications in journal papers have used as much as 100% mutation, it was found that 1.0% gave excellent algorithm performance for the DLBP.

6.10.2.5 DLBP GA Numerical Results Using the Personal Computer Case Study Instance

DLBP GA was used to solve the PC instance. Since one purpose of this exercise was to demonstrate DLBP GA performance with changes in generation size, the data was run varying the generations and without the use of hazard, demand, or direction data to avoid potential complications in the analysis. Also owing to GA's stochastic nature, it was run a minimum of five times and the results were averaged to avoid reporting unusually favorable or unusually poor results.

The GA converged to moderately good solutions very quickly. It was able to find at least one of the four solutions optimal in F (Table 6.7) after no more than 10 generations. One hundred generations consistently provided

TABLE 6.7

The Four Disassembly Sequence Solutions Optimal in F

a)

Part removal sequence	Workstation 1	Workstation 2	Workstation 3	Workstation 4
1	14			
5	23			
3		12		
6		16		
2		10		
8			36	
7				20
4				18
Total time	37	38	36	38
Idle time	3	2	4	2

b)

Part removal sequence	Workstation 1	Workstation 2	Workstation 3	Workstation 4
1	14			
5	23			
3		12		
2		10		
6		16		
8			36	
7				20
4				18
Total time	37	38	36	38
Idle time	3	2	4	2

c)

Part removal sequence	Workstation 1	Workstation 2	Workstation 3	Workstation 4
1	14			
5	23			
2		10		
6		16		
3		12		
8			36	
7				20
4				18
Total time	37	38	36	38
Idle time	3	2	4	2

d)

Part removal sequence	Workstation 1	Workstation 2	Workstation 3	Workstation 4
1	14			
5	23			
2		10		
3		12		
6		16		
8			36	
7				20
4				18
Total time	37	38	36	38
Idle time	3	2	4	2

Time to remove part (in seconds)

three to four of the *F*-optimal solutions, while 1000 generations almost always resulted in the generation of all four optimal solutions. Similarly, the speed for the C++ implemented program searching 1000 generations of 20 chromosomes was just over 1 s each on the previously described workstation.

The balancing aspect of Formula 6.4 worked extremely well with four equivalent (and, in fact, *F*-optimal) solutions being found. All four best solutions consisted of a total of four workstations with 2 to 4 s idle time per workstation. Therefore, the obtained solutions were optimal in the number of workstations and also provided idle times at each workstation of at least 5% but not more than 10% of the total disassembly time of 40 s allocated to each workstation. Note that the search space is 8! or 40,320.

6.10.3   The Ant Colony Optimization Metaheuristic

6.10.3.1   *Ant Colony Optimization Model Description*

ACO is a probabilistic evolutionary algorithm based on a distributed autocatalytic (i.e., positive feedback [9]) process (i.e., a process that reinforces itself in a way that causes very rapid convergence [9]) that makes use of agents called ants (due to these agents' similar attributes to insects). Just as a colony of ants can find the shortest distance to a food source, these ACO agents work cooperatively toward an optimal problem solution. Multiple agents are placed at multiple starting nodes such as cities for the traveling salesman problem (or parts for the DLBP). Each of the m ants is allowed to visit all remaining (unvisited) edges as indicated by a Tabu-type list. Each ant's possible subsequent steps (i.e., from a node p to node q giving edge pq, vertices and arcs in DLBP) are evaluated for desirability and each is assigned a proportionate probability as shown in Formula 6.56 (shown modified for DLBP as described in the next section). The next step in the tour is randomly selected for each ant on the basis of these probabilities. After completing an entire tour (i.e., route or path) from beginning to end, all ants that take a feasible tour are given the mathematical equivalent of additional pheromone (also referred to as "trail" in ACO). The amount of pheromone is proportionate to the tour's desirability (i.e., ants that find better tours receive a larger amount of pheromone). This pheromone is then added to each step that each feasible ant has taken on its individual tour. All paths are then decreased in their pheromone strength according to a measure of evaporation (equal to 1 $- \rho$). This process is repeated for NC_{max} or until stagnation behavior (where all ants make the same tour) is demonstrated. The ant system or ant-cycle model has been claimed to run in $O(NC \cdot n^3)$ [9].

6.10.3.2   *DLBP-Specific Qualitative Modifications and the DLBP ACO algorithm*

The DLBP ACO is a modified ant-cycle algorithm [35]. The DLBP ACO is designed around the DLBP problem by accounting for feasibility constraints and addressing multiple objectives. In DLBP ACO, each part is a vertex on the tour with the number of ants being set equal to the number of parts and having one ant uniquely on each part as the starting position. Each ant is allowed to visit all parts not already in the solution. (In the sequence example solution from Section 6.5—the n-tuple $\langle 5, 2, 8, 1, 4, 7, 6, 3 \rangle$—at time $t = 3$, component p would represent component 8 while component $q \in \{1, 4, 7, 6, 3\}$ and possible partial solution sequences $\langle 5, 2, 8, 1 \rangle$, $\langle 5, 2, 8, 4 \rangle$, $\langle 5, 2, 8, 7 \rangle$, $\langle 5, 2, 8, 6 \rangle$, and $\langle 5, 2, 8, 3 \rangle$ would be evaluated for feasibility and balance.) Each ant's possible subsequent steps are evaluated for feasibility and the measure of balance then assigned a proportionate probability as given by Formula 6.56. Infeasible steps receive a probability of zero and any ants having only infeasible subsequent task options are effectively ignored for the remainder of the cycle. The best solution found in each cycle is saved if it is better than the best found in all the cycles thus far (or if it is the first cycle to initialize the best solution measures for later comparison), and the process is repeated for a maximum designated number of cycles; this process does not terminate for stagnation to enable potential better solutions (due to the ACO's stochastic edge selection capabilities). The best solution found is evaluated as described in Section 6.5.

Owing to the nature of the DLBP (i.e., whether or not a task should be added to a workstation depends on that workstation's idle time and precedence constraints), the probability in Formula 6.56 is not calculated once, at the beginning at each tour as is typical with ACO, but is calculated dynamically, generating new probabilities at each increment of t. The ACO probability calculation formula is modified for DLBP with the probability of ant r taking arc pq at time t during cycle NC calculated as

$$
p_{pq}^r (t) = \begin{cases} \dfrac{[\tau_{pq} (NC)]^\alpha \cdot [\eta_{pq} (t)]^\beta}{\displaystyle\sum_{r \in allowed_r} [\tau_{pr}(NC)]^\alpha \cdot [\eta_{pr}(t)]^\beta} & \text{if } q \in \text{allowed} \\ 0 & \text{otherwise} \end{cases} \tag{6.56}
$$

Also modified for DLBP, the amount of trail on each arc is calculated after each cycle using

$$
\tau_{pq}(NC + 1) = p \cdot \tau_{pq}(NC) + \Delta \tau_{pq} \tag{6.57}
$$

Where (unchanged from the general ACO formulation)

$$
\Delta \tau_{pq} = \sum_{r=1}^{m} \Delta \tau_{pq}^r \tag{6.58}
$$

and (also unchanged, other than the use of arcs versus edges and the recognition that arc $pq \neq qp$ in DLBP—further discussed at the end of this section)

$$
\Delta \tau_{pq}^r = \begin{cases} \dfrac{Q}{L_r} & \text{arc}(p, q) \text{used} \\ 0 & \text{otherwise} \end{cases} \tag{6.59}
$$

As in other DLBP combinatorial optimization methodologies, DLBP ACO seeks to preserve precedence while not exceeding CT in any workstation.

As long as the balance is at least maintained, the DLBP ACO then seeks improvements in hazard measure, demand measure, and direction measure but never at the expense of precedence constraints. For DLBP ACO, the visibility, η_{pq}, is also calculated dynamically at each increment of t during each cycle and is defined as

$$\eta_{pq}(t) = \frac{1}{F_{tr}} \qquad (6.60)$$

where F_{tr} is the balance of ant r at time t (i.e., ant r's balance thus far in its incomplete solution sequence generation). The divisor for the change in trail is defined for DLBP ACO as

$$L_r = F_{nr} \qquad (6.61)$$

where a small final value for ant r's measure of balance at time n (i.e., at the end of the tour) provides a large measure of trail added to each arc. Though L_r and η_{pq} are related in this application, this is not unusual for ACO applications in general; for example, in the traveling salesman problem, L_r is the tour length while η_{pq} is the reciprocal of the distance between cities p and q [9]. However, this method of selecting η_{pq} (effectively a short-term greedy choice) may not always translate into the best long-term (i.e., final tour) solution for a complex problem like the DLBP. Also note that, although this is a multicriteria problem, only a single criterion (the measure of balance F) is used in the basic ACO calculations and trail selection. The other objectives are only considered after balance and only at the completion of each cycle, not as part of the probabilistic vertex selection. That is, an ant's tour solution is produced based on F, while at the end of each cycle the best overall solution is updated based on F, H, D, and R in that order. This process is followed because the balance is considered (for the purpose of this study) to be the overriding requirement as well as to be consistent with previous multicriteria DLBP studies by the authors. (One way to consider other criteria in the n greedy steps taken in the selection of a solution would be a weighting scheme in the probabilistic step selection.)

Three procedural changes were made to the algorithm in Dorigo et al. [9] (Figure 6.10). First, the time counter was reset to zero in each cycle. This resetting has no effect on the software's performance and was performed only for program code readability. Second, "Place the m ants on the n nodes" was moved from step 1 to step 2 (and "nodes" was changed to "vertex" in recognition of DLBP's formulation as a digraph). This change was performed so that the resetting of each ant to each vertex would be repeated in each cycle, prohibiting the potential accumulation of ants on a single (or few) ending vertex (vertices), resulting in the subsequent restarting of all (or many) of the ants from that one (or few) vertex (vertices) in successive cycles; resetting the ants ensures solution diversity. This resetting was also necessary due to the nature of the DLBP where the sequence is a critical and unique element of any solution. A part with numerous precedence constraints will

Procedure DLBP_ACO {
1. Initialize:
        SET NC := 0   {NC is the cycle counter}
        FOR every arc pq **SET** an initial value $\tau_{pq}(NC) := c$ for trail intensity and $\tau_{pq} := 0$

2. Format problem space:
        SET $t := 0$ {t is the time counter}
        Place the m ants on the n vertices
        SET $s := 1$ {s is the Tabu list index}
        FOR $r := 1$ to m **DO**
                Place the starting part of the rth ant in **Tabu**$_r(s)$

3. Repeat until Tabu lists are full   {this step will be repeated ($n-1$) times}
        SET $s := s + 1$
        SET $t := t + 1$
        FOR $r := 1$ to m **DO**
                Choose the part q to move to, with probability $p_{pq}^r(t)$ as given by Formula
                (56)   {at time t the rth ant is on part $p =$ **Tabu**$_r(s-1)$}
                Move the rth ant to the part q
                Insert part q in **Tabu**$_r(s)$

4. **FOR** $r := 1$ to m **DO**
        Move the rth ant from **Tabu**$_r(n)$ to **Tabu**$_r(1)$
        Compute F, H, D, and R for the sequence described by the rth ant
        SAVE the best solution found per procedure BETTER_SOLUTION

    FOR every arc pq
        FOR $r := 1$ to m **DO**

$$\Delta \tau_{pq}^r := \begin{cases} \dfrac{Q}{L_r} & arc\,(p,q)\,used \\ 0 & otherwise \end{cases}$$

$$\Delta \tau_{pq} := \Delta \tau_{pq} + \Delta \tau_{pq}^r$$

5. **FOR** every arc pq compute $\tau_{pq}(NC + 1)$ according to $\tau_{pq}(NC + 1) := \rho \cdot \tau_{pq}(NC) + \Delta \tau_{pq}$
        SET NC := NC + 1
        FOR every arc pq **SET** $\Delta \tau_{pq} := 0$

6. **IF** (NC < NC$_{max}$)
        THEN
                Empty all Tabu lists
                GOTO STEP 2
        ELSE
                PRINT *best_solution*
                STOP

FIGURE 6.10
DLBP ACO procedure.

typically be unable to be positioned in the front of the sequence (i.e., removed early) and will often be found toward the end. In DLBP, not resetting each ant could potentially result in a situation where, for example, all ants choose the same final part due to precedence constraints; when each ant attempts to initiate a subsequent search (i.e., the next cycle) from that last vertex, all fail due to the infeasibility of attempting to remove a final part first, effectively terminating the search in just one cycle. Finally, step 6 normally could also terminate for stagnation behavior but (as discussed in the next section) this is not desired for DLBP and has been deleted.

Finally, it can be inferred from Ref. 9 that *pq* is equivalent to *qp*. Although this is acceptable and desirable in a TSP-type problem, in DLBP, sequence is an essential element of the solution (due to precedence constraints and size constraints at each workstation). For this reason, in DLBP ACO edges *pq* and *qp* are directed (i.e., arcs) and therefore distinct and unique and as such, when trail is added to one, it is not added to the other (i.e., trail is only added traveling in one direction).

6.10.3.3 DLBP-Specific Quantitative Values

In DLBP ACO, the maximum number of cycles is set at 300 (i.e., $NC_{max} = 300$) since larger problems than those studied in this chapter were shown to reach their best solution by that count and as a result of experimentation by the authors. The process was not run until no improvements were shown but, as is the norm with many combinatorial optimization techniques, was run continuously on subsequent solutions until NC_{max} [23]. This also provided the probabilistic component of ACO an opportunity to leave a potential local minimum. Repeating the DLBP ACO method in this way provides improved balance over time. As per the best ACO meta-heuristic performance experimentally determined by Dorigo et al. [9], the weight of pheromone in path selection α was set equal to 1.00; the weight of balance in path selection β was set equal to 5.00; the evaporation rate, $1 - \rho$ was set to 0.50; and the amount of pheromone added if a path is selected Q was set to 100.00. The initial amount of pheromone on all of the paths c was set equal to 1.00.

6.10.3.4 DLBP ACO Numerical Results Using
the 10-Part Case Study Instance

Over multiple runs, DLBP ACO was able to successfully find several (due to the stochastic nature of ACO) feasible solutions to the 10-part instance. Table 6.8 depicts a typical solution sequence. This solution demonstrates the minimum number of workstations while placing the hazardous part (part number 7) and high-demand parts (parts 6, 7, and 9) relatively early in the removal sequence (the exception being part 2, primarily due to precedence constraints and part 2 having a smaller part removal time that is not conducive to the ACO's iterative greedy process) and allowing just six direction changes (due to precedence constraints, the optimum is five while the worst case is eight). The speed for the C++ implemented program on this problem (averaged over five runs) was just over one-tenth of a second on the 1.6 GHz PM x86-family workstation.

The optimality of the NWS minimization and the balancing is demonstrated by the solution consisting of all five workstations operating at

TABLE 6.8

Typical DLBP ACO Disassembly Sequence Solution to the 10-Part Data Set

		Workstation				
		1	2	3	4	5
Part removal sequence	10	10				
	5	23				
	6		14			
	7		19			
	4			17		
	9			14		
	8				36	
	1					14
	2					10
	3					12
Total time		33	33	31	36	36
Idle time		7	7	9	4	4

Time to remove part (in seconds)

78–90% of capacity while meeting all precedence constraints in the exponentially large search space (i.e., 10! or 3, 628, 800).

6.10.4 The Greedy Algorithm and AEHC Heuristic

Two approaches are used sequentially to provide solutions to DLBP. The first rapidly provides a feasible solution to the DLBP and minimum or near-minimum NWS using a greedy algorithm. The greedy algorithm considered here is based on the first-fit-decreasing (FFD) algorithm (developed for the bin-packing problem and effectively used in computer processor scheduling). The second is implemented after the DLBP greedy algorithm to compensate for DLBP greedy's inability to balance the workstations. The AEHC heuristic ensures that the idle times at each workstation are similar. It does this by only comparing tasks assigned in adjacent workstations; this is done both to conserve search time and to only investigate swapping tasks that will most likely result in a feasible sequence.

6.10.4.1 Greedy Model Description and the Algorithm

A greedy strategy always makes the choice that looks the best at the moment. That is, it makes a locally optimal choice in the hope that this choice will lead to a globally optimal solution. Greedy algorithms do not always yield optimal solutions but for many problems, they do [5]. The DLBP greedy algorithm [33] was built around FFD rules. FFD rules require looking at each element in a list, from largest to smallest (PRT in the DLBP) and putting that element into the first workstation in which it fits without violating precedence constraints. When all of the work elements have been assigned to a workstation, the process is complete. The greedy FFD algorithm is further modified with priority rules to meet multiple objectives. The hazardous parts are prioritized to the earliest workstations, greedy ranked large removal time to small. The remaining nonhazardous parts are greedy ranked next, large removal times to small. In addition, selecting the part with the larger demand ahead of those with lesser demands breaks any ties for parts with equal part removal times and selecting the part with an equivalent part removal direction breaks any ties for parts also having equal part removal directions. These selections are undertaken to prevent damage to these more desirable parts. The DLBP greedy algorithm provides an optimal or near-optimal minimum number of workstations; the more constraints, the more likely the optimal number of workstations is found. The level of performance (minimal NWS) will generally improve with the number of precedence constraints.

The specific details for this implementation are as follows. The DLBP greedy algorithm first sorts the list of parts. The sorting is based on part removal times, whether or not the part contains hazardous materials, the subsequent demand for the removed part, and the part removal direction. Hazardous parts are put at the front of the list for selection into the solution sequence. The hazardous parts are ranked from the largest to the smallest part removal times. The same is then done for the nonhazardous parts. Any ties (i.e., two parts with equal hazard typing and equal part removal times) are not randomly broken, but rather ordered based on the demand for the part, with the higher demand part being placed earlier on the list. Any of these parts also having equal demands is then selected based on their part removal direction being the same as the previous part on the list (i.e., two parts compared during the sorting that only differ in part removal directions are swapped if they are removed in different directions—the hope being that subsequent parts and later sorts can better place parts having equal part removal directions).

Once the parts are sorted in this multicriteria manner, the parts are placed in workstations in FFD greedy order while preserving precedence. Each part in the sorted list is examined from first to last. If the part had not previously been put into the solution sequence (as described by ISS_k Tabu* list data structure), the part is put into the current workstation if idle time remains to

---

* Although the name recalls Tabu search as proposed in Refs. [13,14], the ISS_k Tabu list is similar to that used by the ant system [9] including, for example, the absence of any aspiration function.

accommodate it and as long as putting it into the sequence at that position will not violate any of its precedence constraints. ISS_k is defined as

$$ISS_k = \begin{cases} 1, & \text{assigned} \\ 0, & \text{otherwise} \end{cases} \tag{6.62}$$

If no workstation can accommodate it at the given time in the search due to precedence constraints, the part is maintained on the sorted list (i.e., its ISS_k value remains 0) and the next part (not yet selected) on the sorted list is considered. If all parts have been examined for insertion into the current workstation on the greedy solution list, a new workstation is created and the process is repeated. Figure 6.11 shows the DLBP greedy procedure.

While being very fast and generally very efficient, the FFD-based greedy algorithm is not always able to optimally minimize the number of workstations. In addition, there is no capability to balance the workstations; in fact, the FFD structure lends itself to filling the earlier workstations as much as possible, often to capacity, while later workstations end up with progressively greater and greater idle times. This results in an extremely poor balance. This limitation led to the formulation of follow-on methods to fulfill the second objective, one of which is AEHC (see McGovern and Gupta [33] for a 2-opt implementation modified for DLBP by exchanging vertices instead of edges as would be done in a TSP application).

Procedure DLBP_GREEDY {
1. **SET** $j := 0$

2. $\forall\ PRT_k\ |\ 1 \le k \le n$, generate sorted part sequence PSS based on:
$h_k := 1$ parts (i.e., hazardous), large PRT_k to small; then $h_k := 0$ parts, large PRT_k to small; if more than one part has both the same hazard rating and the same PRT_k, then sort by demand, large d_k to small; if more than one part has the same hazard rating, PRT_k, and demand rating, then sort by equal part removal direction

3. $\forall\ p \in PSS\ |\ p := PSS_k,\ 1 \le k \le n$, generate greedy part removal sequence PSG by:
    IF $ISS_p == 1$ (i.e., part p already included in PSG)$\lor$
                PRECEDENCE_FAIL$\lor$
                $I_j < PRT_p$ (i.e., part removal time required exceeds idle time available at ST_j)

    THEN   **IF** $p == n$ (i.e., at last part in sequence)

                THEN   Increment j (i.e., start new workstation)
                                Start again at $p := 1$

                ELSE   Try next p

    ELSE   Assign p to ST_j and set $ISS_p := 1$

4. Using PSG:
    Calculate F, H, D, and R
}

FIGURE 6.11
DLBP greedy procedure.

6.10.4.2 *Hill-Climbing Description and the Heuristic*

A second-phase approach to DLBP was developed to quickly provide a near-optimal and feasible balance sequence using a hill-climbing local search heuristic, AEHC [30]. AEHC was tailored to DLBP (and applicable to any bin-packing-type problem having precedence constraints) to take advantage of the knowledge about the problem's format and constraints to provide a solution that is better balanced than DLBP greedy alone but significantly faster than DLBP 2-opt [33]. Hill climbing is an iterated improvement algorithm, basically a gradient descent or ascent. It makes use of an iterative greedy strategy, which is to move in the direction of increasing value. A hill-climbing algorithm evaluates the successor states and keeps only the best one [23]. AEHC is designed to consider swapping each task in every workstation with each task only in the next adjacent workstation in search of improved balance. It does this while preserving precedence and not exceeding CT in any workstation. Only adjacent workstations are compared to enable a rapid search and therefore it is deemed unlikely that parts several workstations apart can be swapped and still preserve the precedence of all of the tasks in-between. As shown in the Figure 6.12 , a part has a limited number of other parts with which it can be considered for exchange. In the 2-opt methodology from McGovern and Gupta [33] and using the Figure 6.12 data, part number 6 would be considered for exchange by parts 1 through 5, and would consider exchanges with parts 7 through 11; that is, any part could be exchanged with any other part. In AEHC, part 6 would be considered for exchange only by parts 4 and 5, and would consider exchanges only with parts 8, 9, and 10.

The neighborhood definition and search details of AEHC are as follows. After the DLBP greedy algorithm generates a minimum-NWS, feasible solution, the AEHC heuristic is applied to improve the balance. AEHC does this by going through each task element (part) in each workstation and comparing it to each task element in the next adjacent workstation. If the two task elements can be exchanged while preserving precedence, without exceeding either workstations available idle time, and with a resulting improvement in

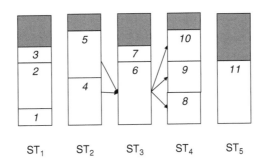

FIGURE 6.12
AEHC example.

Procedure DLBP_AEHC {

1. Initialize:
   SET PST := PSG
   SET TMP := PSG
   SET BST := PSG
   SET *START* := 1

2. $\forall j \mid 1 \le j \le$ NWS – 1, attempt to generate better balanced AEHC part removal sequence PST by performing AEHC part swap (i.e., swap each TMP$_p$ in workstation j with every TMP$_q$ in workstation $j + 1$)

   $\forall p \in$ TMP $\mid p =$ TMP$_k$,*START* $\le k \le$ *START* + NPW$_j$ – 1

   $\forall q \in$ TMP $\mid q =$ TMP$_k$,*START* + NPW$_j$ $\le k \le$ *START* + NPW$_j$ – 1+NPW$_{j+1}$

   Calculate new ST$_j$, and ST$_{j+1}$

   IF ((CT – new ST$_j \ge 0$) $\wedge$
   (CT – new ST$_{j+1} \ge 0$))

   THEN Calculate new *F*, *H*, *D*, and *R*

   IF ((BETTER_SOLUTION(TMP, BST))

   $\wedge$
   (PRECEDENCE_PASS))

   THEN SET BST := TMP

   SET *START* := *START* + NPW$_j$

3. Save results:
   SET PST := BST
   SET TMP := BST

4. Repeat STEP 2 until no more improvements in *F*, *H*, *D*, or *R* can be made
}

FIGURE 6.13
DLBP AEHC procedure.

the overall balance, the exchange is made and the resulting solution sequence is saved as the new solution sequence. This process is repeated until task elements of the last workstation have been examined. The heuristic is given in pseudocode format in Figure 6.13.

In the DLBP greedy/AEHC hybrid heuristic, the DLBP greedy process is run once to generate an initial solution. Hill climbing is typically continuously run on subsequent solutions for as long as is deemed appropriate or acceptable by the user or until it is no longer possible to improve, at which point it is assumed that the local optima has been reached [23]. Repeating the AEHC method in this way provides improved balance with each iteration. The AEHC was tested both ways; run only once after the greedy solution was generated as well as run until the local optima was obtained (a single AEHC iteration has several benefits including the observation that AEHC was seen to typically provide its largest single balance performance improvement in

TABLE 6.9

DLBP Greedy/AEHC Hybrid Heuristic Solution to the Cell Phone Instance

Part ID	1	2	4	3	6	7	8	9	16	5	10	15	18	14	13	17	20	11	12	21	25	19	22	23	24
PRT	3	2	10	3	15	15	15	15	2	10	2	2	3	2	2	2	5	2	2	1	2	18	5	15	2
Workstation	1	1	1	1	2	3	4	5	5	6	6	6	6	7	7	7	7	7	7	7	7	8	9	10	10
Hazardous	1	1	0	0	0	0	0	0	0	0	0	0	0	0	0	0	0	0	0	1	0	1	1	0	1	0
Demand	4	7	1	1	1	1	1	1	1	1	2	1	2	1	1	2	1	1	4	4	4	8	6	7	1
Direction	2	3	5	5	5	5	5	5	2	5	4	0	5	2	5	4	4	4	4	4	4	5	4	4	4

the first iteration). Additionally, a single iteration of AEHC may be recommended for the real-time solution of a very large DLBP instance on a dynamic mixed-model and mixed-product disassembly line.

6.10.4.3   *DLBP Greedy/AEHC Hybrid Heuristic Numerical Results Using the Cell Phone Case Study Instance*

Table 6.9 shows the solution generated by the DLBP greedy/AEHC hybrid heuristic. As per its design, DLBP greedy removed hazardous parts early on, though many of these parts had to wait due to precedence constraints. Four parts with larger part removal times (15 s) were removed very early on (although parts 19 and 23 take as long or longer, precedence constraints prevent earlier removal) but at the expense of using an entire workstation, demonstrating a condition that limits DLBP greedy/AEHC's effectiveness since the greedy portion works to take out large (time) items early (and therefore, potentially adjacent to each other in the sequence) and AEHC is not allowed to look out far enough to add smaller time items into the sequence to better fill those workstations. As can be seen in Figure 6.6, precedence relationships play a large role in determining the allowable solution sequences to this problem.

The DLBP greedy/AEHC heuristic hybrid was run three times to obtain an average computation time (due to the hybrid's deterministic nature, all runs generated the same solution sequence). The solution to the cell phone instance was found extremely quickly, regularly taking less than 1/100th of a second. The hybrid technique was not, however, able to generate the optimal number of workstations or the optimal balance (with the assumption that NWS = 9 and $F = 9$ is optimal [37]).

6.10.5   The H–K General-Purpose Heuristic

A new search approach has been proposed to provide a rapid, near-optimal search for combinatorial optimization applications. This heuristic rapidly provides a feasible solution using a modified exhaustive search technique to provide a data sampling of all solutions.

6.10.5.1 Heuristic Background and Motivation

Exhaustive search is optimal because it looks at every possible answer. In many physical search applications (e.g., antisubmarine warfare, search, and rescue), exhaustive search is not possible due to time or sensor limitations. In these cases, it becomes practical to sample the search space and operate under the assumption that, for example, the highest point of land found in a limited search is either the highest point in a given search area or is reasonably near the highest point. The proposed search technique [31,41] in this section works by sampling the exhaustive solution set; that is, search the solution space in a method similar to an exhaustive search, but in a pattern that skips solutions (conceptually similar to the STEP functionality in a FOR loop as found in computer programming) to significantly minimize the search space (in Figure 6.14, the shading indicates solutions visited, the border represents the search space).

This pattern is analogous to the radar acquisition search pattern known as "spiral scan," the search and rescue pattern of the "expanding square," or the antisubmarine warfare aircraft "magnetic anomaly detector (MAD) hunting circle." Once the solution is generated, the space can be further searched with additional application of the H–K metaheuristic (with modifications from the previous H–K; see the end of Section 6.10.5.2), or the best-to-date solution can be further refined by performing subsequent local searches (such as 2-opt or smaller, localized H–K searches). That is, depending on the application, H–K can be run once, multiple times on subsequent solutions, multiple times from the same starting point using different skip measures (potentially as a multiprocessor or grid computing application), multiple times from a different starting point using the same skip measures (again, potentially as a multiprocessor or grid computing application), or followed up with an H–K or another, differing local search on the best or several of the best suboptimal solutions generated. H–K can also be used as the first phase of a hybrid algorithm or to hot start another methodology (e.g., to provide the initial population in a GA). For the purpose of demonstrating the method and measuring its efficacy, H–K is run alone to a single-phase solution in this chapter. The skip size ψ can be as small as $\psi = 1$, or as large

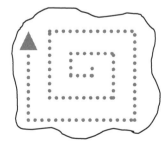

FIGURE 6.14
Exhaustive search space and the H–K search space and methodology.

as $\psi = n$. Since $\psi = 1$ is equivalent to exhaustive search, and $\psi = n$ generates a trivial solution (it returns only one solution, that being the data in the same sequence as it is given to H–K, i.e., $PS_k = \langle 1, 2, 3, \ldots, n \rangle$; also, in the single-phase H–K, this solution is already considered by any value of ψ), in general all skip values can be further constrained as

$$2 \le \psi_k \le n - 1 \tag{6.63}$$

6.10.5.2 H–K Methodology Comparison to Other Solution-Generating Methodologies

The H–K general-purpose heuristic shows a variety of similarities to many currently accepted combinatorial optimization methodologies. For example, like ACO and GA—but unlike Tabu search—H–K generates the best solutions when suboptimal solutions bear a resemblance to the optimal solution. In three dimensions, this would appear as a set with relatively shallow gradients (i.e., no sharp spikes); the more shallow the gradient, the better the chance the found solution is near the optimal solution. These data sets are effective in ACO applications since the ant agents work to reinforce good solution sections by adding trail (similar to pheromones in actual ants). They are effective in GA applications since those algorithms break apart good solutions and recombine them with other good solution sections, again, reinforcing preferred tours. The drawback to these is that if suboptimal solutions do not bear a resemblance to the optimal solution, the optimal solution may not be found. However, in many applications, it is not typical for the data to perform in this manner. Also, it is not a significant drawback since the general H–K takes a deterministic, nonweighted path; therefore, an isolated solution is as likely to be visited as the optimal shallow gradient solution. Isolated or steep gradient solutions are addressed in ACO and GA with probabilistically selected tours, allowing for potential excursions to seemingly poor-performing areas of the search space.

Like SA—and unlike ACO, GA, and Tabu—the multiple phase version of H–K looks at a large area initially, then smaller and smaller areas in more detail.

Like Tabu search (and unlike ACO and GA), the single phase of H–K does not revisit solutions found previously during the same phase of the search.

Unlike many applications of ACO, Tabu, and SA (but like GA), H–K does not generate solutions similar to a branch-and-bound type of method where the best solution is grown by making decisions on which branch to take as the sequence moves from beginning to end. Rather, an entire solution sequence is generated prior to evaluation. Similarly, H–K does not contain any greedy component since the movement through the search space is deterministic and no branching decisions are required, which are typically made based upon the best short-term (greedy) move.

As with each of these heuristics, H–K can be applied to a wide range of combinatorial optimization problems. Like greedy and r-opt [29] heuristics and traditional mathematical programming techniques such as linear programming, H–K selects solutions for consideration based on a deterministic manner, while metaheuristics such as ACO and GA have a strong probabilistic component. As a result, H–K is repeatable, providing the same solution every time it is run. However, probabilistic components can be added in the form of randomized starting point(s), skip size(s), and/or skip types.

Unlike traditional mathematical programming techniques such as linear programming, H–K is not limited to linear problem descriptions and, as shown in this chapter using the DLBP, readily lends itself to multicriteria decision making as well as nonlinear problem formats. However, like most heuristics, H–K cannot consistently provide the optimal solution, while most mathematical programming methods generally do. Also common to heuristics, there is some degree of tuning and design required by the user, for example, selection of the number of generations, mutation rate, and crossover rate and crossover method in GA, or pheromone evaporation rate, number of ants, number of cycles, and visibility description for ACO. H–K decisions include the following:

- *Number of H–K processes*: One or multiple
- *Starting point*: Constant, deterministic varying (i.e., different but known starting points for use in multiple H–K processes only), and random
- *Skip type (type of skipping between solution elements)*: Constant type, deterministic varying type, and random type
- *Skip size (size of data skipped in each solution element)*: Constant size, deterministic varying size, and random size
- *Follow-on solution refinement*: None, different H–K (different starting point, skip type, skip size, etc.), local H–K (H–K search about previous solution), and other (such as 2-Opt or GA)

Depending on these structural decisions, H–K can take on a variety of forms; from a classical optimization algorithm in its most basic form to a general evolutionary algorithm (EA) with the use of multiple H–K processes, to a biological or natural process algorithm by electing random functionality. To demonstrate the method and show some of its limitations, in this chapter the most basic form of the H–K metaheuristic is used; one process, constant starting point of $PS_k = \langle 1, 1, 1, ..., 1 \rangle$ (since the solution set is a permutation, there are no repeated items; therefore the starting point is effectively $PS_k = \langle 1, 2, 3, ..., n \rangle$), constant skip type (i.e., each element in the solution sequence is skipped in the same way), constant skip size of ψ (although different skip sizes are used throughout the chapter to allow for a thorough efficacy analysis, the skip size stays constant throughout each H–K run), and no follow-on solution refinement.

6.10.5.3   The H–K Process and DLBP Application

As far as the H–K process itself, since it is a modified exhaustive search allowing for solution sampling, it searches for solutions similar to depth-first search iteratively seeking the next permutation iteration—allowing for skips in the sequence—in lexicographic order. In the basic H–K and with $\psi = 2$, the first element in the first solution would be 1, the next element considered would be 1, but since it is already in the solution, that element would be incremented and 2 would be considered and be acceptable. This is repeated for all of the elements until the first solution is generated. In the next iteration, the initial part under consideration would be incremented by 2 and, therefore, 3 would be considered and inserted as the first element. Since 1 is not yet in the sequence, it would be placed in the second position, 2 in the third, etc. For DLBP H–K, this is further modified to test the proposed sequence part addition for precedence constraints. If all possible parts for a given solution position fail these checks, the remainder of the positions are not further inspected, the procedure falls back to the previously successful solution addition, increases it by 1, and continues. These processes are repeated until all allowed items have been visited in the first solution position (and by default, due to the nested nature of the search, all subsequent solution positions). For example, with $n = 4$, $P = \{1, 2, 3, 4\}$, and no precedence constraints, instead of considering the $4! = 24$ possible permutations, only five are considered by the single-phase H–K with $\psi = 2$ and using forward-only data: $PS_k = \langle 1, 2, 3, 4 \rangle$; $PS_k = \langle 1, 4, 2, 3 \rangle$; $PS_k = \langle 3, 1, 2, 4 \rangle$; $PS_k = \langle 3, 1, 4, 2 \rangle$; and $PS_k = \langle 3, 4, 1, 2 \rangle$. All of the parts are maintained in a Tabu-type list ISS_k defined by Formula 6.62. Each iteration of the DLBP H–K generated solution is considered for feasibility. If it is ultimately feasible in its entirety, DLBP H–K then looks at each element in the solution and places that element using the next-fit (NF) rule (from the bin-packing problem application; once a bin has no space for a given item attempted to be packed into it, that bin is never used again even though a later, smaller item may appear in the list and could fit in the bin [24]). DLBP H–K puts the element under consideration into the current workstation if it fits. If it does not fit, a new workstation is assigned and previous workstations are never again considered. Although NF does not perform as well as first-fit, best-fit, first-fit-decreasing, or best-fit-decreasing when used in the general bin-packing problem, it is only one of these rules that will work with a DLBP solution sequence due to the existence of precedence constraints (see Section 6.10.4 or McGovern and Gupta [33] for a DLBP implementation of FFD). When all of the work elements have been assigned to a workstation, the process is complete and the balance, hazard, demand, and direction measures are calculated. The best of all of the inspected solution sequences is then saved as the problem solution. Although the actual software implementation for this dissertation consisted of a very compact recursive algorithm, in the interest of clarity, the general DLBP H–K procedure is presented here as a series of nested loops (Figure 6.15).

Procedure DLBP_H–K {
    SET $ISS_k := 0 \; \forall \; k \in P$
    SET FS := 1

    $PS_1 := 1$ to n, skip by ψ_1
        SET $ISS_{PS_1} := 1$

        $PS_2 := 1$ to n, skip by ψ_2
            WHILE $(ISS_{PS_2} == 1 \lor$
                        PRECEDENCE_FAIL$\land$
                        not at n)
              Increment PS_2 by 1

            IF $ISS_{PS_2} == 1$
            THEN SET FS := 0
            ELSE SET $ISS_{PS_2} := 1$
              :
              :

            IF   FS == 1
                $PS_n := 1$ to n skip by ψ_n
                    WHILE $(ISS_{PS_n} == 1 \lor$
                            PRECEDENCE_FAIL $\land$
                            not at n)
                    Increment PS_n by 1

                IF $ISS_{PS_n} == 0$
                THEN evaluate solution PS
              :
              :

        IF   FS == 1
        THEN SET $ISS_{PS_2} := 0$
        ELSE SET FS :=1

    SET $ISS_{PS_1} := 0$
    SET FS :=1

}

FIGURE 6.15
DLBP H–K procedure.

The following section studies how the skip size affects various measures including time complexity. The general form of the skip size-to-problem size relationship is formulated as

$$\psi_k = n - \Delta\psi_k \tag{6.64}$$

While not described in Section 6.10.5.1, early tests of time complexity growth with skip size suggest another technique to be used as part of H–K search. Since any values of ψ that are larger than the chosen skip value for a given H–K problem take significantly less processing time, all larger skip values should also be considered to increase the search space at the expense of a minimal increase in search time. In other words, H–K can be run repeatedly on a given instance using all skip values from a smallest ψ (selected based on time complexity considerations) to the largest (i.e., $n - 1$ as per Formula 6.63) without a significant time penalty. In this case, any ψ_k would be constrained as

$$n - \Delta\psi_k \leq \psi_k \leq n - 1 \quad \text{where } 1 \leq \Delta\psi_k \leq n - 2 \tag{6.65}$$

If this technique is used (as it is in the next sections), it should also be noted that multiples of ψ visit the same solutions; for example, for $n = 12$ and $2 \leq \psi \leq 10$, the four solutions considered by $\psi = 10$ are also visited by $\psi = 2$ and $\psi = 5$.

6.10.5.4 *DLBP A Priori Numerical Results for Varying Skip Size*

The benchmark data set from Section 6.9.4 was then used with DLBP H–K to provide a thorough efficacy analysis. This section also demonstrates proper use of the data set for complexity studies, efficacy analysis, and graphical representation, as well as an overview of the findings with regard to the H–K general-purpose heuristic. The final configuration of the developed benchmark was: $|PRT| = 4$, 19 instances with instance size evenly distributed from $n = 8$ to 80 in steps of $|PRT|$. This provided numerous instances of predetermined, calculable solutions with the largest instance 10 times larger than the smallest instance. The size and range of the instances is considered appropriate and significant for this study, with small ns tested—which decreases the NWS and tends to exaggerate less than optimal performance—as well as large, which demonstrates time complexity growth and efficacy changes with n. To summarize, the test data consisted of $8 \leq n \leq 80$ parts with 4 unique part removal times giving PRT = {3, 5, 7, 11}. Only the last part having PRT = 11 is hazardous; only the last part having PRT = 7 is demanded. The first of each part with PRTs equal to 3, 5, 7, and 11 are removed in direction $+x$; all others are in direction $-x$. The disassembly line is paced and operated at a speed that allows 26 s (CT = 26) for each workstation. Known optimal results include $F^* = 0$, $H^* = 1$, $D^* = 2$, $R^* = 1$.

The DLBP *a priori* instances were first used to determine how skip size affects performance and time complexity, then based on these results, they were used to determine how the heuristic's performance changes with problem size. Because H–K is exceptionally deterministic and performs no preprocessing of the data, the order in which the data is presented can affect the solutions considered. For this reason, the analyses done in the following sections were first run with the data as given by Formula 6.48 and then run again, but this time with the data presented in reverse. The better of the two results were used and the search times (forward and reverse) were added. Both forward and reverse were used to ensure unusually good (or bad) results that were not artificially obtained due to some unknown structural feature of the DLBP *a priori* data sets, since presenting data in both forward and reverse formats would be expected to be the most common application. This is not necessary where an instance contains a relatively large number of precedence constraints, individual parts in an instance do not bear a great deal of resemblance to each other, or where in an actual application one may expect a user of H–K to only run the data exactly as it was presented to them. Even though not covered in this chapter, since this is an introduction to the basic concepts of H–K, note that it could be expected that better results may be achieved with multiple runs of H–K consisting of the data being presented to H–K in differing orders. (This is an effective H–K technique that is not described in Section 6.10.5.1, but one which may be the easiest to implement.)

Skip size was varied to determine if and how solution performance decreased with increases in ψ. All cases are the DLBP *a priori* data set with $n = 12$, and consideration of all skip sizes (including exhaustive search); that is, $\psi = \{1, \ldots, n\}$. This problem size was chosen because it is the largest multiple of |PRT| that could be repeatedly evaluated by exhaustive search in a reasonable amount of time ($n = 12$ took just over 20 min, while $n = 16$ is calculated to take over 2 years). This was essential since exhaustive search provides the time complexity and solution performance baseline. A full range of skip sizes was used to provide maximum and minimum performance measures, and all intermediate skip sizes were used to allow the highest level of detail for the study. Although most of the following results are not optimal, and there exists multiple optimum extreme points, this is more of a reflection of the specially designed DLBP *a priori* data set. Suboptimal solutions are not an atypical result when this data set is evaluated. The data is intended to pose challenges, even at relatively small n values, to a variety of combinatorial optimization techniques.

The first analysis done was a time complexity study. As shown in Figure 6.16, the solution time appears to grow slightly less than factorially to the inverse of ψ, so even a very small change in skip size will yield a significant speed up of any search. (The exhaustive search curve has no relationship to the H–K curve and is depicted only to provide a size and shape reference.) A more detailed view of the H–K time-to-skip size relationship can be seen in Figure 6.17. With an exponential curve (n^n) adjusted for scale and overlaid, it can be seen that the rate of growth is actually exponential to the inverse of ψ.

Next, the balance performance was measured. The measure of balance was seen to remain optimal through $\psi = 4$ (i.e., while ψ stayed within 1/3 of n). As seen in Figure 6.18, performance then decreases as skip size increases but at a relatively slow rate.

The third objective was also seen to decrease in performance with increases in skip size (see Figure 6.19). The hazard measure remains optimal through $\psi = 3$, but eventually drops to worst case at $\psi = 9$.

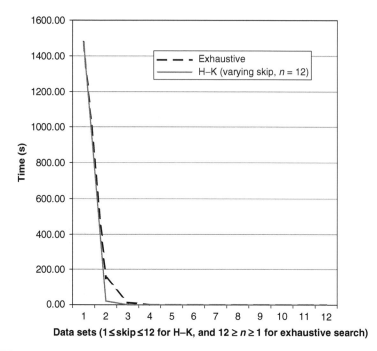

FIGURE 6.16
DLBP H–K time complexity.

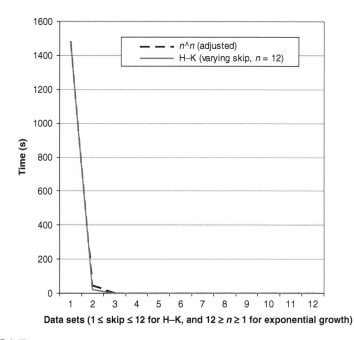

FIGURE 6.17
Detailed DLBP H–K time complexity.

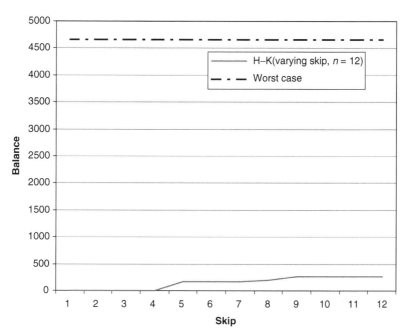

FIGURE 6.18
DLBP H–K F performance with changes in ψ.

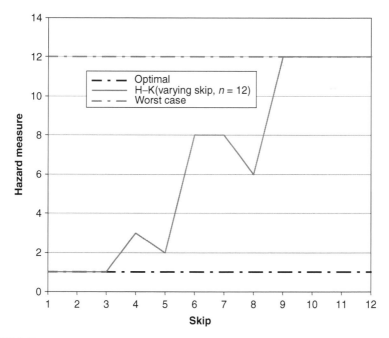

FIGURE 6.19
DLBP H–K H performance with changes in ψ.

The fourth and fifth objectives decreased in performance with increases in skip size as well, each at a slightly steeper angle (attributed to the prioritization of the objectives) (see Figures 6.20 and 6.21). Optimal and better than optimal (i.e., removing the high-demand part first in the situation where the hazardous part is not removed first) D results were seen as far out as $\psi = 8$. The part removal direction measure reaches worst case by $\psi = 6$, but then improves slightly at $\psi = 9$.

These studies show how the heuristic's performance decreases with skip size and should provide at least a starting point for the selection of skip size based on the required level of performance as considered against time complexity. In large problems, time will quickly become the overriding consideration. These and other experiments indicate that a $\Delta\psi$ ranging from 10 to 12 (resulting in a ψ equal to from 10 to 12 less than n) gives time complexity performance on par with many other metaheuristic techniques, regardless of hardware implementation.

6.10.5.5 DLBP A Priori *Numerical Results for Varying* n

Problem size was varied to determine if and how solution performance changed with increases in n. All cases are the DLBP *a priori* problem with size varying between $8 \leq n \leq 80$, $1 \leq \Delta\psi \leq 10$, and the resulting skip sizes of

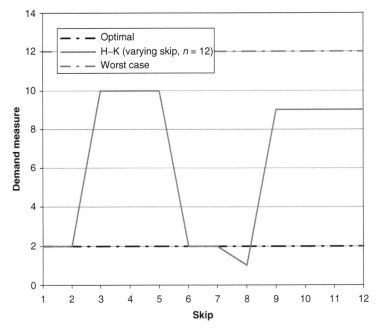

FIGURE 6.20
DLBP H–K D performance with changes in ψ.

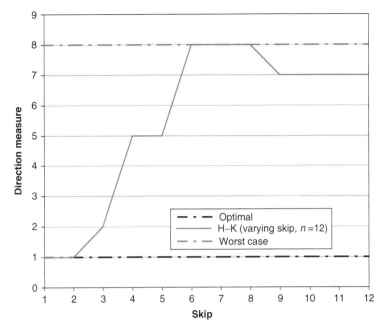

FIGURE 6.21
DLBP H–K R performance with changes in ψ.

$n - 10 \le \psi \le n - 1$ based on the work of the previous section. On the full range of data ($8 \le n \le 80$), DLBP H–K found solutions with NWS* workstations up to $n = 12$, then solutions with NWS* + 1 workstations through data set 11 ($n = 48$), after which it stabilized at NWS* + 2 (Figure 6.22). Increases in the balance measure are seen with increases in data set size. A detailed view of balance performance with problem size can be seen in Figure 6.23.

The hazardous part and the demanded part were both regularly suboptimally placed. Hazard part placement stayed relatively consistent with problem size (though effectively improving as compared to the worst case; see Figure 6.24), although demand decreased in performance when compared to best case and worst case (see Figure 6.25). These results are as expected because hazard performance is designed to be deferential to balance and is affected only when a better hazard measure can be attained without adversely affecting balance, although demand performance is designed to be deferential to balance and hazardous-part placement and is affected only when a better demand measure can be attained without adversely affecting balance or hazardous-part placement.

With part removal direction structured as to be deferential to balance, hazard, and demand, it was seen to decrease in performance when compared to the best case and when compared to the worst case (see Figure 6.26). Again, these results are as expected due the prioritization of the multiple objectives.

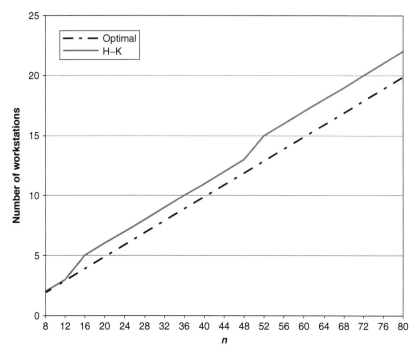

FIGURE 6.22
DLBP H–K workstation calculation.

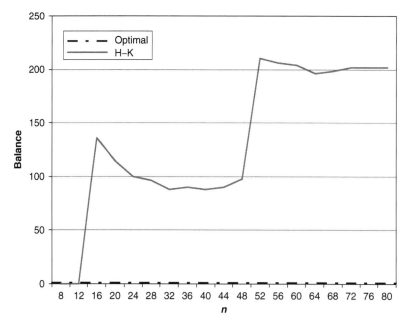

FIGURE 6.23
Detailed DLBP H–K balance measure.

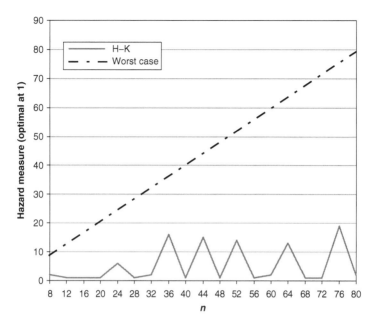

FIGURE 6.24
DLBP H–K hazard measure of performance.

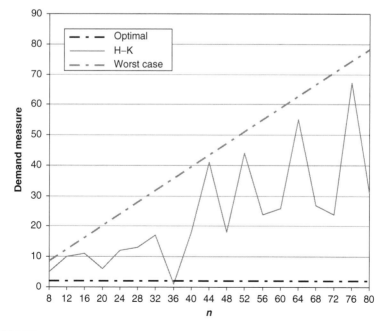

FIGURE 6.25
DLBP H–K demand measure of performance.

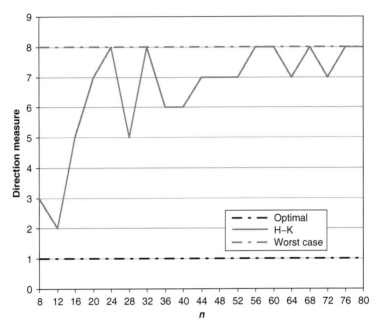

FIGURE 6.26
DLBP H–K part removal direction measure of performance.

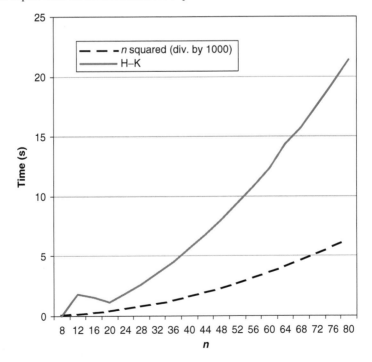

FIGURE 6.27
Detailed DLBP H–K time growth with problem size.

Although less than optimal, these results are not unusual for heuristics run against this data set. The DLBP *a priori* benchmark data is especially designed to challenge the solution-finding ability of a variety of search methods to enable a thorough quantitative evaluation of these methods' performance in different areas. A smaller ψ or, as with other search techniques, the inclusion of precedence constraints will increasingly move the DLBP H–K method toward the optimal solution. As shown in Figures 6.27 and 6.28, the time complexity performance of DLBP H–K provides the tradeoff benefit with the technique's near-optimal performance, demonstrating the moderate increase in time required with problem size that grows markedly slower than the exponential growth of exhaustive search. Exhaustive, n^2 and n^3 curves are shown for comparison. (The anomaly seen in the H–K curve in Figure 6.27 is due to a software rule added by the authors that dictated all ψ be equal to $n - 10$, but no less than $\psi = 3$, to prevent exhaustive or near-exhaustive searches at small n.)

Average-case time complexity using DLBP *a priori* data then appears to be $O(b^b)$ in skip size (where $b = 1/\psi$), while H–K appears to grow at approximately $O(n^2)$ in problem size.

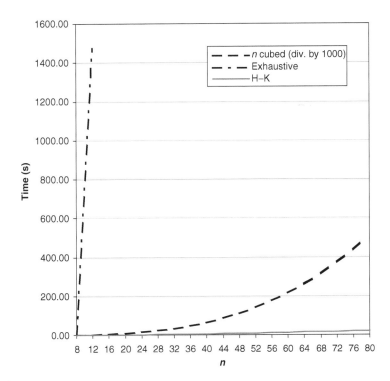

FIGURE 6.28
DLBP H–K time complexity compared to exhaustive search.

6.11   Conclusions

The disassembly line balancing problem was qualitatively introduced and mathematically defined, then proven to be unary NP-complete. Four very fast, near-optimal combinatorial optimization approaches to the problem were presented along with the four primary instances. The heuristics used rapidly provided a feasible solution to the DLBP using a wide variety of search techniques. Each approach demonstrated a near-optimal minimum number of workstations, with the level of optimality increasing with the number of constraints. They were also noted to generate feasible sequences with an optimal or a near-optimal McGovern–Gupta measure of balance while maintaining or improving the hazardous-materials measure, the demand measure, and the part removal direction measure. Although near-optimum methods, the DLBP versions of the ACO and GA metaheuristics, and the H–K general-purpose heuristic along with a purpose-designed greedy or hill-climbing hybrid quickly found near-optimal solutions in DLBP's exponentially large search space. These combinatorial optimization methods are well suited to the multicriteria decision-making problem format as well as for the solution of problems with nonlinear objectives. In addition, these methods are ideally suited to integer problems, a requirement of many disassembly problems, which generally do not lend themselves to rapid or easy solution by traditional optimum solution-generating mathematical programming techniques.

References

1. Aho, A. V., Hopcroft, J. E. and Ullman, J. D., *The Design and Analysis of Computer Programs*, Addison-Wesley, Reading, MA, 1974.
2. Bautista, J. and Pereira, J., Ant algorithms for assembly line balancing, in Dorigo, M. et al. (Eds.), *ANTS 2002, LNCS 2463*, Springer, Berlin, pp. 65–75, 2002.
3. Bierwirth, C., Mattfeld, D. C. and Kopfer, H., On permutation representations for scheduling problems, parallel problem solving from nature, in Voigt, H. M., Ebeling, W., Rechenberg, I. and Schwefel, H. P. (Eds.), *Lecture Notes in Computer Science*, Springer, Berlin, Vol 1141, pp. 3.10–3.18, 1996.
4. Brennan, L., Gupta, S. M. and Taleb, K. N., Operations planning issues in an assembly/disassembly environment, *International Journal of Operations and Production Planning*, 14, 9, 57–67, 1994.
5. Cormen, T., Leiserson, C., Rivest, R. and Stein, C., *Introduction to Algorithms*, The MIT Press, Cambridge, MA, 2001.
6. Dantzig, G. B., Discrete-variable extremum problems, *Operations Research*, 5, 266–277, 1957.
7. Dorigo, M. and Di Caro, G., The ant colony optimization meta-heuristic, in Corne, D., Dorigo, M. and Glover, F. (Eds.), *New Ideas in Optimization*, McGraw-Hill, Maidenhead, UK, pp. 11–32, 1999.

8. Dorigo, M., Di Caro, G. and Gambardella, L. M., Ant algorithms for discrete optimization, *Artificial Life*, 5, 3, 137–172, 1999.

9. Dorigo, M., Maniezzo, V. and Colorni, A., The ant system: optimization by a colony of cooperating agents, *IEEE Transactions on Systems, Man, and Cybernetics–Part B*, 26, 1, 1–13, 1996.

10. Elsayed, E. A. and Boucher, T. O., *Analysis and Control of Production Systems*, Prentice Hall, Upper Saddle River, NJ, 1994.

11. Erel, E. and Gokcen, H., Shortest-route formulation of mixed-model assembly line balancing problem, *Management Science*, 11, 2, 308–315, 1964.

12. Garey, M. and Johnson, D., *Computers and Intractability: A Guide to the Theory of NP Completeness*, W. H. Freeman and Company, San Francisco, CA, 1979.

13. Glover, F., Tabu search—part I, *ORSA Journal on Computing*, 1, 3, 190–206, 1989.

14. Glover, F., Tabu search—part II, *ORSA Journal on Computing*, 2, 1, 4–32, 1990.

15. Gungor, A. and Gupta, S. M., A systematic solution approach to the disassembly line balancing problem, Proceedings of the 25th International Conference on Computers and Industrial Engineering, New Orleans, Louisiana, March 29–April 1, pp. 70–73, 1999.

16. Gungor, A. and Gupta, S. M., Disassembly line balancing, Proceedings of the 1999 Annual Meeting of the Northeast Decision Sciences Institute, Newport, Rhode Island, March 24–26, pp. 193–195, 1999.

17. Gungor, A. and Gupta, S. M., Issues in environmentally conscious manufacturing and product recovery: a survey, *Computers and Industrial Engineering*, 36, 4, 811–853, 1999.

18. Gungor, A. and Gupta, S. M., A solution approach to the disassembly line problem in the presence of task failures, *International Journal of Production Research*, 39, 7, 1427–1467, 2001.

19. Gungor, A. and Gupta, S. M., Disassembly line in product recovery, *International Journal of Production Research*, 40, 11, 2569–2589, 2002.

20. Gupta, S. M. and Taleb, K. N., Scheduling disassembly, *International Journal of Production Research*, 32, 1857–1866, 1994.

21. Gutjahr, A. L. and Nemhauser, G. L., An algorithm for the line balancing problem, *Management Science*, 11, 2, 308–315, 1964.

22. Hackman, S. T., Magazine, M. J. and Wee, T. S., Fast, effective algorithms for simple assembly line balancing problems, *Operations Research*, 37, 6, 916–924, 1989.

23. Hopgood, A. A., *Knowledge-Based Systems for Engineers and Scientists*, CRC Press, Boca Raton, FL, 1993.

24. Hu, T. C. and Shing, M. T., *Combinatorial Algorithms*, Dover Publications, Inc., Mineola, NY, 2002.

25. Kongar, E. and Gupta, S. M., A genetic algorithm for disassembly process planning, Proceedings of the 2001 SPIE International Conference on Environmentally Conscious Manufacturing II, Newton, MA, October 28–29, pp. 54–62, 2001.

26. Koza, J.R., *Genetic Programming: On the Programming of Computers by the Means of Natural Selection*, MIT Press, Cambridge, MA, 1992.

27. Lambert, A. J. D., Disassembly sequencing: a survey, *International Journal of Production Research*, 41, 16, 3721–3759, 2003.

28. Lambert, A. J. D. and Gupta, S. M., *Disassembly Modeling for Assembly, Maintenance, Reuse, and Recycling*, CRC Press, Boca Raton, FL, 2005.

29. Lawler, E. L., Lenstra, J. K., Rinnooy Kan, A. H. G. and Shmoys, D. B., *The Traveling Salesman Problem: A Guided Tour of Combinatorial Optimization*, John Wiley & Sons, New York, NY, 1985.

30. McGovern, S. M. and Gupta, S. M., Greedy algorithm for disassembly line scheduling, Proceedings of the 2003 IEEE International Conference on Systems, Man, and Cybernetics, Washington, D.C., October 5–8, pp. 1737–1744, 2003.

31. McGovern, S. M. and Gupta, S. M., Demanufacturing strategy based upon metaheuristics, Proceedings of the 2004 Industrial Engineering Research Conference, Houston, Texas, May 15–19, CD-ROM, 2004.

32. McGovern, S. M. and Gupta, S. M., Multi-criteria ant system and genetic algorithm for end-of-life decision making, Proceedings of the 35th Annual Meeting of the Decision Sciences Institute, Boston, MA, November 20–23, pp. 6371–6376, 2004.

33. McGovern, S. M. and Gupta, S. M., Local search heuristics and greedy algorithm for balancing the disassembly line, *The International Journal of Operations and Quantitative Management*, 11, 2, 91–114, 2005.

34. McGovern, S. M. and Gupta, S. M., Uninformed and probabilistic distributed agent combinatorial searches for the unary NP-complete disassembly line balancing problem, Proceedings of the SPIE International Conference on Environmentally Conscious Manufacturing V, Boston, MA, October 23–26, pp. 81–92, 2005.

35. McGovern, S. M. and Gupta, S. M., Ant colony optimization for disassembly sequencing with multiple objectives, *The International Journal of Advanced Manufacturing Technology*, 30, 5–6, 481–496, 2006.

36. McGovern, S. M. and Gupta, S. M., Computational complexity of a reverse manufacturing line, Proceedings of the 2006 SPIE International Conference on Environmentally Conscious Manufacturing VI, Boston, MA, October 1–4, CD-ROM, 2006.

37. McGovern, S. M. and Gupta, S. M., Deterministic hybrid and stochastic combinatorial optimization treatments of an electronic product disassembly line, Chapter 13, in Kenneth, D., Lawrence and Ronald, K., Klimberg (Eds.), *Applications of Management Science, Volume 12—Applications of Management Science: In Productivity, Finance, and Operations*, Elsevier Science, New-Holland, Amsterdam, pp. 175–197, 2006.

38. McGovern, S. M. and Gupta, S. M., Performance metrics for end-of-life product processing, Proceedings of the 17th Annual Conference of the Production and Operations Management Society, Boston, MA, April 28–May 1, CD-ROM, 2006.

39. McGovern, S. M. and Gupta, S. M., A balancing method and genetic algorithm for disassembly line balancing, *European Journal of Operational Research*, 179, 3, 692–708, 2007.

40. McGovern, S. M., Gupta, S. M. and Kamarthi, S. V., Solving disassembly sequence planning problems using combinatorial optimization, Proceedings of the 2003 Northeast Decision Sciences Institute Conference, Providence, RI, March 27–29, pp. 178–180, 2003.

41. McGovern, S. M., Gupta, S. M. and Nakashima, K., Multi-criteria optimization for non-linear end of lifecycle models, Proceedings of the Sixth Conference on EcoBalance, Tsukuba, Japan, October 25–27, pp. 201–204, 2004.

42. McMullen, P. R. and Tarasewich, P., Using ant techniques to solve the assembly line balancing problem, *IIE Transactions*, 35, 605–617, 2003.

43. Papadimitriou, C. H. and Steiglitz, K., *Combinatorial Optimization: Algorithms and Complexity*, Dover Publications, Mineola, NY, 1998.

44. Pinedo, M., *Scheduling Theory, Algorithms and Systems*, Prentice Hall, Upper Saddle River, NJ, 2002.

45. Ponnambalam, S. G., Aravindan, P. and Naidu, G. M., A comparative evaluation of assembly line balancing heuristics, *The International Journal of Advanced Manufacturing Technology*, 15, 577–586, 1999.
46. Reingold, E. M., Nievergeld, J. and Deo, N., *Combinatorial Algorithms: Theory and Practice*, Prentice-Hall, Inc., Englewood Cliffs, NJ, 1977.
47. Scholl, A., Balancing and sequencing of assembly lines, *Physica-Verlag*, Heidelberg, Germany, 1995.
48. Suresh, G., Vinod, V. V. and Sahu, S., A genetic algorithm for assembly line balancing, *Production Planning and Control*, 7, 1, 38–46, 1996.
49. Torres, F., Gil, P., Puente, S. T., Pomares, J. and Aracil, R., Automatic PC disassembly for component recovery, *International Journal of Advanced Manufacturing Technology*, 23, 1–2, 39–46, 2004.
50. Tovey, C. A., Tutorial on Computational Complexity, *Interfaces*, 32, 3, 30–61, 2002.
51. Zeid, I., Gupta, S. M. and Bardasz, T., A case-based reasoning approach to planning for disassembly, *Journal of Intelligent Manufacturing*, 8, 97–106, 1997.

7

Multikanban System for Disassembly Line

Gun Udomsawat and Surendra M. Gupta

CONTENTS

7.1   Introduction

The Toyota production system (TPS) is a high-efficiency manufacturing system that generates minimal amount of inventories, involves minimum of nonvalue added processes (such as setup times), and reduced material-transporting times. To achieve high efficiency, the TPS uses the "pull" concept. The pull system plays a major role in controlling and reducing the inventory levels. It is well known that carrying large amounts of inventories is costly and risky. This is especially true in industries where products have short shelf lives, are expensive to produce, or have difficult to predict demands. Manufacturers can benefit greatly by implementing a made-to-order or pull system.

Although several tools are available to implement a pull system, no tool has more widespread implementation than the Toyota kanban system. Based on a "card" concept, a kanban mechanism effectively controls production in an assembly line by signaling a workstation to start producing a target component or processing materials only when the component or the materials are needed. Consequently, the system limits its inventory based on the number of kanbans in the line. Although simple and effective in a deterministic environment, the kanban mechanism is cumbersome when it faces uncertainties in certain situations in a production line. This mechanism relies on a stable supply of materials. For this reason, many companies have put in tremendous effort to ensure that all suppliers are reliable. In other manufacturing environments where the supplier is not very reliable, implementing the kanban mechanism can be problematic. Modification of the kanban mechanism becomes necessary in such situations.

A product reaches its end-of-life (EOL) when it becomes nonfunctional and repair is not feasible or not chosen as an option. However, a significant number of functional products are replaced because newer models that are often more attractive, efficient, and affordable are introduced in the market. This disposal of functional products is witnessed in the case of household appliances, electronic products, communication devices, and personal computers (PCs). Typically, a used automobile can enter and reenter the preowned automobile market until it reaches its EOL. However, only a small number of the other products (such as electronic products) are reused or resold in the second-hand market. Many governmental regulations are now encouraging reuse and recycling of such EOL products to reduce the amount of waste sent to landfills. An additional benefit of reuse and recycling is that these approaches form a source of less expensive raw materials and components in both manufacturing and remanufacturing processes. This emerging market of reusable and recyclable components and materials is a major force behind designing products for remanufacturing and the product recovery system.

Product recovery has become a subject of interest during the last decade. Among the several processes used in product recovery, selective disassembly yields a very high recovery rate of reusable components and recyclable materials. Disassembly line is one of the most common setups in a selective disassembly facility and is suitable for disassembly of products in large quantities. Disassembly line is appropriate for disassembly of many EOL products, especially electronic products, PCs, and household appliances. This appropriateness stems from two major characteristics, similarity in the disassembly process, and similarity in the structure of targeted materials and components. Although the use of a disassembly line offers advantages, the production control mechanism is a complex issue that must be addressed. As in an assembly line setting, there are two types of control mechanisms in a disassembly line setting, the push system and the pull system. A push system is easy to implement but is not efficient in the disassembly environment. By nature, this system tends to generate large amounts of inventories. A pull system, in theory, creates significantly lesser amounts of inventories.

However, most pull system control tools that are used in assembly line settings are not suitable for use in disassembly line settings in their current form. We will discuss the reasons for these difficulties and demonstrate how the existing tools for assembly line settings can be modified for implementation in disassembly line settings.

In this chapter, we study the effect of implementing a kanban mechanism in a disassembly line environment. We identify and discuss some of the complications that are inherent in a disassembly line, namely product arrival, demand arrival, inventory fluctuation, and production control mechanisms. We show how to overcome such complications by implementing a multikanban system (MKS) in the disassembly line setting. The MKS is designed in such a way that its implementation helps reduce the inventory build up in the system, which commonly occurs in a disassembly line when a push system is implemented. The MKS is able to ensure a smooth operation of the disassembly line in which multiple types of EOL products arrive for processing. The modified kanban system relies on the dynamic routing of kanbans according to the state of the system. We investigate the effectiveness of the MKS using simulation. We also explore the effect of product mix on the performance of the traditional push system and the MKS in terms of controlling the system's inventory while attempting to achieve a decent customer service level. We provide a numerical example to illustrate the methodology and obtain results using simulation.

This chapter is organized as follows: (1) Section 7.2 provides a brief literature review; (2) Section 7.3 identifies and discusses some of the complications that are inherent to a disassembly line; (3) Section 7.4 describes the multikanban model; (4) Section 7.5 considers the example of a PC disassembly line to illustrate the concept of an MKS; (5) Section 7.6 presents a numerical example; and (6) Section 7.7 contains the conclusions and the ideas for future research.

7.2   Literature Review

Hopp and Spearman [14] described the traditional kanban control mechanism using the one-card and two-card methodologies. Monden [22] described various types of kanbans, including an emergency kanban, and claimed that a kanban system cannot be adapted to sudden or large variations in demands. However, Gupta and Al-Turki [9] suggested a methodology to adjust the number of kanbans to respond to changing demands. In a subsequent paper, Gupta et al. [12] introduced a flexible kanban system (FKS), which has the ability to adjust the number of kanbans in response to the variations in demand and lead times. Gupta and Al-Turki [10,11] also demonstrated that in the uncertainties and the interruptions caused by preventive maintenance or breakdown of the material handling system, the FKS outperforms the traditional kanban system. Yet another algorithm to adjust the number of kanbans was proposed by Tardif and Maaseidvaag [30]. Their adaptive kanban system monitors the actual inventory level and adjusts the number of kanbans by releasing or capturing the extra kanbans. Takahashi and Nakamura [26–28] proposed a reactive kanban system to detect unstable changes and adjust the number of kanbans and buffer sizes according to instability or switch to the concurrent ordering system. Subsequently, Takahashi et al. [25] presented a reactive just-in-time (JIT) ordering system that responds to changes in both mean and variance of demands. Another way of implementing kanbans in a more diverse environment such as a single-stage hybrid system was suggested by Korugan and Gupta [17]. A hybrid system refers to a combination of two distinct lines, a production line and a disassembly line. Chang and Yih [3] proposed a kanban system called the generic kanban system for job shop systems.

Gungor and Gupta [6] provide a comprehensive survey of issues in environment-conscious manufacturing and product recovery. Within the area of product recovery, many researches address disassembly and its significant domains such as disassembly sequencing [19], disassembly line [8,29], disassembly line balancing [7], and disassembly line scheduling [21]. For more information on disassembly and product recovery, see Refs. 2, 13, and 23. A recent book by Lambert and Gupta [20] is also helpful in understanding the area of disassembly and disassembly modeling. In the area of electronic products and appliance disassembly, Lambert [18] presents a methodology to determine the optimal disassembly sequence. For more information on electronic products disassembly and recycling, see Refs. 1, 4, 5, 15, and 24.

7.3   Dissassembly Line

A disassembly line is composed of a series of workstations functioning in a sequence. These workstations disassemble EOL products into subassemblies or components. EOL products may enter the disassembly line at any workstation, depending upon their type. Similarly, depending upon what is demanded, the demand may occur at any workstation (Figure 7.1). The arrival points, the configuration of EOL products, and the variety of demanded items are the factors that make a disassembly line difficult to manage. These factors also lead to substantial fluctuations in inventory in a disassembly line. In this section, we highlight some of the crucial issues in using a disassembly line for appliance disassembly. We then propose a methodology to overcome these issues by using the pull control principle.

7.3.1   Arrival Pattern of EOL Products

EOL products arriving at a disassembly line may consist of different combinations of components from a given set of components. Generally, from a set

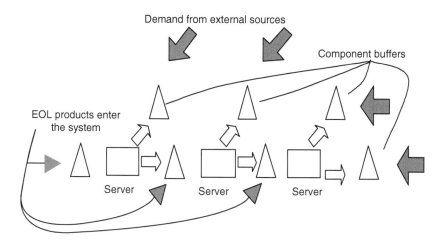

FIGURE 7.1
A disassembly line.

of N components, the total number of possible combinations of components, $Q_{(N)}$ is given by

$$Q_{(N)} = 2^N - N - 1 \qquad (7.1)$$

For example, a set of 4 components (A, B, C, D) can produce up to 11 possible product combinations (AB, ABC, ABCD, ABD, AC, ACD, AD, BC, BCD, BD, and CD). By adding one more component to the set, the number of possible combinations increases to 26. It is, therefore, clear that the number of combinations increases exponentially with the increase in the number of components. Fortunately, in real life, not all combinations exist in every EOL product type. In some EOL products, such as a PC, components are modular and usually come in various combinations. Most of the times, these products have been modified before disposal. Household appliances commonly arrive in fewer combinations because consumers rarely modify such products. Nevertheless, the workstation at which an EOL product enters the disassembly line is determined by the type and the combination of the components that make up the product. Consider a disassembly line with three workstations: if component A is disassembled at workstation 1, component B is disassembled at workstation 2, and components C and D are disassembled at workstation 3, then a product with components B, C, and D arriving at the disassembly line does not have to go to workstation 1 at all. This product can enter the disassembly line directly at workstation 2. Considering the same example, if an arriving product consists of components A, C, and D, it would have to enter workstation 1. However, after getting processed at workstation 1, it could skip workstation 2 entirely and go to workstation 3. Furthermore, EOL products with different precedence relationships must be processed through workstations in different sequences. These three situations destabilize the disassembly line by causing an overflow of materials at one workstation while starving at some other workstations, leading to undesirable fluctuations of inventory in the system. It is, therefore, crucial to balance the line and manage the materials flow of the line.

7.3.2 Demand Fluctuation and Inventory Management

Although a disassembly line has many unique characteristics, the multi-level arrival of demand is one of the most crucial characteristics that makes the disassembly line much more complicated than a typical assembly line. Demand can occur at any station of the disassembly line. In most assembly lines, demand arrives only at the last workstation. However, in that case, even if multilevel arrival of demand were considered, its effect would be benign because the product does not go forward from there as it is taken off the line to fulfill the demand. In a disassembly line setting, however, the arrivals of external demand at workstations other than the last workstation, creates a disparity between the number of demanded components and the number of partially disassembled products. Thus, if the system responds to every request for components, it would end up with a significant amount of extra inventory of components that are in low demand. These issues create chaos in the system. Because service level is important and must be maximized, it is essential to develop a good methodology to control the system and find a way to manage the extra inventory produced.

7.3.3 Applications of Disassembly Line

As mentioned earlier, high technology–related EOL products have become a subject of interest in product recovery in recent years. These products usually have shorter lives and high-volume usage. Examples are electronic products, household appliances, and PCs. These products contain a large number of reusable components and recyclable materials. For example, a majority of components from end-of-life PCs, especially steel covers, adaptors, memories, hard drives, media drives, power supplies, main boards, and steel cases, are reusable or recyclable. Components that are functional can be supplied to the second-hand market, used in refurbished products, or used in PC repairs. Components that are outdated or not functional can be supplied to the recycling industry. Printed circuit boards (PCBs), transformers, and steel covers and cases are composed of recyclable steel, copper, and other precious materials such as silver, gold, and platinum. Household appliances constitute another example. Because these appliances share similar materials and components, they can be disassembled together in the same disassembly facility. These appliances also require similar disassembly processes and disassembly tools. A disassembly facility can benefit from disassembling multiple types of appliances in two ways. First, this disassembly reduces the uncertainty in the supply of materials and components. The disassembly facility has a better chance of receiving a continuous supply of EOL products. Second, the variety of components and materials retrieved from the EOL appliances is much broader, which helps increase customer satisfaction.

However, there are complications while implementing a disassembly line facility. The facility must be flexible enough to accommodate the ensuing variety of products that would enter the system. More notably, it needs a very efficient production control system to deal with disassembly of multiple products, components, and materials. As mentioned earlier, the demand for these components is highly uncertain and varies with the components. This uncertainty and variability result in highly fluctuating inventory levels. Many components can become obsolete within a short period of time. In addition, disassembly is a labor-intensive process. Therefore, components should not be stocked for future demand. Because of these issues, choosing a better production control system is highly desirable.

7.3.4 Production Control System in Disassembly Line

In general, there are two types of control mechanisms: push mechanism and pull mechanism. The push mechanism relies on a predetermined production schedule based on the expected demand of finished products. Raw materials are pushed through the system to meet the future demand. However, the production in pull mechanism is triggered by the actual demand and causes a flow of materials throughout the system. These two mechanisms have been the topics of studies for their superiority in system efficiency, customer service level, and ease of implementation. Conclusions from those studies are diverse. In fact, none of the mechanisms dominates in all situations. The push system has advantages in terms of experience in implementing it and providing higher levels of customer service in certain production scenarios because the system tends to build up inventory. In contrast, the advantage of the pull mechanism is that it does not generate large amounts of inventory. Instead, this system has a mechanism to control the inventory. However, this system relies heavily on the consistency of raw material supplies and the agility of the constant supply of raw materials. The pull mechanism only produces when and where there are needs; therefore, this mechanism is likely to perform better than the push mechanism in a disassembly line.

The TPS is one of the most commonly used pull mechanism tools. However, when implemented in a disassembly line setting, this tool is fraught with numerous uncertainties. A modification of the mechanism is therefore required to improve system's performance by reducing these difficulties and allowing the system to operate at its best. In the next section, we introduce a MKS that is designed for implementation in a disassembly line setting in which supply and demand fluctuate extensively.

7.4. Multikanban Model for Disassembly Line

The multikanban model for disassembly line proposed here offers three major advantages. First, this model implements the pull concept and carries minimal inventory. The model permits initiation of disassembly only when there is an actual demand for components or materials. As a result, the facility does not have to carry a large amount of inventories and can save on the inventory-carrying cost and the disassembly cost of unwanted components. Second, the multikanban pull system can help in keeping track of the inventory, which is very useful in giving out a quote to a customer. Finally, the multikanban mechanism relies on routing rules that direct kanbans to workstations supplying the most needed residuals. In this way, customer satisfaction is retained while the kanban routing takes care of fluctuations in the demand for various components. In this section, we will explore the multikanban mechanism in detail.

7.4.1   Material Types

There are two basic types of materials in the disassembly line, components and subassemblies. A component is a single item that cannot be further disassembled. The component is placed in the component buffer and awaits retrieval by a customer demand. In contrast, a subassembly can be further disassembled and is composed of at least two components. Both types of materials can be further distinguished as regular or overflow items. Regular items are items demanded by customers or downstream workstations. To fulfill this demand, a server must disassemble the demanded component or subassembly. The residual item from this disassembly process that does not fulfill any request is called an overflow item. Because the disassembly process is initiated by a single kanban, the overflow item will not have a kanban attached to it. However, the overflow item is routed in the same way as the regular item. The only difference between these two items is that the overflow item is given priority of being retrieved after it arrives at its buffer. It should be noted that as long as there is an overflow item in the buffer, its demand would not initiate any further disassembly process. This will help the system to eliminate any extra inventory that is caused by unbalanced demand.

7.4.2   Kanban Types

Corresponding to material types, there are two basic types of kanbans in the system, component kanbans and subassembly kanbans. A component kanban is attached to a disassembled component that is placed in the component buffer of the workstation where it was disassembled. Similarly, a subassembly kanban is attached to a residual subassembly that is placed in the subassembly buffer of the workstation where it was separated from the component. A component placed in a component buffer can be retrieved by an external demand. When authorized, a subassembly placed in the subassembly buffer is routed for disassembly to the next workstation based on its disassembly sequence.

At the first workstation, products arrive only from outside sources. However, at any other workstation i, where $1 < i \leq N - 1$, there are two possible types of arrivals. The first type is a subassembly that arrives from an upstream workstation and is called internal subassembly. There is always a subassembly kanban attached to an internal subassembly. The second type is a product (or subassembly) that arrives from outside sources and is called external subassembly. No kanban is attached to an external subassembly. This aspect is also true for the products arriving from external sources to the first workstation. As long as there is an external product or subassembly available at an input buffer, the system will process it first before processing any available internal subassembly. This will avoid unnecessary pulling of an internal subassembly from an upstream workstation. Thus, the number of kanbans attached to the internal subassemblies will remain constant throughout the process. Figure 7.2 illustrates the kanbans and the materials flow in a disassembly line.

7.4.3 Kanban Routing Mechanisms

Consider workstation j, where $1 \leq j \leq N - 1$. When a demand for component j arrives at the component buffer of workstation j, one unit of component j is retrieved and the component kanban j attached to it is routed to the most desirable workstation. The procedure for determining the most desirable workstation to route component kanban j is as follows (this procedure is not applicable to component kanbans $N - 1$ and N; in both cases the kanbans are routed to the input buffer of the last workstation).

A component kanban originating from workstation j will be routed to a workstation i, where $1 \leq i < j$, or workstation j depending on the availability and the desirability of the subassembly that contains component j. Routing component kanban j to workstation i, where $1 \leq i \leq (j - 1)$, will result in an immediate separation of component j from component i. Thus, the only subassembly located at the input buffer of workstation i that would be useful is a subassembly that contains only components i and j. If this type of subassembly exists in the input buffer of workstation i, then workstation i is qualified. Similarly, if there is at least one subassembly in the input buffer of workstation j, then workstation j is qualified.

Next, we need to select the most desirable workstation to route component kanban j to, among the qualified workstations, such that, if chosen, it will cause the least amount of extra inventory in the system. Choosing workstation i will increase the inventory level of component i by an additional unit. Thus, the best workstation i is the one that is most starving for its component. By checking the backorder level for demand i, we could determine the most starving workstation. If there is a tie, we select the most downstream workstation. Choosing workstation j will create a residual subassembly that will be further disassembled at downstream workstations. If we select workstation j, then we must also select a proper subassembly to disassemble. For example, if a backorder exists at the component buffer of workstation k, where $j < k \leq (N - 1)$, then, if available, we might try to disassemble a subassembly that contains only components i and k. If more than one workstation qualifies as

starving workstations, then we select the one that is most starving among them. If there is a tie, then we select the most downstream workstation.

We can now compare the starving levels of workstations i and j. If the highest starving level of workstation i is greater than or equal to the starving level of workstation j, then we will route the component kanban j to workstation i, otherwise we will route it to workstation j. Note that whenever an external subassembly is available, it will always be chosen first. Internal subassemblies will only be used when no external subassembly of the desired kind is available. Subassembly kanbans are routed in a fashion similar to component kanbans. Figure 7.3 shows a concept of the multikanban mechanism.

7.4.4 Selection of Products

As we permit multiple combinations of products, the worker may have several options when selecting the product for disassembly. If the authorization of disassembly is initiated by the subassembly kanban (j_x), which can occur only at workstation i, where $1 \leq i < j$, the workers will have no option but to select the subassembly that results in immediate separation of subassembly (j_x), namely subassembly (ij_x). If the authorization of disassembly is initiated by component kanban j at workstation i, where $1 \leq i < j$, the worker will have to remove subassembly (ij) from the product buffer with no other options because the only subassembly that results in immediate separation of component j is the subassembly (ij). However, if the component kanban j arrives at workstation j, there are multiple options because every subassembly located in the product buffer contains component j and always results in immediate separation of component j. In this case, we determine whether or not the residual that is created by the disassembly will result in an overflow of inventory. We choose the subassembly (j_x), where x is the most desirable residual ranking based on the request of subassembly kanban x at workstation j (existing kanban x at the workstation j) or the current inventory level of subassembly (component) x, respectively.

7.4.5 Determining the Kanban Level

The kanban level plays an important role in the multikanban mechanism as it maintains a proper flow of components and subassemblies at a desired level throughout the system. The kanban level can be determined by considering product arrival rate, demand arrival rate, and disassembly time. The number of kanbans for both the component kanban, k_i and the subassembly kanban, k_j^* can be computed, at any point in the disassembly line, using the following general expressions:

$$k_i = \max\left(1, \frac{R_i}{F_i}\right) \tag{7.2}$$

$$k_j^* = \max\left(1, \frac{R_j^*}{F_j^*}\right) \tag{7.3}$$

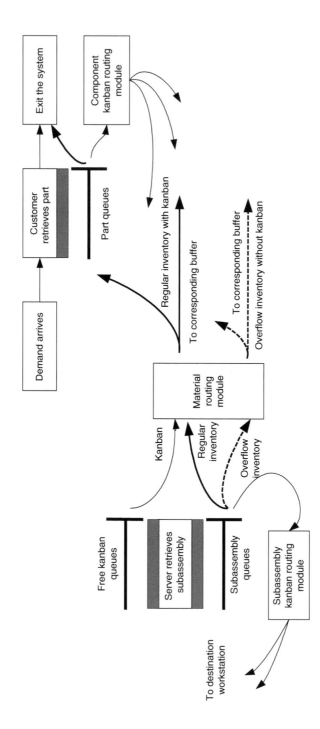

FIGURE 7.3
The multikanban mechanism.

where R_i is the request rate of component i, F_i the furnish rate of component i, R_j^* the request rate of subassembly j, and F_j^* the furnish rate of subassembly j. These request rates and furnish rates can be calculated as follows:

$$R_i = d_i \quad \text{for } 1 \leq i \leq N \tag{7.4}$$

$$F_i = \sum_{w=1}^{i} s_{(i,w)} \quad \text{for } 1 \leq i \leq N \tag{7.5}$$

$$R_j^* = s_i, \quad i \text{ is the next component to be disassembled in the sequence} \tag{7.6}$$

$$F_j^* = a_j^* + \sum_{w=1}^{m-1} s_{(i,w)}, \quad i \text{ is the latest component disassembled in the sequence} \tag{7.7}$$

where d_i is the demand arrival rate of component i, $s_{(i,w)}$ the disassembly rate of component i at workstation w, s_j the disassembly rate of subassembly j, a_j^* the arrival rate of subassembly j (from external source), m the current workstation index, N the maximum number of components, and $N - 1$ the maximum number of workstations. In the case of component kanban, which is requested only from a single source, request rate is equal to the customer demand arrival rate. However, because the component kanban arrives from several sources in the system, the furnish rate is the summation of arrival rates from all possible sources. In the case of subassembly kanban, the furnish rate is influenced by both the disassembly rate and the external subassembly arrival rate. Thus, we take into account all external and internal arrival rates of subassemblies at the buffer. Similarly, the two requesting sources, the demand for target component and the demand for residual subassembly, affect the request rate. The number of kanbans is determined at the beginning of the disassembly process. It is clear that demand, supply, disassembly time, and product structure affect the computation of the number of kanbans.

7.5 Line Description and Assumptions for PC Disassembly

To illustrate the concept of the multikanban mechanism in a disassembly line, we consider a PC disassembly line. Figure 7.4 depicts the complete disassembly sequence for a PC. The PC disassembly line has six workstations (for simplicity, we do not consider removal of cover as a part of the disassembly operation). EOL PCs arrive at the line in various configurations. The input location for EOL PCs depends on their configuration. The input location for a PC is the most upstream workstation that disassembles the first component, according to precedence relationship, from that PC. Only one type of component is disassembled at a given workstation, except the last workstation where motherboard and case are separated.

It takes different amounts of time to disassemble different components. At each workstation, there are two types of output buffers, component buffer

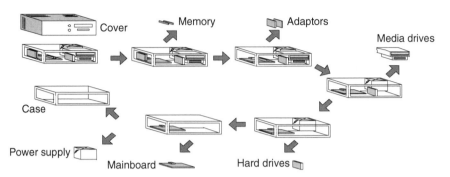

FIGURE 7.4
Disassembly sequence of personal computer.

and subassembly buffer. The component disassembled at a workstation, s_i, is placed in the component buffer, B_i. The rest of the subassembly is routed to the subassembly buffer, B'_i. The subassembly buffer becomes the input buffer for the subsequent workstation to further disassemble the subassembly according to its disassembly sequence. There are multiple sources of demands. A demand can occur at any workstation. The demand at a given workstation is always for the component that is disassembled at that workstation. Regardless of the configuration, a PC must be disassembled in a predefined sequence from the first component to the last component. When a particular component is demanded, this component is retrieved from the output component buffer, B_i, of the workstation where it is disassembled. If no component is available at the component buffer, the demand waits there in the form of a backorder.

In studying the model using simulation, the following assumptions were made

1. Customer backorder is permitted.
2. External demand is for component only and can arrive at any workstation.
3. Components must be disassembled according to their precedence relationships: one type at a time until the last component in the disassembly sequence is disassembled.
4. Products and subassemblies can arrive at any workstation along the line depending on their configurations.

7.6   Numerical Example

A numerical example of a PC disassembly line with six workstations is used to illustrate the application of the multikanban concept. Here, we disregard the removal of the cover to reduce the size of the model. There are seven targeted components, memory, adaptor card, media drive, hard drive, motherboard, power supply, and case. The six most common product configurations are shown in Table 7.1. From this data, disassembly is performed on a disassembly line consisting of six workstations. Component A, representing memory module, is disassembled at workstation 1. Component B, representing adaptor card, is disassembled at workstation 2. Component C, representing media drive, is disassembled at workstation 3. Component D, representing hard drive, is disassembled at workstation 4. Component E, representing motherboard, is disassembled at workstation 5. Finally, component F, representing power supply, is disassembled from component G, representing steel case, at workstation 6. Table 7.1 also provides the mean arrival rates for products, the mean disassembly times for components, the mean demand arrival rates for components, which are all exponentially distributed, and the kanban level, which was calculated using the formulae proposed in Section 7.4.5. Table 7.2

TABLE 7.1

Product Details

	Demand Arrival	Product Configurations (External Arrival)						Disassembly	Kanban
Component	Rate	1	2	3	4	5	6	Time	Level
A Memory	15	X	X	—	—	—	—	2	1
B Adaptors	15	X	X	X	X	—	—	3	1
C Media drives	15	X	X	X	—	X	—	5	2
D Hard drive	10	X	—	—	—	—	—	3	1
E Motherboard	20	X	X	X	X	X	X	3	1
F Power supply	20	X	X	X	X	X	X	3	1
G Case	20	X	X	X	X	X	X	3	1
Arrival Rate		10	5	2	1	1	1		

TABLE 7.2

Subassembly Details

		Subassembly							Disassembly
Component		1	2	3	4	5	6	7	Time
A	Memory	—	—	—	—	—	—	—	2
B	Adaptors	X	X	—	—	—	—	—	3
C	Media drives	X	X	X	X	—	—	—	5
D	Hard drive	X	—	X	—	X	—	—	3
E	Motherboard	X	X	X	X	X	X	—	3
F	Power supply	X	X	X	X	X	X	X	3
G	Case	X	X	X	X	X	X	X	3
External arrival rate (a_j)		—	2	—	1	—	1	—	
Subassembly kanban level		1	1	1	1	2	1	1	

presents the data for all seven subassemblies that appear in the disassembly line from either the disassembly process or external arrival.

We used Arena® software [16] to simulate the model. We conducted two sets of experiments representing the push system and the MKS, respectively. For each experiment, we collected the data over a 2-day period. In the push system, all arriving products were processed continuously in the order of their arrival. The demand was fulfilled as soon as the components were available. In the multikanban pull control system, we used smart routing for a subassembly kanban (as explained in Section 7.4.3) to reduce the inventory build up caused by disparity in demand among components. We also used the product selection method (as explained in Section 7.4.4) for the first three workstations. In these experiments, statistics on the following two performance measures were collected: (1) the system's ability to fulfill the demand and (2) the average inventory levels.

It is clear from Figures 7.5 and 7.6 that the multikanban mechanism significantly reduces average inventory while maintaining the components' demand fulfillment rates. Both systems offer no significant difference in service level during the same period of time. For each component, the MKS does not sacrifice any service to customer demand. Instead, the system targets to service the customer demand by taking the number of waiting customer demands into consideration prior to routing the kanban to the appropriate workstation. Adhering to the product selection rules, only the best choice product is disassembled. These multimodules work in harmony to prioritize the disassembly process only where and when there is a need.

In the push environment, the system builds up inventory to fulfill the demand of customers. The large amount of inventory copes well with any

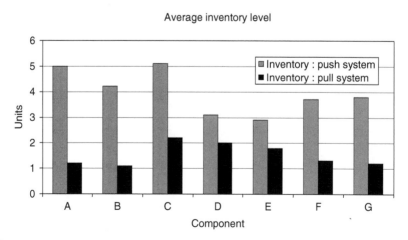

FIGURE 7.5
The average inventory level.

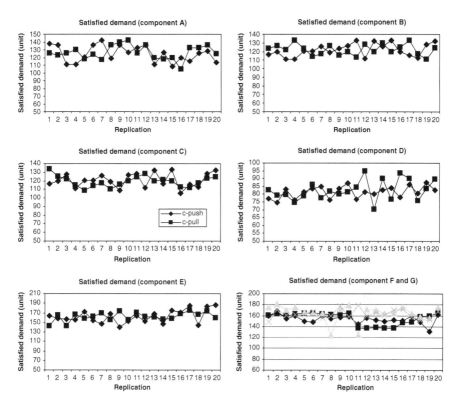

FIGURE 7.6
Number of satisfied demand for multikanban system and push system.

fluctuation in demand. However, this large amount of inventory results in a higher carrying cost and a higher cost of operation. This drawback becomes severe when disassembling products that contain certain components low in demand, have short shelf lives, and incur high costs of disassembly. However, the multikanban mechanism addresses the fluctuations in demand by routing the kanbans to the most suitable workstation. For the example considered, the system was able to reduce the inventory level by an average of 33% while keeping the demand of customers at levels comparable to the push system.

Figure 7.7 shows the average customer waiting time. A customer is considered to be waiting if there is no component stored in the outbound buffer where it arrives. In the MKS, the average waiting time is significantly lower than in the push system. This advantage is the result of the effective routing rules that tend to favor higher demand components. The MKS responds to the backorder faster than the push system does, because a free kanban is routed to the destination where disassembly of subassembly or EOL product yields both the target component and the residual component that are required the most.

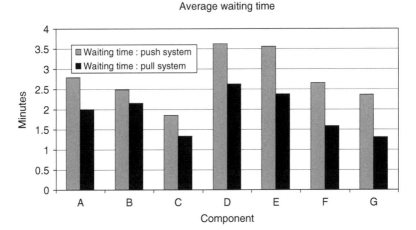

FIGURE 7.7
Customer waiting time for multikanban system and push system.

7.7 Conclusions and Future Research

Despite serious complications in a disassembly line, this chapter demonstrated that the TPS can be adapted to perform effectively in a disassembly line. With the help of an example it was shown that the proposed multikanban mechanism can be effectively implemented. As illustrated by the numerical example, the multikanban mechanism enables the disassembly line to meet the demand of customers and stabilizes the fluctuations in the inventory levels. To these ends, the mechanism does not rely on adjusting the number of kanbans. Instead, the MKS relies on real-time routing adjustment of kanban. In future research, we will study the effect of both sudden and scheduled interruptions in a disassembly line. We will also study the effect of product mix and multiple precedence relationships of components and the effectiveness of the MKS.

References

1. Boon J. E., Isaacs J. E., and Gupta S. M., Economics sensitivity for end of life planning and processing of personal computers, *Journal of Electronics Manufacturing*, 11, 1, 81–93, 2002.
2. Brennan L., Gupta S. M., and Taleb K. N., Operations planning issues in an assembly/disassembly environment, *International Journal of Operations and Production Management*, 14, 9, 57–67, 1994.
3. Chang T.-M. and Yih Y. Generic kanban systems for dynamic environments, *International Journal of Production Research*, 32, 4, 889–902, 1994.

4. Das S., Mani V., Caudill R., and Limaye K., Strategies and economics in the disassembly of personal computers—a case study, Proceedings of the 2002 International Symposium on Electronics and the Environment, San Francisco, CA, May 6–9, pp. 257–262, 2002.

5. Ellis B., Environmental issues in electronics manufacturing: A review, *Circuit World*, 26, 2, 17–21, 2000.

6. Gungor A. and Gupta S. M., Issues in environmentally conscious manufacturing and product recovery: A survey, *Computer and Industrial Engineering*, 36, 4, 811–853, 1999.

7. Gungor A. and Gupta S. M., A solution approach to the disassembly line balancing problem in the presence of task failures, *International Journal of Production Research*, 39, 7, 1427–1467, 2001.

8. Gungor A. and Gupta S. M., Disassembly line in product recovery, *International Journal of Production Research*, 40, 11, 2569–2589, 2002.

9. Gupta S. M. and Al-Turki Y. A. Y., An algorithm to dynamically adjust the number of kanbans in a stochastic processing times and variable demand environment, *Production Planning and Control*, 8, 2, 133–141, 1997.

10. Gupta S. M. and Al-Turki Y. A. Y., Adapting just-in-time manufacturing systems to preventive maintenance interrupts, *Production Planning and Control*, 9, 4, 349–359, 1998.

11. Gupta S. M. and Al-Turki Y. A. Y., The effect of sudden material handling system breakdown on the performance of a JIT system, *International Journal of Production Research*, 36, 7, 1935–1960, 1998.

12. Gupta S. M., Al-Turki Y. A. Y., and Perry R. F., Flexible kanban system, *International Journal of Operations and Production Management*, 19, 10, 1065–1093, 1999.

13. Gupta S. M. and McLean C. R., Disassembly of products, *Computers and Industrial Engineering*, 31, 225–228, 1996.

14. Hopp W. J. and Spearman M. L., *Factory Physics*, Second Edition, McGraw-Hill, New York, 2001.

15. Jung L. B. and Bartel T. J., Computer take-back and recycling: An economic analysis for used consumer equipment, *Journal of Electronics Manufacturing*, 9, 1, 67–77, 1999.

16. Kelton D. W., Sadowski R. P., and Sadowski, D. A., *Simulation with Arena®*, WCB, McGraw-Hill, New York, 1998.

17. Korugan A. and Gupta S. M., Adaptive kanban control mechanism for a single stage hybrid system, Proceedings of the SPIE International Conference on Environmentally Conscious Manufacturing II, Newton, MA, October 28–29, pp. 175–182, 2001.

18. Lambert A. J. D., Determine optimum disassembly sequences in electronic equipment, *Computer and Industrial Engineering*, 43, 553–575, 2002.

19. Lambert A. J. D., Disassembly sequencing: A survey, *International Journal of Production Research*, 41, 16, 3721–3759, 2003.

20. Lambert A. J. D. and Gupta S. M., *Disassembly Modeling for Assembly, Maintenance, Reuse, and Recycling*, CRC Press, Boca Raton, FL, 2005.

21. McGovern S. M. and Gupta S. M., Greedy algorithm for disassembly line scheduling, Proceedings of the 2003 IEEE International Conference on Systems, Man, and Cybernetics, Washington, DC, October 5–8, pp. 1737–1744, 2003.

22. Monden Y., Adaptable kanban system helps Toyota maintain just-in-time production, *Industrial Engineering*, 13, 5, 29–46, 1981.

23. Moyer L. and Gupta S. M., Environmental concerns and recycling/disassembly efforts in the electronics industry, *Journal of Electronics Manufacturing*, 7, 1, 1–22, 1997.

24. Sodhi M. S. and Reimer B., Model for recycling electronics end-of-life products, *OR Spektrum*, 23, 97–115, 2001.

25. Takahashi K., Morikawa K. and Nakamura N., Reactive JIT ordering system for changes in the mean and variance of demand, *International Journal of Production Economics*, 92, 181–196, 2004.

26. Takahashi K. and Nakamura N., Reacting JIT ordering systems to the unstable changes in demand, *International Journal of Production Research*, 37, 10, 2293–2313, 1999.

27. Takahashi K. and Nakamura N., Reactive logistics in a JIT environment, *Production Planning and Control*, 11, 1, 20–31, 2000.

28. Takahashi K. and Nakamura N., Decentralized reactive kanban system, *European Journal of Operational Research*, 139, 3, 262–276, 2002.

29. Tang Y., Zhou M., and Caudill R., A systematic approach to disassembly line design, Proceedings of the 2001 IEEE International Symposium on Electronics and the Environment, Denver, CO, May 7–9, pp. 173–178, 2001.

30. Tardif V. and Maaseidvaag L., An adaptive approach to controlling kanban systems, *European Journal of Operational Research*, 132, 2, 411–424, 2001.

8

Disassembly Sequencing Problem: Resolving the Complexity by Random Search Techniques

Mukul Tripathi, Shubham Agrawal, and M. K. Tiwari

CONTENTS

8.1   Introduction

The past century witnessed both scientific and technological developments as well as the concomitant emergence of environmental threats as an inevitable by-product of this progress. Survival needs and, later, the desire to improve the quality of life motivated technological innovations, which led to the Industrial Revolution. The philosophy of time-based competition (TBC) characterizes a newly emerging manufacturing paradigm. To gain significant competitive advantage and to meet the customized production requirement in TBC, manufacturers are shortening product life cycle (Ceglarek et al. 2004). This shortening of product life cycle has given rise to the alarming problem of waste management, which has prioritized remanufacturing activities and consequently increased the manufacturer's responsibility (Rai et al. 2002; Prakash and Tiwari 2005). Various factors such as legislation, depletion of natural resources, augmented cost of landfill, and enhanced public awareness have forced manufacturers to recycle and remanufacture their postconsumed products after they enter the waste stream (McGovern and Gupta 2007). A significant German reform program, based upon the Closed Substance Cycle and Waste Management Act of 1996, has prioritized remanufacturing activities, thereby increasing the manufacturer's responsibility (Feldmann et al. 1999). The remanufacturing and recycling of a product first requires dismantling to extract reusable and recyclable materials from the used product. Disassembly is a process for carrying out this dismantling and entails the methodical and economical separation of parts and subassemblies from a used or worn-out product or assembly.

In general, products are composed of various modules and submodules that are agglomerated to form the final product. This manifestation of modularity in structural composition has generated interest among the research community to pursue in-depth research in the domain of disassembly planning and sequencing problems. Disassembly has recently received increased attention because of its wide utility in product recovery, and the ease of disassembly after use is becoming a matter of concern while designing new products. The inclusion of disassembly operations as an inevitable element of the impending manufacturing paradigm has given rise to a broad area of research. This research is focused on maximizing the returns obtained and, simultaneously, minimizing the cost incurred during disassembly (Rai et al. 2002). The following are the primary objectives of product disassembly:

1. Extraction of valuable materials from products, which can be reused
2. Exclusion of hazardous components from products
3. Assistance in recycling for the manufacture of new components
4. Aid in the design of new products and support for the analysis of assembly processes
5. Assistance in repair operations

In addition to the extraction of both useful and malignant components to facilitate remanufacturing and recycling of worn-out products, designers and researchers are in search of new tools and techniques to decompose these worn-out products in an environmentally as well as an economically responsible manner. End-of-life (EOL) operations, growing into an issue of considerable industrial relevance, have necessitated this search. In essence, the ease with which parts can be disassembled plays a vital role in determining the commercial viability of reuse of these parts (McGovern and Gupta 2007). The design of fasteners, their location, and their number may have a substantial bearing on product disassembly and, hence, on the product's economic value after EOL. Currently, disassembly is carried out primarily in industries such as computers, automobiles, and defense (for the dismantling of weapons) (Taleb et al. 1997; Gungor and Gupta 1997; Meier 1993), where the quantity of products discarded per year is quite large and poses significant problems during product decomposition. Aerospace, construction, industrial equipment, and electronics are other industries where disassembly is applied during design, repair, and dismantling.

In this chapter, we attempt to optimize (maximize) the net profits from product or part recovery so as to minimize the disassembly cost in a real-world environment where knowledge about the quality of returned products as well as their constituent parts is vague. The ambiguity in the structure and the quality of the returned products, in addition to variations in reliability of the obtained components, poses a serious challenge to the development of a model that can effectively map the real-world disassembly scenario.

The fuzzy disassembly optimization problem (FDOP), proposed in this research, is aimed at determining the optimal disassembly sequence and the depth of disassembly to maximize net profits. Here, the fuzzy control theory has been used to model the variability in quality of the returned product. This problem of variability falls under the category of combinatorial optimization problems, which are computationally complex in nature (Lambert and Gupta 2007). Consideration of the depth of disassembly further complicates the problem of disassembly sequencing by including complete as well as incomplete disassembly sequences. This inclusion leads to a considerable increase in the number of sequences that are to be studied with respect to the number of possible assembly sequences (because a concept such as incomplete assembly does not exist).

The past decade witnessed a significant growth in the use of metaheuristic techniques for the optimization of computationally complex nondeterministic polynomial (NP)-hard problems of real dimensions (Agrawal and Tiwari 2006). Continuous improvements in this field in the past few years have spectacularly reduced the time of response of these metaheuristics without a significant depreciation in solution quality. Considering the complexity inherent in solving the FDOP, the use of artificial intelligence (AI) techniques for its resolution is considered a potential area of research, which is also evident from the increasing number of publications in this area (Agrawal and Tiwari 2006; McGovern and Gupta 2006, 2007). This chapter explores the application potential of several state-of-the-art AI techniques such as genetic algorithm (GA) (Goldberg 1987a,b), simulated annealing (SA) (Kirkpatrick et al. 1983), tabu search (TS) (Glover 1990), particle swarm optimization (PSO) (Kennedy and Eberhart 2001), and ant colony optimization (ACO) (Dorigo et al. 1996) on the proposed FDOP model. The performances of these techniques are bench-marked against a set of carefully generated test instances, and useful insights have been obtained on the performance of these approaches. Further, an anal-ysis is carried out to determine the impact of various factors employed in the modeling of FDOP on the profits associated with disassembly operations.

The rest of the chapter is organized as follows: Section 8.2 contains a literature review of related work, while Section 8.3 discusses FDOP. Section 8.4 illustrates the various random search techniques used for optimi-zation and the implementation procedure of these techniques on the prob-lem concerned. Computational results and insights are developed in Section 8.5, and finally, Section 8.6 concludes the chapter with suggested directions for study and research.

8.2   Literature Review

The literature pertaining to EOL considerations is very rich, and the problem of disassembly sequencing has been studied by various researchers who have made remarkable contributions in this area. The developments in this area can be categorized into three parts: product data modeling to facilitate sequence generation and recovery operations, economical considerations involved with disassembly operations, and the development of solution methodologies and techniques for disassembly sequence optimization. Several reviews of dis-assembly planning and scheduling literature have been published. O'Shea et al. (1998) provided a comprehensive review of 62 relevant papers published in the period from 1993 to 1996. Gungor and Gupta (1999) detail a more com-prehensive review of the field of environmentally conscious manufacturing. Works by Tang et al. (2000) and Lee et al. (2001) are other noteworthy reviews on disassembly sequencing. More recently, Lambert's work (2003) involved a thorough survey of disassembly sequencing literature. His work contained a review of 137 papers on disassembly sequencing. Dong and Arndt (2003) have reviewed current research on disassembly sequence generation and computer-aided design for disassembly. This chapter provides a brief review of the literature on disassembly sequencing; for detailed information, the reader is advised to refer to these resources.

The clear representation of product data facilitates the precise delineation of components, parts, joints, and subassemblies as well as the precedence relation-ships that exist between them. This delineation helps in the development of a disassembly model for analysis and optimization. Initially, in research, it was believed that "the sequence of assembly is the reverse of that of disassembly"

(Takeyama et al. 1983). Bourjault (1984) was the first to provide a comprehensive and systematic methods for assembly planning and introduced tools such as set theory, combinatorial analysis, Boolean algebra, and graph theory to deal with assembly and disassembly problems. Bourjault's work was extended by De Fazio and Whitney (1987) and later by Baldwin et al. (1991).

Homem de Mello and Sanderson (1990) proposed AND/OR graphs for the avoidance of deadlock in assembly optimization. An AND/OR graph (Lambert 2002) is a directed hypergraph, with a node of this graph representing the subassembly. The arcs emanating from the node point to the various modules (either parts or subassemblies) contained within it. According to the authors, this representation provided "a compact representation of all feasible assembly sequences" and became one of the most widely used methods of the relational product data representation for assembly and disassembly planning and sequencing. Jimenez and Torras (2000) presented an application of AND/OR graphs for disassembly.

The concept of AND/OR graph has been extended by many researchers. Lambert (2000) proposed a reduced disassembly graph to avoid the ambiguity that arises out of the concentration of the many entangled hyperarcs. Subramani and Dewhurst (1991) developed the disassembly diagram (DAD). Ishii et al. (1994) suggested a hierarchical network representation scheme called LINKER, which provided an efficient way of representing the physical and geometric connections and the parts. Moorie et al. (1998) proposed the automatic generation of disassembly Petri nets (DPNs) from the disassembly precedence matrix (DPM). Zussman and Zhou (1999) used DPN for the modeling and adaptive planning of disassembly processes. Tiwari et al. (2001) applied Petri nets for calculating indices for serviceability, disassembly, dismantling, recycling, and dumping. Rai et al. (2002), Kumar et al. (2003), and others have also used DPNs for disassembly modeling.

Sarin et al. (2006) proposed a disassembly optimization model (DOM) and an innovative approach for network representation based upon the information derived directly from the bill of material (BOM). Their scheme not only shows the mating relationships between parts and subassemblies but clearly depicts the joints and the fasteners that hold these components together. The problem addressed in this chapter follows the same scheme of network representation proposed by Sarin et al. (2006).

The analysis of EOL alternatives is important for the economic consideration of disassembly processes. It is clear that the benefits that result from reusing, recycling, or remanufacturing the components can outweigh the combined cost of disassembly and disposal of the complete product. Zussman et al. (1994) developed a tool for determining the EOL costs, including the costs of landfill, materials recycling, and component reuse. Harjula et al. (1996) discussed the minimization of disassembly costs. Kriwet et al. (1995) analyzed various recycling options for components. Boothroyd and Alting (1992) proposed the idea of design for assembly and disassembly (DFAD). Corbet (1996) introduced the concept of design for value maximization, and Stuart et al. (1995) described the mathematical modeling approach to evaluate the environmental impacts and yield trade-offs. Ishii et al. (1994) proposed a disassembly cost formula, which incorporates the time required for dismantling components and fasteners from assembly. The analysis of disassembly costs can also be found in the works of Kroll (1996), Kroll et al. (1996), Kroll and Carver (1999), Jovane and Semeraro (1997), and Jovane et al. (1998). The optimization of product design should also consider recycling of materials and the cost associated with recycling should be incorporated into the disassembly costs. In their work, Sodhi and Knight (1998), Knight and Sodhi (2000), and Sodhi and Reimer (2001) discuss the aforementioned aspects for electronic products.

There also exists a need for the discipline of design for environment (DFE) (Hill 1993), which encapsulates several aspects for using clean technologies and recyclable materials. In this context, Sandborn and Murphy (1999) proposed a model that considers both economic and environmental criteria for the optimization of product design in the virtual prototyping phase. Another paper on design for disassembly is by Veerakamolmal and Gupta (2000), who have advocated an integral approach that includes modularity, standardization, and conscious materials selection.

In addition to data modeling and cost analysis, research related to disassembly sequencing primarily focuses on the formulation of tools and techniques to achieve optimal benefits. In general, disassembly sequence optimization problems can be broadly categorized into those problems in which the sequence costs are immaterial and those problems in which these costs are taken into account. The first set of problems can be mapped through linear programs and are less complex to solve. These were addressed by Lambert (1999), Penev and de Ron (1996), and Kuo et al. (2000). The problems that constitute the second group involve sequence-dependent setup costs, and thereby integer variables have been shown to be NP-hard and computationally complex (Moyer and Gupta 1997). The underlying disassembly sequencing problem was treated as price-collecting traveling salesman problem (Balas 1989), with simple variations by Navin-Chandra (1994). Gungor and Gupta (1997) proposed a heuristic procedure for determining disassembly sequence. Lambert and Gupta (2002) proposed a tree network model for this problem, which is based on the disassembly graph approach (Lambert 1997, 1999) and reverse BOM. These authors solved the formulation using a mixed integer program. Recently, Sarin et al. (2006) formulated the disassembly optimization problem as a precedence-constrained asymmetric traveling salesman problem (TSP) and solved this problem using Lagrangian relaxation with a three-stage iterative procedure.

Nature-inspired random search techniques such as SA, GA, and TS have been extensively used in recent times for the optimization of computationally complex problems. These algorithms are marked by their short response times and high-quality solutions to the problems of real dimensions. Li et al. (1995) used SA, while Abe et al. (1999), Dini et al. (1999), Caccia and Pozzetti (1999), De Lit et al. (2001), and others applied GA for disassembly optimization. Seo et al. (2001) used GAs for the economic and environmental analyses of the disassembly process within the framework of a full disassembly chain. McGovern and Gupta (2007) applied GA to balance disassembly lines, and GA proved to be a highly efficient strategy when applied to problems of large dimensions. Ant algorithms have been utilized as optimization tools in disassembly operations by McGovern and Gupta (2006) who employed ACO for disassembly sequencing with multiple objectives. Agrawal and Tiwari (2006) made use of a collaborative ant colony algorithm to solve a stochastic mixed-model U-shaped disassembly line balancing problem. Their approach uses dual colonies of ants, which collaboratively share the information available with each other to identify the interrelated task and model sequences.

This chapter will further explore the implementation of several state-of-the-art random search techniques on the disassembly sequencing problem along with a consideration of ambiguities in the condition of the returned products.

8.3 Fuzzy Disassembly Optimization Problem

This section develops the mathematical model for the disassembly sequence optimization problem. In general, disassembly planning is more complex compared with assembly planning because of the following reasons:

1. Depth of disassembly, that is, disassembly is not performed to its full extent; incomplete disassembly is often preferred, which considerably increases the number of sequences to be studied

2. Uncertainty in the structure of the returned products or the parts or components constituting such products

3. Existence of multiple demand sources

4. Presence of hazardous components in the returned product

5. Variability in the structure of returned products (owing to the existence of different models)

6. Existence of AND/OR precedence constraints compared with AND constraints in assembly.

Therefore, disassembly planning requires the development of efficient models and tools and techniques to resolve the complexities encountered in disassembly sequencing. The proposed model is aimed at determining the disassembly sequence and the depth of disassembly to maximize the net revenue obtained from the recovery of components and the subassemblies minus the cost incurred while disassembling the product and disposing off the components. We have also incorporated the ambiguity in the available data through the use of fuzzy control theory with a view to exemplify the real-world disassembly scenario. The FDOP results in an optimal disassembly sequence of the multisubassembly system, aimed at achieving the following objectives:

1. Maximization of the profits obtained from the optimal disassembly sequencing of the products by simultaneously minimizing the sequence-dependent setup costs and the associated costs required for breaking the joints while preserving the essential precedence relationships

2. Incorporation of the previously accumulated knowledge available in the form of warranty or field service data in the variables of problem domain. The net recovered value of any module is obtained by a fuzzy logic controller (FLC) using the recovered quality parameter and the service area parameter.

8.3.1 Key Concepts

This subsection provides a brief overview of the concept and the key terms involved in the determination of a disassembly sequence.

8.3.1.1 Precedence Relations

Typically, an assembly is composed of modules consisting of parts and subassemblies. Modules may be defined as a group of components that are stable either because of their independence (as individual components) or because they are secured by fasteners (as subassemblies) (Lambert and Gupta 2007). These modules are arranged hierarchically in accordance with the product's structure or the product's BOMs. To extract any module from a subassembly, the latter must first be broken down. Thus, based on hierarchy, certain precedence relationships exist among the joints of a full assembly (FA). In the case of disassembly, such relationships may include the AND, OR, or complex AND/OR type of precedence relations (Gungor and Gupta 2002).

8.3.1.2 Depth of Disassembly

Complete disassembly of a product may include the parts having high disassembly cost and less recovery value. Thus, it is a prolific idea to omit the extraction of such parts from the optimal disassembly sequence and thereby limit the disassembly operation to a certain depth instead of carrying out the complete disassembly of the product. This incomplete disassembly sequencing operation, keeping in mind the methodological retrieval of modules, is termed as depth of disassembly and has been addressed in the proposed model.

8.3.2 The Model Assumptions

The FDOP assumptions include the following:

1. The recovered value of any module is dependent upon two input linguistic variables, that is, service area parameter and recovered quality parameter (discussed in Section 8.3.4).
2. The depth of disassembly is assumed to have a significant effect on the profits related to disassembly operations.

8.3.3 The Mathematical Model

Before stepping into the mathematical formulation, we attempt to explore the various solution sets involved in the formulation. In general, the term "full assembly" (FA) has been used to represent the product as a whole. Eight sets, defined as follows, have been used to formulate the objective function.

1. $P = \{p_1, p_2, ..., p_{n_P}\}$, the set of all the parts (within the FA) and n_P, the cardinality of the set P.
2. $SA = \{sa_1, sa_2, ..., sa_{n_{SA}}\}$, the set of all the subassemblies and n_{SA}, the cardinality of the set SA.
3. $F = \{f_1, f_2, ..., f_{n_F}\}$, the set of all the fasteners and joints and n_F, the cardinality of set F.
4. $FB = \{fb_1, fb_2, ..., fb_{n_{FB}}\}$, the set of fasteners broken and n_{FB}, its number.
5. $PR = \{pr_1, pr_2, ..., pr_{n_{PR}}\}$, the set of parts recovered after disassembly operation and n_{PR}, the cardinality of set PR.
6. $SR = \{sr_1, sr_2, ..., sr_{n_{SR}}\}$, the set of subassemblies recovered and n_{SR}, the cardinality of set SR. PR and SR form total parts and subassemblies recovered.
7. $SB = \{sb_1, sb_2, ..., sb_{n_{SB}}\}$, the set of subassemblies broken and n_{SB}, the cardinality of set SB.
8. $SOL = \{sol_1, sol_2, ..., sol_{n_{FB}+n_{SB}}\}$, a set formed of a particular feasible sequence from numerous possible permutations of $FB \cup SB$.

Consider $P = \{1, 3, 7, 8, 9, 12\}$, then $n_P = 6$ and $p_1 = 1, p_2 = 3, p_3 = 7, p_4 = 8, p_5 = 9, p_6 = 12$. A similar convention stands for the elements of other sets. Section 8.3.3.1 presents the mathematical model based on the above sets.

8.3.3.1 Objective Function

The aim of this chapter is to determine the optimal disassembly sequence, which maximizes the net profit attained from disassembly operations and is mathematically represented as

$$\text{Maximize } Z_P = \sum_{i=1}^{n_{\text{PR}}+n_{\text{SR}}} RC_{psa_i} - \sum_{i'=1}^{(n_P+n_{\text{SA}})-(n_{\text{PR}}+n_{\text{SR}})} DC_{psa_{i'}} - \sum_{j=1}^{n_{\text{FB}}+n_{\text{SB}}} BC_{sol_j}$$
$$- \sum_{k=1}^{n_{\text{FB}}+n_{\text{SB}}-1} W_{sol_k, sol_{k+1}} \tag{8.1}$$

where

$$psa_i \in P \cup \text{SA}, psa_{i'} \in (P \cup \text{SA}) - (\text{PR} \cup \text{SR}), sol_j \in \text{SOL}, sol_k \in \left(\text{SOL} - \left\{sol_{n_{\text{FB}}+n_{\text{SB}}}\right\}\right)$$

The purpose is to maximize the net profits represented by objective function value Z_P. RC_{psa_i} is the composite recovered value of part/subassembly, psa_i, $DC_{psa_{i'}}$ the disposal cost of part/subassembly $psa_{i'}$, BC_{sol_j} the breaking cost of joint/subassembly, sol_j, $W_{sol_k sol_{k+1}}$ the setup cost incurred if the joint/subassembly, sol_k is broken immediately before sol_{k+1}, and i, i', j, k are the counter variables.

The first term of the objective function symbolizes the composite recovered value of all the parts and subassemblies recovered and constitutes the only positive contribution to the profit Z_P. The second term represents the cost of disposal of uneconomical or unrecovered parts. The third term corresponds to the costs involved in breaking joints and subassemblies, while the last term represents the setup costs incurred due to the sequence in which the joints have to be broken. Thus, the total profit attained through disassembly is the sum of profits received from the selling of recovered components minus the costs incurred in disposal, setup, and disassembly operations.

8.3.3.2 Constraints

As previously discussed, certain precedence relationships may exist between the joints of a product. Thus, while determining the optimal disassembly sequence, any violation of precedence relations must be checked. We have used a DPM, which is based upon the BOM of the product. It is represented as R, where

$$R = \left[r_{j_1, j_2}\right], \quad j_1, j_2 \in \{1, 2, \ldots, (n_F + n_{\text{SA}})\} \tag{8.2}$$

where

$$r_{j_1, j_2} = \begin{cases} 1 & \text{joint } j_2 \text{ preceeds joint } j_1 \text{ in the network representation} \\ 0 & \text{otherwise} \end{cases} \tag{8.3}$$

Here, r_{j_1, j_2} is a decision variable that takes the value of 1 when the joint "j_2" precedes joint "j_1" and 0 when the joint "j_2" precedes joint "j_1."

8.3.4 Disassembly Modeling in Fuzzy Environment

We use fuzzy set theory to incorporate the vagueness and the imprecise nature of the recovery cost of the disassembled parts and subassemblies.

Application of fuzzy set theory to represent nonstatistical uncertainty and approximate reasoning encountered in real-life situations can be found in the literature (Gen et al. 1996; Mierswa 2005). The FLC is a conceptual input–output machine, capable of making certain decisions based on the input linguistic variables fed into it. At the end of the useful life of a product, its parts and subassemblies deteriorate as a function of time. These deteriorations, in turn, depend upon the parts and the subassemblies under consideration and vary according to the use of the product as well as the modules within the product. In this research, the FLC is structured as multiple input, single output (MISO) for making relevant decisions based upon the two input linguistic variables discussed below.

The product life cycle management (PLM) is primarily divided into three phases: design, manufacturing, and warranty or field service. The key concern of this research is to effectively and efficiently include the information obtained from the last phase of the PLM into the problem domain of variables associated with disassembly sequencing operations. Analysis of the warranty or field service data reveals vital information about the condition of the product, which is returned from different marketing and distribution areas. This information is based on the expected way of the product's usage and can be categorized in terms of linguistic variables, depending upon the area from which the product returns after its use. Such information constitutes the first input for our FLC and is termed as the service area parameter.

In addition, dissimilar deteriorations can occur for various modules within a product, depending upon the way of the product's usage as well as the manner in which the modules comprising it have been used. Exhaustive use of any single module or the higher degree of sophistication required for a module's use may lead to its early failure and the failure of the product as a whole while keeping other modules in a relatively better condition. For example, frequent use of a panel on the television set may lead to its early degradation compared with other internal circuitry of the product. Furthermore, the expected life of any product depends upon the minimum life expectancy of all the modules constituting it. Thus, we define a new attribute associated with the module, recovered quality parameter, which defines the quality of the modules at the EOL of the product. This parameter has been expressed in terms of linguistic variables and thus acts as the second input for our FLC. Table 8.1 sums up these two inputs and the output of our FLC.

TABLE 8.1

Inputs and Output of FLC

Name	Input/ Output	Minimum Value	Maximum Value
Service area parameter	Input	1	10
Recovered quality parameter	Input	1	10
Recovered cost	Output	$RC_{psa_i}/10$	RC_{psa_i}

An FLC has been designed for the underlying disassembly sequencing problem. This FLC works in accordance with the rule base for the fuzzy parameters. The general fuzzy statement for our problem states: "If the service area parameter for the product is good and the recovered quality parameter for a module under consideration is high then the expected cost of that module is high." Including the two fuzzy inference rules of generalized *modus ponens* and generalized *modus tollens* (Ross 1997) into the aforementioned statement and utilizing the expertise knowledge, we develop the fuzzy rule base for solving the problem. This constitutes a realistic approach, based on the concept of fuzzy systems, developed to approximate the actual cost of the recovered modules within the product.

After formulation of the rule base, the next step is to determine the membership functions of the input and the output parameters. In general, each membership function is defined by a start and an end point and can take either a linear or a nonlinear form, in which triangular, Gaussian, and sigmoid functions are the most prevalent (Ross 1997). In this chapter, the inputs are fuzzified using the triangular membership functions distributed symmetrically across the universe of discourse. Three fuzzy sets are used in each case with an overlapping of about 15% to 25%. If–then rules, which are derived from the rule base, are used to compute the fuzzified output that is ready for defuzzification. In this chapter, we have used the weighted average mean methodology in which the outputs can be defuzzified using the algebraic function as given in Equation 8.4a:

$$Z^* = \frac{\sum m_c(\tilde{z}) \cdot \tilde{z}}{\sum m_c(\tilde{z})} \tag{8.4a}$$

Here, $m_c(\tilde{z})$ is the output membership value of the fuzzified quantity $\tilde{z}$ and Z^* the output defuzzified value. This methodology weighs each membership function in the output by its respective membership value. This approach of defuzzification is applied only to the symmetric output membership functions and thus suits our case.

For instance, the two functions shown in Figure 8.1 would result in the following general form of the defuzzified value:

$$Z^* = \frac{\tilde{z}_1(0.3) + \tilde{z}_2(0.7)}{0.3 + 0.7} \tag{8.4b}$$

Thus, the information from fuzzy theory, based on the service area parameter and the recovered quality parameter, is used to obtain the expected cost of a module and helps in mapping real attributes into the proposed optimization model.

8.4 Random Search Techniques

Recently, random search techniques have gained increasing importance because of their ability to provide highly efficient and competitive solutions

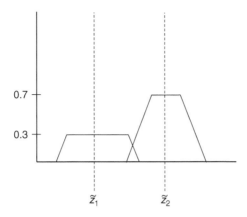

FIGURE 8.1
Illustration of the weighted average mean method for symmetrical and triangular membership functions and two input variables $\tilde{z}_1$ and $\tilde{z}_2$ (on x-axis) having membership function value 0.3 and 0.7 (on y-axis), respectively.

in a very short span of time. The complexities in manufacturing are growing, and the need to provide competitive alternatives at the earliest has motivated the development and the implementation of these search techniques on complex optimization problems. In this chapter, we will briefly illustrate different AI techniques and their implementation on the disassembly sequencing problem.

8.4.1 Genetic Algorithm

It is a powerful optimization tool that imitates the natural process of evolution and Darwin's principle of "survival of the fittest." The works of Goldberg (1989), Holland (1975), and Michaelwicz (1992) are regarded as the representative set of publications on GAs. Some of the typical applications of GAs include TSP (Grefenstette et al. 1985); scheduling problem (Davis 1985, Clevland and Smith 1989); VLSI circuit layout design problem (Fourmann 1985); computer-aided gas pipeline operation problem (Goldberg 1987a, 1987b); communication network control problem (Cox et al. 1991); real-time control problem in manufacturing systems (Lee et al. 1997); cellular manufacturing (Gupta et al. 1996); pattern classification (Bandyopadhyay et al. 1995, Bandyopadhyay and Pal 1998); assembly line balancing (Ponnambalam et al. 2000); and disassembly line balancing (McGovern and Gupta 2007). The efficient implementation of a GA to solve a problem requires that (1) the solution can be expressed as a string (chromosome) and (2) a fitness value corresponding to that string can be calculated. There are three main operators in a GA: selection, recombination, and mutation. After the solution is encoded as a string and a measure of fitness evaluation is chosen, the GA proceeds as shown in the pseudocode (Figure 8.2). Section 8.4.1.1 details the process of GA implementation on the disassembly sequencing problem.

```
Genetic Algorithm
{
        Initialization of population of chromosomes
        Evaluation of chromosomes
        While termination criteria is not satisfied
        {
                Selection of parent chromosomes to form mating pool
                Crossover operation to form offsprings
                Mutation in offsprings
                Evaluation of offsprings generated

        }

}
```

FIGURE 8.2
Pseudocode of genetic algorithm.

8.4.1.1 *String (Chromosome) Representation*

The first step in the implementation of a search technique, to any problem, is the representation of search space in terms of algorithmic parameters. For disassembly of a product containing n_F joints/fasteners and n_{SA} subassemblies, the knowledge-based string is n_t-tuple, where n_t represents the number of segments of the string and is defined as

$$n_t = n_F + n_{SA} + 1$$

Each string segment denotes a joint or a subassembly to be broken. The strings are generated in a manner to prevent the violation of constraints, while the violation of constraints during recombination is checked by the use of PPX crossover (precedence preservative crossover). One extra segment is reserved for saving a special number "0," which determines the depth of disassembly. Integer coding is performed for the string representations in such a way that each joint and each subassembly is assigned the value of a unique positive integer. Each string corresponds to a potential solution to the problem. This string representation can be interpreted as follows:

1. The joints and subassemblies are broken according to the sequence of their appearance in the solution string.
2. The depth of disassembly is ascertained by using the analogy behind the special number "0," which states that only those subassemblies and joints that lie before "0" in the solution string are allowed to be broken down for further disassembly.

As an example, consider the following 9-tuple string representation, <2 3 1 4 8 0 7 6 5>, where integers from 1 to 8 represent joints and subassemblies to be broken, while "0" identifies the depth of disassembly. Thus, from the afore-mentioned discussion, the final sequence of joints to be broken involves those joints that lie ahead of "0," that is, <2 3 1 4 8>. The associated subassemblies and parts recovered can be determined on the basis of the joints broken.

8.4.1.2 Selection

Tournament selection has been utilized in the experiments for the selection of parents. Other selection techniques commonly used with GAs include roulette wheel selection and elite selection, details regarding which can be found in Deb (2004).

8.4.1.3 Crossover

In crossover, two parent strings combine to create new, hopefully, better offspring. A number of crossover methods such as *k*-point crossover, uniform crossover, uniform order-based crossover, order-based crossover, partially matched crossover (PMX), cycle crossover (CX), and PPX have been designed (Goldberg 1989, Spears 1997) to achieve the fusion of genetic materials. Owing to the precedence preservation requirements in the disassembly sequencing problem, the PPX (Biewirth et al. 1996, McGovern and Gupta 2007) operator has been used. This operator preserves ordering within the chromosomes and thereby leads to the generation of feasible offspring. In PPX, first a mask is created. This mask consists of random 1s and 2s corresponding to the parent part from which the information has to be taken. For example, if the mask for child 1 reads 22221211, then the first four genes of second parent would make up the first four genes of child 1; the fifth gene of child 1 will comprise that first gene of parent 1 which is not present in child 1. Figure 8.3 depicts this process.

8.4.1.4 Mutation

Mutation is mainly used to impart exploitation in GAs and thereby prevent them from premature convergence. Mutation involves random changes in the genes comprising the chromosome. Swapping mutation has been used for the problem at hand, and this mutation randomly selects two bits and then swaps them.

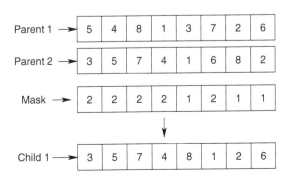

FIGURE 8.3
PPX example.

8.4.2 Simulated Annealing

Originally proposed by Kirkpatrick et al. (1983), SA is a random search technique that derives its inspiration from the physical annealing of solids. In physical annealing, a metal is brought to its lowest energy state by first heating it to a very high temperature (usually recrystallization temperature of metal) and then cooling it at a very slow rate to a very low temperature. If the cooling is not appropriately slow, it may result in quenching, which is not desirable. Based on the iterative improvement concept, the SA algorithm is in fact a heuristic method with the basic idea of generating random displacement from any feasible solution. This process accepts not only the generated solutions, which improve the objective function but also those solutions which do not improve it with the probability exp($-\Delta f/T$), a parameter depending on the objective function and decreasing temperature. The pseudocode of SA (Tiwari and Swarnkar 2004), delineating the sequence of operations, is given in Figure 8.4. For the implementation of SA on the disassembly sequencing problem addressed in this chapter, we use the same aforementioned

```
Simulated Annealing
{
      Initialize a solution save it in X
      Set initial temperature T_U and final temperature T_F
      Initialize current temperature T = T_U
      Initialize temperature counter k = 1
      Initialize iteration counter i = 0
      While  T ≥ T_F
      {
            While  i < i_MAX
            {
                  Generate   a   solution   in   neighborhood   of   current
solution X, the                new string is saved in Y
                  Evaluate fitness associated with point Y, F_Y
                  Find the difference between the fitness ΔF_YX = F_Y - F_X.
                  If ( ΔF_YX > 0 )
                        Y is selected as current solution i.e., X=Y.
                  Endif
                  If ( ΔF_YX < 0 )
                        Generate a random no.  r∈ (0,1)
                        If  r ≤ e^{-|Δz_YX|/T}
                              put X=Y
                        Endif
                  Endif
                  Increment i
            }
            i=0;
            Increment temperature counter k=k+1
            Update temperature as per cooling schedule T(k) = T_U /ln(k)
      }
}
```

FIGURE 8.4
Pseudocode of simulated annealing.

approach (Section 8.4.1.1) for solution encoding. SA avoids being trapped into local optima by selecting the better neighborhood points with certainty and, at the same time, occasionally accepting the comparatively worse points in the neighborhood with some nonzero probability that gradually decreases. This probability is expressed in terms of the acceptance function. In general, the performance of SA search technique is marked by four parameters, that is, choice of initial temperature, the epoch length or number of transitions made before the temperature is reduced, a proper cooling schedule for temperature reduction, and some terminating criteria.

8.4.3   Tabu Search

TS is regarded as a "high-level" iterative improvement procedure that is used for solving various computationally complex problems (Sarma et al. 2002). This procedure avoids terminating into a local optimum by maintaining a tabu list of forbidden moves. The technique is similar to SA as it uses a single particle rather than a population, as used in GA, PSO, and ACO. Many successful implementations of TS for obtaining near-optimal or optimal solutions of the problems pertaining to process planning, scheduling, and set partitioning have been reported (Glover 1990, Taillerd 1990).

The TS process begins with an initial feasible solution and attempts to find a better solution by investigating among a large pool of neighborhood solutions. This method is characterized by inherent simplicity, high adaptability, and a short-term memory via a tabu list. This process avoids recycling and subsequently also enables the backtracking to previous solutions. Aspiration criterion is a checking condition for the acceptance of a solution. The features of tabu list and aspiration make TS a powerful optimization tool for solving many combinatorial optimization problems. In general, the application of TS technique is characterized by (1) generation of initial feasible solution, (2) neighborhood generation, (3) tabu list size, (4) aspiration level, and (5) stopping criterion. Figure 8.5 shows the generic pseudocode of the TS algorithm.

8.4.4   Particle Swarm Optimization

Particle swarm is a metaheuristic proposed by Kennedy and Eberhart (2001) and inspired by the choreography of bird flock. In particle swarm, the particle's position determines its fitness and thereby the particles are inclined to move toward highly favorable positions. The movement of each particle of the swarm depends upon the particle's cognitive and social components. The cognitive component motivates the particle to attain the best position found by it so far, whereas the social component encourages the particle to move toward the global optimum. The combined effect of these two components ensures a better balance between the speed of convergence and the search space exploration, thereby making particle swarm an efficient optimization technique. Figure 8.6 depicts the pseudocode of the particle swarm algorithm.

```
Tabu Search
{
        Generate initial solution
        Evaluate the fitness, F
        While the stopping condition is not met
        {
                Identify Neighborhood set N(s)
                Identify Tabu set T(s,k)
                Identify aspirant Set A(s,k)
                Choose the best s'∈{N(s) - T(s,k)}∪A(s,k)
                Evaluate fitness of s', Fₛ'
                If (Fₛ'> Fₛ)
                        s=s'
                Else
                        Increment Tabu list
                Endif
        }
}
```

FIGURE 8.5
Pseudocode of tabu search.

```
Particle Swarm Optimization
{
        Initialize population of particles
        Evaluate fitness of particles, F
        Initialize particle's memory, Fᵐ
        Determine global best, gb
        While termination criteria is not reached
        {
                For each particle i
                {
                        Update velocity
                        Update position
                        Evaluate fitness of particle i, Fᵢ
                        If (Fᵢᵐ<Fᵢ)
                                Update particle's memory
                        Endif
                        If (Fᵢᵍᵇ<Fᵢ)
                                Update global best
                        Endif
                        Increment i
                }
        }
}
```

FIGURE 8.6
Pseudocode of particle swarm.

The implementation of particle swarm, which usually encodes real variables on the disassembly sequencing problem (which, in contrast, is a combinatorial optimization problem), is not a straightforward task. A generic encoding schema is devised to convert the real-valued solutions provided by the particle swarm into combinatorial solution with precedence preservation as required by the problem at hand. This schema amalgamates the features of directed graph and topological short (TS) algorithm. In a directed graph, vertices

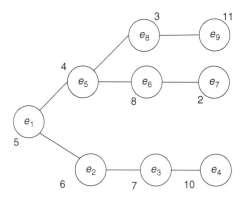

FIGURE 8.7
Directed graph for disassembly operation with precedence relationship.

represent operations, while edges represent precedence relations among dif-ferent operations. A directed edge is represented as $< ei\ ej >$ where vertex ei must be encountered before vertex ej, that is, task i must be accomplished prior to task j. Implementation of the technique initiates with the generation of a string representing priority numbers or ranks. All the feasible operations are listed, and the one with the highest priority number (lowest rank) is selected. To exemplify the process, consider the directed graph of Figure 8.7. The number outside of the vertex represents the priority values. At first, vertex $e1$ is selected as no precedence constraints exist for operation 1 and the edges corresponding to $e1–e2$ and $e1–e5$ are removed. Thereafter, vertex $e5$ is selected owing to its higher priority compared with $e2$, which is the only other feasible alternative. Therefore, vertex $e5$ is selected and the edges corresponding to $e5–e6$ and $e5–e8$ are removed. This procedure is repeated until the selection of the last vertex. Finally, a feasible sequence $\{e1, e5, e8, e2, e3, e6, e7, e4, e9\}$ is uniquely obtained. A major advantage obtained with the proposed schema is the absence of infea-sible sequences in the population. This schema enhances the metaheuristic search potential and maintains the required diversity in the population.

8.4.5 Ant Colony Optimization

Introduced by Dorigo (1992), the ant metaheuristics are inspired by the forag-ing behavior of ant colonies. Ethnologists found that real ants, although blind, construct the shortest path from their colony to the feeding source through the use of pheromone trails. The process is imitated in ACO through the use of a set of simple agents (i.e., artificial ants), which were allocated with computational resources and which exploit stigmergic communication (Stutzle and Dorigo 2002), that is, a form of indirect communication mediated by the environment, to find the solution to the problem at hand. During their motion, ants drop a certain amount of pheromone, a chemical substance used by ants to communi-cate with one another and exchange information about which course to follow. Ants mark their chosen path with a trail of this substance. An approaching

```
Ant Colony Optimization
{
      Initialize ants
      Initialize pheromone matrix
      Initialize visibility matrix
      While the termination criteria is not reached
      {
            For each ant i
            {
                  For each node n
                  {
                        Probabilistically select the next node to move
                        Update the Tabu list
                  }
                  Empty the Tabu list
            }
            Evaluate the solution
            Update the pheromone counts on path
      }
}
```

FIGURE 8.8
Pseudocode of ant colony optimization.

ant senses the pheromone on different paths and probabilistically selects a path in accordance with the probability that is directly proportional to the amount of pheromone on it. To ensure that the ants visit all the nodes, a special data structure called tabu list is associated with each ant. The tabu list saves the node already visited and thereby forbids the ants to visit them again. The precedence preservation in ACO has been performed by selecting only that node which did not belong to a node within the tabu list and which had no precedence constraint with any of the nodes contained within the tabu list as the next probable node. Figure 8.8 shows the pseudocode of the ACO.

During the start of the new tour, tabu list is emptied and the updated pheromone values serve to guide the future path of ants. This process is iteratively repeated for a designated maximum number of cycles, specified CPU time limit, or maximum number of cycles between two improvements of the global best solution. The ACO has been successfully applied to solve many complex NP-hard combinatorial optimization problems such as the TSP (Dorigo et al. 1996, Dorigo and Gambardella 1997), quadratic assignment problem (Stutzle and Hoos 1998, Maniezzo et al. 1999), vehicle routing problems (Bullnheimer et al. 1998), and disassembly line balancing problem (McGovern and Gupta 2006).

8.5 Experimental Design and Results

This section contains the experimental design used and the results obtained by the implementation of the aforementioned AI techniques on the problem at hand. The algorithmic performance is studied on a set of carefully generated test beds. The results portray the performance of various algorithms on the problem concerned; however, the superior performance of one approach over another cannot be ascertained by the results alone as rigorous parameter settings with suitable selection of operators need to be performed for getting optimum performance from any approach. These results demonstrate the efficacy and the applicability of all these metaheuristics, in real time, on the disassembly sequencing problem.

8.5.1 Test-Bed Formulation

In general, a product is composed of numerous parts that have a complex structure. This complexity is achieved by modularizing the product into smaller subassemblies. This modular assembly provides ease of manufacturing of the product as well as in the analysis of the product. The problem discussed in this chapter involves numerous factors. Out of these factors, seven factors were identified as having a major impact on the results. These factors varied at two different levels (low and high), and their effect on the objective function value was studied.

The factors considered for this study are as follows:

1. *Number of parts* (n_p). This factor directly reflects the total number of parts that could be separated out of the worn-out products. This includes the parts contained either directly within the FA or within any subassembly lying underneath it. The factor is varied at two levels of 20 and 40 parts.

2. *Number of subassemblies* (n_{SA}). The problem instances are varied at two different levels of this factor, that is, 3 (low) and 6 (high).

3. *Number of joints* (n_J). The total number of joints obtained by the union of joints within all the subassemblies as well as those joints within the FA constitutes the parameter n_J. This term forms a deciding factor for the determination of problem size and is kept at 20 for low and 40 for high level.

4. *Setup costs* ($W_{sol_k, sol_{k+1}}$). The sequence-dependent costs between the two joints are randomly generated at the interval (1, 5) for low level and (10, 15) for high level.

5. *Joint-breaking costs* (BC_{sol_j}). The costs for breaking the joints are usually dissimilar for different joints and are generated uniformly at an interval of (10, 20) for low level and (30, 40) for high level.

6. *Recovery value* (RC_{psa_i}). It is the actual value of the part or the subassembly after recovery from the FA at its EOL and is assumed to be uniformly distributed over the range of (50, 70) and (90, 110) for the two levels during the analysis.

7. *Disposal costs* (DC_{psa_r}). It represents the costs involved in the disposal of unextracted or unrecovered parts and components. In the experimental design used, these costs are generated at two different levels at the uniform interval of (5, 10) and (20, 25).

8.5.2 Parameter Settings

The aforementioned algorithms were coded in C++ and compiled on GNU C++ compiler (GCC) using KDevelop on a Linux platform. Further, the compiled program was run on a system specification of 1.5 GHz Pentium IV processor and 768 MB RAM. The parameters chosen to perform the experiments were taken from the relevant research results reported in the literature. In the case of GA, a population size of 100 chromosomes, a crossover probability of 0.7, and a mutation probability of 0.1 were used. For SA, the choice of initial and final temperature was set equal to 1000°C and 0.7°C, respectively. The convergence rate was controlled by the selection of a suitable cooling schedule, which we took as the Boltzman function ($T(k) = T_U/ln(k)$) in the present case. Moreover, before making a transition from $T(k)$ to $T(k + 1)$, 50 neighbors of the current solution were evaluated by the acceptance function. A tabu list of 50 particles was used for TS.

A swarm size of 50 particles was used for the PSO, for each problem instance, and each particle represented a vector containing $n_F + n_{SA} + 1$ elements. The number of cities and the ants in the ACO system were both posited to be equal to $n_F + n_{SA} + 1$. Other matching parameters in the pseudorandom proportional rule were adapted from those used by Dorigo and Gambardella (1997). The maximum number of functional evaluations were set as the termination criteria to evaluate both types of approaches (population-based and nonpopulation-based) used for testing on the same grounds.

8.5.3 Performance Comparison

Eight problem instances were generated with due consideration to the Taguchi's $L_8(2^7)$ orthogonal arrays as shown in Table 8.2. The characteristics and the search capability of the five algorithms were compared, and the results are provided in Table 8.3 for detailed analysis. The results clearly reveal that ACO and PSO outperform other algorithms in terms of maximizing the net profits

TABLE 8.2

L_8 Orthogonal Array Design

Experiment No.[a]	n_P	n_{SA}	n_F	$W_{sol_k, sol_{k+1}}$	BC_{sol_j}	RC_{psa_i}	DC_{psa_i}
1	1	1	1	1	1	1	1
2	1	1	1	2	2	2	2
3	1	2	2	1	1	2	2
4	1	2	2	2	2	1	1
5	2	1	2	1	2	1	2
6	2	1	2	2	1	2	1
7	2	2	1	1	2	2	1
8	2	2	1	2	1	1	2

[a] All symbols are detailed in the text.

TABLE 8.3

Performance Comparison

Experiment No.	Algorithms				
	ACO	PSO	SA	Tabu	GA
1	289.6	282.21	280.77	277.94	245.99
2	248.85	210.95	200.94	198.73	195.23
3	481.37	439.19	436.78	426.71	418.14
4	−541.65	−563.32	−591.89	−594.74	−610.22
5	−396.13	−405.65	−437.87	−442.24	−458.81
6	1890.07	1841.41	1834.21	1833.2	1764.66
7	2022.32	1974.46	1977.09	1971.87	1956.99
8	291.67	282.6	282.16	278.24	272.23

attained in disassembly operation, and ACO was found to be better than the other four in all the circumstances. The success of the algorithm lies in the fact that the pseudorandom proportional rule of ACO provides a balance between exploitation and exploration, and, at the same time, a minimum pheromone count on the paths ensures exploration on all paths throughout the process and avoids premature convergence. Another reason for ACO's success on the disassembly sequencing problem is the graphical representation of the problem as TSP, over which ACO is known to perform significantly well.

Further, to provide detailed insights, the best fitness values obtained from ACO, PSO, GA, SA, and TS algorithm were plotted against the total number of functional evaluations (Figure 8.9) to estimate the convergence trends of the algorithms. It can be seen from Figure 8.9 that all the algorithms have good convergence property and portray competitive performance.

8.5.4 Analysis of Results

From Table 8.3, it is clear that ACO outperforms the remaining approaches when applied to the underlying problem. To draw insights on the effect of different parameters, which are estimated to have some influence on the final objective, a statistical technique, ANOVA (analysis of variance), was applied over the results obtained after the implementation of ACO algorithm on each problem instance of the orthogonal array. F-test was carried out at a relatively high significance level of 95%, which clearly showed that the value of F obtained for the model distinctly surpassed the $F_{critical}(2.75)$, indicating high accuracy and repeatability in the performance of the algorithm. The results of the ANOVA test are detailed in Table 8.4.

The ability of a factor to affect the objective function value depends on its contribution to total sum of squares (Phadke 1989). It is clearly visible from Table 8.5 that the percentage contribution of the recovery cost (RC_{psa_i}) is highest (about 50% of the total contribution to the objective function value) followed by the number of parts (n_p) and disposal cost (DC_{psa_i}), which also have

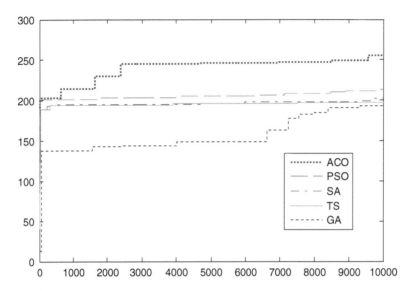

FIGURE 8.9
Convergence curve for the five algorithms on the problem (FDOP). x-axis represents number of fitness evaluations and y-axis represents the global best value attained.

some significant contribution to the same. The number of subassemblies (n_{SA}) and the setup costs ($W_{sol_k, sol_{k+1}}$) seem to have a negligible effect on the formulated model of the FDOP. In a nutshell, the aforementioned analysis not only statistically validates the efficiency of our algorithm but also explores new dimensions that can aid in designing for disassembly.

8.6 Summary and Conclusion

This chapter addressed a real-world product disassembly sequence optimization problem. The proposed FDOP model has a wider applicability and works with the real systems encountered in daily life. The model makes use of the warranty or field service information for prediction of recovered quality parameter. Moreover, the same source of information also predicts the type of use of the product based on the area from which it is received during reverse logistic activities. This information, in the form of linguistic variables, is amalgamated using the concept of fuzzy logic to get the resultant effect on the recovery cost of the modules.

In this chapter, we considered the fuzzified model of the disassembly optimization problem that aims at minimizing the disassembly costs while maximizing the benefits associated with the component recovery. Random search techniques were applied to test its efficacy over a set of problem instances of varying size and complexity. The results from the preliminary experiments revealed that the AI techniques such as ACO, PSO, GA, TS, and

TABLE 8.4

Results of ACO and Experimental Design

Experiment No.	n_P	n_{SA}	n_F	$W_{sol_k, sol_{k+1}}$	BC_{sol_j}	RC_{psa_i}	DC_{psa_i}	Run1	Run2	Run3	Run4	Run5	Average
1	1	1	1	1	1	1	1	288.92	292.32	278.64	292.45	295.67	289.60
2	1	1	1	2	2	2	2	257.29	255.17	231.11	260.62	240.06	248.85
3	1	2	2	1	1	2	2	480.02	488.73	460.62	492.26	485.22	481.37
4	1	2	2	2	2	1	1	−547.83	−550.51	−535.63	−543.37	−530.91	−541.65
5	2	1	2	2	2	1	2	−388.6	−399.81	−388.82	−400.21	−403.21	−396.13
6	2	1	2	2	1	2	1	1894.09	1902.78	1896.32	1845.22	1911.94	1890.07
7	2	2	1	1	2	2	1	2028.46	2011.31	2036.48	2016.74	2018.61	2022.32
8	2	2	1	2	1	1	2	295.3	295.59	292.29	287.68	287.49	291.67

Note: 1 = low and 2 = high.

TABLE 8.5

The ANOVA Analysis

	Sum of Squares	DF	Mean Square	F	Percent Contribution
No. of parts	6929563.536	1	6929563.536	43741.01	22.06918
No. of subassemblies	30614.089	1	30614.089	193.2432	0.096997
No. of joints	1258085.43	1	1258085.43	7941.326	4.006321
Setup costs	161429.7302	1	161429.7302	1018.982	0.513626
Joint breaking costs	1638873.289	1	1638873.289	10344.95	5.219077
Recovery costs	15619500.48	1	15619500.48	98593.9	49.74541
Disposal costs	5755422.36	1	5755422.36	36329.56	18.32971
Model	31393488.92	7	4484784.131	28309	–
Residual	5069.5226	32	158.4225813	–	–
Total	31398558.44	39	805091.2	–	–

SA, which were used on the test bed, gave a satisfactory performance. This performance indicates the need for further research and exploration. The experiment was designed for statistically studying the obtained results as well as validating the repeatability and efficiency of the results and thereby the algorithm. Insights regarding the contribution of various factors on the objective function value were acquired through the ANOVA analysis.

The following directions for future research are suggested to interested readers: (i) inclusion of certain other disassembly considerations such as removal of hazardous parts or high demand parts early in the disassembly process and (ii) use of the multiobjective techniques for solving the problem.

Acknowledgment

Professor M.K. Tiwari is thankful to Professor S.M. Gupta and Professor A.J.D. Lambert for their encouragement to pursue this research. In fact, the authors learned the basics of disassembly modeling through their significant contributions published in different journals.

References

Abe, S., Murayama, T., Oba, F., and Narutaki, N., Stability check and reorientation of subassemblies in assembly planning. *Proceedings of the IEEE International Conference on Systems, Man, and Cybernetics*, 1999, **2**, 486–491.

Agrawal, S., and Tiwari, M. K., A collaborative ant colony algorithm to stochastic mixed-model U-shaped disassembly line balancing and sequencing problem. *International Journal of Production Research*, 2006, Available online, DOI: 10.1080/00207540600943985.

Balas, E., The prize collecting traveling salesman problem. *Networks*, 1989, **19**(6), 621–636.

Baldwin, D. F., Abell, T. E., Lui, M. M., De Fazio, T. L., and Whitney, D. E., An integrated computer aid for generating and evaluating assembly sequences for mechanical products. *IEEE Transactions on Robotics and Automation*, 1991, **7**, 78–94.

Bandyopadhyay, S., Murthy, C. A., and Pal, S. K., Pattern classification using genetic algorithms, *Pattern Recognition Letters*, 1995, **16**, 801–808.

Bandyopadhyay, S. and Pal, S. K., Incorporating chromosome differentiation in genetic algorithms, *Information Science*, 1998, **104**, 293–319.

Biewirth, C., Mattfeld, D. C., and Kopfer, H., On permutation representations for scheduling problems, parallel problem solving from nature. In: Voigt, H. M., Ebeling, I., Rechenberg, I., Schwefel, H. P. (Eds.), *Lecture Notes in Computer Science, 1141*. Springer, Berlin, Germany, 1996, pp. 3.10–3.18.

Boothroyd, G. and Alting, L., Design for assembly and disassembly. *Annals of the CIRP*, 1992, **41**(2), 625–636.

Bourjault, A., Contribution aune approcheme thodologique de l'assemblage automatise: elaboration automatique des sequences operatoires (Besanc̣ on, France: Faculty of Science and Technology, Universite de Franche-Comte), Ph.D. Thesis, November 12, 1984 (in French).

Bullnheimer, B., Hartl, R. F., and Strauss, C., Applying the ant system to the vehicle routing problem. In: Voss, S., Martello, S., Osman, I. H., Roucairol, C. (Eds.), *Meta-Heuristics: Advances and Trends in Local Search Paradigms for Optimization*, Kluwer, Boston, 1998, pp. 109–120.

Caccia, C. and Pozzetti, A., A genetic algorithm for disassembly strategy definition. Proceedings of 1999 SPIE Conference on Environmentally Conscious Manufacturing, 1999, pp. 68–77.

Ceglarek, D., Huang, W., Zhou, S., Ding, Y., Kumar, R., and Zhou, Y., Time-based competition in multistage manufacturing: stream-of-variation analysis (SOVA) methodology—review. *The International Journal of Flexible Manufacturing Systems*, 2004, **16**, 11–44.

Clevland, G. A. and Smith, S. F., Using genetic algorithms to scheduling flow shop releases. *Proceedings of Third International Conference on Genetic Algorithm and Their Applications*. Morgan Kaufmann, Palo Alto, CA, 1989, pp. 160–169.

Corbet, K. S., Design for value maximization: putting a business lens on environmental activities. Proceedings of the 1996 IEEE International Symposium on Electronics and the Environment, 1996, pp. 81–86.

Cox, L. A., Davis, L., and Qiu, Y., Dynamic anticipatory routing in circuit-switched telecommunications networks. In: *Handbook of Genetic Algorithms.* Van Nostrand Reinhold, New York, 1991.

Davis, L., Job shop scheduling with genetic algorithms. *In Proceedings of the First International Conference on Genetic Algorithms and their Applications*, Morgan Kaufmann, 1985, pp. 136–140.

Deb, K., *Optimization For Engineering Design: Algorithms and Examples*, Prentice Hall of India, New Delhi, 2004.

De Fazio, T. L. and Whitney, D. E., Simplified generation of all mechanical assembly sequences. *IEEE Journal of Robotics and Automation*, 1987, **3**(6), 640–658.

De Lit, P., Latinne, P., Rekiek, B., and Delchambre, A., Assembly planning with an ordering genetic algorithm. *International Journal of Production Research*, 2001, **39**(16), 3623–3640.

Dini, G., Failli, F., Lazzerini, B., and Marcelloni, F., Generation of optimized assembly sequences using genetic algorithms. *CIRP Annals*, 1999, **48**(1), 17–20.

Dong, J. and Arndt, G., A review of current research on disassembly sequence generation and computer aided design for disassembly. *Proceedings of the Institution of Mechanical Engineers, Part B. Journal of Engineering Manufacturer*, 2003, **217**(3), 299–312.

Dorigo, M., Optimization, Learning and Natural Algorithms. Ph.D. Thesis, Politecnico di Milano, Italy, [in Italian], 1992.

Dorigo, M. and Gambardella, L. M., Ant colony system: a cooperative learning approach to the travelling salesman problem. *IEEE Transactions on Evolutionary Computation*, 1997, **1**(1), 53–66.

Dorigo, M., Maniezzo, V., and Colorni, A., The ant systems: optimization by a colony of cooperative agents. *IEEE Transactions on Man, Machine and Cybernetics—Part B*, 1996, **26**(1), 29–41.

Feldmann, K., Trautner, S., and Meedt, O., Innovative disassembly strategies based on flexible partial destructive tools. *Annual Reviews in Control*, 1999, **23**, 159–164.

Fourmann, M. P., Comparison of symbolic layout using genetic algorithm. *Proceedings of the First International Conference on Genetic Algorithms and their Applications*, Lawrence Erlbaum Associate, Hillsdale, NJ, 1985, 141–155.

Gen, M., Tsujimura, Y., and Li, Y., Fuzzy assembly line balancing using genetic algorithms. *Computers and Industrial Engineering*, 1996, **31**(3/4), 631–634.

Goldberg, D. E., Computer aided gas pipeline operation using genetic algorithms and rule learning, part 1: genetic algorithms in pipeline optimization. *Engineering with Computers*, 1987a, **3**(1), 35–45.

Goldberg, D. E., Computer aided gas pipeline operation using genetic algorithms and rule learning, part 2: rule learning control of a pipeline under normal and abnormal conditions. *Engineering with Computers*, 1987b, **3**(1), 47–58.

Goldberg, D. E., *Genetic Algorithms in Search, Optimization and Machine Learning*. Addison-Wesley, Reading, MA, 1989.

Glover, F., Tabu search: a tutorial. *Interfaces*, 1990, **20**(4), 74–94.

Grefenstette, J. J., Gopal, R., Rormatia, B., and Vangucht, D., Genetic algorithms for travelling salesman problem In: *Proceedings of the First International Conference on Genetic Algorithms and their Applications*. Lawrence Erlbaum Associates, Hillsdale, NJ, 1985.

Gupta, M., Gupta, Y., Kumar, A., and Sundram, C., A genetic algorithm-based approach to cell composition and layout design problems. *International Journal of Production Research*, 1996, **34**(2), 447–482.

Gungor, A. and Gupta, S. M., An evaluation methodology for disassembly processes. *Computers & Industrial Engineering*, 1997, **33**(1/2), 329–332.

Gungor, A. and Gupta, S. M., Issues in environmentally conscious manufacturing and product recovery: a survey. *Computers and Industrial Engineering*, 1999, **36**, 811–853.

Gungor, A. and Gupta, S. M., Disassembly line in product recovery. *International Journal of Production Research*, 2002, **40**(11), 2569–2589.

Harjula, T., Rapoza, B., Knight, W. A., and Boothroyd, G., Design for disassembly and the environment. *CIRP Annals*, 1996, **45**(1), 109–114.

Hill, B., Industry's integration of environmental product design. *IEEE International Symposium on Electronics and the Environment*, 1993, 64–68.

Holland, H. H., *Adaptation in Natural and Artificial Systems*. University of Michigan Press, Detroit MI, 1975.

Homem de Mello, L. S. and Sanderson, A. C., AND/OR graph representation of assembly plans. *IEEE Transactions on Robotics and Automation*, 1990, **6**(2), 188–198.

Ishii, K., Eubanks, C. F., and Di Marco, P., Design for product retirement and material life-cycle. *Materials and Design*, 1994, **15**(4), 225–233.

Jimenez, P. and Torras, C., An efficient algorithm for searching implicit AND/OR graphs with cycles. *Artificial Intelligence*, 2000, **124**(1), 1–30.

Jovane, F. and Semeraro, Q., Computer-aided disassembly planning as a support to product redesign. Proceedings of 4th CIRP International Seminar on Life Cycle Engineering, 1997, 388–399.

Jovane, F., Semeraro, Q., and Armillotta, A., On the use of the profit rate function in disassembly process planning. *The Engineering Economist*, 1998, **43**(4), 309–330.

Kennedy, J. and Eberhart, R. C., *Swarm Intelligence*, Morgan Kaufmann, San Mateo, CA, 2001.

Kirkpatrick, F., Gelatte, C. D., and Vecchi, M. P. (Eds.), Optimization by simulated annealing. *Science*, 1983, **220**, 671–680.

Knight, W. A. and Sodhi, M., Design for bulk recycling: analysis of materials separation. *CIRP Annals: Manufacturing Technology*, 2000, **49**(1), 83–86.

Kriwet, A., Zussman, E., and Seliger, G., Systematic integration of design-for-recycling into product design. *International Journal of Production Economics*, 1995, **38**(1), 15–32.

Kroll, E., Application of work-measurement analysis to product disassembly for recycling. *Concurrent Engineering*, 1996, **4**(2), 149–156.

Kroll, E., Beardsley, B., and Parulian, A., A methodology to evaluate ease of disassembly for product recycling. *IEE Transactions*, 1996, **28**, 837–845.

Kroll, E. and Carver, B. S., Disassembly analysis through time estimation and other metrics. *Robotics and Computer Integrated Manufacturing*, 1999, **15**, 191–200.

Kumar, S., Kumar, R., Shankar, R., and Tiwari, M. K., Expert enhanced coloured stochastic Petri net and its application in assembly/disassembly. *International Journal of Production Research*, 2003, **41**(12), 2727–2762.

Kuo, T. C., Zhang, H. C., and Huang, S. H., Disassembly analysis for electromechanical products: a graph-based heuristic approach. *International Journal of Production Research*, 2000, **38**(5), 903–1007.

Lambert, A. J. D., Optimal disassembly of complex products. *International Journal of Production Research*, 1997, **35**, 2509–2523.

Lambert, A. J. D., Linear programming in disassembly/cluster sequence generation. *Computers & Industrial Engineering*, 1999, **36**, 723–738.

Lambert, A. J. D., Optimum disassembly sequence generation. Proceedings of 2000 SPIE Conference of Environmentally Conscious Manufacturing, 2000, pp. 56–67.

Lambert, A. J. D., Determining optimum disassembly sequences in electronic equipment. *Computers and Industrial Engineering*, 2002, **43**(3), 553–575.

Lambert, A. J. D. and Gupta, S. M., Demand-driven disassembly optimization for electronic products. *Journal of Electronics Manufacturing*, 2002, **11**(19), 121–135.

Lambert, A. D. J., Disassembly sequencing: a survey. *International Journal of Production Research*, 2003, **41**(16), 3721–3759.

Lambert, A. D. J. and Gupta S. M., *Disassembly Modeling for Assembly, Maintenance, Reuse, and Recycling*, CRC Press, Boca Raton, FL, 2005.

Lee, C. Y., Piramuthu, S., and Tsai, Y. K., Job shop scheduling with genetic algorithm and machine learning. *International Journal of Production Research*, 1997, **35**(4), 1171–1191.

Lee, D. H., Kang, J. G., and Xirouchakis, P., Disassembly planning and scheduling: review and further research. *Proceedings of Institute of Mechanical Engineers*, 2001, **215**(5), 695–710.

Li, W., Zhng, C., Wang, H. P. B., and Awoniyi, S. A., Design for disassembly analysis for environmentally conscious design and manufacturing. Proceedings of ASME International Mechanical Engineering Congress and Exposition, 1995, **2**, 969–976.

Maniezzo, V., Colorni, A., and Dorigo, M., The ant system applied to the quadratic assignment problem. *IEEE Transaction on Knowledge and Data Engineering*, 1999, **11**(50), 769–778.

McGovern, S. M. and Gupta, S. M., A balancing method and genetic algorithm for disassembly line balancing. *European Journal of Operational Research*, 2007, **179**(3), 692–708.

McGovern, S. M. and Gupta, S. M., Ant colony optimization for disassembly sequencing with multiple objectives. *International Journal of Advanced Manufacturing and Technology*, 2006, **30**(5–6), 481–496.

Meier, B., Breaking Down an Arms Buildup: Dismantling and Recycling Weapons. *New York Times* (Late New York Edition), 1993, pp. D1–D2.

Michaelwicz, Z., *Genetic Algorithm + Data Structure = Evaluation Programs*. Springer, New York, 1992.

Mierswa, I., Incorporating Fuzzy Knowledge into Fitness: Multiobjective Evolutionary 3D Design of Process Plants. *Genetic and Evolutionary Computation Conference, Proceedings of the 2005 conference on Genetic and Evolutionary Algorithm*, ACM Press, 2005, pp. 1985–1992.

Moorie, K. E., Gungor, A., and Gupta, S. M., A Petrinet approach to disassembly process planning. *Computers & Industrial Engineering*, 1998, **35**(1), 165–168.

Moyer, L. and Gupta, S. M., Environmental concerns and recycling/disassembly efforts in the electronics industry. *Journal of Electronics Manufacturing*, 1997, **7**(1), 1–22.

Navin-Chandra, D., The recovery problem in product design. *Journal of Engineering Design*, 1994, **5**(1), 65–86.

O'Shea, B., Grewal, S. S., and Kaebernick, H., State of the art literature survey on disassembly planning. *Concurrent Engineering*, 1998, **6**(4), 345–357.

Penev, K. D. and de Ron, A. J., Determination of a disassembly strategy. *International Journal of Production Research*, 1996, **34**(2), 495–506.

Phadke, M. S., *Quality Engineering using Robust Design*, Prentice Hall, Englewood Cliffs, NJ, 1989.

Ponnambalam, S. G., Aravindan, P., and Naidu, G. M., A multi-objective genetic algorithm for solving assembly line balancing problem. *International Journal of Advanced Manufacturing Technology*, 2000, **16**(5), 341–352.

Prakash, and Tiwari, M. K., Solving a disassembly line balancing problem with task failure using a psycho-clonal algorithm, International Design Engineering Technical Conferences and Computers and Information in Engineering Conference, DETC2005, Long Beach, CA, 2005.

Rai, R., Rai, V., Tiwari, M. K., and Allada, V., Disassembly sequence generation: a Petri net based heuristic approach. *International Journal of Production Research*, 2002, **40**(13), 3183–3198.

Ross, T. J., *Fuzzy Logic with Engineering Applications, International Editions*, McGraw-Hill, New York, NY, 1997.

Sandborn, P. A. and Murphy, C. F., A model for optimizing the assembly and disassembly of electronic systems. *IEEE Transactions on Electronics Packaging Manufacturing*, 1999, **22**(2), 105–117.

Sarin, S. C., Sherali, H. D., and Bhootra, A., A precedence-constrained asymmetric travelling salesman model for disassembly optimization. *IIE Transactions*, 2006, **38**, 223–237.

Sarma, U. M. B. S., Kant, S., Rai, R., and Tiwari, M. K., Modeling the machine loading problem of FMSs and its solution using a tebu-search-based heuristic. *International Journal of Computer Integrated Manufacturing*, 2002, **15**(4), 285–295.

Seo, K. K., Park, J. H., and Jang, D. S., Optimal disassembly sequence using genetic algorithms considering economic and environmental aspects. *International Journal of Advanced Manufacturing Technology*, 2001, **18**, 371–380.

Sodhi, M. and Knight, W. A., Product design for disassembly and bulk recycling. *CIRP Annals: Manufacturing Technology*, 1998, **47**(1), 115–118.

Sodhi, M. and Reimer, B., Models for recycling electronics end-of-life products. *OR Spektrum*, 2001, **23**, 97–115.

Spears, W., Recombination parameters. In: Back, T., Fogel, D. B., Michalewicz, Z. (Eds.), *The Handbook of Evolutionary Computation*. Chapter E1.3, IOP Publishing and Oxford University Press, Philadelphia, PA, 1997, pp. E1.3:1–E1.3:13.

Stuart, J. A., Ammons, J. C., Turbini, L. J., Saunders, F. M., and Saminathan, M., Evaluation approach for environmental impact and yield tradeoffs for electronics manufacturing product and process alternatives. Proceedings of the 1995 IEEE International Symposium on Electronics and the Environment, IEEE, Piscataway, NJ, 1995, pp. 166–170.

Stutzle, T. and Dorigo, M., A short convergence proof for a class of ant colony optimization algorithms. *IEEE Transactions on Evolutionary Computation*, 2002, **6**(4), 358–365.

Stützle, T. and Hoos. H. H., MAX-MIN ant system and local search for combinatorial optimization problems. In: Voss, S., Martello, S., Osman, I. H., Roucairol, C. (Eds.), *Meta-Heuristics: Advances and Trends in Local Search Paradigms for Optimization*, Kluwer Academic Publishers, Dordrecht, The Netherlands, 1998, pp. 137–154.

Subramani, A. K. and Dewhurst, P., Automatic generation of product disassembly sequence, *Annals of the CIRP*, 1991, **40**(1), 115–118.

Taillerd, E., Some efficient heuristic methods for flow shop sequencing problems. *European Journal of Operational Research*, 1990, **47**, 65–74.

Takeyama, H., Sekiguchi, H., Kojima, T., Inoue, K., and Honda, T., Study on automatic determination of assembly sequence. *CIRP Annals*, 1983, **32**(1), 371–374.

Taleb, K. N., Gupta, S. M., and Brennan, L., Disassembly of complex product structures with parts and materials commonality. *Production Planning and Control*, 1997, **8**(3), 255–269.

Tang, Y., Zhou, M. C., Zussman, E., and Caudill, R., Disassembly modelling, planning, and application: a review. Proceedings of 2000 IEEE International Conference on Robotics and Automation, 2000, **3**, 2197–2202.

Tiwari, M. K., Sinha, N., Kumar, S., Rai, R., and Mukhopadhyay, S. K., A Petri net based approach to determine the disassembly strategy of a product. *International Journal of Production Research*, 2001, **40**(5), 1113–1129.

Tiwari, M.K., and Swarnkar, R., Modeling machine loading problem of FMSs and its solution methodology using a hybrid tabu search and simulated annealing-based heuristic approach. *Robotics and Computer-Integrated Manufacturing*, 2004, **20**, 199–209.

Veerakamolmal, P. and Gupta, S. M., Design for disassembly, reuse and recycling. In: Goldberg, L. (Ed.), *Green Electronics/Green Manufacturing: Environmentally Responsible Engineering*, Butterworth-Heinemann Newnes, Chapter 5, 2000, pp. 69–82.

Zussman, E., Kriwet, A., and Seliger, G., Disassembly-oriented methodology to support design for recycling. *CIRP Annals*, 1994, **43**(1), 9–14.

Zussman, E. and Zhou, M., A methodology for modeling and adaptive planning of disassembly processes. *IEEE Transactions on Robotics and Automation*, 1999, **15**(1), 190–194.

9

Human-in-the-Loop Disassembly Modeling and Planning

Ying Tang and Meng-Chu Zhou

CONTENTS

9.1 Introduction

As the average life cycle of products continues to decrease with rapid technological advancement and global competition, outdated consumables fill up the available landfill space with hazardous or valuable components and materials. Ravi et al. (2005) had indicated that about 500 million computers would be rendered obsolete by 2007 in the United States alone. The Environmental Protection Agency's (EPA) *Municipal Solid Waste FactBook* reports that 29 states in the United States have 10 years or more of landfill capacities remaining, 15 states between 5 and 10 years, and 6 states less than 5 years (Rogers and Tibben-Lembke 1998). The high level of product turnover has, in turn, precipitated the

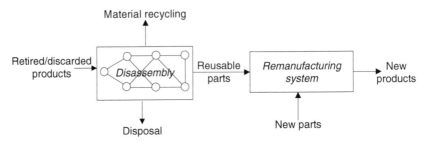

FIGURE 9.1
The material flow in reverse logistics systems.

influx of more toxins into the waste stream. If this influx is not contained, these toxins can take the form of air, food, and water contaminants. Failure to treat mounting piles of obsolete products poses significant environmental threats.

Reverse logistics, as one natural solution, has received growing attention in view of the fast depletion of landfill capacity and raw material. Several countries have enforced environmental legislation, charging manufacturers with the responsibility for the whole product life cycle. This responsibility includes end-of-life (EOL) product retrieval and, if possible, reprocessing and reuse of the constituent components and materials of the product (Fleischmann et al. 1997). As depicted in Figure 9.1, disassembly plays an important role in the entire reverse logistics process. Disassembly involves separation of a reclaimed product into a number of components and subassemblies and subsequently redirecting them for remanufacturing, material recycling, or disposal. However, this process is very expensive because of the increasing complications of reclaimed products and the uncontrollable element, that is, human behavior, involved. The unknown state of recovered parts and human performance involved may lead to stochastic routing, varying disassembly lead time, and much uncertainty in process planning (Beullens 2004). A study of some surveys held over the past 10 years (Tang et al. 2002; Lambert 2003) indicates that many researchers and industry executives have started to realize that disassembly can be performed in a cost-effective manner. Researchers have developed some methodologies to identify the extent to which disassembly of a product should be conducted to keep the process profitable and environment-friendly (Meacham et al. 1999; Zussman and Zhou 2000; Gao et al. 2004). Several heuristic approaches have been proposed to generate the optimal disassembly sequence that would minimize the cost of disassembly (assuming that a certain level of disassembly is required) or obtain the best cost–benefit ratio for disassembly (Moore et al. 2001; Seo et al. 2001; Lambert 2002; Tang et al. 2002; Tiwari et al. 2002; Hula et al. 2003).

While a significant amount of research is being conducted in this area, not much effort has been focused on human factors in disassembly. In fact, disassembly is currently manual or semiautomatic and labor-intensive. The large amount of human intervention in disassembly provides extra flexibility but also gives rise to much uncertainty in the process. For instance, disassembly processing time, quality of disassembled components, and disassembly cost may vary significantly with the varying skill levels of human operators. Moreover, this uncertainty contains fuzzy or imprecise information that is sometimes described by linguistic terms only. Thus, there is an emerging need for an effective analytical model to manage the inherent uncertainty and assist strategic decision making in the operation of human-in-the-loop disassembly systems.

The competence, motivation, and availability of human workers play a major role in the performance of manufacturing systems (Mikler et al. 1999). A lesson learned from process automation is that if human factors are not taken into consideration, even state-of-the-art systems can be more problematic than beneficial (Goodrich and Boer 2003). Bidanda et al. (2005) raise several engineering issues that are related to the human component of manufacturing cell design, implementation, and operation. Baines et al. (2005) extensively review theoretical frameworks for human performance modeling within manufacturing system design. Fuzzy set theory is applied to explore imprecise human cognitive boundaries and judgmental behaviors. Jahan-Shahi et al. (2001) propose multivalued fuzzy sets to model nonprocess factors (e.g., operators' condition and working environment) in a time and cost estimation of flat plate processing. Taking fuzzy human behavior into account, Utkin et al. (1997) propose a method involving probability and possibility measures to analyze computer-integrated manufacturing systems. Considering the involvement of human operators in flexible manufacturing systems (FMS), Xu (1996) presents a fuzzy Petri nets (PNs) model and a two-stage fuzzy reasoning algorithm to derive the optimum schedule for the proposed FMS. In a highly automated manufacturing environment, human operators are recognized as decision makers or problem solvers, who interact with the hardware and the software. In the disassembly process, human operators are involved not only in decision-making but also in the dismantling tasks in many situations since disassembly automation level is still very low compared with manufacturing and assembly automation. Coupled with the uncertainty in product condition, this low level of automation makes disassembly a very complex and challenging process. Considering the similarities and the differences between assembly and disassembly, in this chapter, we present formal models that analyze the influence of human intervention in disassembly and incorporate the disturbances into disassembly planning and control.

9.2 Problem Statement

From an operational standpoint, a disassembly process can be perceived as a sequence of decisions. Starting from an originally returned product unit, this sequence of decisions is selected from a set of feasible options that take the product apart into its constituent subassemblies and components, which optimize the expected revenue and minimize the cost involved. In general, such decision-making is contingent upon the unit's condition as well as other operational constraints (e.g., human factors) and is repeated on newly generated items until none of the artifacts need further disassembly. A set of data is required for the detailed computation of the optimal disassembly plan (i.e., a sequence of decisions). These data must characterize the quality attributes of the end product unit itself and of the various subassemblies and components extracted during the process as well as the impact of operational constraints on the process.

9.2.1 Impact of Human Factors on Disassembly

Disassembly involves two types of human operations. In heavy-duty tasks (e.g., shredding and processing contaminated materials), usually handled by machines, the human operator is involved in the process by way of interacting with the hardware and the software (e.g., set up, loading, and monitoring). The human's performance of these operators depends to a large extent on their understanding of the instructions. In contrast, human operators participating in simple dismantling tasks (e.g., taking a part out from a product or a subassembly) may make decisions regarding which tool to use and which orientation to follow.

In this chapter, variations in the output of different operators working under identical conditions are attributed to skill, defined as "proficiency at following a given method." The time that an operator needs to complete a task, the quality of the product or the decision resulting from this manual operation, and the labor cost for this operation vary with skill levels. This chapter focuses on these aspects.

- *Disassembly time.* A disassembly task, referred to as "a task" in this chapter, is defined as a process handled by a machine or an operator to dismantle a product or a subassembly into two or more subassemblies or components. It is assumed that each task can be completed by a machine or an operator, on an average, within a fixed time called operation time. This operation time depends on the condition of the product or the subassembly. As a human operator is involved in the task, either to set up a machine for the process or to completely deal with the process, extra time called extended time is added to the operation time, depending on the skill level of the operator. Disassembly time, defined as the time taken to complete a particular task, is thus the sum of the operation time and the extended time.

- *Labor cost.* Labor cost is an important factor that partially determines the final cost for a disassembly task. The higher skill level an operator has, the more salary the operator receives. An operator with a higher skill level spends less time on a task than an operator with a lower skill level who works on the same task.

- *Quality of disassembled subassemblies and components.* An unskilled worker may damage disassembled components by using inappropriate tools or by following an incorrect disassembly orientation due to lack of experience. This performance may further degrade disassembly revenue in view of the resale or reuse values of disassembled components. The possibility of obtaining a quality subassembly or component from a skilled operator is generally higher than that from an unskilled operator.

9.2.2 Challenges and Solutions for the Disassembly Process

As stated in Section 9.2, the optimal disassembly plan might be quite straightforward to obtain in principle. However, the challenge arises because much of the information necessary for disassembly decision making is not available during the process initiation and can only be provided through the real-time observation of the process itself. For instance, obsolete products exhibit a high level of uncertainty in their structure and condition due to the uncontrollable element of human behavior involved in product usage. Furthermore, such data or information might not be constant but vary with time. As the physical, psychological, personal, and other characteristics of humans and social and environmental conditions evolve over time, the impact of human factors on disassembly is hard to estimate *a priori*.

As discussed in Reveliotis (2003), the problem of dealing with the lack of data necessary for disassembly planning and control falls under the broader area of system identification and parameter estimation. Figure 9.2 depicts the two general approaches to this problem.

- *Deterministic approach.* Starting from a detailed representation of all feasible disassembly options, the first approach seeks to estimate the

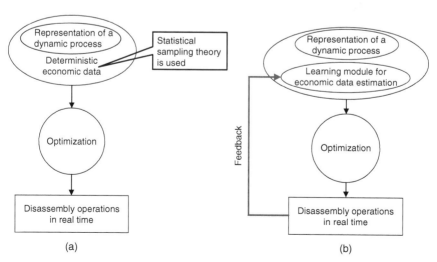

FIGURE 9.2
(a) Deterministic approach. (b) Dynamic approach.

unknown data using statistical sampling theory and subsequently uses the information obtained for the disassembly optimization process. The resulting decisions guide disassembly in real time. As represented by Figure 9.2a, the data necessary for the decision-making process is obtained in advance, and any additional information that might become available during the process cannot be used for optimization.

- *Dynamic approach.* Compared with the first approach, the dynamic approach engages a feedback mechanism that enables the system to monitor its own behavior and then integrates and exploits the newly obtained "knowledge" for future decision making. The intention of this setting is to bring the estimates maintained closer to the estimates of the observed disassembly system.

Obviously, the deterministic approach is computationally easy to implement when the unknown data pertaining to product condition and human performance exhibit high uniformity. In contrast, the dynamic approach uses an adaptation strategy in which the disassembly decision-making mechanism is guided by an expert system or a knowledge-based system. Although a large amount of computation is required, this method can deal with nonstationary process in which unknown parameters present significant variation.

9.3   Models for Human-in-the-Loop Disassembly

Modeling and analysis are very critical in the design and operation of disassembly. It is widely known that flaws in the modeling process can substantially affect the development time and cost as well as the operational efficiency (Zurawski and Zhou 1994). This section focuses on two important models for human-in-the-loop disassembly. Considering Petri nets (PNs) as the common ground for these two models, the fundamental definition of PNs is first given in Section 9.3.1. Subsequently, we present the detailed features of these two extended PN models and their deployment in disassembly process planning.

9.3.1   Petri Nets

PNs, as a graphical and mathematical tool, provide a uniform environment for modeling, formal analysis, and design of discrete-event systems (Zhou and Venkatesh 1998; Hruz and Zhou 2007). A PN may be identified as a particular kind of bipartite-directed graph populated by three types of objects. These objects are places, transitions, and directed arcs connecting places to transitions and transitions to places. Pictorially, places are depicted by circles that represent the locations where objects await processing or the conditions that objects are in and transitions by bars that model processes or activities.

A place is an input (output) place to a transition if a directed arc connects the place (transition) to the transition (place). To study the dynamic behavior of a modeled system, each place may potentially hold either none or a positive number of tokens, pictured by small, solid dots. The presence or the absence of tokens in a place indicates the number of objects available in that place or the true or false condition associated with that place. Formally, a PN can be defined as follows:

> **Definition 9.1:** *A PN is defined as a 5-tuple: PN = (P, T, I, O, M), where*
>
> - $P = \{p_1, p_2, \ldots, p_n\}$ *is a finite set of places.*
> - $T = \{t_1, t_2, \ldots, t_s\}$ *is a finite set of transitions,* $P \cup T \neq \varnothing, P \cap T = \varnothing$.
> - $I: P \times T \rightarrow \mathcal{N}$ *is the input function that defines the set of ordered pairs* (p_i, t_j), *where* $I(p_i, t_j)$ *represents the number of direct arcs from place* p_i *to* t_j, *and* $\mathcal{N} = \{0, 1, 2, 3, \ldots\}$.
> - $O: P \times T \rightarrow \mathcal{N}$ *is an output function that defines the set of ordered pairs* (p_i, t_j), *where* $O(p_i, t_j)$ *represents the number of direct arcs from place* t_j *to* p_i.
> - $M: P \rightarrow \mathcal{N}$ *is a function that defines a marking vector, where* $M(p_i)$ *represents the number of tokens in* p_i. *The initial marking and the final marking are denoted by* m_{i0} *and* m_f, *respectively.*

$^\bullet p$ is the set of input transitions to the place p. Mathematically, $^\bullet p = \{t, O(p, t) \neq 0\}$. And $p^\bullet$ is the set of output transitions to the place p. Mathematically, $p^\bullet = \{t, I(p, t) \neq 0\}$. Following the same fashion, $^\bullet t$ is the set of input places to the transition t. Mathematically, $^\bullet t = \{p, I(p, t) \neq 0\}$. And $t^\bullet$ is the set of output places to the transition t. Mathematically, $t^\bullet = \{p, O(p, t) \neq 0\}$.

If $I(p, t) = k(O(p, t) = k)$, then k-directed arcs connecting place p to transition t (transition t to place p) exist. Graphically, parallel arcs connecting a place (transition) to a transition (place) are represented by a single directed arc labeled with weight k.

The occurrence of events or the execution of operations can be modeled by the flow of tokens on places, using the following rules:

> *Enabling rule.* A transition t is said to be enabled if the number of tokens in each input place of t is equal to or larger than the weight of the directed arc connecting p to t. Mathematically, t is enabled if $\forall p \in \, ^\bullet t$, $M(p) \geq I(p, t)$.
>
> *Firing rule.* A firing of an enabled transition t removes from each input place p the number of tokens equal to the weight of the directed arc connecting p to t. It also deposits in each output place p the number of tokens equal to the weight of the directed arc connecting t to p. Mathematically, firing at M leads to M' such that $\forall p \in P, M'(p) = M(p) + O(p, t) - I(p, t)$.

9.3.2 Deterministic Model for Human-in-the-Loop Disassembly

Usually, qualitative linguistic terms are used to evaluate the impact of human intervention. A linguistic variable differs from a numerical variable in that its values are not numbers but words or sentences in a natural or an artificial language. For example, the skill level of an operator is a linguistic variable and the values of this level can be numerical or high, fair, low, and so on. Fuzzy set theory provides a good tool to represent such vague input data by formulating the values using membership functions. By taking advantage of both fuzzy logic and well-formed formalism of PNs, this section describes a fuzzy attributed petri net (FAPN) model to analyze human-in-the-loop disassembly planning. In this model, the operation time for each disassembly task and the expected profit through the resale or the reuse of a discarded product, subassembly, or component are assumed to be deterministic. The model's deployment for disassembly process planning is presented later.

9.3.2.1 Fuzzy Attributed Petri Net

Definition 9.2: *An FAPN is defined as 8-tuple* (Tang et al. 2006):

FAPN = $(P, T, I, O, M, \tau, \alpha, \lambda)$, where T, I, and O are defined the same as in PN and

- P is a nonempty set, where $P = \{p_1, p_2, ..., p_n\} = W \cup Q$. The set of places $W = \{w_1, w_2, ..., w_r\}$ represents operators and the set of places $Q = \{q_1, q_2, ..., q_s\}$ stands for a product, subassembly, or component. $r + s = n$.
  1. $W \cap Q = \varnothing$.
  2. There is a unique $p \in Q$ such that $\cdot p = \varnothing$. This place is usually denoted by p_1 and named as a root or a product place.
  3. $\exists\, Q' \subset Q$ such that $Q' \neq \varnothing$ and $\forall p \in Q'$, $p^\bullet = \varnothing$. Places in Q' are called leaves.
  4. $I(p, t) = O(p, t)$, $\forall t \in T$, $\forall p \in W$
- $M: P \to \mathcal{N}$ is a function that defines a marking vector, where $M(p_i)$ represents the number of tokens in p_i. The initial and the final markings satisfy
  1. $m_0(p_1) = 1$, $m_0(p) = 0$, $\forall p \in Q - \{p_1\}$, and $m_0(p) = 1$, $\forall p \in W$
  2. $m_f(p) = m_0(p)$, $\forall p \in W$.
- $\tau: Q \to \{0\} \cup \mathfrak{R}^+$ is a revenue function assigned to a place, which is the average profit earned by the resale or the reuse of the product, subassembly, or component represented by p. $\mathfrak{R}^+$ is the set of positive real numbers.
- $\alpha(w) = \cup_i\{(\rho_i(w), \sigma_{p_i}(w)\}$, $i = 1, 2, ..., i_k$ is a mapping function that maps a w to a union of 2-tuples: $\rho_i(w)$, a fuzzy proposition and $\sigma_{p_i} \in [0, 1]$, the

corresponding truth degree. i_k is the number of propositions applicable to w.

- $\lambda: T \to \mathfrak{R}^+$ is an operation time associated with a transition. Its unit is hour.

The human factors discussed in this model are limited to disassembly time, labor cost, and quality of disassembled subassemblies or components. Therefore, $\forall w \in W$, three attributes are associated with it, which are the fuzzy propositions regarding the impact of the operator w on disassembly (e.g., the impact on disassembly time is small, $\rho_1(w)$; the labor cost is high, $\rho_2(w)$; and the bad influence on disassembly quality is small, $\rho_3(w)$). The corresponding truth degrees represent the extent to which the proposition is true. Furthermore, for simplicity, this model assumes that each disassembly operation takes only one component away from an assembly or a subassembly. Figure 9.3 depicts a simple example of an FAPN that models the decomposition of a product into two components processed by an operator.

$FAPN = (P, T, I, O, M, \tau, \alpha, \lambda)$

$P = \{q_1, q_2, q_3, w_1\}$, where q_1 is the root place

$T = \{t_1\}$

$m_0 = \{1, 0, 0, 1\}$

$\tau = \{0, 0.2, 0.5\}$

$\alpha(w_1) = \{$(the effect on disassembly time is small, 0.75), (the labor cost is high, 0.7), (the bad effect on disassembly quality is small, 0.8)$\}$

$\lambda = \{1/6\}$

9.3.2.2 Disassembly Process Planning

In FAPN, an operator place is introduced as an input place to a transition. Considering the impact of a human operation on disassembly cost and revenue, not only are fuzzy attributes $\rho_i(w_j)$ introduced to represent human factors but also a truth degree is associated with each attribute for the strength of the impact.

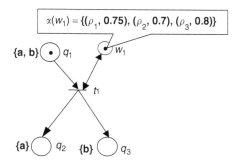

FIGURE 9.3
A simple example of an FAPN. (From Tang, Y. et al., *IEEE Transactions on Systems, Man & Cybernetics: Part A*, 36, 718–726, 2006.)

For a particular disassembly operation t, the disassembly time is calculated as the sum of the operation time and the extended time, where the latter is a function of an operator's skill. The higher is the skill of an operator, the shorter is the extended time. As the truth degree drops with increasing extended time, the relationship between the truth degree and the extended time is assumed to be exponential (Xu 1996). The extended time $T_{ext}(w_j)$ is then calculated as follows:

$$T_{ext}(w_j) = \frac{-\ln(\sigma_{p_1}(w_j))}{\beta} \tag{9.1}$$

where β is a constant whose value is chosen based on statistics and $\sigma_{p_1}(w_j)$ is the truth degree of the attribute $p_1(w_j)$. Based on Equation 9.1, the disassembly time for the disassembly operation t, $T_{dis}(t)$, is

$$T_{dis}(t) = \frac{-\ln(\sigma_{p_1}(w_j))}{\beta} + \lambda(t), \; w_j \in {}^{\bullet}t \cap w_j \in W \tag{9.2}$$

Labor cost is also an important factor in determining the final disassembly cost. The more experienced operators are, higher the salary they get. This model assumes that the fuzzy proposition—the labor cost is high—is characterized as the function of operator's salary. For example, as shown in Figure 9.4, the maximum and the minimum salaries are $18/h and $6/h, respectively.

Therefore, the final disassembly cost for a particular disassembly operation t, $c(t)$, is calculated as follows (assuming the charge for disassembly tools is ignored):

$$c(t) = T_{dis}(t) \times (6 + 12\sigma_{p_2}(w_j)) \tag{9.3}$$

where $\sigma_{p_2}(w_j)$ is the truth degree of the attribute $p_2(w_j)$.

Considering the damage caused by an unskilled operator to a disassembled subassembly or component, the adjusted revenue for each place p, $p \in Q$ is also introduced and denoted as $r(p)$.

$$r(p) = \begin{cases} \sigma_{p_3}(w_j) \times \tau(p), & p \in (w^{\bullet})^{\bullet} \\ \tau(p), & p = p_1 \end{cases} \tag{9.4}$$

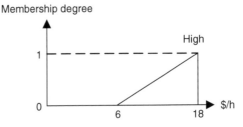

Membership degree

High

1

0

6   18

$/h

FIGURE 9.4
Membership function for an operator's salary. (From Tang, Y. et al., *IEEE Transactions on Systems, Man & Cybernetics: Part A*, 36, 718–726, 2006.)

where $\sigma_{p_3}(w_j)$ is the truth degree of the attribute, $\rho_3(w_j)$, $\forall p \in Q - \{p_1\}$, $r(p)$ may take different values as p results from separate disassembly operations processed by different operators w.

Finally, the disassembly planning algorithm with the consideration of human factors is described in Algorithm 9.1. The aim is to identify the order of disassembly operations (i.e., transition firing in an FAPN) that maximizes the expected return from the processed items. This objective is then formally expressed by introducing the disassembly value function, $d : Q \rightarrow \{0\} \cup \mathfrak{R}^+$, as detailed in the following definition:

> **Definition 9.3:** *The disassembly value of place p is denoted as $d(p)$ and calculated in a recursive manner using the following equations:*
>
> 1. $\forall p \in Q'$, $d(p) = r(p)$
> 2. $\forall p \in Q-Q'$, $d(p) = \max\{r(p), \max_{t \in p^{\bullet}} \Delta(t)\}$

where

$$\Delta(t) = \sum_{q \in t^{\bullet} \cap q \in Q} d(q) - c(t)$$

is the margin benefit of transition t, $t \in T$, which represents the cost or revenue when place $^{\bullet}t \in Q - Q'$ takes t for further disassembly. $c(t)$ and $r(p)$ are calculated using Equations 9.3 and 9.4, respectively.

This function maps each place, $p \in Q$, to a nonnegative value, characterizing the expected return obtained from processing the product, subassembly, or component represented by place p. For a leaf place, the disassembly value is determined by its predicted resale revenue multiplied by the truth degree. For a nonleaf place, the value results from either a processing option or a resale option that maximizes the expected return. The former is defined as the cumulative returns of all the disassembled items reduced by the corresponding processing cost. The disassembly plan is then calculated by determining the disassembly sequence, which maximizes $d(p_1)$ through a recursive computation that starts from the leaf places and derives the best d-values for every place $p \in Q$. Figure 9.5 summarizes the overall logic of the approach. The detailed algorithm is given as follows:

Algorithm 9.1:

> *Step 1.* For $p \in Q'$ the set of leaves according to Definition 9.2, set $d(p) = r(p)$;
>
> *Step 2.* Set $L = Q - Q'$;
>
> While $(L \neq \varnothing)$ do:
>
> Find a place $p \in L$, such that $\forall q \in (p^{\bullet})^{\bullet}$, $q \notin L \cap q \in Q$
>
> Calculate:
>
> $\Delta(t) = \sum_{q \in t^{\bullet}} d(q) - c(t)$
>
> $d(p) = \max\{r(p), \max_{t \in p^{\bullet}} \Delta(t)\}$

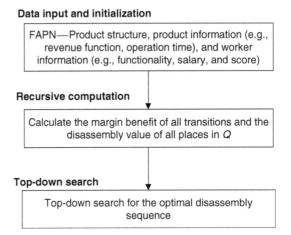

FIGURE 9.5
Logic of optimal disassembly planning. (From Tang, Y. et al., *IEEE Transactions on Systems, Man & Cybernetics: Part A*, 36, 718–726, 2006.)

Let $L = L - \{p\}$, i.e., removes p from L;

End

Step 3. Set $Z = \{p_1\}$ (assuming that p_1 is the root node for every incoming product), DP $= \emptyset$ and $C = \emptyset$ (DP represents final disassembly path and C represents the last components after disassembly).

While $(Z \neq \emptyset)$ do:

For each node p in Z:

a. Find a transition t, where $t \in p^\bullet$, $\Delta(t) = d(p) \neq r(p)$

b. If (a) succeeds, DP $=$ DP $\cup \{t\}$ and $Z = Z \cup \{t^\bullet\}$; otherwise, $C = C \cup \{p\}$;

c. $Z = Z - \{p\}$;

End

Owing to the deterministic features of the FAPN (i.e., the operation time associated with each transition, the revenue function associated with each place in Q, and the truth degree of each attribute associated with a worker place are assumed to be known *a priori*), the exhausted search in Algorithm 9.1 takes $O(n)$ and guarantees the optimal disassembly planning where n is the number of transitions in an FAPN (Tang et al. 2006).

9.3.3 Dynamic Model for Human-in-the-Loop Disassembly

The negative aspect of the FAPN model is the assumption that the associated revenue function and the time function are deterministic. In reality, these costs and profits involved in disassembly are determined dynamically because of the prevailing condition of discarded products and other operational constraints. Taking this factor into consideration, this section presents a dynamic model. Instead of presuming that the pertinent data is already known, a self-adaptive disassembly process planner and associated computationally effective algorithms are designed in a way as to (1) accumulate the past experience of predicting such data and, simultaneously, (2) exploit the "knowledge" captured in the data to choose the best disassembly plan and improve the overall disassembly performance.

9.3.3.1 Fuzzy Petri Net

Definition 9.4: A fuzzy Petri net (FPN) is defined as a 10-tuple (Turowski and Tang 2005):

FPN = $(P, T, I, O, M, \theta, \alpha, \delta, \tau, \lambda)$, where $P = W \cup Q, T, I, O, M$ are defined the same as in FAPN and

- $\theta: W \to \mathfrak{R}^+$ is operator wage rate associated with a place in W.
- $\varepsilon(w) = \cup_i\{v_i, \iota_{v_i}(w)\}, \forall w \in W$ is a mapping function that maps w to a union of 2-tuples: v_i, the fuzzy variables on human skill level and $\iota_{v_i} \in [0, 1]$, the corresponding truth degree.
- $\delta(q) = \cup_i\{(\mu_i, \varsigma_{\mu_i}(q)\}, \forall q \in Q$ is a mapping function that maps q to a union of 2-tuples: μ_i, the fuzzy condition variables and $\varsigma_{\mu_i} \in [0, 1]$, the corresponding truth degree.
- $\tau(q) = \cup_i\{(\beta_i, \gamma_{\beta_i}(q)\}, \forall q \in Q$ is a mapping function that maps q to a union of 2-tuples: β_i, the fuzzy revenue variables and $\gamma_{\beta_i} \in [0, 1]$, the corresponding truth degree.
- $\lambda(t) = \cup_i\{(\pi_i, \eta_{\pi_i}(t)\} \forall t \in T$ is a mapping function that maps t to a union of 2-tuples: π_i, the fuzzy disassembly time variables and $\eta_{\pi_i} \in [0, 1]$, the corresponding truth degree.

The value of θ is given and assumed to be reasonably correlated to the skill of an operator represented by w, and the values of τ and λ are decided through adaptive learning, indicating the resale or the reuse revenue of the disassembled unit q and the disassembly time of transition t, respectively.

9.3.3.2 Adaptive Disassembly Process Planning

Based on the FPN (Section 9.3.3.1), a learning technique is engaged in each disassembly process, which enables the disassembly system to make good decisions by observing its own behavior and to continuously improve its decision making through learning. Figure 9.6 presents the basic architecture. Disassembly operations run in real time based on a sequence of plans controlled by a learning-based process planner. This planner consists of three interactive components: fuzzy learning module, plan selection module, and database. The fuzzy learning module functions as a black box. With the inputs available for each disassembly operation (e.g., the condition of the unit

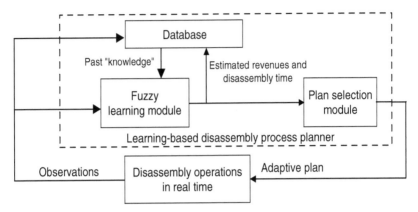

FIGURE 9.6
The basic architecture of the learning-based disassembly process planner (Turowiski, M. and Tang, Y. in Proceedings of the IEEE International Conference on Networking, Sensing and Control, Tucson, AZ, 2005, 141–146).

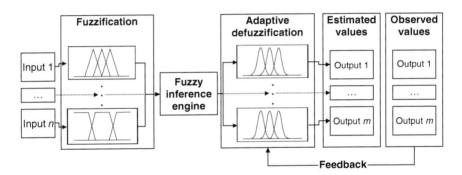

FIGURE 9.7
Fuzzy learning module (Turowski, M. and Tang, Y. in Proceedings of the IEEE International Conference on Networking, Sensing and Control, Tucson, AZ, 2005, 141–146).

that the operation targets and the skill level of the human operator for this operation), the module predicts the corresponding outputs (i.e., disassembly time and revenues of disassembled parts). Such predications are adjusted every time a feedback is obtained from the execution of each plan in real-time disassembly. Meanwhile, the feedback is maintained in the database and is accumulated as the past "knowledge" to improve future decision making. The estimated parameters are fed into the plan selection module in which an algorithm determines the disassembly plan that maximizes the expected returns during the process operation.

9.3.3.2.1 Fuzzy Learning Module

The learning module, as depicted in Figure 9.7, consists of three fundamental elements just like any other traditional fuzzy logic system: fuzzification, inference engine, and defuzzification.

Fuzzification is the process of transforming a "crisp" input (e.g., x) into one or more fuzzy sets through membership functions. A fuzzy set, A_j, defined on a universe of discourse, X, is characterized as $A_j = \{(x, \varsigma_{A_j}(x)) | x \in X, \varsigma_{A_j}(x) \in [0, 1]\}$. The membership function, $\varsigma_{A_j}(x)$, is also known as the "truth degree of membership of x in A_j," which specifies the extent to which x can be regarded as belonging to the set A_j. Therefore, the output of fuzzification passed to inference engine is a union of 2-tuples of a fuzzy set and its corresponding truth degree (e.g., $x \rightarrow \{(\text{class } 1, 0.25), (\text{class } 2, 0.75), (\text{class } 3, 0.00)\}$).

The input data considered in this model are limited to operator wage and product inspection scores to characterize operator skill level and product condition. An operator place, w_i, is introduced as an input of a transition, t_j, in an FPN, where $I(w_i, t_j) = 1, w_i \in W$. A wage rate parameter $\theta(w_i)$ is then associated with w_i and classified through the fuzzification process to reflect the corresponding fuzzy skill level of w_i, $\varepsilon(w_i)$. It is assumed that the wage rate of an operator is proportional to the operator's skill level. In an FPN, a fuzzy condition variable $\delta(q_i)$ is also associated with each product place q_i, whose membership function varies with the different items that q_i represents. The condition of a processing item is characterized by an inspection score. Quality testing of a discarded product is assumed to be conducted before disassembly, and different inspection procedures and scales may be used for various products. For the subassemblies and components of a product, their corresponding inspection scores are given *a priori* based on the previous benchmark experience.

A set of IF–THEN rules is then used in the fuzzy inference engine to describe the impact of the aforementioned factors on disassembly. Consequently, the processing time of a disassembly operation, t_j, and the revenues of disassembled units, for example, q_i, $O(q_i, t_j) = 1, q_i \in Q$, are fuzzy and are characterized by the union of fuzzy output sets and their corresponding truth degrees. These are denoted by $\lambda(t_j)$ and $\tau(q_i)$ in an FPN, respectively. For instance, there are two fuzzy inputs to the inference engine—human operator skill: {(low, 0.00), (medium, 0.75), (high, 0.25)} and product condition: {(bad, 0.30), (good, 0.70)}. According to the following two rules, the output of the inference engine is {(disassembly time is low, 0.25), (disassembly time is normal, 0.30)}.

Rule A. IF operator skill is high AND product condition is good THEN disassembly time is low.

Rule B. IF operator skill is medium AND product condition is bad THEN disassembly time is normal.

Compared with the traditional defuzzification process, the defuzzifier is designed in a fashion that membership functions are not fixed but continuously adapted to the observed data corresponding to the real disassembly behavior. In this application, Gaussian functions are chosen to map outputs of the inference engine to "crisp" data in predicting disassembly time and revenues of disassembled units for each transition. Such functions are initialized with some arbitrary mean and deviation and then dynamically shaped whenever observations on real disassembly operations are fed back into the process. Finally, the centroid function, Ω, is used to derive crisp data with a certainty degree, ψ, both of which are defined as follows:

$$\Omega(\cup_j(A_j, \varsigma_{A_j}(x))) = \frac{\sum_j(\text{CENTER}(A_j) \cdot \varsigma_{A_j}(x))}{\sum_j \varsigma_{A_j}(x)} \qquad \Psi(\Omega(\cup_j(A_j, \varsigma_{A_j}(x)))) = \max_j(\varsigma_{A_j}(x))$$

where $\cup_j(A_j, \varsigma_{A_j}(x))$, the output of the inference engine, is a union of 2-tuples: A_j, the fuzzy set and $\varsigma_{A_j}(x) \in [0, 1]$, the corresponding truth degree. $\text{CENTER}(A_j)$ is the center of mass of the fuzzy set A_j. The detailed learning is given in Algorithm 9.2.

Algorithm 9.2   Adaptive Learning:

1. Let $\Lambda_{A_j} = \{x_1, x_2, ..., x_n\}$ be the set of crisp data (both observed and predicted) classified into the fuzzy set A_j, and initially $\Lambda_{A_j} = \varnothing$.

2. For each fuzzy set A_j, initialize its Gaussian membership function with an arbitrary mean $\hat{\phi}$ and standard deviation σ (e.g., $\hat{\phi} = \sigma = 0$). Let $\hat{V}(\hat{\phi})$ be the standard error.

3. Iterate the following learning procedure:

   a. Predict final outputs of the fuzzy learning module (e.g., $\Omega(\lambda(t))$, $\Psi(\lambda(t))$ based on the given inputs, for example, operator's salary $\theta(w_i)$, reasoning rules, and present Gaussian membership functions. Each output includes the crisp defuzzified data, $x_k = \Omega$ $(\cup_j(A_j, \varsigma_{A_j}(x)))$, its associated certainty degree, $\Psi(x_k)$, and the fuzzy set, A_l, where $\Psi(x_k) = \varsigma_{A_l}(x_k)$.

   b. Collect the observed value that corresponds to x_k and feed it back to the module. Let x_k^* be the corresponding observed value, then $\Lambda_{A_l} = \Lambda_{A_l} \cup \{x_k^*\}$.

   c. Calculate the mean and standard deviation to shape the membership function of A_l using the updated data set Λ_{A_l}.

Although simple, the aforementioned approach can theoretically approximate the behavior of any statistic according to the central limit theorem (Devore 2004). As stated in the theorem, the sample mean of every probability distribution is normally distributed, even if the probability distribution itself is not normally distributed. Outputs of the proposed adaptive fuzzy system can reasonably estimate any probability distribution, as long as a large enough sample set is obtained. Unlike a universal approximator (e.g., multilayer perceptron), the proposed system adapts to find a statistical description of any factor under varying input conditions.

9.3.3.2.2   Plan Selection Module

The objective of the plan selection module is to identify the best disassembly sequence, that is, transition firing in an FPN, which maximizes the expected returns from the processed items. With the information estimated from the fuzzy learning module, this process is essentially similar to the process described in Section 9.3, except for the different calculation of the disassembly value function, $d : Q \to \{0\} \cup \Re^+$, where

$$\forall p \in Q', \quad d(p) = \psi(\tau(p)) \cdot \Omega(\tau(p)) \tag{9.5}$$

$$\forall p \in Q - Q', \quad d(p) = \max \left\{ \psi(\tau(p)) \cdot \Omega(\tau(p)), \right.$$

$$\left. \max_{t \in p^\bullet} \left(\sum_{q \in t^\bullet, w \in t^\bullet} d(q) - \Omega(\lambda(t)) \cdot \psi(\lambda(t)) \cdot \theta(w) \right) \right\} \tag{9.6}$$

9.3.3.3 Example

This section presents a case study to demonstrate the adaptive process planning through the disassembly of a batch of obsolete flashlights. The components of a flashlight include cover (C), glass (G), head housing (H), bulb (B), spring (S), and main housing (M). The flashlight's connectivity graph and FPN are presented in Figures 9.8a and 9.8b, respectively. The dynamic method is simulated in MATLAB, in which a set of training data and test instances that characterize variations in product condition and human operator skill as well as their impact on disassembly are generated randomly. Several sets of experiments are then conducted to illustrate different aspects of this method.

9.3.3.3.1 Illustration of Learning Procedure

In this set of experiments, the transition t_1 of the sample FPN, shown in Figure 9.8b, is chosen to exemplify the adaptive learning procedure. The fuzzy membership functions for the skill level of w_1, $\varepsilon(w_1)$, and the condition of q_1, $\delta(q_1)$ are shown in Figures 9.9a and 9.9b, respectively. As the number of training data increases, the Gaussian membership functions of $\lambda(t_1)$, $\tau(q_2)$, and $\tau(q_7)$ with regard to $\delta(q_1)$ and $\varepsilon(w_1)$ evolve. Figure 9.10 depicts the evolution of $\tau(q_2)$ in three different categories (i.e., revenue is low, medium, or high). Note that the three initial Gaussian membership functions are overlapped when training starts. As more knowledge is exploited through iterations, these functions are dynamically reshaped by Algorithm 9.2. Ninety-five percent confidence analysis (Banks et al. 2000) for the mean of $\tau(q_2)$ is also conducted. The approximate $100(1 - \alpha)\%$ confidence interval for ϕ is defined as follows. As shown in Figure 9.11, the prediction of $\tau(q_2)$ classified into the "the revenue is high" set converges to the mean (i.e., 2) of training data with high confidence (e.g., 95%).

$$\hat{\phi} \pm g_{\alpha/2,(n-1)} \hat{V}(\hat{\phi}) \quad \text{or} \quad \hat{\phi} - g_{\alpha/2,(n-1)} \hat{V}(\hat{\phi}) \leq \phi \leq \hat{\phi} + g_{\alpha/2,(n-1)} \hat{V}(\hat{\phi}) \tag{9.7}$$

where

$\hat{\phi}$ = sample mean of ϕ based on a sample of size n

$\hat{V}(\hat{\phi})$ = standard error of $\hat{\phi}$

$g_{\alpha/2, (n-1)}$ = $100(1 - \alpha)\%$ percentage point of a t-distributed with $n - 1$ degrees of freedom

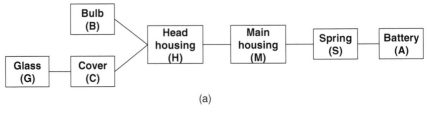

(a)

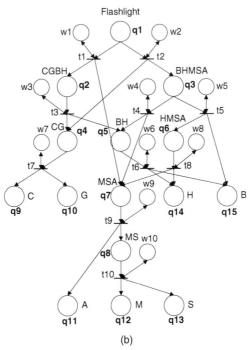

(b)

FIGURE 9.8
(a) The connectivity graph of a flashlight (Tang, Y. et al., *J. Manuf. Syst.*, 21(3), 200–217, 2002) and (b) the FPN of a flashlight.

9.3.3.3.2 Comparison Results

This set of experiments compares two cases applied to the same test instances. In the baseline case, the system completely disassembles the incoming flashlight, following a static sequence, regardless of the flashlight's condition and the human factors involved. As shown in Figure 9.6b, the three complete disassembly paths are $\{t_1, t_3, t_7, t_6, t_9, t_{10}\}$, $\{t_2, t_4, t_7, t_6, t_9, t_{10}\}$, or $\{t_2, t_5, t_7, t_8, t_9, t_{10}\}$. In the proposed case, each flashlight is disassembled on the basis of the sequence derived by the learning-based process planner in which Algorithm 9.2

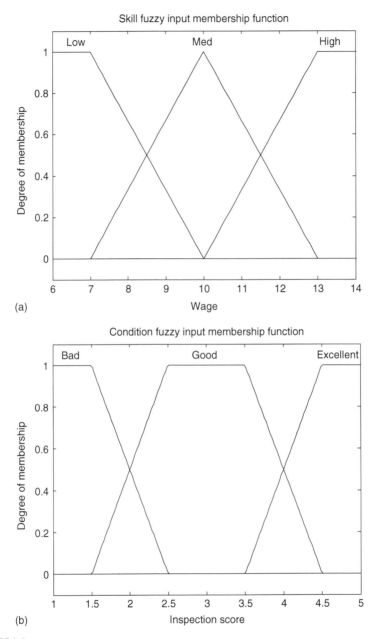

FIGURE 9.9
(a) Fuzzy membership function for operator skill and (b) fuzzy membership function for product condition.

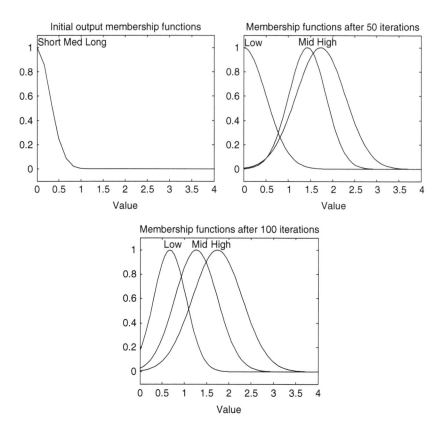

FIGURE 9.10
Learning the fuzzy membership functions of $\tau(q_2)$.

is applied. To show the difference between these two cases, the value gain in a disassembly plan (DP) is adapted (Tang et al. 2001).

$$f(DP) = \sum_{p \in C} \text{the total revenue of the resulting}$$

$$p - \sum_{\substack{t \in DP \\ w \in \cdot t \cap w \in W}} (\theta(w) \times \text{the disassembly time of } t) \qquad (9.8)$$

As the number of instances changes from 500 to 5000, the average value gain is obtained and compared between the baseline and the proposed cases. Figure 9.12 shows the average value gain for the proposed case and two static disassembly paths. The average value gain of the third static disassembly path has a very high variability and obscures the other two plots; so, this average value gain is removed for clarity. A 95% confidence interval for the true mean difference in average value gain is calculated using the data in Table 9.1, where only the best static disassembly path is chosen for the baseline case.

A 95% confidence interval for average value gain is given by

$$0.6900 \pm (2.26)0.0259$$

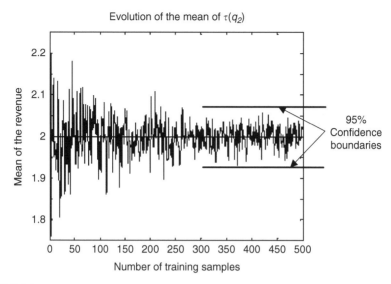

FIGURE 9.11
The evolution of the mean of $\tau(q_2)$ in "the revenue is high" set.

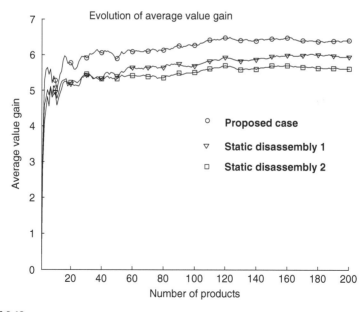

FIGURE 9.12
The average value gain.

TABLE 9.1

Comparison of the Average Value Gain

| | Average $f(DP)$ ($) | | Observed |
Number of Products	Baseline Method	Proposed Method	Difference
500	5.638	6.223	0.585
1000	5.688	6.384	0.696
1500	5.619	6.349	0.730
2000	5.681	6.388	0.707
2500	5.716	6.324	0.608
3000	5.634	6.366	0.732
3500	5.646	6.342	0.696
4000	5.697	6.384	0.687
4500	5.785	6.274	0.489
5000	5.573	6.323	0.750
Sample mean	5.6678	6.3360	0.6680
Sample variance	0.0035	0.0028	0.0067
Standard error			0.0259

or

$$0.7485 \leq \phi_2 - \phi_1 \leq 0.6315$$

The 95% confidence interval for average value gain lies completely above zero, which provides strong evidence that $\phi_2 - \phi_1 > 0$. Specifically, the proposed method is better than the baseline method because the value gains of products are increased in the proposed case.

9.4 Conclusion

In this chapter, we discussed two formal models for uncertainty management in disassembly process planning owing to the large amount of human intervention. The deterministic model mathematically represented the influence of human factors on disassembly (e.g., disassembly time, quality of disassembled components, and labor cost) as fuzzy membership functions. To derive the cost-effective disassembly sequence of a discarded product, an algorithm was then developed on the basis of this model. However, the assumption of *a priori* availability of the information that characterized the randomness in the cost data limited the application of this model. Taking this limitation into consideration, the dynamic model, which is derived from the first model, incorporated fuzzy learning strategy into the disassembly process. This incorporation enabled the system to exploit the past "experience" of the aforementioned uncertainty, remain alert to any variations in real disassembly, and predict the relevant data associated with the model, for example, disassembly time and revenue of a disassembled unit. In this model, data acquisition was considered a part of disassembly planning and control. The learning process ensured that the maintained estimates were closer to the data corresponding to the observed disassembly behavior. In short, the concepts and the methods outlined in this chapter provide design engineers with efficient modeling, analysis, and management tools for uncertainty management in human-in-the-loop disassembly.

Acknowledgments

We are grateful for the support provided by A. Charles and Anne Morrow Lindbergh Foundation.

References

Banks, J., Carson, J. S. II, Nelson, B. L. and Nicol, D. M., 2000, *Discrete-Event System Simulation*. Englewood Cliffs, NJ: Prentice-Hall.

Baines, T. S., Asch, R., Hadfield, L., Mason, J. P., Fletcher, S. and Kay, J. M., 2005, Towards a theoretical framework for human performance modeling within manufacturing systems design. *Simulation Modeling Practice and Theory*, 13, 486–504.

Beullens, P., 2004, Reverse logistics in effective recovery of products from waste materials. *Review in Environmental Science and Bio/Technology*, 3, 283–306.

Bidanda, B., Ariyawongrat, P., Needy, K. L., Norman, B. A. and Tharmmaphornphilas, W., 2005, Human related issues in manufacturing cell design, implementation, and operation: a review and survey. *Computers and Industrial Engineering*, 48, 507–523.

Devore, J. L., 2004, *Probability and Statistics for Engineering and the Sciences*. Belmont, CA: Thomson Brooks/Cole.

Fleischmann, M., Bloemhof-Ruwaard, J. M., Dekker, R., Van der Laan, E., van Nunena, J. A. E. E. and Van Wassenhove, L. N., 1997, Quantitative models for reverse logistics: a review. *European Journal of Operational Research*, 103(1), 1–17.

Gao, M., Zhou, M. C. and Tang, Y., 2004, Intelligent decision making in disassembly process based on fuzzy reasoning Petri nets. *IEEE Transactions on Systems, Man, and Cybernetics: Part B*, 34(5), 2029–2034.

Goodrich, M. A. and Boer, E. R., 2003, Model-based human-centered task automation: a case study in ACC system design. *IEEE Transactions on Systems, Man, and Cybernetics—Part A: Systems and Humans*, 33(3), 325–336.

Hula, A., Jalali, K., Hamza, K., Skerlos, S. J. and Saitou, K., 2003, Multi-criteria decision-making for optimization of product disassembly under multiple situations. *Environmental Science and Technology*, 37(23), 5303–5313.

Hruz, B. and Zhou, M.C., 2007, *Modeling and Control of Discrete Event Dynamic Systems*. London: Springer.

Jahan-Shahi, H., Shayan, E. and Masood, S. H., 2001, Multivalued fuzzy sets in cost/time estimation of flat plate processing. *International Journal of Advanced Manufacturing Technology*, 17, 751–759.

Lambert, A. J. D., 2002, Determining optimum disassembly sequences in electronic equipment. *Computers and Industrial Engineering*, 43(3), 553–575.

Lambert, A. J. D., 2003, Disassembly sequencing: a survey. *International Journal of Production Research*, **41(16)**, 3721–3759.

Meacham, A., Uzsoy, R. and Venkatadri, U., 1999, Optimal disassembly configurations for single and multiple products. *Journal of Manufacturing Systems*, **18(5)**, 311–322.

Mikler, J., Hådeby, H., Kjellberg, A. and Sohlenius, G., 1999, Towards profitable persistent manufacturing human factors in overcoming disturbances in production systems. *International Journal of Advanced Manufacturing Technology*, **15(10)**, 749–756.

Moore, K. E., Gungor, A. and Gupta, S. M., 2001, Petri net approach to disassembly process planning for products with complex AND/OR precedence relationships. *European Journal of Operational Research*, **135(2)**, 428–449.

Ravi, V., Shankar, R. and Tiwari, M. K., 2005, Analyzing alternatives in reverse logistics for end-of-life computers: ANP and balanced scorecard approach. *Computers and Industrial Engineering*, **48**, 327–356.

Reveliotis, S., 2003, Uncertainty management in optimal disassembly planning through learning-based strategies. *Advances in Manufacturing, Logistics and Supply Chain Management*, N. R. S. Raghavan, Y. Narahari, P. Luh and R. Akella (Eds.). Bangalore, India: IIS, pp. 135–141.

Rogers, D. S. and Tibben-Lembke, R. S., 1998, *Going Backwards: Reverse Logistics Trends and Practices*. Pittsburgh, PA: Reverse Logistics Executive Council, Center for Logistics Management.

Seo, K. K., Park, J. H. and Jang, D. S., 2001, Optimal disassembly sequence using genetic algorithms considering economic and environmental aspects. *International Journal of Advanced Manufacturing Technology*, **18(5)**, 371–380.

Tang, Y., Zhou, M. C. and Caudill, R., 2001, An integrated approach to disassembly planning and demanufacturing operation. *IEEE Transaction on Robotics and Automation*, **17(6)**, 773–784.

Tang, Y., Zhou, M. C. and Gao, M., 2006, Fuzzy-Petri-net based disassembly planning considering human factors. *IEEE Transactions on Systems, Man, and Cybernetics: Part A*, **36(4)**, 718–726.

Tang, Y., Zhou, M. C., Zussman, E. and Caudill, R., 2002, Disassembly modeling, planning, and application. *Journal of Manufacturing Systems*, **21(3)**, 200–217.

Tiwari, M. K., Mukhopadhyay, S. K., Sinha, N., Kumar, S. and Rai, R., 2002, A Petri net based approach to determine the disassembly strategy of a product. *International Journal of Production Research*, **40(5)**, 1113–1129.

Turowski, M. and Tang, Y., 2005, Adaptive fuzzy system for disassembly process planning. Proceedings of the IEEE International Conference on Networking, Sensing and Control, Tucson, AZ, March 19–22, pp. 141–146.

Utkin, L. V., Gurov, S. V. and Shubinsky, I. B., 1997, Analysis of computer integrated manufacturing systems by fuzzy human operator behavior. *Journal of Quality in Maintenance Engineering*, **3(3)**, 189.

Xu, J. X., 1996, Fuzzy Petri net-based optimum scheduling of FMS with human factors. Proceedings of the 35th conference on Decision and Control, Kobe, Japan, Dec. 11–13, pp. 4451–4452.

Zhou, M. C. and Venkatesh, K., 1998, *Modeling, Simulation and Control of Flexible Manufacturing Systems: A Petri Net Approach*. Singapore: World Scientific.

Zurawski, R. and Zhou, M. C., 1994, Petri nets and industrial applications: a tutorial. *IEEE Transactions on Industrial Electronics*, **41(6)**, 567–583.

Zussman, E. and Zhou, M.C., 2000, Design and implementation of an adaptive process planner for disassembly processes. *IEEE Transactions on Robotics & Automation*, **16(2)**, 171–179.

10

Planning Disassembly for Remanufacture-to-Order Systems

Karl Inderfurth and Ian M. Langella

CONTENTS

10.1 Introduction

10.1.1 Motivation

The previous decade starting in 1990, which started with the fall of communism and the resulting emergence of capitalism as the dominant economic system, was one of amazing economic growth. This era was characterized by increased consumption of goods and services, which compounded by the shorter product life cycles observed, has led to a rapid depletion of natural resources and available landfill space. Where a generation ago few would have given thought to environmental impact of operations, nowadays many are concerned with the ramifications for us all. This is evidenced by the recent push toward sustainability in operations (see e.g., Kleindorfer et al. 2005), which calls for an expanded view of profitability to include effects on people and the planet.

Within the realm of sustainability, reverse logistics and closed-loop supply chain management fit quite well. Companies are motivated to participate not only to comply with environmental legislation and to woo the so-called green customers, but also because they profit economically from doing so (see deBrito and Dekker, 2004). When deciding on what to do with returned products, there are many options available, for example, recycling, refurbishing, and remanufacturing (see Thierry et al., 1995). Remanufacturing results in products assumed "good as new" (in difference to refurbishing) and avoids energy consumption required in material recycling and maximizes the recovery of value added during manufacturing. Thus, it is seen as an advantageous option for products for which a demand for remanufactured items exists or can be made to exist. Remanufacturing can be seen with a diverse set of products, from single-use cameras to photocopiers to automotive engines.

To remanufacture items, parts must be harvested through the disassembly of the returned products. These parts are then cleaned and inspected, and provided they meet the "good as new" quality standard, are reassembled into remanufactured items. Therefore, parts can be seen as the requisite input for the remanufacturing process, and this chapter will focus on how disassembly can be planned to meet this demand for parts.

10.1.2 Problem Description

We can start by defining the terminology used in this area, using the sample product structures given in Figure 10.1. First, returned products (represented by the boxes 1 and 2 in Figure 10.1) are referred to as *cores*. Once completely disassembled, the resulting parts (W, X, Y, and Z in the Figure 10.1) are called

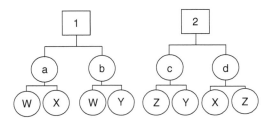

FIGURE 10.1
Exemplary product structures.

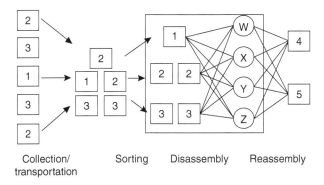

Collection/   Sorting   Disassembly   Reassembly
transportation

FIGURE 10.2
The remanufacturing process from collection to reassembly.

leaves. Items that are neither cores nor leaves (a, b, c, and d in Figure 10.1) are referred to as *intermediate* items.

Next, we will turn our attention to understanding remanufacturing as a process, using Figure 10.2 to illustrate. Starting on the left-hand side of the figure, the cores (depicted by the boxes numbered 1 through 3) must first be collected and transported to the remanufacturing facility. Once the cores arrive at the facility, they must normally be sorted as the return stream usually contains cores of many different types (e.g., model, year, and variant).

The sorted cores then wait to be disassembled, in what Guide et al. (2003) refer to as a *remanufacture-to-order* system, where demand for remanufactured product triggers disassembly. This is often seen in practice, particularly with more complex product examples such as automotive engines (see Seitz and Wells, 2006). We can see that a specified demand for remanufactured products would result in demand for parts through a bill-of-materials explosion of the remanufactured items. A *disassemble-to-order* problem (sometimes alternatively called *demand-driven* disassembly) is represented inside the box, and has the task of determining how many of each type of core to disassemble (and related decisions) to meet this given demand.

The following costs are relevant in this problem setting. *Core procurement costs* include compensating the customer and transporting the core to the remanufacturing facility. More generally, this cost term would include any routine operations performed on a core prior to the decision to disassemble it, for example, receiving and handling costs. *Separation costs* result from separating a core or an intermediate item. Disassembly is notoriously labor-intensive and requires much more manual labor (with up to three times the labor cost according to Seitz and Wells, 2006) than manufacturing in mass scale. Owing to the multitude of models encountered in a remanufacturing facility (about 400 core variants were encountered by the engine remanufacturer in Seitz and Wells, 2006), personnel must be familiar with and proficient at disassembling many different core types and disassemble cores using general-purpose tools. This extreme need for flexibility also impedes the ability to automate the process for most products. *External procurement costs* result from obtaining leaves from either an external supplier or internal supply source as an alternative to harvesting them from disassembly. Having an alternative supply for parts is advantageous for two reasons. First, at times when a part is being mass produced for manufacturing, this usually represents an inexpensive option in the presence of substantial scale economies. Second, this source is all the more attractive when parts from disassembly fail the stringent quality conditions for remanufacturing, and procured parts have to compensate for this loss. *Disposal costs* for leaves and intermediate items may represent actual disposal through landfilling or incineration. Leaves can also be "disposed of" through material recycling, leading to (marginal) revenue. This is attractive as the landfilling costs are saved and replaced with a (albeit small) revenue stream. This was often seen in our experience with several industrial partners. Finally, *holding costs* are relevant for cores, intermediates, and leaves. A released leaf, which is not immediately demanded, can be held in stock to satisfy future periods demand. Intermediate items, which contain leaves not immediately demanded, might be held in stock perhaps due to constrained disassembly capacity. Cores are most often held in stock to protect against scantiness, and one commonly observes very large core stocks (about 1 year of production) in practice, particularly in automotive engines. The engines are very often stored outside exposed to the elements, whereas leaves and intermediate items are stored in a dedicated storage space where one would expect higher physical storage costs.

Demand-driven disassembly planning is made difficult by several aspects. First, it is often so that a certain leaf can be harvested from several types of cores, an aspect referred to in the disassembly literature as *commonality*. Referring back to Figure 10.1, leaves X and Y were *common* to both cores, whereas, leaves W and Z were *unique* to one core. Using common parts to manufacture different models of a product makes a lot of sense from an economic point of view, allowing for scale economies in both manufacturing as well as product recovery. From a planning point of view, however, commonality increases the possible ways in which the leaf may be obtained. This aspect, particularly in combination with the next two aspects, makes it more difficult to judge which core to disassemble.

Second, disassembling a certain core results in the release of several leaves, some of which might be immediately needed while others might not. Leaves not immediately needed can either be held to satisfy future periods demand or disposed of. Thus, the optimal decision to this problem must weigh the two extremes. On the one hand, if very few cores are disassembled, separation and disposal costs are avoided, but external procurement costs have to be accepted. On the other hand, disassembling too many leads to high (or unnecessarily early) separation and disposal and holding costs. The optimal decisions must strike the perfect balance between these two extremes, depending on the specified costs. This is quite similar to joint production systems, where several products result from a certain production process (a prominent example being petrochemical production).

Third, the input–output relation, i.e., the yields from disassembly—how many good-quality parts resulting from the disassembly of a core—is subject to uncertainty. The reason for this is that the quality of the core returned is often not known before the inspection of the parts following disassembly. The combination of these three factors (which is both plausible and observed in our experience) is particularly challenging, not only for practitioners but also for researchers.

10.1.3 Literature Review

We will now review the relevant literature for the disassemble-to-order problem. In doing so, we choose to limit our review to approaches that do not incorporate setup costs. The stream of literature including setup costs is relevant for product examples where significant setup costs are incurred when starting to disassemble a certain product model, for example, when using robots to disassemble. In the automotive setting, disassembly is commonly very manual and uses general-purpose tools, which do not have to be changed between models. A more complete review also including other types of disassembly research is afforded in Langella (2007b). We will start by examining the literature for planning the disassembly of a single core.

Gupta and Taleb (1994) pioneered research in this area and were the first to note that disassembly planning is not simply the reverse of assembly. In their heuristic approach, demand for leaves in a single core with no inner-core commonality were specified and must be met with disassembly, i.e., without external leaf procurement. Other approaches in the single-core problem would include the following two, which include capacity considerations. Lee et al. (2002) present an integer programming (IP) model, which minimizes the sum of purchase, holding, and separation costs subject to a capacity constraint in a multi-period setting, also including lead times. Cao and Tang (2005) consider aspects of sequencing and more detailed treatment of capacity, presenting an IP model leading to a heuristic approach. Taleb et al. (1997) built on the work of Gupta and Taleb (1994) including inner-core commonality in the single-core case and developing a heuristic solution approach. Neuendorf et al. (2001) suggest an alternative timed Petri net solution algorithm for the same problem.

We can now turn our focus to multiple core problems with commonality. Verrakamolmal and Gupta (1998) provide a mathematical programming model, which satisfies demand for leaves purely from disassembly including a constraint on the amount of cores available for disassembly (a *return constraint*). Next, Gupta et al. (2000) extended this incorporating a deterministic yield rate (a fraction of good quality parts that are suitable for remanufacturing), with the discarded parts generating recycling revenue. Both of these models have nonlinear objective functions due to the way that separation costs were modeled, a disadvantage when seeking an optimal solution. Lambert and Gupta (2002) developed the *tree network model*, based on Veerakamolmal and Gupta (1998) and another approach which included sequencing, resulting in a linear objective function for the single-period problem. Veerakamolmal and Gupta (2000) extended the model of Gupta et al. (2000) into a multi-period setting with no external leaf procurement but included a return constraint and a shelf life for leaves.

Kongar and Gupta (2002) provide a single-period goal programming formulation that optimizes several goals (e.g., maximize profit and minimize disposed and held items) using aspiration levels for each goal. This was extended to a fuzzy goal programming approach of Kongar and Gupta (2006), who allow for uncertain revenue from recycled material harvested from the cores, although the leaf demand and yield rates remained deterministic.

Taleb and Gupta (1997) provided the first heuristic approach to the multi-core multi-period disassemble-to-order problem. Their heuristic approach contained two algorithms, which resulted in planning the disassembly to meet the entire demand for the leaves. This approach was extended to include external leaf procurement and holding in Langella (2007a). Langella and Schulz (2006) add refinements to the heuristics and examine performance over a rolling multi-period planning horizon. Schulz (2006) extends the heuristics for settings with constrained disassembly capacity, whereas Langella (2007b) allows additionally for constrained returns.

The remaining approaches additionally consider yield uncertainty in planning. The first work here was by Inderfurth and Langella (2003) and the work was further extended in Inderfurth and Langella (2006). In both papers, the authors consider a single-period problem with stochastic yields for leaves harvested from several cores including commonality and develop and test heuristic solution approaches. Langella (2007b) provides the adaptation of the heuristics to the multi-period problem. The multi-criteria work of Kongar and Gupta (2002) was extended for stochastic yields using elements of the heuristics of Inderfurth and Langella (2006) by Imtanavanich and Gupta (2004). Imtanavanich and Gupta (2005) then extended their previous work for a multi-period planning horizon.

10.1.4 Roadmap

This chapter is organized as follows. In the next section, we first start by examining the planning problem faced when the yields of disassembly are deterministic, formulated as an integer linear program. Section 10.3 then relaxes this assumption and we explore the problem through recourse models from stochastic linear programming. Both Sections 10.2 and 10.3 will include a discussion of solution methods. Section 10.4 provides our conclusions and an outlook on further research.

10.2 Planning Disassembly with Deterministic Yields

This section will put forth methods for planning disassembly when the yield of disassembly is known with certainty, an assumption that will later be relaxed. Within this section, we will start by examining a single-period planning problem and later extend it to a general multi-period problem, highlighting the differences brought about by the inclusion of future periods.

10.2.1 Planning for a Single Period

First, we will put forth a basic model, which is subject to assumptions (e.g., complete disassembly and no capacity constraints) and later extend this model to a more general case by relaxing some of the assumptions.

10.2.1.1 Basic Model

In the basic model setting, we make the following assumptions:

- The demand for leaves is deterministic and must be fulfilled using either leaves gained through disassembly or external procurement.
- Leaves can be obtained from external procurement (without limits) at given costs, and excess leaves may be disposed of (also without limits) at given costs.
- There are no lead times for core procurement or separation.
- Yields from disassembly are deterministic (relaxed in Section 10.3).
- Cores are completely disassembled, releasing all the leaves contained therein. As such, we make no mention of intermediate items (relaxed in extended model).
- There are no limits on the amount of cores that can be procured from the market (return availability) or disassembled (disassembly capacity) (relaxed in extended model).

With these assumptions in mind, we now define the following notation:

i: Core index
k: Leaf index
I: Core set
K: Leaf set
c_i^z: Core cost, i.e., the cost to acquire, transport, and completely disassemble the core and clean and inspect the leaves for core i
c_k^p: External procurement cost for leaf k
c_k^d: Disposal cost for leaf k
$\pi_{i,k}$:   Yield rate, i.e., the amount of leaf k resulting from the disassembly of core i
D_k: Demand for leaf k
x_i^z: Amount of core i to acquire and disassemble
x_k^p: Amount of leaf k to externally procure
x_k^d: Amount of leaf k to dispose of

Using this notation, we can formulate an integer linear programming model as

$$\min C = \sum_{i \in I} c_i^z \cdot x_i^z + \sum_{k \in K} \left(c_k^p \cdot x_k^p + c_k^d \cdot x_k^d \right) \tag{10.1}$$

subject to

$$\sum_{i \in I} x_i^z \cdot \pi_{i,k} + x_k^p - x_k^d = D_k \quad \forall k \in K \tag{10.2}$$

$$x_i^z, x_k^p, x_k^d \geq 0 \quad \text{and integer} \quad \forall i \in I, k \in K \tag{10.3}$$

The objective function (10.1) minimizes the total relevant costs, which consist of summed core cost and summed leaf procurement and disposal costs. The first constraint (10.2) ensures that the demand is satisfied by disassembly and external leaf procurement, with any superfluous leaves being disposed of. The last constraint (10.3) forces the decision variables to be nonnegative integers. We can point out, however, that when all the yield rates and leaf demands are integer valued, the leaf procurement and disposal decisions will automatically be integer-valued, making the integrality constraints (for these two variables, not the core decisions) superfluous.

10.2.1.2   Extended Model

Several of the assumptions made for the previously introduced model can be relaxed, leading to a more general formulation. First, the complete disassembly assumption can be relaxed by incorporating *intermediate* items (those items, which are neither cores nor leaves) in the planning. These intermediate items may be obtained through the separation of cores, and are either further disassembled to the leaf items (in case of separation) or disposed of in the single-planning period. Second, the amount of cores available from the market for purchase can be constrained. Finally, disassembly capacity can be expressly incorporated into the model. This disassembly capacity limits core and intermediate item disassembly. All other assumptions will remain unchanged. We quickly note that although we have chosen to display two levels (cores and one level of intermediates), this can be easily generalized. To extend the model, we will need some additional notation:

j: Intermediate item index
J: Intermediate set
c_j^s: Separation cost for intermediate item j
c_j^d: Disposal cost for intermediate item j
$\pi_{i,j}$:   The amount of intermediate item j resulting from the disassembly of core i
$\pi_{j,k}$:   The amount of leaf k resulting from the disassembly of intermediate item j

A: Amount of disassembly capacity available to separate cores and intermediate items

a_i: Amount of disassembly capacity absorbed by the separation of core i to the intermediate items contained therein

a_j: Amount of disassembly capacity absorbed by the separation of intermediate item j to the leaf items contained therein

R_i: Amount of core i available for acquisition and disassembly

x_j^s: Amount of intermediate item j to separate further into leaf items

x_j^d: Amount of intermediate item j to dispose of

The model would now have the following objective and constraints:

$$\min C = \sum_{i \in I} c_i^z \cdot x_i^z + \sum_{j \in J} \left(c_i^s \cdot x_i^s + c_j^d \cdot x_j^d \right) + \sum_{k \in K} \left(c_k^p \cdot x_k^p + c_k^d \cdot x_k^d \right) \quad (10.4)$$

subject to

$$\sum_{i \in I} x_i^z \cdot \pi_{i,j} - x_j^d = x_j^s \quad \forall j \in J \quad (10.5)$$

$$\sum_{i \in J} x_j^s \cdot \pi_{j,k} + x_k^p - x_k^d = D_k \quad \forall k \in K \quad (10.6)$$

$$\sum_{i \in I} a_i \cdot x_i^z + \sum_{j \in J} a_j \cdot x_j^s \leq A \quad (10.7)$$

$$x_i^z \leq R_i \quad \forall i \in I \quad (10.8)$$

$$x_i^z, x_j^s, x_j^d, x_k^p, x_k^d \geq 0 \quad \text{and integer} \quad \forall i \in I, j \in J, k \in K \quad (10.9)$$

As can be seen, the objective function (10.4) now also contains the relevant costs for intermediate items in addition to those for cores and leaves. The first constraint (10.5) tracks intermediate items ensuring that all items harvested from disassembly are either separated further or disposed of. The second constraint (10.6) ensures that demand for leaves is met exactly. The third constraint (10.7) ensures that the available disassembly capacity (consumed by both core and intermediate separation) is not exceeded. The fourth constraint (10.8) limits our core decision to the amount of cores returned. Finally, as was the case earlier, all decision variables must be nonnegative integers.

10.2.2 Planning for More than One Period

We will now turn our attention to how the situation changes when there are more than a single planning period and an inventory can be used to hold leaves between the periods. As was the case in the single-period problem, we will start with a basic model formulation and later relax some of the assumptions.

10.2.2.1 Basic Model

In the basic model setting, we make the following assumptions:

- The demand for leaves is deterministic and must be fulfilled using either leaves gained through disassembly or external procurement.
- An inventory can be used to hold leaves to satisfy demand in future periods. Holding leaves results in a holding cost per part period, and there are no limits on the amount of leaves that may be held in stock.
- Leaves can be obtained from external procurement (without limits) at given costs, and excess leaves may be disposed of (also without limits) at given costs.
- There are no lead times for core procurement or separation.
- Yields from disassembly are deterministic (relaxed in Section 10.3).
- Cores are completely disassembled, releasing all the leaves contained therein (relaxed in extended model).
- There are no limits on the amount of cores that can be procured from the market (return availability) or disassembled (disassembly capacity) (relaxed in extended model).

Some new notation is also needed, so we hereby define t as the time index with $t = 1, \ldots, T$, using T to denote the number of periods in the planning horizon. The decisions will now be time-phased, that is, $x_{i,t}^z$ refers to the amount of core i collected and disassembled in period t. Holding cost (per part period) for leaves will be denoted c_k^h, and $y_{k,t}$ will represent the inventory of leaf k at the end of period t. The starting inventory (the leaf inventory available at the start of period 1) is exogenous and will be denoted by $\bar{y}_{k,0}$. Additionally, an end-of-horizon inventory could be fixed, which we denote by $\bar{y}_{k,T}$.

The objective function and constraints can be given by

$$\min C = \sum_{t=1}^{T} \left[\sum_{i \in I} c_i^z \cdot x_{i,t}^z + \sum_{k \in K} \left(c_k^p \cdot x_{k,t}^p + c_k^d \cdot x_{k,t}^d + c_k^h \cdot y_{k,t} \right) \right] \quad (10.10)$$

subject to

$$y_{k,1} = \bar{y}_{k,0} + \sum_{i \in I} x_{i,1}^z \cdot \pi_{i,k} + x_{k,1}^p - x_{k,1}^d - D_{k,1} \quad \forall k \in K \quad (10.11)$$

$$y_{k,t} = y_{k,t-1} + \sum_{i \in I} x_{i,t}^z \cdot \pi_{i,k} + x_{k,t}^p - x_{k,t}^d - D_{k,t} \quad \forall k \in K, t = 2, \ldots, T \quad (10.12)$$

$$y_{k,T} = \bar{y}_{k,T} \quad \forall k \in K \quad (10.13)$$

$$x_{i,t}^z, x_{k,t}^p, x_{k,t}^d, y_{k,t} \geq 0 \quad \text{and integer} \quad \forall i \in I, k \in K, t = 1, \ldots, T \quad (10.14)$$

The objective function (10.10) now contains relevant costs for all of the periods, and separate inventory balance constraints are provided for the first (10.11) and following (10.12) periods in the problem. The third constraint (10.13) ensures that the specified end-of-horizon inventory is realized.

10.2.2.2 Extended Model

We can now work at relaxing some of the assumptions, which were made in the previous model. Specifically, we will allow for partial disassembly by incorporating intermediate items and incorporate return availability and disassembly capacity into the planning model. This will necessitate several changes. First, we will additionally allow for inventories of cores (using $y_{i,t}$ to denote the inventory of core i at the end of period t) and intermediate items (analogously represented by $y_{j,t}$). Core inventories become necessary when dealing with return availability, decoupling the core procurement and separation decisions. Inventories of intermediate items will allow for scarce disassembly capacity to be better utilized. Starting inventories and end-of-horizon will be expressed conventionally, for example, with $\bar{y}_{i,0}$ and $\bar{y}_{i,T}$, respectively, for the core stock. Second, we will decouple the core procurement and separation decisions and costs by introducing $x_{i,t}^p$ and $x_{i,t}^s$ to denote the core procurement and separation decisions, respectively. The previously utilized core cost c_i^z will likewise be separated into procurement, (c_i^p), and separation, (c_i^s), cost components, i.e., $c_i^z = c_i^p + c_i^s$. Now, the disassembly capacity available for each period (A_t) and the amount of returns available (R_t) are specified.

The objective function and constraints are given as

$$
\min C = \sum_{t=1}^{T} \left[\begin{array}{l} \sum_{i \in I} \left(c_i^p \cdot x_{i,t}^p + c_i^s \cdot x_{i,t}^s + c_i^h \cdot y_{i,t} \right) \\[2mm] + \sum_{j \in J} \left(c_j^s \cdot x_{j,t}^s + c_j^d \cdot x_{j,t}^d + c_j^h \cdot y_{j,t} \right) \\[2mm] + \sum_{k \in K} \left(c_k^p \cdot x_{k,t}^p + c_k^d \cdot x_{k,t}^d + c_k^h \cdot y_{k,t} \right) \end{array} \right] \tag{10.15}
$$

subject to

$$
y_{i,1} = \bar{y}_{i,0} + x_{i,1}^p - x_{i,1}^s \quad \forall i \in I \tag{10.16}
$$

$$
y_{i,t} = y_{i,t-1} + x_{i,t}^p - x_{i,t}^s \quad \forall i \in I, t = 2, \ldots, T \tag{10.17}
$$

$$
y_{i,T} = \bar{y}_{i,T} \quad \forall i \in I \tag{10.18}
$$

$$y_{j,1} = \bar{y}_{j,0} + \sum_{i \in I} x_{i,1}^s \cdot \pi_{i,j} - x_{j,1}^d - x_{j,1}^s \quad \forall j \in J \tag{10.19}$$

$$y_{j,t} = y_{j,t-1} + \sum_{i \in I} x_{i,t}^s \cdot \pi_{i,j} - x_{j,t}^d - x_{j,t}^s \quad \forall j \in J, t = 2, \ldots, T \tag{10.20}$$

$$y_{j,T} = \bar{y}_{j,T} \quad \forall j \in J \tag{10.21}$$

$$y_{k,1} = \bar{y}_{k,0} + \sum_{j \in J} x_{j,1}^s \cdot \pi_{j,k} + x_{k,1}^p - x_{k,1}^d - D_{k,1} \quad \forall k \in K \tag{10.22}$$

$$y_{k,t} = y_{k,t-1} + \sum_{j \in J} x_{j,t}^s \cdot \pi_{j,k} + x_{k,t}^p - x_{k,t}^d - D_{k,t} \quad \forall k \in K, t = 2, \ldots, T \tag{10.23}$$

$$y_{k,T} = \bar{y}_{k,T} \quad \forall k \in K \tag{10.24}$$

$$\sum_{i \in I} a_i \cdot x_{i,t}^s + \sum_{j \in J} a_j \cdot x_{j,t}^s \leq A_t \quad t = 1, \ldots, T \tag{10.25}$$

$$x_{i,t}^p \leq R_{i,t} \quad \forall i \in I, t = 1, \ldots, T \tag{10.26}$$

$$x_{i,t}^s, x_{i,t}^i, y_{i,t}, x_{j,t}^s, x_{j,t}^d, y_{j,t}, x_{k,t}^p, x_{k,t}^d, y_{k,t} \geq 0 \quad \text{and integer} \quad \forall i \in I, j \in J, k \in K,$$
$$t = 1, \ldots, T \tag{10.27}$$

The objective function (10.15) contains all relevant costs in both periods. The first three constraints (10.16–10.18) deal with the core stock, defining it as the stock at the end of the previous period plus any procured minus any separated in the current period. Next, the inventory of intermediates (10.19–10.21) is specified as the previous value plus that which is harvested from the separation of the core items minus any disposed of and the amount further separated into leaf items. Finally, the leaf inventory (10.22–10.24) is given as its previous value plus the amount harvested from the disassembly of intermediate items and gained through external procurement minus any disposed of or demanded in the respective period. The disassembly capacity constraint (10.25) and return availability constraint (10.26) have both been modified to hold in each of the periods.

10.2.3   Solution Methods

Even the most basic single-period formulation can be shown to be NP-hard, a term that indicates intractability and thereby the fact that such problems cannot be solved exactly within polynomial time. Examining the model

(10.28), we can see that it can be transformed into an unbounded knapsack problem (see Martello and Toth, 1990, p. 3) by specifying certain procurement and disposal costs for the leaves. Since the unbounded knapsack is known to be NP-hard (see Martello and Toth 1990, p. 6), its special case is therefore also NP-hard. An exact solution for the IP models presented in this chapter can be obtained using a commercial linear programming solver, for example, XPRESS. Owing to the problem complexity described previously, solving large industrial-sized problem (which contain many decisions that must be integer valued) instances will probably require a prohibitively large amount of computational time.

To illustrate this, a more empirical way of examining complexity of problems is to formulate several instances of the problem at different problem sizes (numbers of decisions) and examine the average computational time required to solve them as the problem size is increased. This average computational time can be thought of as a proxy for the average complexity (see Coffin and Saltzman, 2000). To this end, Langella (2007b) generated instances of the multiple period problem (10.10) for different problem sizes as depicted in Table 10.1. The sizes (labeled A through E) have different numbers of cores ($|I|$), leaves ($|K|$), and time periods (T), resulting in increasing amounts of integer decisions (IDV). For each size level, 20 problem instances were generated and solved on the same computer using XPRESS, and the table provides the minimum, maximum, and average computation times in seconds over the 20 instances.

As one can see by examining the running times in the table, at least one instance in each size group was solved within one second. However, other instances of the same size took considerably longer to solve, particularly when looking at the last three size levels C through E. When looking at the average computational time for these levels, we can see that the average time needed increases *dramatically* when compared with the *gradual* increase in integer decisions. One can note that industrial-sized problems would probably contain many more cores and leaves than even in the highest (E) level. This motivates the development and use of heuristic solution methods, which offer "good and fast" solutions to the problem.

The pioneering work on heuristics for this multiple core problem with commonality is credited to Taleb and Gupta (1997). In their approach, leaves could not be externally procured and the entire demand had to be fulfilled

TABLE 10.1

Problem Sizes and Computational Time Results

Size	$|I|$	$|K|$	T	IDV	Minimum	Mean	Maximum
A	2	3	4	8	0.2	0.3	1.2
B	3	7	6	18	0.2	0.5	2.3
C	4	10	8	32	0.3	145.0	1864.4
D	4	10	9	36	0.2	1930.0	37232.4
E	5	11	8	40	0.7	11448.8	71979.0

by harvesting leaves from disassembly. The approach resulted in infeasible solutions to certain instances, and so was corrected and extended for leaf procurement and holding costs (for the multi-period setting). The resulting heuristics in Langella (2007b) represent the best-performing heuristics to date for this problem.

For single-period problems, Langella (2007b) suggests an iterative heuristic, where in each iteration one core is selected for disassembly to fulfill leaf demand. To decide which core is disassembled, different forms of profitability measurements (e.g., ratio or absolute) can be utilized in the heuristic. By profitability, we mean the difference between the revenue from disassembly (i.e., the foregone external procurement costs of contained leaves) and the cost (the core procurement and separation cost). An examination of the heuristics' performance, measured by the percent penalty and based on both randomly generated instances as well as random variants of real-world data, indicates that in the single-period realm (where admittedly heuristics will be less necessary than in the multi-period setting), the heuristics work well. Table 10.2 provides results for the ratio profitability measure, whereas, Table 10.3 contains penalties for the absolute profitability measure, differentiating between two factors of experimentation. The first factor was profitability with high profitability (HP) containing instances, where the core cost was less than half the summed leaf procurement costs and low (LP) where the leaves cost less than twice the core cost. The second factor was the problem size, with small problems containing 3 cores and 10 leaves, where the larger problem sizes twice as many of each. Each table presents average and maximum penalties for each factor combination where 32 instances were generated per combination. The performance was noticeably

TABLE 10.2

Percent Penalties for Ratio Profitability Measure Heuristic for Single-Period Problem

	Small Problem		Large Problem	
	Mean	Maximum	Mean	Maximum
LP	0.6	3.5	0.5	3.2
HP	2.3	17.6	1.9	6.1

TABLE 10.3

Percent Penalties for Absolute Profitability Measure Heuristic for Single-Period Problem

	Small Problem		Large Problem	
	Mean	Maximum	Mean	Maximum
LP	0.5	7.7	0.6	5.1
HP	2.6	11.6	2.4	16.5

TABLE 10.4

Percent Penalties for Look-Ahead Heuristic for Multiple-Period Problem

	Small Problem		Large Problem	
	Mean	Maximum	Mean	Maximum
ST	3.5	12.4	7.5	19.6
LT	3.4	25.3	7.3	28.4

good e.g, generated using real product structures, specifically those given in Veerakamolmal and Gupta (1998) and Lambert and Gupta (2002). The heuristics were also adapted to cases where returns are constrained, another area where performance was particularly good.

For multi-period problems, Langella (2007b) put forth different heuristic approaches, the best of which turned out to be a so-called look-ahead heuristic. This heuristic will hold leaves that will be demanded in the coming periods (the number is determined by a cost ratio which can be interpreted as an economic run out time) risking that the leaves will also be availed coincidentally in the future. Performance of this heuristic was examined using a test battery consisting of both randomly generated instances and random variants of real-world data. The average and maximum penalties obtained by the look-ahead heuristic for the randomly generated instances are given in Table 10.4. In the results, we differentiate between small and large instances again as well as instances with a short (ST) and long (LT) horizon with four and eight periods, respectively. The heuristic's performance was considerably better when applied to random variants of real-world product structures obtained from both Veerakamolmal and Gupta (1998) and Imtanavanich and Gupta (2005). When the return stream is constrained, the performance of heuristics generally increases with a tightening of the return constraint. In situations where the disassembly capacity is constrained, Schulz (2006) put forth a heuristic solution method, which shows much promise. Nevertheless, the performance results in Table 10.4 indicate that in the multi-period case some more effort should be exerted in improving the present heuristics.

10.3 Planning Disassembly with Stochastic Yields

Owing to the fact that the quality of the cores is generally not known beforehand, it is often the case that the amount of leaves harvested from the disassembly of a particular core is subject to considerable uncertainty. This further complicates planning, as we will soon see.

There are several competing approaches to modeling yield randomness as presented in the excellent overview provided by Yano and Lee (2003). Within our disassembly setting, we revert to modeling yield uncertainty using the stochastically proportional yield approach where the number of good leaves obtained is the product of the number of cores disassembled (input) multiplied by a random fraction. In other words, the yield rates ($\pi_{i,k}$ for the complete disassembly models) are stochastic. This chapter will provide the so-called recourse models from the realm of stochastic linear programming, starting for a single-period and later extending this to a two-period problem, under the assumption that the yield rates are stochastic. Owing to this complicated setting, we will restrict our attention to complete disassembly and to two periods in the multi-period case in this chapter. A generalization is straightforward, but makes the presentation of the material far more complicated.

Recourse models were first introduced by Dantzig (1955), and represent one approach to extend linear programming into settings containing certain stochastic parameters. Recourse models work (see, e.g., Sen and Hingle, 1999 for a recommendable tutorial) by bifurcating decisions into two groups, proactive ones made before the realization of the random variables and reactive decisions which are made after some realization. The uncertainty is incorporated by introducing different *scenarios*, with each scenario containing joint realizations for each stochastic variable and having a defined probability of occurrence. Reactive decisions (sometimes also called recourse decisions) depend on the realized scenario and are charged with compensating for the random influence. Optimal proactive decisions bring about an optimal sum of proactive costs and expected reactive costs, assuming for the moment that a cost minimization problem is at hand.

10.3.1 Planning for a Single Period

We can start by framing the planning situation we face, and dissecting the decision variables into proactive and reactive. Following the work of Inderfurth and Langella (2003), we base the modeling on the following sequence of decisions:

- First, we decide (proactively) how many of each core is to be disassembled (x_i^z).
- Then, the random yields are realized. We will use $q \in Q$ as the scenario index and set, with $\pi_{i,k,q}$ representing the amount of leaf k resulting from the disassembly of core i in scenario q.
- Finally, leaf procurement decisions and disposal decisions will be made reactively. Leaf decisions will be denoted $x_{k,q}^p$ and $x_{k,q}^d$ as they now depend on the yield rate realization.

A scenario q contains joint realizations for each of the $\pi_{i,k}$ yield rates, and its probability is denoted P_q, noting quickly that as one would expect $\Sigma_{q \in Q} P_q = 1$. The number of elements in the scenario set increases dramatically as the number of possible yield rates per core–leaf relationship (outcomes) is raised. To illustrate this, we can imagine a problem containing two cores and three leaves, one of which is common to both cores (see Figure 10.3). In this problem, there are four core–leaf relationships, the common leaf having one with each

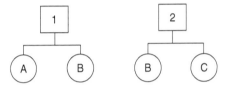

FIGURE 10.3
Exemplary product structure.

core, and the unique leaves having one each with their respective core. If each relationship had three possible outcomes (i.e., each yield rate could, say, be 0, 1/2, or 1), there would be $3^4 = 81$ scenarios in the scenario set, as each scenario must contain combined outcomes for all variables. Increasing the number of outcomes in this simple problem from 3 to 10 increases the number of scenarios from 81 to 10,000, underscoring the fact that this method also cannot be relied upon for solving realistic-sized problems. We note that both independent and dependent yield rates may be modeled using the scenario method, which we feel may indeed prove useful for some practical applications.

The objective function and constraints are given as

$$\min C = \sum_{i \in I} c_i^z \cdot x_i^z + \sum_{k \in K} \sum_{q \in Q} P_q \cdot \left(c_k^p \cdot x_{k,q}^p + c_k^d \cdot x_{k,q}^d \right) \tag{10.28}$$

subject to

$$\sum_{i \in I} x_i^z \cdot \pi_{i,k,q} + x_{k,q}^p - x_{k,q}^d = D_k \quad \forall k \in K, q \in Q \tag{10.29}$$

$$x_i^z \geq 0 \quad \text{and integer} \quad \forall i \in I \tag{10.30}$$

$$x_{k,q}^p, x_{k,q}^d \geq 0 \quad \forall k \in K, q \in Q \tag{10.31}$$

As we can see, the objective function (10.28) now contains both proactive and reactive cost terms. The latter is contained in the second term, where the scenario-dependent leaf costs are multiplied with the probability that the scenario results. When summed over all scenarios and leaves, this provides an expectation of the cost resulting from the recourse decisions. Demand-fulfillment constraints (10.29) are now provided for each leaf and scenario, linking yield rate realizations with their corresponding recourse decisions to satisfy the demand. Core decisions are restricted to nonnegative integers as before, while the leaf decisions are merely constrained by nonnegativity. The reason for this is that if the scenarios contain yield rate outcomes that are not entirely integer-valued, non-integer leaf harvests may result, necessitating disposal and procurement decisions to be treated as continuous to satisfy the integer-valued demand. In practice, of course, the harvest will always be integer-valued, and the respective decisions must be rounded to implement the policy.

10.3.2 Planning for Two Periods

When adapting the recourse model for two planning periods, the model becomes more complex than in the single period for a couple of reasons. First, leaf inventories must now be incorporated into the environment. Second, there are yield realizations in each of the two planning periods, which must also be accounted for. To illustrate the environment we now face (see Figure 10.4) and to introduce the modified notation, we will now describe the sequence of decisions and information.

1. The first-period core disassembly decisions $(x^z_{i,1})$ are chosen before any yield revelation.

2. The yield rates of the first period become known. For the first-period yield rates (which are denoted by $\pi_{i,k,1,q}$), we use the same scenario index and set $q \in Q$ as in the single-period formulation.

3. Once the yields of disassembly from the first period are known, the first-period leaf recourse decisions (now including inventory and denoted by $y_{k,1,q}$, $x^p_{k,1,q}$, $x^d_{k,1,q}$ as they are now scenario-dependent) are taken.

4. The second-period disassembly decisions are made after the revelation of the first period's yields but before the second period. Therefore, they will also be scenario-dependent and denoted by $x^z_{i,2,q}$. Intuitively, one would expect that the smaller the first period's yield rates turn out to be, the more cores will be disassembled in the second period.

5. The second period's yield rates realize, denoted by $\pi_{i,k,2,r}$ using $r \in R$ as the second-period scenario index and set.

6. Finally, once the second period's yield rates are known, the second-period leaf recourse decisions $(y_{k,2,q,r}, x^p_{k,2,q,r}, x^d_{k,2,q,r})$ are made. Each decision depends on both the first and the second period's yield revelation.

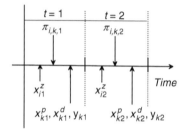

FIGURE 10.4
The decisions and information in a two-period stochastic problem.

The objective function and constraints are given as

$$\min C = \sum_{i \in I} c_i^z \cdot x_{i,1}^z + \sum_{q \in Q} \sum_{i \in I} P_q \cdot c_i^z \cdot x_{i,2,q}^z$$

$$+ \sum_{q \in Q} \sum_{k \in K} P_q \cdot \left(c_k^p \cdot x_{k,1,q}^p + c_k^d \cdot x_{k,1,q}^d + c_k^h \cdot y_{k,1,q} \right)$$

$$+ \sum_{q \in Q} \sum_{r \in R} \sum_{k \in K} P_q \cdot P_r \cdot \left(c_k^p \cdot x_{k,2,q,r}^p + c_k^d \cdot x_{k,2,q,r}^d + c_k^h \cdot y_{k,2,q,r} \right) \qquad (10.32)$$

subject to

$$y_{k,1,q} = \bar{y}_{k,0} + \sum_{i \in I} x_{i,1}^z \cdot \pi_{i,k,1,q} + x_{k,1,q}^p + x_{k,1,q}^d - D_{k,1} \quad \forall k \in K, q \in Q \qquad (10.33)$$

$$y_{k,2,q,r} = y_{k,1,q} + \sum_{i \in I} x_{i,2,q}^z \cdot \pi_{i,k,2,r} + x_{k,2,q,r}^p + x_{k,2,q,r}^d - D_{k,2} \quad \forall k \in K, q \in Q, r \in R \qquad (10.34)$$

$$y_{k,2,q,r} = \bar{y}_{k,2} \quad \forall q \in Q, r \in R \qquad (10.35)$$

$$x_{i,1}^z, x_{i,2,q}^z \geq 0 \quad \text{and integer} \quad \forall i \in I, q \in Q \qquad (10.36)$$

$$x_{k,1,q}^p, x_{k,1,q}^d, y_{k,1,q}, x_{k,2,q,r}^p, x_{k,2,q,r}^d, y_{k,2,q,r} \geq 0 \quad \forall k \in K, q \in Q, r \in R \qquad (10.37)$$

The objective function is grouped by decisions. As can be seen, the first term contains the first-period core decisions, made before any yield revelation. The second term includes the scenario-dependent cost resulting from disassembly decisions for the second period multiplied by the probability that the scenario realizes. The third term contains the scenario-specific recourse decisions for the first period, which just as the previous term, are multiplied with their respective probabilities resulting in an expectation. The last term includes recourse decisions for the second period. As these decisions depend on both scenarios, we multiply the costs by the product of both scenario's probabilities and sum over all combinations. Hereby, we implicitly assume independence of scenarios across the two periods. In case of dependence, conditional probabilities for P_r, i.e., $P_r(q)$ would have to be used.

The first constraint (10.33) specifies the inventory in the first period as the starting inventory plus the harvest and amount externally procured minus the amount disposed of or demanded. This constraint will hold for all leaves and scenarios, effectively linking first-period recourse decisions with their corresponding realized value. The second constraint (10.34) handles the second-period inventory, and is equal to the inventory available at the end of the first period (which depends on q) plus the harvest (depends on both q and r, the former because the disassembly decision depends on it, while the yield rate is dependent on the latter) and the amount externally procured minus the amount disposed of (both leaf decisions depend also on both realizations) and demanded. The third constraint ensures that the given end-of-horizon inventory is adhered to, which must hold for all leaves and combinations of both scenarios. As was the case previously, the core decisions are constrained to nonnegative integers, while other decisions must be nonnegative but may be treated as continuous, for the same reasons as in the single-period recourse model.

10.3.3 Solution Methods

As was the case in the previous chapter, the mixed-integer linear recourse models can also be solved using commercial solvers such as XPRESS. Unfortunately, this carries with it an identical disadvantage in that larger instances will probably require a prohibitively large amount of time to arrive at an optimal solution. The situation is even worse than in the deterministic setting, since the scenario set Q will increase exponentially with an increasing number of yield outcomes. To illustrate the growth of the number of integer decisions, we can see that the single-period model contained $|I|$ integer decisions (one for every core) and the two-period model has $|I|+|I|\cdot|Q|$, and adding additional periods will obviously increase this even more. So the models in Sections 10.3.1 and 10.3.2 are only useful for generating benchmark results to test the performance of heuristics, which must be developed for this problem. In our experience with industrial partners, it seems that they often take the mean yield rate and plan using this as the deterministic yield rate. This heuristic we will refer to as the mean yield rate method.

Inderfurth and Langella (2006) suggest two more-sophisticated heuristics where adjusted yield factors are developed by examining the solution to decomposed stochastic problems and comparing this with the solution to the deterministic problem. Thus, using adjusted yield factors (a deterministic equivalent) in the deterministic model would lead to decisions optimal in the (decomposed) stochastic problems. The adjusted yield factors are then inserted into the deterministic model and solved, providing decisions which one hopes will work well in the many core, many leaf problem.

The first, simpler heuristic (called one-to-one) separates the problem into a single-core single-leaf problem, yielding an adjusted yield rate to be used as a deterministic equivalent. In decomposing the many-to-many problem into many one-to-one problem, the core cost must be split among the leaves it contains. This results from the joint production like aspect of disassembly, and there are several alternative ways in which the core cost can be allocated among the leaves. Inderfurth and Langella (2003) use a yield-proportional split where Inderfurth and Langella (2006) use cost-proportional split using external procurement costs as a proxy for value. It can be said, however, that this rule does not critically impact the performance of the heuristic. Inderfurth and Langella (2006) show that its performance is often quite good. Specifically, in most of the tested cases it outperforms the mean yield rate method preferred by practitioners and obtains much smaller average and maximum penalties. This relatively simple heuristic performs best when the demand for leaves matches the bills of materials of the returned cores (an aspect referred to as *demand symmetry*). We can presume then that it will probably work well when returned cores match the demanded remanufactured products. In some instances in the study, however, performance was less impressive. This was the case when the demand was less symmetric, yield variation was high, and disassembly profitability was high. For these cases, a heuristic with more sophistication is needed.

Such a more complicated heuristic (one-to-many) examines a single-core multi-leaf problem, thus, ignoring commonality among cores in the many-to-many problem. To do this, the demand for common leaves is first split among the cores, which contain the leaf. Since splitting demand for leaves among the cores is basically the answer sought by the problem (an output, rather than an input of optimization), and since the initial demand split has proven itself critical to the performance of the heuristic, a sophisticated split method is advocated. In this initial split, the cores' profitability is modeled as a normal random variable, and the demand is split according to the probability that one core's profitability surpasses the other. Once the demand is split, adjusted yield rates are obtained by comparing the deterministic and stochastic solution to one-to-many problems resulting from decomposing the many-to-many problem. These adjusted yield rates are then inserted into a deterministic model providing a solution and thereby an updated demand split, if you will. With this updated demand split, new adjusted yield factors are calculated and used within the deterministic model to arrive at a new solution. This procedure repeats itself until two consecutive demand splits are identical. As shown in both Inderfurth and Langella (2006) and Langella (2007b), the heuristic works well in all conditions of demand symmetry, yield variability, and profitability. Specifically, the penalty of using the heuristic in these single-period problems never exceeded a maximum deviation of 3% in all tested cases.

The results of the performance study are partly recapitulated in Tables 10.5 through 10.7 for the mean yield rate, one-to-one, and one-to-many heuristics, respectively. The tables discern between instances with low-yield variation (LYV) and high-yield variation (HYV) given by coefficients of yield rate variation of 0.1 and 0.25, respectively, in addition to the factor of core profitability introduced in Section 10.2. As can be seen, the mean yield rate (Table 10.5) performs decently on average in cases of low-yield variability. The performance

TABLE 10.5

Percent Penalties for Mean Yield Rate Heuristic in Single-Period Stochastic Yield Problem

	LYV		HYV	
	Mean	Maximum	Mean	Maximum
LP	2.3	10.4	6.1	18.1
HP	2.3	6.7	4.5	23.4

TABLE 10.6

Percent Penalties for One-to-One Heuristic in Single-Period
Stochastic Yield Problem

	LYV		HYV	
	Mean	Maximum	Mean	Maximum
LP	1.0	4.2	1.9	7.7
HP	2.4	9.9	2.9	17.1

TABLE 10.7

Percent Penalties for One-to-Many Heuristic in Single-Period
Stochastic Yield Problem

	LYV		HYV	
	Mean	Maximum	Mean	Maximum
LP	0.4	1.6	0.6	2.3
HP	0.7	2.4	0.6	2.9

deteriorates as the yield variation increases, with elevated average and maximum penalties. The one-to-one heuristic (Table 10.6) outperforms the mean yield rate most of the time, with lower averages and maximum penalties. It's own maximum penalties, however, particularly in cases with high profitability and yield variability, motivate the more sophisticated one-to-many heuristic well. The one-to-many heuristic (Table 10.6) performs well in all the tested cases with low maximum values overall, although revealing a slightly better performance in cases with lower profitability and yield uncertainty.

The work of Langella (2007b) extends these heuristics for application to multi-period problems. To do this, the one-to-one and one-to-many heuristics are applied period by period in each of the periods in the planning problem, subject to a small modification to account for the changed costs in the multi-period problem. Additionally, in the multi-period context, heuristics must allow for inventories (by netting demand) and decide what to do with leaves resulting from higher-than-planned yields (those not immediately demanded, which may be held for future demand or immediately discarded). Owing to the computational effort required to arrive at solutions for even the smallest instances of the multi-period stochastic problem, a relatively small battery of instances was used in the performance study. That being said, it appears that the one-to-one heuristic's performance remains good in the multi-period setting for problems with low profitability and symmetric demand. Similarly, the results indicate that the one-to-many heuristic may be sensitive to high profitability coupled with less symmetric demand. This remains an area for further examination and we expect that a more sophisticated adaptation to the multi-period problem will probably lead to heuristic performance improvements.

10.4 Conclusion and Outlook

As we have seen, remanufacturing fits quite well within sustainable operations management, leading to additional revenue for the firm and a decrease in landfill space and virgin material consumption. Unfortunately, this requires some complex planning as the main input for remanufacturing, the parts, are harvested through the disassembly of cores. We recall that disassembling a core results in the release of several parts, some of which are common to other cores, and that the input–output relationship is as uncertain as the quality of the cores themselves. This makes real-world problems (containing perhaps dozens of core variants and hundreds of leaves over a planning horizon consisting of several time periods) impossible to solve exactly.

In the first part of this chapter, we have put forth linear programming models, which can be used to plan disassembly when yields are known. Starting with a single-period problem, we developed a simple model containing assumptions and then subsequently relaxed many of these assumptions in an extended model. We then turned our attention to multi-period models and formulated in turn a basic and extended model. Solution methods were also discussed, paying particular attention to available heuristics and providing information on their performance.

Later, we allowed for yields of disassembly to be subject to uncertainty, and suggested recourse models to incorporate the randomness. Unfortunately, it is doubtful that exact solutions can be obtained in acceptable amounts of computational time. As such, the developed heuristics were particularly highlighted in the discussion of solution methods. These heuristics have been shown to outperform the mean yield rate method, a common practitioner's heuristic.

Several extensions will prove useful as product returns are used to fulfill demand for spare part requirements, a topic which has been receiving ever-increasing attention. First, in spare parts settings, demand uncertainty will play more of a role, necessitating the relaxation of the deterministic demand assumption. Second, return uncertainty can also be included and approaches can be modified to incorporate this. Finally, solution methods can be developed for settings where selective (or incomplete) disassembly is more often found, perhaps additionally incorporating disassembly labor capacity explicitly.

References

Cao, D. and Tang, O. (2005). Modeling a multi-period disassembly planning problem with capacity constraints. Paper presented at the 2005 INFORMS Conference, San Francisco, CA.

Coffin, M. and Saltzman, M.J. (2000). Statistical analysis of computational tests of algorithms and heuristics. *INFORMS Journal of Computing*, 12: 24–44.

Dantzig, G.B. (1955). Linear programming under uncertainty. *Management Science,* 1: 197–206.

de Brito, M.P. and Dekker, R. (2004). A framework for reverse logistics. In: Dekker, R., Fleischmann, M., Inderfurth, K., and Van Wassenhove, L.N., editors, *Reverse Logistics: Quantitative Models for Closed-Loop Supply Chains,* pp. 3–27, Springer, Berlin.

Guide, V.D.R., Jayaraman, V. and Linton, J. (2003). Building contingency planning for closed-loop supply chains with product recovery. *Journal of Operations Management,* 21: 259–279.

Gupta, S.M., Lee, Y.J., Xanthopulos, Z. and Veerakamolmal, P. (2000). An optimization approach for a reverse logistics supply chain. Proceedings of the 2000 World Symposium on Group Technology/Cellular Manufacturing, San Juan, Poerto Rico, March 27–29, pp. 227–232.

Gupta, S.M. and Taleb, K.N. (1994). Scheduling disassembly. *International Journal of Production Research,* 32(8): 1857–1866.

Imtanavanich, P. and Gupta, S.M. (2004). Multi-criteria decision making for disassembly-to-order system under stochastic yields. Proceedings of the SPIE International Conference on Environmentally Conscious Manufacturing IV, Philadelphia, PA.

Imtanavanich, P. and Gupta, S.M. (2005). Multi-criteria decision making approach in multiple periods for a disassembly-to-order system under stochastic yields. Paper presented at Sixteenth Annual Conference of POMS, Chicago, IL.

Inderfurth, K. and Langella, I.M. (2003). An approach for solving disassemble-to-order problems under stochastic yields. In: Spengler, T., Voss, S., and Kopfer, H., editors, *Logistik Management: Prozesse, Systeme, Ausbildung,* pp. 309–331, Physica, Heidelberg.

Inderfurth, K. and Langella, I.M. (2006). Heuristics for solving disassemble-to-order problems with stochastic yields. *OR Spectrum,* 28: 73–99.

Kleindorfer, P.R., Singhal, K. and Van Wassenhove, L.N. (2005). Sustainable operations management. *Production and Operations Management,* 14: 482–492.

Kongar, E. and Gupta, S.M. (2002). A multi-criteria decision making approach for disassembly-to-order systems. *Journal of Electronics Manufacturing,* 11: 171–183.

Kongar, E. and Gupta, S.M. (2006). Disassembly to order system under uncertainty. *Omega,* 34: 550–561.

Lambert, A.J.D. and Gupta, S.M. (2002). Demand-driven disassembly optimization for electronic products. *Journal of Electronics Manufacturing,* 11(2): 121–135.

Langella, I.M. (2007a). Heuristics for demand driven disassembly planning. *Computers & Operations Research,* 34: 552–577.

Langella, I.M. (2007b). Planning Demand-Driven Disassembly for Remanufacturing. DUV, Wiesbaden.

Langella, I.M. and Schulz, T. (2006). Effects of a rolling schedule environment on the performance of disassemble-to-order heuristics. Proceedings of the Fourteenth International Working Seminar on Production Economics, Vol I, Innsbruck, Austria, pp. 241–256.

Lee, D.H., Xirouchakis, P. and Züst, R. (2002). Disassembly scheduling with capacity constraints. *Annals of the CIRP (College International pour la Recherche en Productique),* 51(1): 387–390.

Martello, S. and Toth, P. (1990). *Knapsack Problems: Algorithms and Computer Implementations.* Wiley, Chichester.

Neuendorf, K.-P., Lee, D.-H., Kiritsis, D. and Xirouchakis, P. (2001). Disassembly scheduling with parts commonality using Petri nets with timestamps. *Fundamenta Informaticae*, 47: 295–306.

Schulz, T. (2006). A disassemble-to-order heuristic for use with constrained disassembly capacities. In: Waldmann, K.-H., and Stocker, U.M. editors, *OR 2006 Proceedings*, Springer, Berlin.

Seitz, M.A. and Wells, P.E. (2006). Challenging the implementation of corporate sustainability: The case of automotive engine remanufacturing. *Business Process Management Journal*, 12(6): 822–836.

Sen, S. and Hingle, J. (1999). An introductory tutorial on stochastic linear programming models. *Interfaces*, 29: 33–61.

Taleb, K.N. and Gupta, S.M. (1997). Disassembly of multiple product structures. *Computers and Industrial Engineering*, 32(4): 949–961.

Taleb, K.N., Gupta, S.M. and Brennan, L. (1997). Disassembly of complex product structures with parts and materials commonality. *Production Planning and Control*, 8(3): 255–269.

Thierry, M., Salomon, M., Van Nunen, J. and Van Wassenhove, L. (1995). Strategic issues in product recovery managment. *California Management Review*, 37(2): 114–135.

Veerakamolmal, P. and Gupta, S.M. (1998). Optimal analysis of lot-size balancing for multiproducts selective disassembly. *International Journal of Flexible Automation and Integrated Manufacturing*, 6(3–4): 245–269.

Veerakamolmal, P. and Gupta, S.M. (2000). Optimizing the supply chain in reverse logistics. Paper presented at the 2000 Environmentally Conscious Manufacturing Conference, Boston, MA.

Yano, C.A. and Lee, H.L. (1995). Lot sizing with random yields: a review. *Operations Research* 43(2): 311–334.

11

Facility and Storage Space Design Issues in Remanufacturing

Aysegul Topcu, James C. Benneyan, and Thomas P. Cullinane

CONTENTS

11.1 Introduction

Over the last two decades, supply chain management has received remarkable attention from both the world of business and academia. The Council of Supply Chain Management Professionals (2007), formerly known as the Council of Logistics Management, defines supply chain management as "the process of planning, implementing and controlling the efficient, cost-effective

flow of raw materials, in-process inventory, finished goods and related information from the point of origin to the point of consumption for the purpose
of conforming to customer requirements." Most research has focused on the
forward movement of materials from the supplier or the manufacturer to the
end customer. Reverse supply chain management, however, has not received
as much attention. Reverse supply chain management is defined as "the effective and efficient management of the series of activities required to retrieve a
product from a customer and either dispose it or recover value (Prahinski and
Kocabasoglu 2006)." According to the Reverse Logistics Executive Council, the
cost of handling, transporting, and determining the disposition of returned
products is $35 billion annually for U.S. firms (Meyer 1999). Reverse supply
chain management includes activities such as remanufacturing, reuse, reconditioning, recycling, and repair. Remanufacturing includes the processes of
disassembling a used product, assessing the condition of the components of
this product, repairing or reworking these components into refurbished components to satisfy exactly the same or higher quality standards as new components, and using these units in the manufacture of a new product (Topcu
and Cullinane 2005, Lund 1984).

Since 1955 an estimated 8% of the U.S. gross national product has been
annually spent on new facilities (Rosenblatt 1986). Tompkins et al. (2003)
estimate that approximately $250 billion is spent to plan or re-plan facilities
in the United States. Therefore, effective facilities of planning and design
can reduce these expenses significantly and improve productivity, efficiency,
responsiveness, and profitability. Conversely, a poor design will prove costly
and time consuming and cause delays, damaged product, congestion, and
lost product.

Remanufacturing is becoming very important in the economies of the
world and is attractive to both manufacturers and consumers. This trend
has stemmed from the rapid development in technology that has resulted
in an increased demand for new consumer goods, a shorter use time for
many products, and an increasing quantity of salvageable, used, and scrap
products. The anticipation of environmental and end-of-life take-back laws
in the United States and the European Union over the last decade has also
increased the importance of remanufacturing and has made it necessary for
businesses to manage the entire life of the product (Toffel 2003). In recent
years, the environment has been seriously threatened by the continuous
growth in consumer waste. According to the U.S. Environmental Protection Agency (1994), 196 million tons of waste were generated by the United
States in 1990. Growing consumer awareness and preference for environment-conscious products and concern about the exhaustion of nonrenewable
resources have also induced many firms to design products for reuse and
more environmentally benign products and processes (Canon 2005, Xerox
2005, Eastman Kodak 2005, Sharp 2001). Finally, sales opportunities in secondary and global markets have increased revenue generation for returned
products (Meyer 1999). With remanufacturing, manufacturers can realize
significant savings on raw materials and the time required to obtain the

component parts of a product. In summary, the major benefits of remanufacturing include significant reductions in raw materials, human resources, energy, and pollution.

Logistically, however, remanufacturing can be labor intensive and a source of high variability in the overall supply chain. Processing times often depend upon the age, amount of wear, and uncertain condition of returned products. When a returned item arrives at a remanufacturing facility, the condition of its parts is usually not known until the item is disassembled. After disassembly, the parts are typically tested or inspected to determine their functionality and then put into storage. Uncertainty as to when these parts will be needed creates uncertainty about how much storage space should be allocated for each part type, which in turn has an impact on the layout of a remanufacturing facility. Other factors that influence the layout are the characteristics (e.g., dimensions or size, volume, weight, and fragility) of the recovered and remanufactured units, work to be provided on the returned products, storage requirements (e.g., temperature or humidity range), and demand for remanufactured units.

The objective of this chapter is to provide insights into the structure of an effective coordination of facilities planning and remanufacturing decisions. Identifying the optimal facility layout and storage space design assumes a particular relevance since the layout itself must accommodate efficient operations under varying conditions. To address the inherent uncertainty and variability due to (i) the number of returned products, (ii) the type and number of parts reclaimed from each returned product, (iii) the type of processes required to remanufacture a part, (iv) the flow of parts and materials, and (v) the demand for the remanufactured part or the final product, there is a need for designs that minimize warehouse space and inventory holding costs while also facilitating effective coordination of facilities planning and remanufacturing decisions.

11.2 Important Issues in Remanufacturing Facility Layout

11.2.1 Traditional Facility Design

The key attribute of a successful facilities plan is its adaptability and ability to become suitable for some new use. Good facilities ideally have the characteristics of flexibility, modularity, upgradeability, adaptability, and selective operability (Tompkins et al. 2003). Three of the most popular types of layouts are product layouts, process layouts, and hybrid layouts. The selection of layout type is based on the volumes and the types of products to be remanufactured. Product layouts (or production line product layouts) are based on the operation sequence and the capacity requirements of the units being produced. An automated soft drink bottling plant is a good example of the flow in a product layout since the routing of bottles of soda must proceed from washing to filling to capping, such that washing and filling should be placed next to each other in the layout. This type of layout makes it easy to decide where to locate stations and is therefore appropriate for high-volume, low-variety production.

Although product layouts are efficient in terms of faster processing rates, lower inventories, and less unproductive time lost to changeovers and materials handling, they are less flexible and may require redesign, especially for products with short or uncertain lives (Krajewski and Ritzman 1999). In contrast, process layouts (or functional layouts or job shops) group machines according to function, such that equipment of the same type is grouped together. Process layouts are best for low-volume and high-variety production such as job shops. Hybrid layouts are a combination of product and process layouts and are found in situations where production lines are routed through automated machining and assembly cells. More information on all these types of layouts can be found in Krajewski and Ritzman (1999) and Tompkins et al. (2003).

Figure 11.1 illustrates the primary processes in a traditional manufacturing facility. These processes typically include assembly, subassembly, inspection for certification, and shipping. The first steps involve the acquisition of raw materials, components, and parts, which are then assembled into a finished product, inspected, and designated as repairable or good products. Good products are stored in warehouse and then shipped to customers, whereas repairable products are cycled through repair processes.

11.2.2   Remanufacturing Considerations

In contrast, in a remanufacturing facility, the inputs also include components and products recovered from the field. As shown in Figure 11.2, when used products are returned from the consumer to the remanufacturing plant, they are inspected and sorted into three categories: repairable, nonrepairable, and acceptable products. Repairable products can be reused without disassembly or any major operations performed on them. After the necessary repair, they are inspected for recertification. In turn, after acceptable products are disassembled, the resulting components are classified as usable, nonusable, or recycle-only parts (nonusable). Usable parts are inspected and, if acceptable, cleaned, processed, and sent to a parts inventory. Nonrepairable products are directly disposed off. The traditional manufacturing processes are then carried out using the cleaned parts or new parts from the parts supplier, or in some cases combining both kinds of parts. The inspection takes place for recertification rather than certification and therefore may be slightly different in nature. If the parts are not acceptable, that is they require some repair, they may first be put in an inventory unrepaired. After the demand emerges for these parts, they are repaired and sent to parts inventory. When replacement parts are not readily available in repaired inventory for a subsequent product being remanufactured, these parts are acquired from a parts supplier to complete the assembly process.

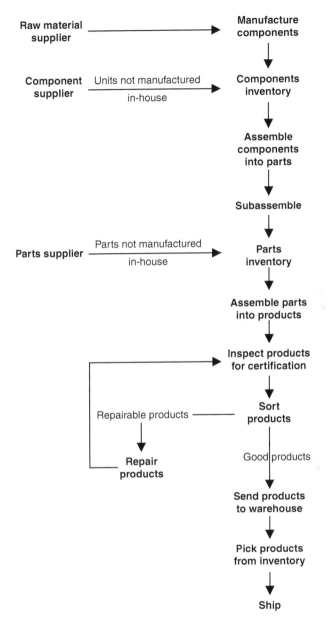

FIGURE 11.1
Typical operations in manufacturing processes.

A comparison of the operations in Figure 11.1 with those in Figure 11.2 clearly reveals that the disassembly of returned products increases the number of operations required to produce a saleable product. This disassembly process can be viewed as the breakdown of a product and the

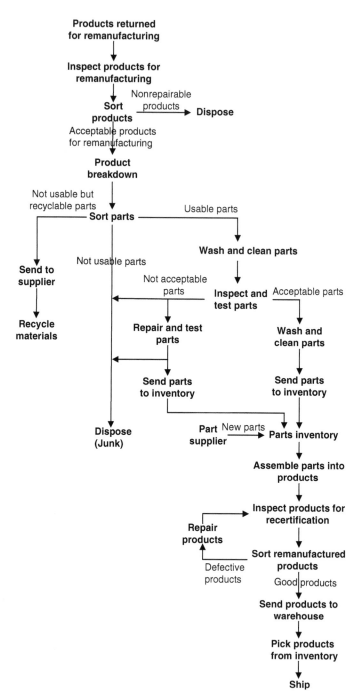

FIGURE 11.2
Typical operations in remanufacturing processes.

FIGURE 11.3
Typical disassembly processes.

subsequent processing, cleaning, inspecting, testing, and repairing of the resultant parts (see Figure 11.3). When products are returned, the condition of the parts is uncertain until the products have been disassembled and tested. After cleaning and testing, the parts are stored. At what point in time these parts are usually needed is not precisely known. This issue creates uncertainty about how much inventory space should be allocated for each part type. A good storage system can significantly affect the material flow in terms of minimizing the travel time between stations and material handling equipment.

11.2.3 Literature Review

11.2.3.1 Facility Layout in Traditional Manufacturing

For an overview on facility layout problems in manufacturing, Meller and Gau (1996) presented the recent trends as well as the trends emerging over the past 10 years. These trends included new methodologies, objectives, and algorithms. These authors also compared the state of the art in facility layout software to the state of the art in facility layout research. Singh and Sharma (2006) reviewed different approaches to facility layout by formulations, solution methodologies, and current as well as emerging trends, including a detailed review of layout software packages. In early research, the objective of these layout problems was to minimize the material handling cost, whereas several more recent studies have addressed the design of layouts in dynamic environments.

Balakrishnan and Cheng (1998) provide a comprehensive review of papers on the dynamic facility layout problem. Benjaafar et al. (2002) provide a comprehensive list of papers that are pertinent to the design of layouts in dynamic environments and define the dynamic facility layout problem as follows: "Assuming demand information for each period is available at the initial design stage, the objective is to identify a layout for each period such that both the material handling and re-layout costs are minimized over the planning horizon." Shore and Tompkins (1980) were among the first researchers to consider the design of layouts under uncertainty. These authors used a facility penalty function that assessed the operation of a facility at levels different than the current one. A change in the departmental areas was not considered. Rosenblatt (1986) developed a formal model and an optimal solution procedure for determining optimal layouts for multiple periods that deal with the dynamic nature of plant layout. Both material handling cost and the cost of relocating departments from one period to the other are taken into account in a different manner from Shore and Tompkins' approach. Later, a number of researchers such as Batta (1987), Urban (1992, 1998), and Balakrishnan (1993) improved on Rosenblatt's solution procedure.

Afentakis and Millen (1990), Kouvelis and Kiran (1991), and Balakrishnan et al. (1992) studied variations of the basic dynamic layout problem. Montreuil and Venkatadri (1991) developed a methodology for designing dynamic layouts for the expansion of manufacturing systems. The stochastic plant layout problem has also been addressed by Montreuil and Laforge (1992), Yang and Peters (1998), and other researchers. Montreuil and Laforge (1992) assume that future production scenarios and their probability of occurrence are known, and these authors propose a method for developing multiple-period layouts. Like Montreuil and Venkatadri's approach (1991), a limitation of this method is that the relative positions of departments are fixed for all periods and only their sizes and shapes can vary. Yang and Peters (1998) developed a model that assumes the flow matrices, and their probability of occurrence is known for multiple periods.

Several authors studied the robustness approach to the stochastic plant layout problem in which the most robust layout is the one with the highest frequency of being closest to the optimal solution for the largest number of scenarios. Rosenblatt and Lee (1987) were among the first researchers to introduce this concept in analyzing single-period layouts. These authors proposed a robust approach to the stochastic plant layout problem, which was further elaborated by Rosenblatt and Kropp (1992), who presented an optimal solution procedure for the single-period stochastic plant layout problem. This procedure only requires solving a deterministic flow-to-flow matrix, where the deterministic matrix is a weighted average of all possible flow matrices.

Because of the computational cost and solution quality, some researchers turned away from mathematical programming techniques to heuristics (Palekar et al. 1992; Kouvelis et al. 1992; Lacksonen and Enscore 1993; Urban 1993; Conway and Venkatramanan 1994; Meller and Bozer 1996; Kaku and Mazzola 1997; Kochhar and Heragu 1999; Rawabdeh and Tahboub 2006). Palekar et al. (1992) focused on the issue of modeling uncertainties in the plant layout problem. These researchers solved such models using dynamic programming for small problems and heuristics for large problems. The proposed heuristics were able to generate good solutions in a reasonable amount of time for problems with up to 40 departments. Simulation studies indicate that a rolling horizon approach yields better results than a fixed horizon approach. Kouvelis et al. (1992) present heuristic strategies for developing robust layouts for multiple planning periods. Kochhar and Heragu (1999) describe a genetic-based algorithm for single- and multiple-period dynamic layout problems that consider layout changeover costs. Ferrari et al. (2003) classify such heuristics into construction, improvement, and hybrid types. Additionally, multicriteria models are proposed by Rosenblatt (1979), Waghodekar and Sahu (1986), Housyar (1991), and Welgama and Gibson (1995).

11.2.3.2 *Material Handling Systems in Facility Layouts*

As stated by Tompkins and Reed (1976), facilities design is the joint selection of an integrated materials handling system and plant layout. Previously existing models based on the cost of the materials handling system criterion to evaluate alternative layouts include those by Hillier (1963), Armour et al. (1964), Hillier and Connors (1966), and Pritsker and Ghare (1970). Webster and Tyberghein (1980) considered the layout with the lowest material handling cost over a number of demand scenarios to be the most flexible layout. Bullington and Webster (1987) extended this definition to the multiperiod case and presented a method for evaluating layout flexibility based on estimating the costs of future re-layout. They recommended that these costs could be used as an additional criterion in determining the most flexible layout. Gupta (1986) presented a simulation approach for measuring layout flexibility based on material flow. Askin et al. (1997) evaluated the layout alternatives for agile manufacturing based on material flow.

Lacksonen and Meller (2000) developed an IDEF0 functional model to describe the facility layout design process. These authors state that there are two key concepts in this process:

1. Iteration
2. Simultaneous designing of the material handling system with the facility layout as also stated by Tompkins and Reed (1976)

The layout design process is considered in three steps: conceptual design, preliminary design, and detailed design. The iterative nature of the process is explained through these steps. A comprehensive understanding of the relationships between material handling system design and layout design as well as other critical activities is found by working through each individual step of the model.

Ferrari et al. (2003) support the design activity of plant layout by means of an integrated approach that takes into account many criteria, both quantitative and qualitative, and present a global approach based on material flow and activity relationships. This approach is carried out using a software called LRP (layout and re-layout program), which has a modular architecture and enables a continuous improvement with the design or in the testing phase. These authors conclude that the integration of LRP with modules of warehouse design and manufacturing cells design appears very interesting. More recently, Jaramillo and McKendall (2004) generated a dynamic extended facility layout problem by simultaneously allocating machines to the plant floor, assigning part or product flow to machines, and defining department boundaries while minimizing total material handling cost. Braglia et al. (2005) present an extended formulation of the layout flexibility concept, assuming that uncertainty in material handling costs may be described by the expected values and standard deviations of demand forecasts. The definition, analysis, and properties of layout flexibility are introduced with reference to a previous work (Braglia et al. 2003). The concept is discussed and a procedure is formulated that is capable of characterizing the configuration that enables the best reduction in handling cost fluctuations.

11.2.3.3 Facility Layout in Remanufacturing

Remanufacturing systems are more dynamic, variable, and complex than traditional manufacturing systems as a result of the variability associated with the routings, processing times, and demand. Ferrer and Whybark (2001) describe the first fully integrated material planning system to facilitate the management of a remanufacturing facility. Kekre et al. (2003) developed a simulation-based line configuration model that simultaneously considers line balancing and line length to maximize the remanufacturing system's effective throughput. More recently, Lim and Noble (2006) evaluated the performance of different layout alternatives such as job shop, cellular, fractal, and holonic layouts in remanufacturing. These authors state that it is possible to improve overall system performance through logistical issues such as facility layout rather than through operational-based approaches using appropriate production planning and control techniques. The primary performance measures included were throughput time and work in progress. These two measures were used to examine how each layout scheme affects the performance and the efficiency of a remanufacturing system. Based on the experimental and operational results, the authors found that each layout has unique performance characteristics that make it most suitable for different operating scenarios. Consequently, a multicriteria perspective is useful to determine which layout organization should be selected based on the criteria chosen by the decision maker. Franke et al. (2006) introduced a model that allows the continuous adaptation of remanufacturing facilities under quickly changing product, process, and market constraints by means of combinatorial optimization and discrete-event simulation. Uncertainties regarding quantity and conditions of mobile phones, reliability of capacities, processing times, and demand are considered. Capacity and remanufacturing program planning are determined by the optimization model, while the simulation enables the planner to determine the required transport, storage capacities, and performance of the remanufacturing system.

11.3 Handling Uncertainty

11.3.1 Sources of Randomness in the Remanufacturing Supply Chain

A fundamental characteristic of a remanufacturing environment is the inbound flow of used products, where most of the components or subassemblies will have probabilistic yields in the sense that not all items will be suitable for reuse. Some salvaged components ultimately may cause a remanufactured unit to fail a subsequent reliability or functionality test. This stochastic nature of returned products affects predictability, safety stock, production targets, and rework and waste (such as additional cleaning, testing, inspection, and reassembly).

The flow of parts through a remanufacturing facility is dependent on the necessary disassembly, cleaning, and testing processes, the age and the amount of wear, and the demand for immediate assembly or for inventory. Required materials or equipment such as cleaning solvents or particular tools depend on the condition and the nature of a part, resulting in product flow and equipment needs that are uncertain until the unit arrives at the facility and substantially impacting the layout of a remanufacturing facility. Both remanufactured units and reclaimed components need to be stored in the inventory until they are needed to satisfy demand. All these sources of uncertainty make the design of a remanufacturing facility more complex. Moreover, with multiple types of products or model variations being returned, some products will share common components such that a recaptured part from one product may be used in another product. Different component yields within and between products further complicates effective inventory planning and warehouse design.

Consider the remanufacturing of a product consisting of three components, denoted by A, B, and C, and with disassembly yield rates (γ) of 0.85, 0.65, and 0, respectively. That is, for any given unit received for remanufacturing, after disassembly the probabilities that parts A, B, and C are salvageable are $\gamma_A = 0.85$, $\gamma_B = 0.65$, and $\gamma_C = 0$, respectively. If a remanufacturing facility receives a batch of $n = 100$ products per period for remanufacturing, the expected number of salvaged parts A, B, and C are $n\gamma_A = 85$, $n\gamma_B = 65$, and $n\gamma_C = 0$, respectively, although there will be variability in each observed yield described by binomial probabilities batch to batch. Figure 11.4 illustrates the flow of parts through this remanufacturing system. The processes are denoted by rectangles, and the storage requirements are denoted by triangles. The numbers outside the parentheses on the top of each flow arrow indicate the expected number of parts that flow through each process and

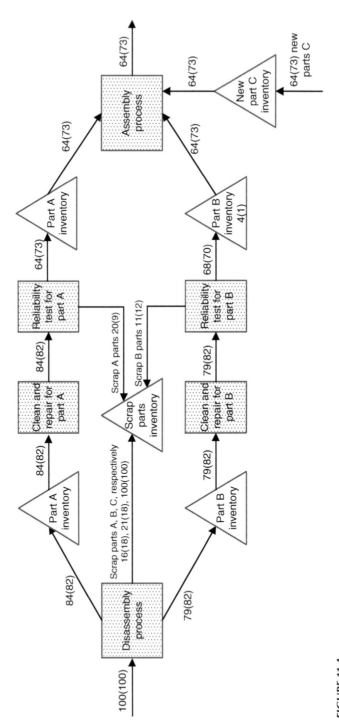

FIGURE 11.4
Remanufacturing product flow example.

storage space in the first period, and the numbers inside the parentheses denote the expected number of parts moving from one step in the process to another step in the second period.

To remanufacture these 100 units, the nonsalvageable parts need to be either replaced from inventory or purchased from an outside vendor, with an expected number of replacements of part i equal to $n(1 - \gamma_i)$, i = A, B, C. Again, for any given batch the actual number of scrapped and replaced parts will be random with binomial probabilities and variances. In this illustration, for example, the observed yield of part B will have a standard deviation of $[n\gamma_B(1 - \gamma_B)]^{0.5} = [100(0.65).35]^{0.5} \approx 4.77$, implying that roughly 95% of batches will yield between 56 and 74 usable pieces of part B, with the remaining 26–44 pieces drawn from inventory or sourced externally. Moreover, in many realistic scenarios, the incoming batch size would be random variables as well, resulting in compound probability distributions and greater variability in the component yields (where n has a prior distribution) such as a Poisson-binomial model.

After subsequent reliability testing of the salvageable A and B parts, a subset of these are available for reassembly, with expectations $n\gamma_A\alpha_A$ and $n\gamma_B\alpha_B$ and variances $n\gamma_A\alpha_A(1 - \gamma_A\alpha_A)$ and $n\gamma_B\alpha_B(1 - \gamma_B\alpha_B)$, respectively, where α_i is the reliability failure rate for part i. Although the appropriate probability distributions, means, and variances can be derived for this simple example, in more complex and realistic scenarios the logic may become intractable. If, for illustration, we assume that initially there are no parts of either type in the inventory, then ultimately the number of remanufactured units from the original batch depends on the smaller number of A and B parts that pass the reliability test. This number of C parts would need to be forecasted, ordered from an outside manufacturer, and on hand to then remanufacture the units, or given long part C lead times all A and B parts would need to be stored until they are received. Either way, the unused A or B parts would be stored in inventory for possible use in a subsequent period.

11.3.2 A Simulation Model Illustration

The following simulation model has been designed to provide an insight into how the number of remanufactured products and the inventory builds of each part are affected by the disassembly and reliability yield rates, and how this affects appropriate warehouse planning. The goal of the simulation example shown below is to simply show how the inventory of the remanufactured product and the inventory build of parts A and B vary as the disassembly and reliability yield rates vary. (The simulation of a full-scale scenario would be very complex and is beyond the intended scope here but could follow similar logic.) In the given scenarios, the batch size is held constant at 100 returned products per period and the model is run for 100 replications of 50 periods each. Both the disassembly and reliability testing yield rates follow binomial distributions. An extension of this model would allow the batch size to vary from period to period.

Some of the key assumptions in the model are as follows:

1. All returned products are disassembled immediately, meaning that no storage is necessary after the products are returned to the remanufacturing facility at the end of their life cycle.

2. Other than disassembly, cleaning, repairing, and reliability testing, no other operation is performed on any part or remanufactured product.

3. Scrap part inventory is fed by disassembly and reliability testing stored in the same storage area.

4. All other nonserviceable parts are disposed off immediately without storage. Only part C is purchased from an outside supplier, with the number of remanufactured items in any period equal to the minimum number of available (in inventory from past periods or recaptured in the current period) parts of type A or B.

5. Inventory holding costs for returned products, disassembled products, and remanufactured products are constant during each period.

6. The machines that perform remanufacturing operations are assumed to operate without any repair or maintenance interruption.

7. No inspection before shipping the product to the customer is required.

Figure 11.5 shows one replication of inventory builds of parts A and B over 50 time periods, assuming equal disassembly and reliability yield rates for both parts. It illustrates the complementary nature of part accumulation and the variability in storage requirements over time. The mean and the maximum of 100 replications per period are shown in Figure 11.6.

Figures 11.7 and 11.8 illustrate the distribution of the average and maximum, respectively, of the number of parts in inventory at any time over each of 100 replications. As shown, the inventory behaviors of both parts are very

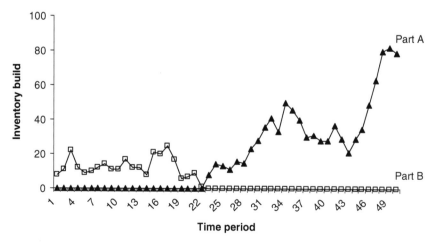

FIGURE 11.5
Sample of the replications for inventory build of parts A and B ($\gamma_A = \gamma_B = 0.80$; $\alpha_A = \alpha_B = 0.80$).

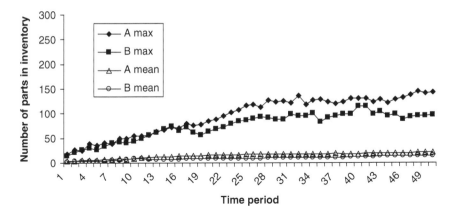

FIGURE 11.6
Inventory builds ($\gamma_A = \gamma_B = 0.80$; $\alpha_A = \alpha_B = 0.80$), 100 replications.

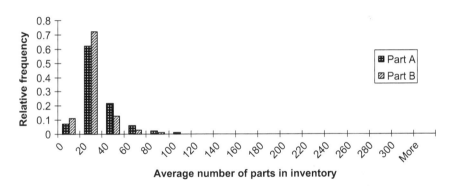

FIGURE 11.7
Average number of parts in inventory ($\gamma_A = \gamma_B = 0.80$; $\alpha_A = \alpha_B = 0.80$).

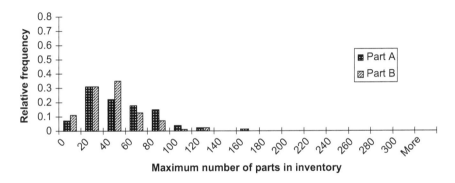

FIGURE 11.8
Maximum number of parts in inventory ($\gamma_A = \gamma_B = 0.80$; $\alpha_A = \alpha_B = 0.80$).

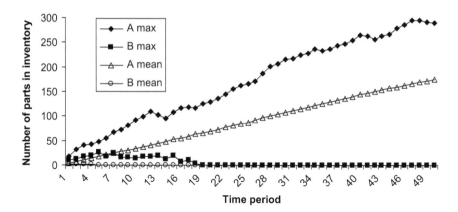

FIGURE 11.9
Inventory builds ($\gamma_A = 0.82$, $\gamma_B = 0.78$; $\alpha_A = 0.80$, $\alpha_B = 0.80$), 100 replications.

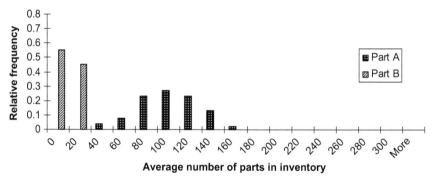

FIGURE 11.10
Average number of parts in inventory ($\gamma_A = 0.82$, $\gamma_B = 0.78$; $\alpha_A = 0.80$, $\alpha_B = 0.80$).

similar, due to their similar yields. Conversely, Figures 11.9 through 11.11 illustrate the same information for an example with slightly different yield rates for parts A and B ($\gamma_A = 0.82$, $\gamma_B = 0.78$). As illustrated, results change significantly for even slightly different yields with storage requirements quickly becoming more dissimilar and variable. In particular, the average and the maximum number of storage spaces are dramatically impacted by several folds, thus having implications on fixed and flexible design planning. Because the inventory build of parts occurs when the demand for one is less than the other in any given period, unused parts are stored until there is demand in some following period. This process can cause a sudden build and depletion of inventory for some parts over a few periods, and the storage space needs to be adaptable to accommodate this phenomenon.

Finally, Figures 11.12 and 11.13 illustrate the impact of inventory availability on the number of remanufactured products produced per period. This remanufacturing rate is almost the same for both examples but will be lower if any given part has a lower yield rate.

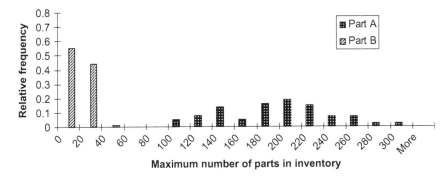

FIGURE 11.11
Maximum number of parts in inventory ($\gamma_A = 0.82$, $\gamma_B = 0.78$; $\alpha_A = 0.80$, $\alpha_B = 0.80$).

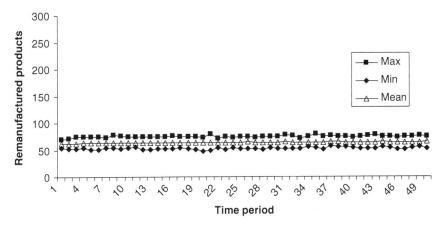

FIGURE 11.12
Number of remanufactured products ($\gamma_A = \gamma_B = 0.80$; $\alpha_A = \alpha_B = 0.80$), 100 replications.

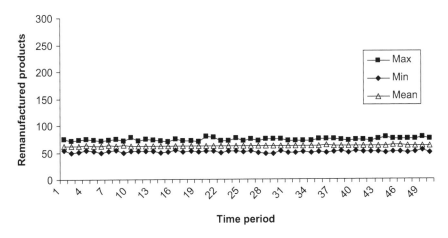

FIGURE 11.13
Number of remanufactured products ($\gamma_A = 0.82$, $\gamma_B = 0.78$; $\alpha_A = 0.80$, $\alpha_B = 0.80$), 100 replications.

11.4   Reconfigurable and Flexible Designs

11.4.1   Reconfigurable Facility Layout

Rapid changes in technology and markets require production systems that are themselves easily upgraded and into which new technologies and new functions can be integrated. These changes include (Koren et al. 1999) (1) increasing frequency of introduction of new products, (2) changes in parts for existing products, (3) large fluctuations in product demand and mix, (4) changes in government regulations (safety and environment), and (5) changes in process technology. In Koren and Ulsoy (2002), a reconfigurable manufacturing system is defined as "one designed at the outset for rapid change in its structure, as well as its hardware and software components, in order to quickly adjust its production capacity and functionality within a part family in response to sudden market changes or intrinsic system changes." For a manufacturing system to be reconfigurable, certain characteristics must exist.

Koren and Ulsoy define these characteristics as (1) modularity (the ability to design all system components, both software and hardware to be modular), (2) scalability (the ability to easily change existing production capacity by rearranging an existing production system and changing the production capacity of reconfigurable components within that system), (3) integrability (the ability to design systems and components for both ready integration and future introduction of new technology), (4) convertibility (the ability to allow quick changeover between existing products and quick system adaptability for future products), (5) customizability (the ability to design the system capability and flexibility to match the applications), and (6) diagnosibility (the ability to quickly identify the sources of quality and reliability problems that occur in large systems). Mehrabi et al. (2000a) examine and identify the key interrelated technologies that should be developed and implemented to achieve these characteristics. These authors explain the issues related to technology requirements of reconfigurable manufacturing systems at the system and machine design levels and ramp-up time reduction.

Mehrabi et al. (2000b) present an overview of available reconfigurable manufacturing techniques, their key drivers and enablers, and their impacts, achievements, and limitations. These authors state that reconfigurable manufacturing systems go beyond the objectives of mass, lean, and flexible manufacturing by (1) the reduction of lead time for launching new systems and reconfiguring existing systems and (2) the rapid manufacturing modification and quick integration of new technology and new functions into existing systems. The authors explain how reconfigurability differs from agility but also conclude that agile manufacturing complements reconfigurable manufacturing and define reconfigurability as "the set of methodologies and techniques that aid in design, diagnostic, and ramp-up of reconfigurable manufacturing systems and machines that give corporations the engineering tools that they need to be flexible and respond quickly to market opportunities and changes." Reconfigurable manufacturing systems are defined as "... designed for rapid adjustment of production capacity and functionality, in response to new circumstances by rearrangement or change of its components." The core idea is to design the production system around a part family to achieve a lower cost and therefore a higher production rate with the required flexibility (Mellor 2002).

Elmaraghy (2005) outlines and compares the characteristics of reconfigurable manufacturing systems and flexible manufacturing systems, linking the concept of a manufacturing system life cycle with aspects of manufacturing system flexibility and reconfigurability. There is little research, however, that focuses specifically on reconfigurable facility layout. Benjaafar et al. (2002) provide examples that support the migration of next generation systems toward highly adaptable and quickly reconfigurable systems. Meng et al. (2004) apply the concept of reconfigurability in layout problems. These researchers state that the reconfigurable layout problem differs from traditional robust and dynamic layout problems in two aspects: (1) it assumes that production data are available only for the current and the upcoming production period and (2) it considers queuing performance measures such as work in progress inventory and product lead time in the objective function of the layout problem.

11.4.2 Reconfigurable Warehouse and Inventory Design

As the simulation results in Section 11.3.2 illustrated, significant variability in inventory can occur in remanufacturing facilities. A reconfigurable warehousing design system should be able to adjust its capacity to accommodate these fluctuations beyond the objectives of dynamic or flexible facility layout problems by resolving the configurability problem more frequently in any real time. As an example, the number of usable parts from returned products changes from period to period, and a company may desire to modify its layout to store the extra parts until they are needed.

11.4.3 Modeling Approaches

No modeling approaches that address reconfigurable remanufacturing facility designs that adapt to inherent uncertainties have appeared in the literature. To illustrate how such a model might work and the associated issues, we describe a preliminary stochastic programming model that optimizes the changing storage space requirements for multiple components over multiple time periods. Stochastic programming is one potential approach that has promise for modeling reconfigurable facility layout problems. This approach has been applied previously to a wide variety of fields such as agriculture, capacity planning, energy, finance, fisheries management, forestry, military, production control, scheduling, sport, telecommunications, transportation, and water management (The Committee on Stochastic Programming 2007).

In this type of formulation, some of the data in the objective or constraints are uncertain and therefore characterized by a probability distribution. A central idea in stochastic programming is that instead of replacing uncertain parameters simply with a deterministic expected value, these parameters are modeled with all possible outcomes. Two general types of stochastic programming problems are chance-constrained linear programs (probabilistic constraints) and stochastic programming with recourse (SPR). The latter was first introduced independently by Dantzig (1955) and Beale (1955). The fundamental idea behind this type of problem is the concept of recourse, which is the ability to take corrective action after a random event has occurred. Chance-constrained stochastic programming models were first introduced by Charnes and Cooper (1959) and focus on scenarios in which a decision is made, such that some constraints are met with specified probability levels.

The given SPR model assumes that the decision maker is the supplier of remanufactured parts. The goal is to determine the storage space required for any remaining remanufactured parts after demand is met. This space assignment should minimize all associated costs. For example, the units for storage might be skid spaces, with each remanufactured part requiring one skid space. If the number of extra remanufactured parts exceeds the capacity of the in-house storage space, then the two options are (i) reconfigure available space to store the extra parts in-house or (ii) pay for temporary outside storage space. If the number of extra remanufactured parts is less than the capacity of the in-house storage space, an opportunity cost is associated with the unused storage space.

Figure 11.14 illustrates a simple SPR model in which the decision maker needs to make storage planning decisions in the current (first) period and then reconfigure storage space based on yield rates in the next period with three possible yield rates and their corresponding probabilities. For example, the remanufacturer will retrieve 60 usable antennas out of 100 returned cell phones with 0.1 probability, 90 with 0.6 probability, and 75 with 0.3

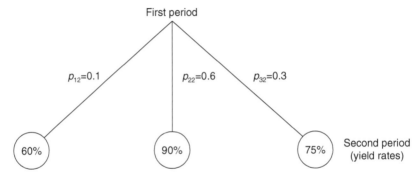

FIGURE 11.14
Probability of possible yield rates for simple SPR model.

probability. Looking forward one period into the future, initially a decision is made about how much in-house storage space to have in the first stage, S_1. After the demand is met in the second period, $X_{k,2}$ is the number of unneeded remanufactured pieces that now need to be stored.

The decision variables in this model are as follows:

S_1 = In-house storage space for all extra remanufactured parts after demand is met

$Z_{k,2}$ = Reconfigured storage space for period 2 under scenario k $(k = 1, 2, 3)$

$Y_{k,2}$ = Outside storage space for period 2 under scenario k $(k = 1, 2, 3)$

$W_{k,2}$ = Unused storage space for period 2 under scenario k $(k = 1, 2, 3)$

Here the $k = 3$ scenarios are the three possible yields (illustrated in Figure 11.14). Note that if the variable is scenario-independent, then the scenario subscript is dropped, such as in S_1 and T_1. The roles of the following cost, demand, capacity, and other variables in this model are shown in Figure 11.15.

$X_{k,l}^{RT}$ = Number of returned products under scenario k in period l

$RYR_{k,l}$ = Remanufacturing yield rate

D_l = Demand in period l

S_l = In-house storage space for the extra remanufactured parts after demand is met

$X_{k,l}$ = Number of extra remanufactured parts under each scenario that one has on hand after demand is met

c_{in} = Cost of storing extra remanufactured parts in-house

c_{rec} = Cost of reconfiguring storage space to accommodate the extra remanufactured parts

c_{out} = Cost of outside storage space if the number of extra remanufactured parts exceeds the capacity one has on hand

c_{loss} = Cost of unused space (space that one had devoted, but did not use)

$p_{k,l}$ = Probability of each outcome

$Z_{k,l}$ = Reconfigured storage space to store extra remanufactured parts if the in-house storage space is exceeded

$Y_{k,l}$ = Outside storage space to store extra remanufactured parts if the in-house storage space is exceeded

$W_{k,l}$ = Unused in-house storage space (when the number of extra remanufactured parts is less than the in-house storage space)

CAP_l^{in} = Capacity for in-house storage space in period l

CAP_l^{rec} = Capacity for reconfigured storage space in period l

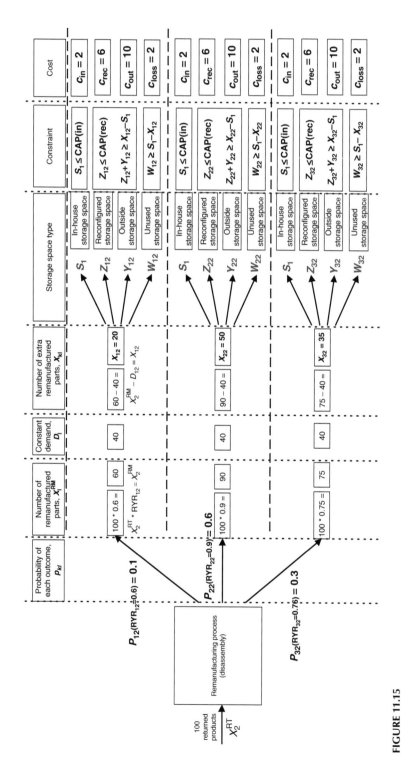

FIGURE 11.15
Stochastic programming representation assuming single-part, two-period, and constant demand.

Given the preceding notation, an SPR two-period optimization model can be written as

$$\text{minimize } (c_{in} * S_1) + \sum_{k=1}^{3} (p_{k,2} * c_{rec} * Z_{k,2}) + \sum_{k=1}^{3} (p_{k,2} * c_{out} * Y_{k,2})$$
$$+ \sum_{k=1}^{3} (p_{k,2} * c_{loss} * W_{k,2})$$

subject to

$$(RYR_{k,2} * X_2^{RT}) - D_2 = X_{k,2} \tag{11.1}$$

$$(RYR_{k,2} * X_2^{RT}) \geq D_2 \tag{11.2}$$

$$Z_{k,2} + Y_{k,2} \geq X_{k,2} - S_1 \tag{11.3}$$

$$W_{k,2} \geq S_1 - X_{k,2} \tag{11.4}$$

$$S_1 \leq CAP_1^{in} \tag{11.5}$$

$$Z_{k,2} \leq CAP_2^{rec} \tag{11.6}$$

$$S_1, Z_{k,2}, Y_{k,2}, W_{k,2} \geq 0 \quad k = 1, 2, 3 \tag{11.7}$$

The objective function minimizes the expected total cost after two periods. Constraint 1 calculates the number of extra parts after demand is met in the second period. Constraint 2 ensures that the demand is met during the second period. Constraint 3 ensures that parts are stored either in reconfigurable storage space or outside if the in-house storage space does not satisfy the storage needs. Constraint 4 ensures the unused storage space incurs a loss if the in-house storage space exceeds the number of extra parts. Constraints 5 and 6 are the capacity limits for the in-house and reconfigured storage spaces. The last set of equations is nonnegativity constraints.

As an example, suppose we have the following model inputs:

$$p_{1,2}(RYR_{1,2} = 0.6) = 0.1,\ p_{2,2}(RYR_{2,2} = 0.9) = 0.6,\ p_{3,2}(RYR_{3,2} = 0.75) = 0.3,$$

$$X_2^{RT} = 100,\ D_2 = 40,\ CAP_1^{in} = 50,\ CAP_2^{rec} = 10,$$

$$c_{in} = 2,\ c_{out} = 6,\ c_{loss} = 10,\ c_{rec} = 2$$

The results on the left-hand side of Table 11.1 indicate that the minimum expected total storage cost is \$115, with an optimal in-house storage space of 50. Out of these 50, 30 will be unused if scenario 1 occurs and 15 will be unused if scenario 3 occurs. Solving the same problem using the expected number of unused remanufactured parts instead ($p_{1,2*RYR_{1,2}+p_{2,2}}$ * $RYR_{2,2+p_{3,2}*RYR_{3,2}}$) * $X_2^{RT} - D_2$, produces the results shown on the right-hand side of Table 11.1. In the expected value formulation, S_1 is not a decision variable and is approximately 43. In particular, note that the expected value

TABLE 11.1

Comparison of SP with Expected Value Formulation (Two-Period Model)

	Period	Stochastic Programming Scenario			Expected Value Scenario		
		1	2	3	1	2	3
In-house S.S.[a]	1	50 (independent of scenario)			Constant at 43		
Number of returns	2	Constant at 100			Constant at 100		
Reconfigured S.S.	2	—	—	—	—	8	—
Outside S.S.	2	—	—	—	—	—	—
Unused S.S.	2	30		15	23	—	8
Total cost		$115			$123		

[a] Storage space.

formulation produces a solution with a higher total cost than the SPR formulation due to the assumption of perfect information about the future. The SPR model, however, minimizes the total cost over a number of possible scenarios, each multiplied by their respective probabilities.

11.4.3.1 Model Extensions

The same general approach can be followed to extend this type of model to more realistic scenarios with multiple incoming product streams, each with multiple components and multiple planning periods. The example shown in Figure 11.16 considers three periods, now with uncertain yield rates in periods 2 and 3. The costs of in-house, reconfigured, outside, and unused storage spaces are held the same in the third period, and the demand is assumed to be constant for both periods. The number of returned products is constant at 100 in the second period but is a "pull" decision variable in the third period, depending on the excess number of remanufactured parts from the previous period and the remanufacturing yield rate in period 3.

In the final period, an initial decision is made about how much in-house storage space is needed based on all available information and scenario probabilities. After the second period, the remanufacturing yield rate for that period is known and a second decision is made as to the values of the recourse variables, namely the amounts of reconfigured storage space, outside storage space, and unused storage space. The decision maker now also makes a decision as to how much in-house storage space to have to be best prepared for the next period (the third period) to accommodate the extra remanufactured parts. After the remanufacturing yield rate for the third period is observed, a decision is made again regarding the values of the recourse variables and the

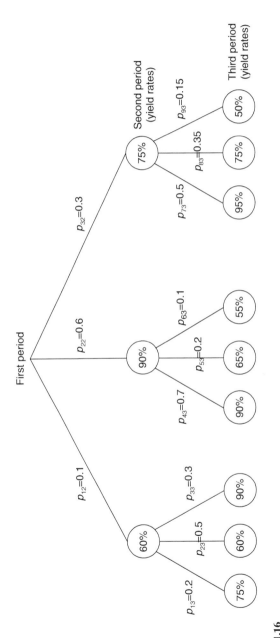

FIGURE 11.16
Probability of possible yield rates for three-period SPR model.

TABLE 11.2

Possible Three-Period Yield Combinations

Scenario (k)	Second Period ($l = 2$) (%)	Third Period ($l = 3$) (%)	Probability ($p_{k,3}$)
1	60	75	0.02
2	60	60	0.05
3	60	90	0.03
4	90	90	0.42
5	90	65	0.12
6	90	55	0.06
7	75	95	0.15
8	75	75	0.105
9	75	50	0.045

number of returned products. In the third period, again three possible yield rates are possible, which, depending upon yield in the second period, result in the $3^2 = 9$ possible scenarios as shown in Table 11.2.

Given the above assumptions, a three-period SPR optimization model can be written as

$$\text{minimize } c_{\text{in}} * S_1 + \sum_{k=1}^{9}(p_k * c_{\text{in}} * S_{k,2}) + \sum_{l=2}^{3}\sum_{k=1}^{9}p_k * (c_{\text{rec}} * Z_{k,l} + c_{\text{out}} * Y_{k,l} + c_{\text{loss}} * W_{k,l})$$

subject to

$$(\text{RYR}_{k,2} * X_2^{\text{RT}}) - D_2 = X_{k,2} \tag{11.8}$$

$$(\text{RYR}_{k,2} * X_2^{\text{RT}}) \geq D_2 \tag{11.9}$$

$$X_{k,2} - D_3 = X_{k,3} \quad \text{if } X_{k,2} - D_3 \geq 0 \tag{11.10}$$

$$[(\text{RYR}_{k,3} * X_{k,3}^{\text{RT}}) + X_{k,2}] - D_3 = X_{k,3} \quad \text{if } X_{k,2} - D_3 < 0 \tag{11.11}$$

$$(\text{RYR}_{k,3} * X_{k,3}^{\text{RT}}) + X_{k,2} \geq D_3 \tag{11.12}$$

$$X_{k,2} \geq D_3 \tag{11.13}$$

$$Z_{k,2} + Y_{k,2} \geq X_{k,2} - S_1 \tag{11.14}$$

$$Z_{k,3} + Y_{k,3} \geq X_{k,3} - S_{k,2} \tag{11.15}$$

$$W_{k,2} \geq S_1 - X_{k,2} \tag{11.16}$$

$$W_{k,3} \geq S_{k,2} - X_{k,3} \tag{11.17}$$

$$S_1 \leq CAP_1^{in} \quad S_{k,2} \leq CAP_l^{in} \quad Z_{k,l} \leq CAP_l^{rec} \tag{11.18}$$

$$S_{1,2} = S_{2,2} = S_{3,2} \quad S_{4,2} = S_{5,2} = S_{6,2} \quad S_{7,2} = S_{8,2} = S_{9,2} \tag{11.19}$$

$$Z_{1,2} = Z_{2,2} = Z_{3,2} \quad Z_{4,2} = Z_{5,2} = Z_{6,2} \quad Z_{7,2} = Z_{8,2} = Z_{9,2} \tag{11.20}$$

$$Y_{1,2} = Y_{2,2} = Y_{3,2} \quad Y_{4,2} = Y_{5,2} = Y_{6,2} \quad Y_{7,2} = Y_{8,2} = Y_{9,2} \tag{11.21}$$

$$W_{1,2} = W_{2,2} = W_{3,2} \quad W_{4,2} = W_{5,2} = W_{6,2} \quad W_{7,2} = W_{8,2} = W_{9,2} \tag{11.22}$$

$$S_1, S_{k,2}, Z_{k,l}, Y_{k,l}, W_{k,l}, X_{k,3}^{RT} \geq 0 \quad k = 1, 2, 3 \quad l = 2, 3 \tag{11.23}$$

As previously mentioned, the objective function minimizes the expected total cost, now over three periods. Constraints 1, 3, and 4 calculate the number of extra parts in periods 2 and 3. Constraints 2, 5, and 6 ensure that demand is met during both periods. Constraints 7 and 8 determine the number of reconfigured storage spaces, whereas Equations 11.16 and 11.17 determine the unused storage space. Constraint 11 is a set of constraints that puts capacity limits for the in-house and reconfigured storage space areas. Constraints 12 through 15 are "nonanticipativity" constraints, which means that one cannot anticipate the future and ensure that scenarios with a common history must have the same set of decisions. The last set of equations is the nonnegativity constraints.

For example, we have the following model inputs:

$$RYR_{k,2} = [0.6 \quad 0.6 \quad 0.6 \quad 0.9 \quad 0.9 \quad 0.9 \quad 0.75 \quad 0.75 \quad 0.75]$$

$$RYR_{k,3} = [0.75 \quad 0.6 \quad 0.9 \quad 0.9 \quad 0.65 \quad 0.55 \quad 0.95 \quad 0.75 \quad 0.5]$$

$$p_k = [0.02 \quad 0.05 \quad 0.03 \quad 0.42 \quad 0.12 \quad 0.06 \quad 0.15 \quad 0.105 \quad 0.045]$$

$$X_2^{RT} = 100, \quad D_1 = D_2 = 40, \quad CAP_2^{in} = CAP_3^{in} = 50, \quad CAP_2^{rec} = CAP_3^{rec} = 10$$

$$c_{in} = 2, \quad c_{out} = 6, \quad c_{loss} = 10, \quad c_{rec} = 2$$

The optimal solution is shown in Table 11.3. For this example, in the first period, the optimal in-house storage space is always 50. Of this figure, 30 will be unused if scenarios 1–3 occur in the second period, and 15 will be unused if scenarios 7–9 occur in the second period. If scenarios 1–3 occur in the second period, then the optimal in-house storage space in the third period is 1. If scenarios 4–6 occur in the second period, then the optimal in-house storage space in the third period is 10. If scenarios 7–9 occur in the second period, then the optimal in-house storage space in the third period is 14. An additional six reconfigured storage spaces will be needed if scenario 3 occurs in the third period given that scenarios 1–3 occurred in the second period. The number of returned products to be remanufactured is a decision variable in the third period, meaning that the decision maker has

TABLE 11.3

SP Results with X_{RT} (Number of Returned Products) as a Decision Variable (Three-Period Model)

	Period	Scenario								
		1	2	3	4	5	6	7	8	9
In-house S.S.[a]	1	50 (Independent of scenario)								
	2	1			10			14		
Number of returns	2	Constant at 100								
	3	28	35	30	–	–	–	20	24	38
Reconfigured S.S.	2	–			–			–		
	3	–	–	6	–	–	–	–	–	–
Outside S.S.	2	–			–			–		
	3	–	–	–	–	–	–	–	–	–
Unused S.S.	2	30			–			15		
	3	–	–	–	–	–	–	–	–	–
Total cost	$137									

[a] Storage space.

the flexibility to determine the number of returned products, which will go into the remanufacturing process (i.e., pull-control inventory system).

As previously mentioned, a simpler expected value formulation for this problem results in a much higher minimum cost solution, underscoring the potential value in this type of modeling approach. This model can also be extended further to multiple products with interdependent parts, random return volumes, and random demand in a multiperiod setting (significantly expanding the number of possible scenarios). However, the number of variables in this type of SPR model will increase significantly, especially in multiproduct and multiperiod cases, because of the explosion of possible scenarios and size of the model. Even this simple model, for example, results in 113 constraints and 79 integer variables. After a large-scale model is formulated somewhere along the lines mentioned, then heuristics such as tabu search, genetic algorithms, or simulated annealing might be used to overcome the resulting problem of solving large combinatorial optimization models.

11.5 Conclusions

Remanufacturing environments are affected by more stochastic events and uncertainty than traditional manufacturing. Therefore, identifying the optimal facility layout and storage space requirements becomes especially important since these aspects must accommodate efficient operations under varying conditions. Examples of stochastic characteristics include (i) the number of returned products, (ii) the type and number of parts acquired from each returned product, (iii) the type of processes required to remanufacture a part, (iv) the flow of parts and materials, and (v) the demand for the remanufactured part and the final product. These considerations have implications on the optimal storage capacity requirements, on the advantages of designing flexible policies, and on modeling approaches to understand and optimize overall design performance.

Possible modeling approaches that can address the unique characteristics of remanufacturing include simulation, stochastic programming, chance constraints, and option-based models. Two approaches were used in this chapter to illustrate the implications of remanufacturing on facility design performance and space requirements and to illustrate generalized modeling approaches. Reconfigurable designs that adjust storage capacity to accommodate the changes in recapture rates, product condition, and codependent part inventories may be particularly effective. A simple simulation model was used to demonstrate the effects of reliability and yield rates on storage space requirements and variability. Similar models can be used to gain further insights into the design complexity of remanufacturing facilities. Stochastic programming models can also help the decision maker to optimize facility designs in the presence of future uncertainty and to design efficient multiperiod flexible layouts.

References

Afentakis, P. and R. A. Millen (1990). Dynamic layout strategies for flexible manufacturing systems. *International Journal of Production Research* **28**(2): 311–323.

Armour, G. C., E. S. Buffa and T. E. Vollmann (1964). Allocating facilities with craft. *Harvard Business Review* **42**(2): 136–158.

Askin, R. G., N. H. Lundgren and F. Ciarallo (1997). A material flow based evaluation of layout alternatives for agile manufacturing. In *Progress in Material Handling Research*. R. J. Graves, L. F. McGinnis, D. J. Medeiros and R. E. Ward. (Eds.), Material Handling Institute, Ann Arbor, MI, pp. 71–90.

Balakrishnan, J. (1993). Notes: "The dynamics of plant layout". *Management Science* **39**(5): 654–655.

Balakrishnan, J. and C. H. Cheng (1998). Dynamic layout algorithms: A state of the art survey. *Omega* **26**(4): 507–521.

Balakrishnan, J., F. R. Jacobs and M. A. Venkataramanan (1992). Solutions for the constrained dynamic facility layout problem. *European Journal of Operational Research* **57**(2): 280–286.

Batta, R. (1987). Comment on "The dynamics of plant layout". *Management Science* **33**(8): 1065–1065.

Beale, E. M. L. (1955). On minimizing a convex function subject to linear inequalities. *Journal of the Royal Statistical Society, Series B* **17**: 173–184.

Benjaafar, S., S. S. Heragu and S. A. Irani (2002). Next generation factory layouts: research challenges and recent progress. *Interfaces* **32**(6): 58–76.

Braglia, M., S. Zanoni and L. Zavanella (2005). Robust vs. stable layout design in stochastic environments. *Production Planning and Control* **16**(1): 71–80.

Braglia, M., S. Zanoni and L. E. Zavanella (2003). Layout design in dynamic environments: strategies and quantitative indices. *International Journal of Production Research* **41**(5): 995–1016.

Bullington, S. F. and D. B. Webster (1987). Evaluating the flexibility of facility layouts using estimated relayout costs. Proceedings of the IXth International Conference on Production Research, Cincinnati, OH, pp. 2230–2236.

Canon Inc (2005). Sustainability Report. http://www.canon.com/environment/report/report2005e.pdf, last accessed on February 8, 2007.

Charnes, A. and A. A. Cooper (1959). Chance-constrained programming. *Management Science* **6**: 73–79.

Conway, D. G. and M. A. Venkataramanan (1994). Genetic search and the dynamic facility layout problem. *Computers and Operations Research* **21**(8): 955–960.

Dantzig, G. B. (1955). Linear programming under uncertainty. *Management Science* **1**: 197–206.

Eastman Kodak Company (2005). Health, safety, environment and responsible growth Annual Report. Rochester, NY. http://www.kodak.com/US/plugins/acrobat/en/corp/environment/05CorpEnviroRpt/HSE2005AnnualReport.pdf, last accessed on February 8, 2007.

Elmaraghy, H. A (2005). Flexible and reconfigurable manufacturing systems paradigms. *International Journal of Flexible Manufacturing Systems* **17**(4): 261–276.

Ferrari, E., A. Pareschi, A. Persona and A. Regattieri (2003). Plant layout computerised design: logistics and relayout program (LRP). *International Journal of Advanced Manufacturing Technology* **21**(12): 917–922.

Ferrer, G. and D. C. Whybark (2001). Material planning for a remanufacturing facility. *Production and Operations Management* **10**(2): 112–124.

Franke, C., B. Basdere, M. Ciupek and S. Seliger (2006). Remanufacturing of mobile phones—capacity, program and facility adaptation planning. *Omega* **34**(6): 562–570.

Gupta, R. M (1986). Flexibility in layouts: a simulation approach. *Material Flow* **3**(4): 243–250.

Hillier, F. S (1963). Quantitative tools for plant layout analysis. *Journal of Industrial Engineering* **14**(1): 33–40.

Hillier, F. S. and M. M. Connors (1966). Quadratic assignment problem algorithms and the location of indivisible facilities. *Management Science* **13**(1): 42–57.

Housyar, A. (1991). Computer-aided facility layout: an interactive multi-goal approach. *Computers and Industrial Engineering* **20**(2): 177–186.

Jaramillo, J. R. and J. Alan R. McKendall (2004). Dynamic extended facility layout roblem. Proceedings of the IIE Annual Conference, Houston, TX.

Kaku, B. K. and J. B. Mazzola (1997). A tabu-search heuristic for the dynamic plant layout problem. *INFORMS Journal on Computing* **9**(4): 374–384.

Kekre, S., U. S. Rao, J. M. Swaminathan and J. Zhang (2003). Reconfiguring a remanufacturing line at Visteon, Mexico. *Interfaces* **33**(6): 30–43.

Kochhar, J. S. and S. S. Heragu (1999). Facility layout design in a changing environment. *International Journal of Production Research* **37**(11): 2429–2446.

Koren, Y., F. Jovane, T. Moriwaki, G. Pristchow, G. Ulsoy and H. V. Brusel (1999). Reconfigurable manufacturing systems. *Annals of CIRP* **48**: 527–539.

Koren, Y. and A. G. Ulsoy (2002). Vision, principles and impact of reconfigurable manufacturing systems. *Powertrain International* **5**(3): 14–21.

Kouvelis, P. and A. S. Kiran (1991). Single and multiple period layout models for automated manufacturing systems. *European Journal of Operational Research* **52**(3): 300–314.

Kouvelis, P., A. A. Kurawarwala and G. J. Gutierrez (1992). Algorithms for robust single and multiple period layout planning for manufacturing systems. *European Journal of Operational Research* **63**(2): 287–303.

Krajewski, L. J. and L. P. Ritzman (1999). *Operations Management, Strategic and Analysis*, Addison-Wesley Publishing Company, Inc. Reading, MA.

Lacksonen, T. A. and E. E. Enscore (1993). Quadratic assignment algorithms for the dynamic layout problem. *International Journal of Production Research* **31**(3): 503–517.

Lacksonen, T. A. and R. D. Meller (2000). An iterative layout design process. Industrial Engineering Research Conference, Cleveland, OH.

Lim, H.-H. and J. S. Noble (2006). The impact of facility layout on overall remanufacturing system performance. *International Journal of Industrial and Systems Engineering* **1**(3): 357–371.

Lund, R. T. (1984). Remanufacturing. *Technology Review* **87**(2): 19–29.

Mehrabi, M. G., A. G. Ulsoy and Y. Koren (2000a). Reconfigurable manufacturing system and their enabling technologies. *International Journal of Manufacturing Technology and Management* **1**(1): 113–130.

Mehrabi, M. G., A. G. Ulsoy and Y. Koren (2000b). Reconfigurable manufacturing systems: key to future manufacturing. *Journal of Intelligent Manufacturing* **11**(4): 403–419.

Meller, D. and K. Y. Gau (1996). The facility layout problem: recent and emerging trends and perspective. *Journal of Manufacturing Systems* **15**(5): 351–366.

Meller, R. and Y. Bozer (1996). A new simulated annealing algorithm for the facility layout problem. *International Journal of Production Research* **34**(6): 1675–1692.

Mellor, C. (2002). Quick-change artists: why techs must get ready for reconfigurable manufacturing. *The Ontario Technologist* **44**(1): 12–15.

Meng, G., S. S. Heragu and H. Zijm (2004). Reconfigurable layout problem. *International Journal of Production Research* **42**(22): 4709–4729.

Meyer, H. (1999). Many happy returns. *Journal of Business Strategy* **20**(4): 27–31.

Montreuil, B. and A. Laforge (1992). Dynamic layout design given a scenario tree of probable futures. *European Journal of Operational Research* **63**(2): 271–286.

Montreuil, B. and U. Venkatadri (1991). Strategic interpolative design of dynamic manufacturing systems layouts. *Management Science* **37**(6): 682–694.

Palekar, U. S., R. Batta, R. M. Bosch and S. Elhence (1992). Modeling uncertainties in plant layout problems. *European Journal of Operational Research* **63**(2): 347–359.

Prahinski, C. and C. Kocabasoglu (2006). Empirical research opportunities in reverse supply chains. *Omega* **34**(6): 519–532.

Pritsker, A. A. B. and P. M. Ghare (1970). Locating facilities with respect to existing facilities. *AIIE Transactions* **2**: 290.

Rawabdeh, I. and K. Tahboub (2006). A new heuristic approach for a computer-aided facility layout. *Journal of Manufacturing Technology Management* **17**(7): 962–986.

Rosenblatt, M. J. (1979). The facilities layout problem: a multi-goal approach. *International Journal of Production Research* **17**(4): 323–332.

Rosenblatt, M. J. (1986). The dynamics of plant layout. *Management Science* **32**(1): 76–86.

Rosenblatt, M. J. and D. H. Kropp (1992). The single period stochastic plant layout problem. *IIE Transactions* **24**(2): 169–176.

Rosenblatt, M. J. and H. L. Lee (1987). A robustness approach to facilities design. *International Journal of Production Research* **25**(4): 479–486.

Sharp Corporation (2001). Environmental, Health and Safety Report: Sharp Corporation North American Facilities. http://www.sharpusa.com/files/ envactivity2001.pdf, last accessed on February 8, 2007.

Shore, R. H. and J. A. Tompkins (1980). Flexible facilities design. *AIIE Transactions* **12**(2): 200–205.

Singh, S. P. and R. R. K. Sharma (2006). A review of different approaches to the facility layout problems. *International Journal of Advanced Manufacturing Technology* **30**(5–6): 425–433.

The Committee on Stochastic Programming. http://www.stoprog.org, last accessed on February 8, 2007.

The Council of Supply Chain Management Professionals. http://www.cscmp.org, last accessed on February 8, 2007.

Toffel, M. W (2003). The growing strategic importance of end-of-life product management. *California Management Review* **45**(3): 102–131.

Tompkins, J. A. and R. J. Reed (1976). An applied model for the facilities design problem. *International Journal of Production Research* **14**(5): 583–595.

Tompkins, J. A., J. A. White, Y. A. Bozer and J. M. A. Tanchoco (2003). *Facilities Planning*. Wiley, New York.

Topcu, A. and T. P. Cullinane (2005). Understanding facilities design parameters for a remanufacturing system. Proceedings of the SPIE, Boston, MA.

Urban, T. L. (1992). Computational performance and efficiency of lower bound procedures for the dynamic facility layout roblem. *European Journal of Operational Research* **57**(2): 271–279.

Urban, T. L. (1993). A heuristic for the dynamic layout problem. *IIE Transactions* **25**(4): 57–63.

Urban, T. L. (1998). Solution procedures for the dynamic facility layout problem. *Annals of Operations Research* **76**(0): 323–342.

U.S. Environment Protection Agency (1994). EPA Proposes to Reduce Air Pollutants from Municipal Waste Incinertors. http://yosemite.epa.gov/opa/admpress. nsf/621960f958b27cf18525701c005d8428/1eb9b464ebfc3ba58525644c002fd60d! OpenDocument, last accessed on February 7, 2007.

Waghodekar, P. and S. Sahu (1986). Facilities layout with multiple objectives: MFLAP. *Engineering Costs and Production Economics* **10**(2): 105–112.

Webster, D. B. and M. B. Tyberghein (1980). Measuring flexibility of job shop layouts. *International Journal of Production Research* **18**(1): 21–29.

Welgama, P. and P. Gibson (1995). Computer-aided facility layout—a status report. *International Journal of Advanced Manufacturing Technology* **10**(1): 66–77.

Xerox Corporation (2005). Environment, Health and Safety Progress Report. Stamford, CT. http://www.xerox.com/downloads/usa/en/e/ehs_2005_progress_report. pdf, last accessed on February 8, 2007.

Yang, T. and B. A. Peters (1998). Flexible machine layout design for dynamic and uncertain production environments. *European Journal of Operational Research* **108**(1): 49–64.

12

Some Studies on Remanufacturing Activities in India

Kampan Mukherjee and Sandeep Mondal

CONTENTS

12.1 Introduction

Remanufacturing is a form of product recovery process in which used and discarded products, components, or parts are subjected to a sequence of activities to make them reusable. Since used products (returns) form the basic raw material in this set of activities, the first step is acquiring returns from the disposer market (returns acquisition management). Next, these returns are transported to the remanufacturing plant through a logistics chain (reverse logistics).

Increasing concerns over environmental issues, take-back obligations, and disposal bans along with growing environmental awareness among customers, the economic benefits of remanufacturing, creation of stock of components or parts from disassembly operation, and demand for spare parts in the postproduct life cycle period are the prime reasons that motivated the industry toward reuse and remanufacturing of used products (Thierry et al. 1995). Japan, various European countries, and several states of the United States have stringent laws on reuse and product recovery (Doppelt and Nelson 2001; Spicer and Johnson 2004). A remanufactured item is often cheaper than a new item as the processing and manufacturing expenditures (time, energy, cost, etc.) are avoided. In the United States alone, there are over 73,000 remanufacturing firms with total sales of $53 billion (Lund 1998). These firms directly employ 350,000 workers and the average profit margins of these firms exceed 20% (Nasr et al. 1998). Further, many corporates have adopted voluntary take-back programs. FujiFilm, for example, purchases back its cameras, and Kodak has initiated design for the environment in its cameras and reuses 86% (by weight) of all the cameras. These measures have been adopted to build a green corporate image and thus attract environment-conscious customers, which would ultimately contribute to growth in sales. IBM Europe, Digital Europe, and Xerox could also create a positive image among customers through product recovery activities, resulting in better economic benefits (Guide et al. 2000). Today, a wide range of products such as automotive parts, locomotives, buses, aircrafts, engines, machine tools, photocopiers, cellular phones, computers, and consumer durables are remanufactured worldwide. But this practice is still in its nascent stage in many African, South American, and Asian countries, including India.

In this chapter, we describe the current remanufacturing business scenario in India, followed by identification of the critical factors that hinder the modest success of this business as an organized sector. Subsequently, we critically study the remanufacturing process of a photocopier remanufacturer in India and depict the interrelationships among managerial issues through a structural model. Managerial issues relating to returns acquisition are identified as one of the key sets of issues demanding critical study for cost–benefit analysis. Section 12.6 presents this economic analysis and formulation of a returns acquisition model. Analysis of the model leads to some significant conclusions.

12.2 Literature Review

Researchers and management scientists have also focused on remanufacturing as one of the emerging areas, and various studies have been completed on relevant business issues, reverse logistics networks, production planning, and inventory decisions. In this context, the contribution of the team members of the European Union–sponsored REVLOG project is noteworthy (Brito de and Dekker 2004). Rochester Institute of Technology's National Center for Remanufacturing and Resource Recovery is a leading center for applied research in remanufacturing. Many universities around the globe have initiated research on similar lines.

We refer to the works of Gungor and Gupta (1999) (331 articles and books) and de Brito et al. (2002) (121 articles, books, and Web sites) for their extensive review of literature on product recovery and reverse logistics. Guide et al. (1999) present a thorough discussion of the literature relevant to production planning and control. In our review of the remanufacturing literature, we primarily focus on contributions that attempt to establish the fundamentals of remanufacturing from a managerial standpoint. These fundamentals include a general body of principles that explain remanufacturing issues and address a variety of decision-making problems. Case examples and empirical investigations into managerial insights and strategy formulations are also covered. Table 12.1 provides the general taxonomy of the remanufacturing literature with the classification and the subclassification. Each of these groups and subgroups highlights certain sets of characteristics and research directions that lead to a conceptual framework of the remanufacturing process and capture the uniqueness of this business process.

12.3 Remanufacturing Sector *vis-à-vis* Indian Economy

Interestingly, in developing countries, remanufacturing is yet to gain acceptance as an organized business sector. We are, however, well aware of India's growth potential in the industrial sector. As one of the Asian giants, since the early 1990s India has been effectively implementing its plan for massive technological and societal development in the globalized economy. Indian market is growing and thus most of the remanufacturable products are produced today at a faster rate than a decade earlier. For example, passenger car penetration is 9 for every 1000 Indians and this figure is growing at the rate of 7% per annum. In the case of personal computers (PC), market penetration is 11 for every 1000 people and by 2010 PC sales would reach 14.3 million. The projected number of mobile cell phone subscribers by 2010 is estimated to be 500 million from the existing user strength of 44 million (with a constant growth rate of 1.5 million per month). We can perceive

TABLE 12.1

Taxonomy of the Remanufacturing Literature

A. Strategic and environmental issues

   1. Strategic and economic related

     Recovery options in product recovery management (Thierry et al. 1995)

     Life cycle and end-of-life aspects (Nagel and Meyer 1999, Stuart et al. 1999)

     Overall cost–benefit application of remanufacturing (Guide and Wassenhove 2001, Amini et al. 2005)

     Macroeconomic importance of remanufacturing (Enviromental Protection Agency, 1998)

     Empirical and survey-related study (Seitz and Peatlie 2004, Daugherty et al. 2005, Mondal and Mukherjee 2006b)

   2. Product design

     Design issues facilitating remanufacturing (Amezquita et al. 1995, Kriwet and Seliger 1995, Hammond et al. 1998)

     Design for remanufacturing (DFR) (Ishii et al. 1995, Gungor and Gupta 1999)

     Computer software that supports the design of a product in terms of disassemblability (Hesselbach and Kuhn 1996)

     Metrics for assessing the remanufacturability of a product design (Bras and Hammond 1996)

     Environment related

     Impact of environmental legislation on remanufacturing (Doppelt and Nelson 2001)

     Environment strategy (Sarkis 1998, Maslennikova and Foley 2000, White et al. 2003)

     Green supply chain management (Bloemhof-Ruwaard et al. 1995, Rao 2002)

     Impact of environmental issues in operations strategy (Sarkis 1999, Georgiadis and Dimitrion 2004)

B. Operational issues in remanufacturing

   1. Production planning and control

     Exploring issues in production planning and control (Guide 2000, Guide et al. 2000)

     Disassembly planning (Johnson and Wang 1995, Guide et al. 1997b, Lambert 2002)

     Scheduling (Gupta and Taleb 1994)

     Capacity planning techniques (Guide et al. 1997c)

     Reverse Wagner/Whitin's dynamic production planning (Richter Sombrutzki 2000)

     Performance study of a remanufacturing shop (Guide et al. 2005)

   2. Reverse logistics

     General framework (Fleischmann et al. 1997, Krumwiede and Sheu 2002, de Brito et al. 2004a)

     Overview of scientific literature and cases in reverse logistics activities (de Brito et al. 2002)

     Facility location model (Jayaraman et al. 1999, Berman et al. 2000, Fleischmann et al. 2001, Listes and Dekker 2005)

     Reverse logistics network design (Johnson 1998, Shiu 2001, Jayaraman et al. 2003, Nagurney and Toyasaki 2005)

     Routing problem (Dethloff 2001, Beullens et al. 2004)

     Reverse logistics operating costs (Tung-Lai Hu et al. 2002, Savaskan et al. 2004)

     Necessity of information technology in managing reverse logistics (Kokkinaki et al. 2004)

   3. Inventory management

     Overview and issue based (Dekker et al. 2000)

TABLE 12.1 (continued)

Taxonomy of the Remanufacturing Literature

> Warehousing (de Brito et al. 2004b)
>
> MRP system (Guide and Srivastava 1997a, Ferrer and Whybark 2001)
>
> Optimal lot-size policies (Ferrer 2003, Minner and Linder 2004)
>
> Model considering stochastic demand and return flow (Inderfurth 1997, Heisig and Fleischmann 2001, Fleischmann and Roelof 2003, Mahadevan et al. 2003, Kiesmuller and Scherer 2003)
>
> Dynamic demands and returns taking into consideration seasonality and product life cycle (Kleber et al. 2002)
>
> PUSH and PULL controlled systems (Van der Laan et al. 1999)
>
> EOQ models (Teunter 2001, Dobos and Richter 2004, Mostard and Teunter 2006)
>
> Safety stock planning (Minner 2001)
>
> Average cost inventory models (Teunter and Van der Laan 2002)
>
> Optimal periodic policy (Inderfurth et al. 2004)
>
> Stochastic hybrid manufacturing/remanufacturing problems (Inderfurth 2004)
>
> Opportunity cost rates (Teunter et al. 2000)
>
> Life cycle inventory analysis (Daniel et al. 2003)
>
> 4. Returns acquisition management
>
> Development of a framework (Guide and Wassenhove 2001)
>
> Approaches of implementation of returns take back (Spicer and Johnson 2004)
>
> Formulation of decision rules on handling of returns (Krikke et al. 1999)
>
> Quantitative model for buy-back program (Klausner and Hendrickson 2000, Guide et al. 2003, Mondal and Mukherjee 2006a)
>
> 5. Market competition for remanufactured products (Majumder and Groenevelt 2001)
>
> 6. Profitability analysis (Hoshino et al. 1995, Ferrer 1997, Guide and Wassenhove 2001, Guide et al. 2003, Heese et al. 2005)

the growth of business activities more realistically if we consider the huge population base of India. The expected benefits may be gleaned from the following facts:

- Material cost savings from remanufacturing, in general, lies between 40% and 65%.
- Energy consumption of a remanufactured product is only 15% of the new product.
- Remanufacturing is a labor-intensive operation (50% casual labor) and thus India may be an appropriate destination because of the availability of cheap labor.
- Profit from remanufacturing activities is almost 20%, although for the automotive sector this figure may go up to $30\pm10\%$.
- Price of a remanufactured product is 30–40% of a new product (particularly in developing countries).

- One million potential customers may be created in India with an attempt to cutting the car price to Rs. 75,000 (the currency is Indian rupee here and in subsequent occurrences) (around 25–30% of the current new car price).

- Automakers in India are worried about the increase in input costs and hence car sales price.

- Indian machine tools manufacturers require raw materials that are expensive with 30% import duty.

- PCs normally become obsolete in 2–3 years, causing disposal and environmental problems.

In the Indian economy, remanufacturing is almost nonexistent as an organized sector. We consulted the Centre for Monitoring Indian Economy (CMIE) database as the authentic source of listed business organizations. Out of nearly 1000 manufacturers, only six organizations could be identified as involved in any form of product recovery. These six organizations are (i) Xerox India Limited (XIL), remanufacturing photocopier under the project named "asset recovery management"; (ii) United Van Der Horst Limited (UVDHL), remanufacturing marine, oilfield, and industrial products; (iii) Soft-AID Computers Private Limited, Mumbai; (iv) Kores Printer Technology Limited; (v) Transdot Electronic Private Limited, Trivandrum, refurbishing printer heads; and (vi) Timkin India Limited, refurbishing large industrial and rail bearings. Among these six firms, XIL and UVDHL have been engaged in remanufacturing for quite some time and have gained sufficient experience and knowledge about remanufacturing activities. In XIL, the scale of remanufacturing activities is much lower than that of manufacturing business, but UVDHL is a full-fledged remanufacturing company with an annual turnover of Rs. 2.75 crores. The remaining four organizations maintain a small scale of business for product recovery.

Furthermore, unlike Western countries, environmental and take-back laws are still not adequately stringent enough in India, although some legislative rules such as the Batteries (Management and Handling) Rules, 2001 and the Recycled Plastics Manufacture and Usage Rules, 1999 exist. However, in India we find product recovery practices existing in the form of an unorganized business sector. There also exists a strong second-hand market segment. Moreover, we must admit that India is a developing country with a price-sensitive market. This means that if remanufactured products were available, people would definitely plan to buy a remanufactured product from the market as the price is comparable to a second-hand product and the quality is comparable to a new product.

Global trends, expected benefits from remanufacturing, and the aforementioned Indian industrial scene strongly justify the logical conclusion that India is one of the prospective countries for initiating and maintaining the remanufacturing sector.

12.4 Empirical Study on Acceptability of Remanufacturing

The current status of remanufacturing business in India, as described in Section 12.3, clearly highlights India's massive potential for remanufacturing. But this sector is still almost nonexistent as an organized sector in the Indian economy. Against the backdrop of this discussion, an empirical investigation is carried out to seek answers to the following sequentially related questions (Mondal and Mukherjee 2006b):

i. Why are Indian manufacturers not interested in remanufacturing their products?
ii. Does any industry-wide commonality exist among these demotivating factors?

12.4.1 Development of the Survey Instrument

Referring to the literature, we identified several issues as prime hindrances to the modest success of remanufacturing. Fourteen such issues were considered (shown in Table 12.2) for constructing the survey instrument. The questionnaire was simple and closed-ended. Data were collected on a five-point Likert scale to rate the 14 issues causing nonpopularity of remanufacturing in India. On this scale, "not important" or "0" value corresponds to respondent's denial to consider an issue in the remanufacturing decision, whereas 1 and 5 correspond to "least importance" and "most importance" of an issue, respectively. These values represent degrees of importance of an issue considered for rejecting a decision to take up the remanufacturing business.

12.4.2 Population and Sample Profile

12.4.2.1 Selection of Population

The following characteristics were considered to define the remanufacturability of a product (according to Lund 1998): (i) stability of product technology, (ii) product durability, (iii) functional failure of a product, (iv) standardization of a product and interchangeability of its parts, and (v) high scope of value additivity on the product. With the help of the CMIE database we prepared a complete list of Indian companies, which are manufacturing products that are currently remanufactured worldwide or are remanufacturable. Of these, 972 companies constituted the survey population. For economical reasons, we adopted simple random sampling technique to decide on the sample size from a finite population.

12.4.2.2 Determination of Sample Size

A total of 110 companies were randomly selected from the aforementioned population, and questionnaires were mailed to these companies. Repeated

TABLE 12.2

Issues Considered in the Empirical Investigation

Issues	Notation	References
No specific market for remanufactured products	V_1	Thierry et al. 1995, Ferrer and Whybark 2003
Relatively few customers in the market	V_2	
Disorganised business sector is already active	V_3	Majumder and Groenevelt 2001
Second-hand market is thriving	V_4	
Customer thinks that refurbished goods are inferior	V_5	Gungor and Gupta 1999, Guide et al. 2000
Mindset of people is not like the Western world	V_6	
Economically not profitable	V_7	Guide and Srivastava 1997a, Dekker et al. 2000
Technically infeasible	V_8	Ferrer and Whybark 2003, Giuntini and Gaudette 2003
Expertise not available in this area	V_9	Giuntini and Gaudette 2003, Seitz and Peatlie 2004
No environmental compulsion for remanufacturing	V_{10}	Doppelt and Nelson 2001, Spicer and Johnson 2004
Uncertainty in timing, quantity, and quality of returns	V_{11}	Thierry et al. 1995, Guide 2000, de Brito et al. 2002, Ferrer and Whybark 2003, Spicer and Johnson 2004
Acquisition of returns is difficult	V_{12}	
Reverse distribution of used products is difficult	V_{13}	
High cost is associated with logistics	V_{14}	

reminders were sent out subsequently. The survey responses were received between February 2004 and August 2004. A total of 41 valid responses were finally received and considered for the study. Table 12.3 lists out the industry type, sales turnover, quantity of business, and other aspects of these responses.

12.4.3 Data Analysis

Table 12.4 shows the descriptive statistics and the correlations among the 14 variables.

The correlation matrix indicated the existence of high correlation among some of the variables. This motivated us to conduct factor analysis and identify the group of variables representing a single underlying construct or factor. The analysis was conducted using the statistical software, SPSS 10.0.

TABLE 12.3

Sample Characteristics

Industry Type	Company Size (Turnovers in billion rupees)				Numbers in Each Group			Turnover (Billion rupees)		
	<1	1–4	4–8	>8	Respondent	Sample	Population	Respondent	Sample	Population
Automobile	2	10	4	2	18	33	307	206.49	278	1006.87
Electronics and computers	1	6	1	1	9	24	244	36.84	104.48	239.68
Consumer durables	1	0	1	0	2	11	68	7.73	13.62	174.54
Industrial machinery	6	4	1	1	12	42	353	31.93	85.76	468.35
Total	10	20	7	4	41	110	972	282.99	481.86	1789.44

The factor loadings for the six factors (approximately explain 90% of the total sample variance) were as follows:

Factor 1 (F_1): V_3 (0.770), V_4 (0.774), V_{11} (0.454), V_{12} (0.763), V_{13} (0.577), and V_{14} (0.780)

Factor 2 (F_2): V_8 (−0.822), V_9 (−0.898), V_{11} (0.659), V_{12} (0.401), V_{13} (0.716), and V_{14} (0.489)

Factor 3 (F_3): V_1 (0.914), V_2 (0.903), and V_5 (0.572)

Factor 4 (F_4): V_5 (0.693) and V_6 (0.803)

Factor 5 (F_5): V_7 (0.877)

Factor 6 (F_6): V_{10} (0.860)

12.4.3.1 Interpretation of Factors

We attempt to explain these factors for getting some insight into the remanufacturing scenario of India and its physical interpretation. The variables V_{12} and V_{13} relate to the difficulty in the process of acquisition of returns and the complexity involved in reverse logistics. These issues are primarily influenced by uncertainty (reflected by V_{11}) in timing, quantity, and quality of returns. Variables V_3 and V_4 represent implications of shortage of returns due to the existence of a strong secondary market. Logically, because of the nonavailability of returns and the complex processes of acquisition and reverse logistics, the logistics cost increases. This increase is represented by variable V_{14}. Thus, issues related to difficulty in acquiring used products and to the reverse distribution (i.e., inbound logistics) process along with the market factors affecting availability of used products may be explained by a single factor, that is, factor 1. This factor may thus be termed as acquisition factor.

TABLE 12.4

Descriptive Statistics and Correlation of the Study Variables

Variable	Mean	Standard Deviation	Correlation													
			V_1	V_2	V_3	V_4	V_5	V_6	V_7	V_8	V_9	V_{10}	V_{11}	V_{12}	V_{13}	V_{14}
V_1	3.07	1.47	1.00													
V_2	2.95	1.36	0.91	1.00												
V_3	2.12	1.19	-0.38	-0.34	1.00											
V_4	1.85	0.96	-0.59	-0.52	0.74	1.00										
V_5	3.17	1.28	0.71	0.74	-0.33	-0.47	1.00									
V_6	2.80	1.42	0.62	0.61	-0.46	-0.50	0.80	1.00								
V_7	2.66	1.32	0.48	0.55	-0.05	-0.49	0.42	0.39	1.00							
V_8	2.22	1.53	-0.01	0.04	-0.42	-0.42	0.07	0.24	0.11	1.00						
V_9	1.93	1.35	0.03	0.05	-0.32	-0.22	0.11	0.22	-0.06	0.69	1.00					
V_{10}	1.90	1.11	0.39	0.41	0.14	-0.04	0.50	0.37	0.44	-0.19	-0.17	1.00				
V_{11}	3.05	1.26	-0.27	-0.32	0.60	0.56	-0.27	-0.44	-0.22	-0.73	-0.57	0.29	1.00			
V_{12}	3.05	1.14	-0.18	-0.19	0.62	0.67	-0.21	-0.40	-0.24	-0.55	-0.44	0.22	0.76	1.00		
V_{13}	3.10	1.07	-0.20	-0.20	0.48	0.53	-0.14	-0.25	-0.19	-0.69	-0.60	0.05	0.74	0.67	1.00	
V_{14}	3.15	1.13	-0.26	-0.26	0.55	0.66	-0.12	-0.26	-0.22	-0.60	-0.45	0.09	0.61	0.77	0.84	1.00

Factor 2 explains the combined effect of six variables (V_8, V_9, V_{11}, V_{12}, V_{13}, and V_{14}) as per the factor analysis. A close scrutiny of these six variables shows that these variables are in fact a combination of two sets of variables— technology-related (V_8, V_9) and reverse logistics–related (V_{11}, V_{12}, V_{13}, V_{14}) variables. Interestingly, the correlation matrix shows that there is a clear-cut negative relationship between these sets of technology-related and reverse logistics–related variables (refer Table 12.4). This negative relationship reflects some intrinsic conflict between these two sets of variables in terms of their importance in assessing the viability of remanufacturing as a business proposition. For example, if the remanufacturing option is rated low (considered to be rejected) because of the strong role of technology issues, the role of reverse logistics issues in this decision-making weakens. However, as per the factor loadings, the technology issues are negatively related to factor 2, whereas the reverse logistics issues are related positively. If we combine these two phenomena, we may note that low value of factor 2 reflects a situation in which remanufacturing as a business proposition may be rejected because of the high importance of technology-related issues. In this situation, the reverse logistics issues may lose their importance as the determining factor in the remanufacturing decision. Similarly, high value of factor 2 justifies the existence of a situation where acquisition and inbound distribution seem to be more important in remanufacturing decision than availability of technology know-how. We term this factor as technology factor, which is primarily governed by the technical feasibility and design-related issues of the remanufacturing process.

Factor 3 addresses the issues relating to the market demand for remanufactured products. The demand for remanufactured products is possibly dictated by the existence of a proper market structure for the remanufactured product and awareness among the customers. Let us term this factor as market factor.

Factor 4 represents customer's attitude toward the remanufactured product. This factor is called attitudinal factor. Factors 5 and 6 are trivial factors as each of these factors has only one variable, V_7 and V_{10}, respectively. Factor 5 relates to profitability and factor 6 to the legislation and the laws imposed by the government.

12.4.3.2 Clustering of Companies Based on the Responses

Based on the structure of the responses obtained from the various companies, we developed a belief that, perhaps, "natural" groupings existed among the respondents. Members of any group are supposed to exhibit similar perception on the viability of remanufacturing. Hence, using cluster analysis we further extended our analysis to identify whether any pattern exists within the data set or not. Finally, a dendrogram showing the distances for all the respondents was obtained (shown in Figure 12.1). Three clusters were distinguished from the dendrogram.

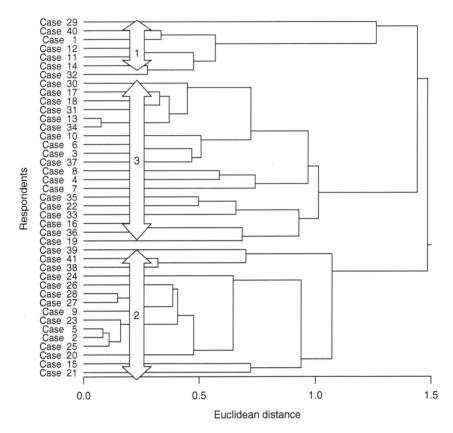

FIGURE 12.1
Dendrogram used to cluster the respondents.

A critical analysis of the respondents' profiles revealed that each of the clusters represented a particular type of industry. These clusters are defined as the following industrial sectors:

Cluster I:	Computers and electronics industry
Cluster II:	Industrial machinery industry
Cluster III:	Automotive industry

12.4.3.3 Criticality of Factors for Each Industry Type

After the clusters were identified, our next step was to identify the criticality of the factors influencing the remanufacturing decision industry-wise. The factor scores for each of the three clusters were separated and then Friedman's test was applied to test whether any significant difference

TABLE 12.5

Industry-Wise Criticality of Factors

Factors	Cluster I (Computers and Electronics)	Cluster II (Industrial Machinery)	Cluster III (Automobile)
Acquisition	1.86	5.40	2.11
Technology	1.43	4.73	2.74
Market	4.29	2.47	4.37
Attitude	5.14	2.80	3.79
Profitability	4.86	2.60	3.79
Legislation	3.43	3.00	4.21
Friedman's Test Statistics	24.306	33.210	21.466
p value	0.000	0.000	0.001

existed among the means of the six factors. Table 12.5 shows the test results and the mean ranks for the six factors for computer and electronics industry, industrial machinery industry, and automotive industry, respectively. The test result also shows that significant differences exist among the means.

The situation under each industry type may now be explained on the basis of mean rank values. In cluster 1, low value of mean rank for technology factor clearly indicates that compared with reverse logistics issues, technological issues are quite critical in making the remanufacturing process infeasible. Today, hardware and electronic products are primarily assembled in India and hardly a few of these items are manufactured here. Therefore, India lacks manufacturing expertise in hardware, IC chips, electronic equipment, and other products. Even if technical know-how and expertise is acquired from other countries, the remanufacturing process may not be quite economical. In addition, customer response toward remanufactured products is not very encouraging. These two aspects are reflected by the market, attitudinal, and profitability factors. It may be noted that nonprofitability is a deterrent for remanufacturing of PCs even in the developed countries. With the help of data, Boon et al. (2001) have demonstrated that remanufacturing of obsolete PCs is not profitable (profitability is measured without including environmental costs).

In the industrial machinery segment, acquisition of returns is quite critical, as high market competition exists for returns in the disorganized and second-hand market. Moreover, high value of mean rank for technology factor indicates that though remanufacturing feasibility is not a serious issue, reverse logistics issues are of prime concern. Retail concept is lacking in this industrial sector. Perhaps, this is the prime reason behind the difficulty in tracking the returns. Further, the high cost associated with transportation of the used machines and use of special handling equipment in reverse logistics, perhaps, makes remanufacturing business unviable. It is also true that in India, technological development is taking place at a slow pace in the manufacturing sector. Consequently, industrial machines are purchased or upgraded after long intervals. The acquired returns may either be quite old or may be technologically obsolete for their use in remanufacturing. In the United States, where technological changes are rapid, remanufacturing of capital goods such as complex military equipment and mining, agricultural, and industrial machines provides the highest profit (Giuntini and Gaudette 2003), unlike the situation in India.

In the automotive sector, poor market structure owing to the lack of customer awareness and the nonexistence of legislative compulsion are found to be the crucial reasons behind the modest success of remanufacturing. Lack of customer awareness about the economic benefits accruing from remanufactured products causes poor market demand for such products in the automotive sector. However, this problem may be overcome by campaigning on the "as-good-as-new" nature of remanufactured products and on the advantage of price-quality trade-off compared with that of the second-hand automotive products. Hammond et al. (1998) found that in the U.S. automotive remanufacturing sector, market demand, availability of parts, and profit potential were the three key criteria considered for determining whether or not to remanufacture a product. Furthermore, legislative compulsion plays the role of a prime driver for remanufacturing business apart from creating customer awareness about benefits.

Table 12.6 summarizes the critical factors from the aforementioned discussion. Interestingly, contrary to common belief, the legislation factor is not a critical factor in the Indian industry, except in the automotive sector. While environmental legislation is considered to be the genesis of remanufacturing business in the Western countries, market economy and financial incentives are the prime driving forces in India. Guide (2000) mentioned that legislation is the main motivation behind remanufacturing in Europe and that economics or profitability plays a similar role in the United States. Thus, it appears that Indian business perception toward remanufacturing is closer to that of the United States.

TABLE 12.6

Dominant Factors in Each of the Three Clusters

Computers and Electronics	Industrial Machinery	Automotive
Technology (design focus)		
Attitude	Acquisition	Market
Market	Technology (reverse logistics focus)	Legislation
Profitability		

12.5 Managerial Issues Relating to Remanufacturing Business—a Case Study

From the empirical investigation in Section 12.4, we identified some critical reasons for the poor existence of remanufacturing business in India. Moreover, the investigation revealed the varied perception of different industries toward remanufacturing. In this section, we extend our research to critically study a particular remanufacturing process in India as a case example to show how a company manages its remanufacturing business. This section also aims at studying and understanding the relationships among key issues in the management of the remanufacturing process, so as to obtain some meaningful insights into managerial decision making.

The company (an original equipment manufacturer or OEM) under study is a well-known photocopier manufacturer in India. This company was incorporated in 1983 and is part of a Fortune 500 document management company. In 1991, the parent global company set a goal of becoming a waste-free business house. This parent company recognized a number of essential characteristics of a waste-free company: financial, competitive advantage, compliance with legislative regulation, and meeting customer requirements. With the adoption of the end-of-life equipment take-back program, the company obtained savings of over $80 million in Europe in 1997. Keeping this strategic goal in view, the OEM in India also followed the practice of the parent company by setting up a remanufacturing center in 1998 as Asset Management Business Centre. This center is engaged in remanufacturing photocopier machines under various brand names (currently three in number). The remanufacturing plant is around 2 km away from the company's manufacturing plant located in the northern region of India. The remanufacturing plant is managed independently and treated as a profit center, separate from the company's manufacturing business. In 2003, with production facility for 60–70 units of monthly production, the sales turnover of the company from remanufacturing business was around Rs. 30 million ($1 U.S. = Rs. 42 approximately). This company is the only Indian company that pursues remanufacturing business of photocopiers in India and thus enjoys a monopolistic market for this product. The market for remanufactured photocopiers and the new products of the OEM are identified as two separate segments. The OEM manages all the basic remanufacturing operations, including acquisition of the used products and reverse logistics activities, through a well-established supply chain network. The company workforce involves both casual (or temporary) and permanent employees. However, the company opts for hiring more temporary workers than permanent workers, particularly for operations such as disassembly, cleaning, sorting, and inspection, which are highly labor-intensive. This recruitment policy is possibly dictated by cost advantage and better workforce management considerations. Currently, only 11 permanent workers are on roll.

Although the primary motivation for remanufacturing business was the waste-free policy or environment friendliness of the parent global company, adoption of remanufacturing by this Indian company was also driven by another factor. This factor is the company's unique service policy to customers regarding the maintenance of newly installed machines. Based on this policy, the company provides free service and maintenance with every new machine that is sold, excluding the cost of consumables, until the product's end-of-life (10 years approximately). Thus, the company not only maintains a close relationship with the customers but also maintains a record of all installed machines through a strong network of service centers across the country. However, the company bears an additional service cost for offering free service and for maintenance of such a vast service network. As the machine ages, the service cost on a unit also increases. So, high cost is borne for older machines. Keeping these points in view, the company has initiated the "exchange offer" and "buy-back" schemes. Customers are given the option of getting their old units replaced by new and upgraded models through exchange offers or of simply selling the old units to the OEM through the buy-back scheme. The company is thus directly benefited by the economic advantage of discontinuing the service to old machines. Instead of being disposed off, these collected returns are remanufactured by the company. The price of remanufactured photocopiers is generally 30% of the new photocopiers and quality is "as good as new." This cost benefit has created sufficient demand for these units in the Indian market. Photocopy shops, small business centers, and other units are regular customers of this remanufactured product. The company has been engaged in remanufacturing for the last 7 years and has found remanufacturing to be an economically viable business proposition.

12.5.1 Issues Pertaining to Remanufacturing

Managing any remanufacturing business requires consideration of various issues that primarily influence decision-making at different levels of remanufacturing activities. After an extensive literature survey, we identified nine sets of issues (shown in Table 12.7) as the key factors controlling remanufacturing management. Interpretive structural model (ISM) methodology is used to critically examine relevant issues in managing the remanufacturing process. ISM methodology helps in establishing order and direction in complex relationships among various elements of a system (Warfield 1974). The decision-making environment of the remanufacturing management is affected by a set of enablers that are directly or indirectly interrelated. ISM methodology can be used to structurally depict the role of these enablers and the impact of their collective influence.

12.5.2 Interpretive Structural Model

Structural self-interaction matrix, depicting the contextual relationship among the sets of issues, is constructed in consultation with the management of

TABLE 12.7

Issues Relevant to Remanufacturing Business

1. Product design issues relevant to remanufacturing process (Amezquita et al. 1995, Gungor and Gupta 1999)
2. Impact of workplace environment and use pattern of returns (Guide, 2000)
3. Level of technology and tools for remanufacturing (Hammond et al. 1998)
4. Factors relevant to the process of returns acquisition (Klausner and Hendrickson 2000, Mondal and Mukherjee 2006a)
5. Factors relevant to the reverse distribution process (Fleischmann et al. 2001)
6. Issues relevant to disassembly and reassembly planning (Gupta and Taleb 1994)
7. Role of skill and expertise of workforce (Hammond et al. 1998)
8. Issues related to inventory management (Van der Laan et al. 1999, Ferrer and Whybark 2001)
9. Issues related to marketing of remanufactured products (Majumder and Groenevelt 2001)

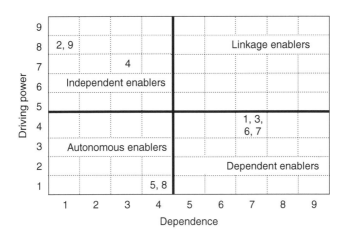

FIGURE 12.2
Driving power and dependence diagram.

the remanufacturing company (Mukherjee and Mondal 2006). Subsequently, the issues are classified into four groups, that is, autonomous, dependent, linkage, and driver enablers based on driving power and dependence. This classification, as depicted in Figure 12.2, shows that issues 5 and 8 are autonomous enablers and are relatively disconnected from the system. Issues 1, 3, 6, and 7 are dependent enablers and 2, 9, and 4 are independent enablers. There are no linkage enablers. Finally, the ISM-based model (Figure 12.3) is constructed using the identified levels of the issues and the information from the driving power–dependence diagram.

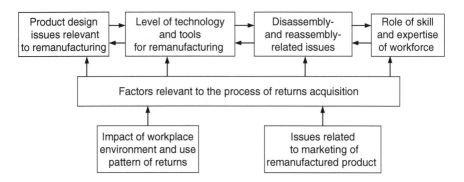

FIGURE 12.3
ISM-based model for the issues related to remanufacturing process.

12.5.3 Discussion on the Interpretive Structural Model

Impact of the workplace environment, use pattern of returns, and issues related to marketing of the remanufactured product are the strongest drivers in the decision-making framework of this remanufacturing business. The existence of a market for the remanufactured product and inflow of returns with better recovery rates substantially influence the economic benefit from any remanufacturing business. This is perhaps the reason why issues relating to marketing of the remanufactured product are in the level of the strongest drivers. Thus, sufficient managerial effort is required to create awareness among customers, particularly in countries like India. Cheaper price, "as-good-as-new" quality and "green image" can be suitably used to develop appropriate marketing strategy. Moreover, the proportion of parts or components recovered in good condition depends on the workplace environment and the use pattern of the parts or components. Better workplace environment results in higher recovery rates and thus leads to more economic benefits. Since the OEM maintains a strong database of the customers through its unique service policy, the company can somewhat manage to control the uncertainty in recovery rates.

The aforementioned two prime sets of issues trigger the next level of issues, which includes the factors relevant to the process of acquisition. These issues are referred to as sensitive issues as they are affected by issues 2 and 9 and also influence the other four types of issues. An acquisition decision is primarily affected by the quality and the quantity of returns. Workplace environment once again influences the quality of returns, while market demand guides the quantity of returns to be procured in the acquisition process as per pull policy of remanufacturing.

The next level of issues, as shown in the ISM in Figure 12.3, includes issues related to product design, remanufacturing technology, disassembly and reassembly, and skill and expertise of the workforce. These issues occur at the operational level of the remanufacturing business. In this case, since the company is an OEM, product design issues relevant to remanufacturing are found to be at the lowest level in terms of driving power. However, experience shows that in most of the reported cases of remanufacturing, factors related to product design play a very important role and trigger activation of other issues.

12.6 Returns Acquisition Policy Decision in Managing Remanufacturing

The ISM revealed that returns acquisition is the most sensitive issue in managing the OEM company's remanufacturing business and perhaps guides the profitability of this business. Thus, the management requires an effective returns acquisition policy for the success of its remanufacturing business. While planning for acquisition of returns, the decision maker tries to identify the optimal buy-back period of a used product to maximize the economic benefit from remanufacturing. For certain products, which are under the free servicing policy of the company, acquisition planning is supposed to be dictated by the efficiency of service management of the company and also by working environment at the user's end (Mondal and Mukherjee 2006a). In this section, we studied this hypothesis and tried to express these indices in terms of various economic parameters related to the remanufacturing process.

12.6.1 Role of Service Management by Original Equipment Manufacturer and Workplace Management by Users

Experience indicates that proper service management by the OEM prolongs the useful life of a product, with a higher possibility of obtaining recoverable parts from a used product till the later part of the product's life. So, remanufacturing is expected to remain viable for a longer period. In contrast, if the OEM spends significantly on improvement of its service, it is quite logical that the unit should be bought back early so as to save cost of servicing. It may thus be concluded that OEM's management of services substantially influences acquisition planning of the returns.

Further, from a general observation it may be noted that the recovery rate decreases as the product ages, where recovery rate is the proportion of content (parts or components) remaining healthy or in good condition or the proportion of parts or components that can be reused (Guide and Srivastava 1997a). Recovery rate is mainly controlled by working conditions and workplace environment at the user's end. Therefore, in a poorly managed workplace, the recovery rate decreases at a faster rate over time. In this case, the used products are to be bought back early so as to obtain advantage of lower remanufacturing cost. Thus, we may logically postulate that the workplace environment also influences acquisition planning. From this discussion, we hypothesize the following statement:

> **Hypothesis:** *Under free-servicing policy, acquisition planning of returns is essentially controlled by service management of the OEM and workplace management at the user's end.*

Now, we first analyze the economic issues determining acquisition decision making. Subsequently, we study the model for determining the optimal period for buying back the used products. The objective is to critically analyze the decision-making situation to get sufficient insight and relevant information for logical testing of the hypothesis.

12.6.2 Acquisition Decision Structure and Modeling

We make the following assumptions about the recovery system to keep the problem within manageable limits:

- A single model of the product is considered in the decision analysis.
- Various costs and revenues are not influenced by interest rates.
- Inventory carrying costs are ignored.
- The buy-back price of the used product is determined solely by the OEM.
- Acquisition plan is constructed keeping in view the interest of the OEM only, not that of the seller of the used product from the disposer market.
- Remanufacturing cost and revenue collected from the recyclers by selling the nonrecovered parts or components are stochastic parameters due to the inherent randomness of recovery rate.

Let us consider that a unit is bought back after time t from the date of its sale, where the scale of t is chosen proportional to the life of the product. Thus, t here is a continuous variable and lies between 0 and 1, where $t = 0$ refers to the time when a new unit is sold and $t = 1$ refers to the end-of-life of the unit. The relevant cost and benefit parameters either directly depend on the age of the product or are influenced by the recovery rate, which in turn is a function of the age itself. Table 12.8 shows the trade-off factors influencing

TABLE 12.8

Cost and Benefit Parameters

Costs	Benefits
• Buy-back price of a unit $C_u(t) = K_1 - K_2 t$, where K_1 represents the price of a new unit and K_2 is the rate of depreciation of value of the product. • Remanufacturing cost per unit $C_r(\cdot)$ $K_f + K_p(1-\alpha)$, here K_f is the fixed remanufacturing cost per unit remanufactured and $K_p(1-\alpha)$ is the variable cost depending on the amount of recovery. K_p represents the total cost of parts/components required to manufacture a new unit.	• Scrap Sales, $P_d(\cdot) = K_d(1-\alpha)$, where K_d refers to the revenue generated from selling all the parts/components in a unit. • Service Cost, $C_s(t) = K_s \cdot t$, where K_s is the rate of expenditures for servicing a unit. The savings from service cost in the period $[t, 1]$ is $$S_s(t) = \int_t^1 C_s(x)dx$$

Recovery rate $\alpha = 1 - \beta t$ where β is a random variable affected by the working environment of the unit and the random factors associated with failure possibilities.

acquisition planning. The difference of benefits and costs at any instant t is thus given as $D(t) = P_d(\cdot) + S_s(t) - C_u(t) - C_r(\cdot)$. Incorporating the functions as proposed in Table 12.8 and maximizing the expected value of $D(t)$, the optimal t is obtained as

$$t_{opt} = \frac{K_2 - E(\beta)(K_p - K_d)}{K_s}$$

12.6.3 Economic Analysis

The value of t_{opt} demands a special attention for some meaningful interpretation. As expressed by $C_u(t)$ function, the salvage value at the end-of-life may be expressed as $(K_1 - K_2)$ where K_1 is the market price of a new unit. Thus, the amount of depreciation during the life of its use is K_2, or in other words, it is the economic value of the product, which has been made available to the user during the product's life. K_p represents the cost of procuring all components of a unit, whereas K_d is the scrap value of the components (not to be reused) by selling them off to the recyclers. Logically, it can be stated that better servicing by the manufacturer and provision of efficient workplace environment by the user lengthens the life of a product. In other words, these factors intensify the use of the product till a very low value of $(K_1 - K_2)$ is achieved and also lead to more recovery of components for reuse. Thus, better service effectiveness may be reflected by higher value of K_2 and more recovery from $(K_p - K_d)$. This phenomenon leads to establishing an index, which may be termed as service effectiveness, expressed as

Service Effectiveness (SE) = Recoverability Index (RI) × Service Index (SI)

where

$$SE = \frac{K_2 - E(\beta)(K_p - K_d)}{K_s}; \quad RI = \frac{K_2 - E(\beta)(K_p - K_d)}{K_2}; \quad SI = \frac{K_2}{K_s}$$

RI shows the expected recovered value from the available economic value of the unit (K_2) during its useful life, whereas SI is the ratio of the economic value and the rate of service expenditure (K_s) meant for creation of this value. SI is practically under the control of the company, showing the efficiency of servicing activities during the useful life of the product. However, a better user's management perhaps reduces the nonrecovery factor, $E(\beta)$, and hence improves the RI. Here, it is to be noted that all these indicators of effectiveness are meant for achieving better remanufacturability.

Clearly, higher level of SE causes higher recovery even in the later part of the product life. So, in that case, remanufacturing remains viable till the older age of the unit, which means, because of the trade-off between remanufacturing and buy-back price, the buy-back decision may be delayed further. In contrast, an early buy-back may be due to the poor management practice of the user or primary customer and improper or inefficient service management of the OEM.

12.7   Conclusion

Remanufacturing has already been accepted worldwide as a viable business proposition because of its ecological and economic benefits. However, in India, remanufacturing is not yet an organized business sector. Empirical study and subsequent analysis revealed that the modest success of remanufacturing in India is perhaps guided by six factors, namely acquisition of returns, technology for remanufacturing, market, customer's attitude, profitability, and legislative compulsion. The analysis further shows that the major reason for nonacceptance of remanufacturing in computers and electronics industry is technological infeasibility; in the industrial machinery segment, the major reason is the complex reverse logistics system; and in the automotive sector, the reason is legislation and customer's negative attitude toward the remanufactured products. It is further noted that Indian manufacturers do not consider existence of stringent legislative measures or take-back policies on remanufacturing as important for a remanufacturing decision. Environmental awareness among Indian customers is almost negligible perhaps as a result of the low literacy rate (65.38% as per 2001 census) and the nonexistence of opportunities for proper environmental education. Lack of awareness of the benefits accruing from remanufacturing also contributes to the low demand for remanufactured products. Before undertaking any remanufacturing business, firms should attempt to popularize the benefits (ecological and economical) of remanufacturing.

This research project was further extended to critically study a remanufacturing process conducted in the Indian business environment. The interrelationship of the various issues relevant to remanufacturing was critically examined. Some sensitive issues were identified as a result of this structural analysis. A returns acquisition model was developed incorporating these sensitive issues. Analysis of the model led to the development of certain indices, which govern the buy-back policy of returns. This analysis also supports the hypothesis that in case of free servicing policy, acquisition planning is directly influenced by service management of the OEM and workplace management of the user.

References

Amezquita, T., Hammond, R., Salazar, M. and Bras, B., Characterizing the remanufacturability of engineering systems, Proceedings of ASME, Advances in Design Automation Conference, Boston, MA, 1995, 82, 271–278.

Amini, M.M., Retzlaff-Roberts, D. and Bienstock, C.C., Designing a reverse logistics operation for short cycle time repair services, *International Journal of Production Economics*, 2005, 96(3), 367–380.

Berman, O., Drezner, Z. and Wesolowsky, G.O., The collection depots location problem on networks, *Naval Research Logistics*, 2000, 49, 15–24.

Beullens, P., Oudheusden, D.V. and Wassenhove, L.N.V., Collection and vehicle routing issues in reverse logistics, In *Reverse Logistics: Quantitative Models for Closed-Loop Supply Chains*, Dekker, R., Fleischmann, M., Inderfurth, K. and Wassenhove, L.N.V. (Eds.), 2004, Springer, Berlin, Germany.

Bloemhof-Ruwaard, J.M., Van Beek, P., Hordijk, L. and Wassenhove, L.N.V., Interactions between operational research and environmental management, *European Journal of Operational Research*, 1995, 85, 229–243.

Boon, J.E., Isaacs, J.A. and Gupta, S.M., Economics of PC recycling, environmentally conscious manufacturing, Proceedings of SPIE, 2001, pp. 29–35.

Bras, B. and Hammond, R., Design for remanufacturing metrics, Proceedings of the First International Working Seminar on Reuse, Eindhoven, The Netherlands, 11–13 Nov., 1996, pp. 35–51.

Brito de, M.P., Flapper, S.D.P. and Dekker, R., Reverse Logistics: A review of case studies, Economic Institute Report EI 2002-21, Erasmus University, Rotterdam, 2002.

Brito de, M.P. and Dekker, R., A framework of reverse logistics, In *Reverse Logistics: Quantitative Models for Closed-Loop Supply Chains*, Dekker, R., Fleischmann, M., Inderfurth, K. and Wassenhove, L.N.V. (Eds.), 2004, Springer, Berlin, Germany.

Brito de, M.P. and Koster, de Rene M.B.M., Product returns: Handling and warehousing issues, In *Reverse Logistics: Quantitative Models for Closed-Loop Supply Chains*, Dekker, R., Fleischmann, M., Inderfurth, K. and Wassenhove, L.N.V. (Eds.), 2004, Springer, Berlin, Germany.

Daniel, S.E., Pappis, C.P. and Voutsinas, T.G., Applying life cycle inventory to reverse supply chains: A case study of lead recovery from batteries, *Resources Conservation and Recycling*, 2003, 37, 251–281.

Daughterty, P.J., Richey, R.G., Genchev, S.E. and Chen, H., Reverse logistics: Superior performance through focused resource commitments to information technology, *Transportation Research Part E*, 2005, 41, 77–92.

Dekker, R., Van der Laan, E. and Inderfurth, K., A review on inventory control for joint manufacturing and remanufacturing, *Conference in Management and Control of Production & Logistics, IEEE*, France, July 5–8, 2000.

Dethloff, J., Vehicle routing and reverse logistics: The vehicle routing problem with simultaneous delivery and pick-up, *OR Spektrum*, 2001, 23, 79–96.

Dobos, I. and Richter, K., An extended production/recycling model with stationary demand and return rates, *International Journal of Production Economics*, 2004, 90(3), 311–323.

Doppelt, B. and Nelson, H., *Extended Producer Responsibility and Product Take-Back: Application for Pacific Northwest*, 2001, The Center for Watershed and Community Health, Portland State University, Portland, Oregon.

Environmental Protection Agency, Macroeconomic importance of recycling and remanufacturing, Office of Solid Waste, United States, October 28, 1998.

Ferrer, G., The economics of personal computer remanufacturing, *Resources, Conservation and Recycling*, 1997, 21, 79–108.

Ferrer, G. and Whybark, D.C., Material planning for remanufacturing facility, *Production and Operations Management*, 2001, 10(2), 112–124.

Ferrer, G. and Whybark, D.C., The economics of remanufacturing, In *Business Aspects of Closed-Loop Supply Chains*, Guide, V.D.R., Jr. and Wassenhove, L.N.V. (Eds.), 2003, Carnegie Bosch Institute, Carnegie Mellon University Press, Pittsburg, PA.

Fleischmann, M., Beullens, P., Bloemhof-Ruwaard, J.M. and Wassenhove, L.N.V., The impact of product recovery on logistics network design, *Production and Operations Management*, 2001, 10(2), 156–173.

Fleischmann, M., Bloemhof-Ruwaard, J.M., Dekker, R., Van der Laan, E., Van Nunen, J.A.E.E. and Wassenhove, L.N.V., Quantitative models for reverse logistics: A review, *European Journal of Operational Research*, 1997, 103, 1–17.

Fleischmann, M. and Roelof, K., On optimal inventory control with independent stochastic item returns, *European Journal of Operational Research*, 2003, 151, 25–27.

Georgiadis, P. and Dimitrion, V., The effect of environmental parameters on product recovery, *European Journal of Operational Research*, 2004, 157, 449–464.

Giuntini, R. and Gaudette, K., Remanufacturing: The next great opportunity for boosting US productivity, *Business Horizons* (Nov–Dec), 2003, 46(6), 41–48.

Guide, V.D.R., Jr. Production planning and control for remanufacturing: Industry practice and research needs, *Journal of Operations Management*, 2000, 18, 467–483.

Guide, V.D.R., Jr. Jayaraman, V. and Srivastava, R., The effect of lead time variation on the performance of disassembly release mechanisms, *Computers & Industrial Engineering*, 1999, 36(4), 759–779.

Guide, V.D.R., Jr. Jayaraman, V., Srivastava, R. and Benton, W.C., Supply chain management for recoverable manufacturing systems, *Interfaces*, 2000, 30(3), 125–142.

Guide, V.D.R., Jr. Kraus, M.E. and Srivastava, R., Scheduling policies for remanufacturing, *International Journal of Production Economics*, 1997b, 48, 187–204.

Guide, V.D.R., Jr. Souza, G.C. and Van der Laan, E., Performance of static priority rules for shared facilities in a remanufacturing shop with disassembly and reassembly, *European Journal of Operations Research*, 2005, 164, 341–353.

Guide, V.D.R., Jr. and Srivastava, R., Buffering from material recovery uncertainty in recoverable manufacturing environment, *Journal of Operational Research Society*, 1997a, 48, 519–529.

Guide, V.D.R., Jr. Srivastava, R. and Spencer, M.S., An evaluation of capacity planning technique in a remanufacturing environment, *International Journal of Production Research*, 1997c, 35(1), 67–82.

Guide, V.D.R., Jr. Teunter, R.H. and Wassenhove, L.N.V., Matching demand and supply to maximize profits from remanufacturing, Working Paper, Economic Institute, Rotterdam School of Economics, Rotterdam, 2003.

Guide, V.D.R., Jr. and Wassenhove, L.N.V., Managing product returns for remanufacturing, *Production and Operations Management*, 2001, 10(2), 142–155.

Gungor, A. and Gupta, S.M., Issues in environmentally conscious manufacturing and product recovery: A survey, *Computers & Industrial Engineering*, 1999, 36(4), 811–853.

Gupta, S.M. and Taleb, K.N., Scheduling disassembly, *International Journal of Production Research*, 1994, 32(8), 1857–1866.

Hammond, R., Amezquita, T. and Bras, B., Issues in the automotive parts remanufacturing industry—a discussion of results from surveys performed among remanufacturers, *International Journal of Engineering Design and Automation*, 1998, 4(1), 27–46.

Heisig, G. and Fleischmann, M., Planning stability in product recovery system, *OR Spektrum*, 2001, 23, 25–50.

Heese, H.S., Cattani, K., Ferrer, G., Gilland, W. and Roth, A.V., Competitive advantage through take-back of used products, *European Journal of Operations Research*, 2005, 164, 143–157.

Hesselbach, J. and Kuhn, M., Disassembly assessment and planning for electronic consumer appliances. Proceedings of the First International Working Seminar on Reuse, Eindhoven, The Netherlands, 11–13 Nov., 1996, pp. 136–169.

Hoshino, T., Yura, K. and Hitomi, K., Optimisation analysis for recycle-oriented manufacturing systems, *International Journal of Production Research*, 1995, 33(8), 2069–2078.

Inderfurth, K., Simple optimal replenishment and disposal policies for a product recovery system with leadtimes, *OR Spektrum*, 1997, 19, 111–122.

Inderfurth, K., Optimal policies in hybrid manufacturing/remanufacturing systems with product substitution, *International Journal of Production Economics*, 2004, 90(3), 325–343.

Inderfurth, K., Flapper, S.D.P., Lambert, A.J.D., Pappis, C.P. and Voutsinas, T.G., Production planning for product recovery management, In *Reverse Logistics: Quantitative Models for Closed-Loop Supply Chains*, Dekker, R., Fleischmann, M., Inderfurth, K. and Wassenhove, L.N.V. (Eds.), 2004, Springer, Berlin, Germany.

Ishii, K., Lee, B.H. and Eubanks, C.F., Design for product retirement and modularity based on technology life-cycle, *Manufacturing Science and Engineering*, 1995, 921–933.

Jayaraman, V., Guide, V.D.R. Jr. and Srivastava, R., A closed-loop logistics system for remanufacturing, *Journal of Operational Research Society*, 1999, 50, 497–508.

Jayaraman, V., Patterson, R.A. and Rolland, E., The design of reverse distribution networks: Models and solution procedures, *European Journal of Operational Research*, 2003, 150, 128–149.

Johnson, P.F., Managing value in reverse logistics systems, *Transportation Research Part E*, 1998, 34(3), 217–227.

Johnson, M.R. and Wang, M.H., Planning product disassembly for material recovery opportunities, *International Journal of Production Research*, 1995, 33(11), 3119–3142.

Kiesmuller, G.P. and Scherer, C.W., Computational issues in a stochastic finite horizon one product recovery inventory model, *European Journal of Operational Research*, 2003, 146, 553–579.

Klausner, M. and Hendrickson, C.T., Reverse logistics strategy for product take-back, *Interfaces*, 2000, 30(3), 156–165.

Kleber, R., Minner, S. and Kiesmuller, G., A continuous time inventory model for a product recovery system with multiple options, *International Journal of Production Economics*, 2002, 79, 121–141.

Kokkinaki, A., Zuidwijk, R., Nunen, J.V. and Dekker, R., ICT enabling reverse logistics, In *Reverse Logistics: Quantitative Models for Closed-Loop Supply Chains*, Dekker, R., Fleischmann, M., Inderfurth, K. and Wassenhove, L.N.V. (Eds.), 2004, Springer, Berlin, Germany.

Krikke, H.R., Harten van, A. and Schuur, P.C., Business case Roteb: Recovery strategies for monitors, *Computers & Industrial Engineering*, 1999, 36(4), 739–757.

Kriwet, Z.E. and Seliger, G., Systematic integration of design-for recycling into product design, *International Journal of Production Economics*, 1995, 38, 15–22.

Krumwiede, D.W. and Sheu, C., A model for reverse logistics entry by third-party providers, *Omega*, 2002, 30, 325–333.

Lambert, A.J.D., Determining optimal disassembly sequences in electronic equipment, *Computers and Industrial Engineering*, 2002, 43, 553–575.

Listes, O. and Dekker, R., A stochastic approach to a case study for product recovery network design, *European Journal of Operational Research*, 2005, 160, 268–287.

Lund, R., Remanufacturing: An American resource, Proceedings of the Fifth International Congress Environmentally Conscious Design and Manufacturing, June 16 and 17, 1998, Rochester Institute of Technology, Rochester, NY.

Mahadevan, B., Pyke, D.F. and Fleischmann, M., Periodic review, push inventory policies for remanufacturing, *European Journal of Operational Research*, 2003, 151, 536–551.

Majumder, P. and Groenevelt, H., Competition in remanufacturing, *Production and Operations Management*, 2001, 10(2), 125–141.

Maslennikova, I. and Foley, D., Xerox's approach to sustainability, *Interfaces*, 2000, 30(3), 226–233.

Minner, S., Strategic safety stocks in reverse logistics supply chains, *International Journal of Production Economics*, 2001, 71, 417–428.

Minner, S. and Linder, G., Lot sizing decisions in product recovery management, In *Reverse Logistics: Quantitative Models for Closed-Loop Supply Chains*, Dekker, R., Fleischmann, M., Inderfurth, K. and Wassenhove, L.N.V. (Eds.), 2004, Springer, Berlin, Germany.

Mondal, S. and Mukherjee, K., Buy-back policy decision in managing reverse logistics, *International Journal of Logistics Systems and Management*, 2006a, 2(3), 255–264.

Mondal, S. and Mukherjee, K., An empirical investigation on feasibility of remanufacturing activities in Indian economy, *International Journal of Business Environment*, 2006b, 1(1), 70–88.

Mostard, J. and Teunter, R.H., The newsboy problem with resalable returns: A single period model and case study, *European Journal of Operations Research*, 2006, 169(1), 81–96.

Mukherjee, K. and Mondal, S., Analysis of issues relating to remanufacturing technology—a case of an Indian company, Proceedings of the Second European Conference on management of Technology, Aston Business School, Birmingham, UK, 2006, 510–518.

Nagel, C. and Meyer, P., Caught between ecology and economy: End-of-life aspects of environmentally conscious manufacturing, *Computers and Industrial Engineering*, 1999, 36(4), 781–792.

Nagurney, A. and Toyasaki, F., Reverse supply chain management and electronic waste recycling: A multitiered network equilibrium framework for e-cycling, *Transportation Research Part E*, 2005, 41, 1–28.

Nasr, N., Hughson, C., Varel, E. and Bauer, R., *State-of-the-Art Assessment of Remanufacturing Technology*, 1998, Rochester Institute of Technology, National Center for Remanufacturing, Rochester, NY.

Rao, P., Greening the supply chain: a new initiative in South East Asia, *International Journal of Operations & Production Management*, 2002, 22(6), 632–655.

Richter, K. and Sombrutzki, M., Remanufacturing planning for the reverse Wagner/ Whitin models, *European Journal of Operations Research*, 2000, 121, 304–315.

Sarkis, J., Evaluating environmentally conscious business practices, *European Journal of Operational Research*, 1998, 107, 159–174.

Sarkis, J.A., Methodological framework for evaluating environmentally conscious manufacturing programs, *Computers & Industrial Engineering*, 1999, 36(4), 793–810.

Savaskan, R.C., Bhattacharya, S. and Wassenhove, L.N.V., Closed-loop supply chain models with product remanufacturing, *Management Science*, 2004, 50(2), 239–252.

Seitz, M.A. and Peatlie, K., Meeting the closed-loop challenge: The case of remanufacturing, *California Management Review*, 2004, 46(2), 74–89.

Shiu, Li-Hsing, Reverse logistics system planning for recycling electrical appliances and computers in Taiwan, *Resources, Conservation and Recycling*, 2001, 32, 55–72.

Spicer, A.J. and Johnson, M.R., Third party demanufacturing as a solution for extended producer responsibility, *Journal of Cleaner Production*, 2004, 11, 445–458.

Stuart, J.A., Ammons, J.C. and Turbini, L.J., A product and process selection model with multidisciplinary environmental considerations, *Operations Research*, 1999, 47(2), 221–234.

Teunter, R.H., Economic ordering quantities for recoverable item inventory systems, *Naval Research Logistics*, 2001, 48, 484–495.

Teunter, R.H. and Van der Laan, E., On the non-optimality of the average cost approach for inventory models with remanufacturing, *International Journal of Production Economics*, 2002, 79, 67–73.

Teunter, R.H., Van der Laan, E. and Inderfurth, K., How to set the holding cost rates in average cost inventory models with reverse logistics? *Omega*, 2000, 28, 409–415.

Thierry, M.C., Salomon, M., Van Nunen, J.A.E.E. and Wassenhove, L.N.V., Strategic production and operations management issues in product recovery management, *California Management Review*, 1995, 37(2), 114–135.

Tung-Lai, Hu, Sheu Jiuh-Bung and Huang, K., A reverse logistics cost minimization model for the treatment of hazardous wastes, *Transportation Research Part E*, 2002, 38, 457–473.

Van der Laan, E., Salomon, M., Dekker, R. and Wassenhove, L.N.V., Inventory control in hybrid systems with remanufacturing, *Management Science*, 1999, 45(5), 733–747.

Warfield, J.N., Developing interconnection matrices in structural modeling, *IEEE Transactions on Systems, Man and Cybernetics*, 1974, 30(7/8), 710–716.

White, C.D., Masanet, E., Rosen, C.M. and Beckman, S.L., Product recovery with some byte: An overview of management challenges and environmental consequences in reverse remanufacturing for computer industry, *Journal of Cleaner Production*, 2003, 11, 445–448.

13

Optimal Control Policy for Environment-Conscious Manufacturing Systems

Kenichi Nakashima

CONTENTS

13.1 Introduction

The continuous growth in consumer waste in recent years has posed a serious threat to the environment. According to the U.S. Environmental Protection Agency (EPA), in 1990, the amount of waste generated in the United States reached a whopping 196 million tons, up from 88 million tons in the 1960s (Nasr 1997). Wann (1996) reports that on an average, an American consumes 20 tons of materials every year. Environment-conscious manufacturing and product recovery (ECMPRO) is primarily driven by the escalating deterioration of the environment. ECMPRO has become an obligation toward the environment and the society. Therefore, many countries are contemplating regulations that would force manufacturers to take back the used products from consumers so that components and materials retrieved from those products may be reused or recycled. For example, Germany has passed a regulation that requires companies to remanufacture products until the product is obsolete. Japan has passed a similar legislation requiring design and assembly methodologies that facilitate recycling of durable goods (Guide et al. 1996). The United States is also contemplating regulations on remanufacturing. Within the next few years, the United States is expected to legislate two acts in this regard—the Automotive Waste Management Act (which will enforce the complete reclamation of automobiles) and the Polymers and Plastic Control Act (which will enforce the complete reclamation of polymers and plastics) (Gungor and Gupta 1999).

Disassembly is the first step to retrieve components and materials (for reuse, recycling, and remanufacturing) from consumer products. Disassembly is the process of systematic removal of desirable constituents from the original assembly so that there is no impairment to any useful component (Gupta and Taleb 1994). Disassembly can be partial (product not fully disassembled) or complete (product fully disassembled) and may use a destructive (focusing on materials recovery) or a nondestructive (focusing on components recovery) methodology. In this chapter, we will only discuss complete and nondestructive disassembly in the remanufacturing system.

We focus on the operational aspect of product recovery in the remanufacturing environment. Fleischmann et al. (1997) define remanufacturing as a process of bringing used products back to "as new" condition by performing the necessary operations such as disassembly, overhaul, and replacement. Remanufacturing is also referred to as recycling-integrated manufacturing (Hoshino et al. 1995). As in conventional production systems, in remanufacturing systems also, decisions related to operations, manufacturing, inventory, distribution, and marketing need to be made (Kopicky et al. 1993; Stock 1992). In general, the existing methods for conventional production systems cannot be used for the remanufacturing systems. Remanufacturing environments are characterized by their highly flexible structures. Flexibility is required to handle the uncertainties involved in remanufacturing. In addition, we should recognize that products used by customers would be used to produce new products after collection and recovery.

With regard to periodic review models, Cohen et al. (1980) developed the product recovery model in which collected products are used directly. In a different model, Inderfurth (1997) discussed the effect of nonzero lead times for orders and recovery. In continuous review models, Muckstadt and Isaac (1981) dealt with a model for a remanufacturing system with nonzero lead times and a control policy with the traditional (Q, r) rule. Van der Laan and Salomon (1997) suggested push-and-pull strategies for the remanufacturing system. These authors, however, consider that demand and procurement are independent of each other in the inventory systems.

Product recovery is aimed at minimizing the amount of waste sent to landfills by recovering materials and parts from old or outdated products. Recovery is achieved through recycling and remanufacturing. According to Gungor and Gupta (1999), product recovery should be considered in designing and managing manufacturing systems. In this case, product recovery includes collection, disassembly, cleaning, sorting, repairing bad components, reconditioning, reassembling, and testing. Recovered parts or products are used in repair and in remanufacturing of other products and components. We focus on a product recovery system in a remanufacturing system under stochastic variability.

This chapter deals with new analytical approaches toward evaluating and optimizing environment-conscious manufacturing systems with stochastic variability such as customer demand, recovery rate, and disposal rate using Markov model analysis. In the proposed models, we consider two types of inventories—the actual product inventory in a factory and the virtual inventory that is used by the customer. We apply the traditional inventory theory to this model with consideration for disposal and return. In Section 13.2, the system is formulated into a general model using a discrete-time Markov chain (MC). This model is composed of states that are denoted by the number of the inventory, the transition probabilities between states, and the costs associated with the transitions. Using Markov analysis, we can calculate the total expected average cost per period exactly and show some properties of the system. We also consider some scenarios in the system and discuss the efficient management of the system. In Section 13.3, we optimize a single-item remanufacturing system with consideration for product lifetime under stochastic demand. The system is formulated into an undiscounted Markov decision process (UMDP) to determine the optimal control policy that minimizes the expected average cost per period. We also show the numerical results of controlling the remanufacturing system under various conditions. This section also deals with the cost management problem of a remanufacturing system using design of experiments under stochastic variability.

13.2 Performance Evaluation of Environment-Conscious Manufacturing

13.2.1 Discrete-Time Markov Chain

A discrete-time MC is based on periodical observation of the system, which is recognized as the state at each epoch. It has a Markov property in that the previous state is independent of the future state or the present state. Therefore, information about the present state of the system is enough for future analysis, and we do not need the history of the system. The MC models can be applied to various problems in economics, linguistics, and so on. Each model consists of states, transition probability, and reward or cost associated with the transition. The state space S is a set of all the states of the system. Let us consider a state $i(i = 1, 2, ..., N)$ at the beginning of period n. The state i moves to state j by next period with transition probability p_{ij}. We assume here that the probability is stationary. When the state transits from i to j, it receives a reward r_{ij}. We can calculate the steady-state distribution of the system by solving the linear equation with the same number of the dimensions as that of the states. This distribution leads to the expected average reward or cost per period. The approach is used for controlling the production system in engineering (Deleersnyder et al. 1989).

13.2.2 Model Description

Using a discrete-time Markov model we formulate a product recovery system with stochastic variability such as demand (Nakashima et al. 2002). Let us consider a single process that produces a single-item product. The finished products are stocked in the factory and phased out according to customer demand. Traditional inventory management focuses only on the inventory in the factory. In the remanufacturing system, however, we should focus on the outdated products that are collected from customers. Specifically, a producer involved in remanufacturing has to consider the products in use as a part of future inventory. Here, we consider the products used by customers as virtual inventory. Managing the virtual inventory until the products are collected and used in the remanufacturing process is as important as controlling the inventory at hand.

Figure 13.1 shows the product recovery system in the remanufacturing environment. Remanufacturing preserves the identity of the product or its part and performs the required disassembly and refurbishing operations to bring the product to a desired level of quality with remanufacturing cost. In contrast, in normal manufacturing, new resources are used to manufacture products. The number of products by normal manufacturing at period t is

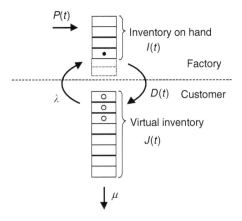

FIGURE 13.1
A product recovery system.

defined by $P(t)$. Products are manufactured through normal manufacturing or through remanufacturing with parts taken back from customers. Both forms of production begin at the start of the period t, and all products are completed by the end of this period. At the end, customers buy new products. We assume that the number of finished products and the number of products bought by customers are $I(t)$ and $J(t)$, respectively. If backlog occurs, $I(t)$ is negative. Demand in successive periods $D(t)$ constitutes the independent random variables with known identical distributions and densities. When sold, products are recovered at the recovery rate λ and products in use are discarded at the discarded rate μ with out-of-date cost. The out-of-date cost is assumed as $\lambda + \mu \leq 1$.

13.2.3 Formulation

The state of the system is denoted by

$$s(t) = (I(t), J(t)) \tag{13.1}$$

The transition of each inventory is given by

$$I(t + 1) = I(t) + P(t) + \lambda J(t) - D(t) \tag{13.2}$$

and

$$J(t + 1) = J(t) - \lambda J(t) - \mu J(t) + D(t) \tag{13.3}$$

It is assumed that $P(t) = \max\{0, I_{max} - I(t) - \lambda J(t)\}$.

The transition probability is defined as

$$P_{s(t)s(t+1)} = \begin{cases} \Pr\{D(t) = d\} & \text{if } S(t + 1) = (I(t) + P(t) + \lambda J(t) - d, \\ & \quad J(t) - \lambda J(t) - \mu J(t) + d) \\ 0 & \text{otherwise} \end{cases} \tag{13.4}$$

Total cost per period, $Q(t)$ is given by

$$Q(t) = cP(t) + \theta \lambda J(t) + h[I(t)]^+ + b[-I(t)]^+ + \delta \mu J(t) \tag{13.5}$$

where the parameters are as follows:

c = Normal manufacturing cost of a new product
θ = Remanufacturing cost of a product
h = Holding cost per unit
b = Backlog cost per unit
δ = Out-of-date cost per unit

We can calculate the stationary distribution of the system to solve a linear equation of the steady-state distribution. In this system, we obtain the total expected average cost per period from Equation 13.5.

13.2.4 Numerical Results

In this section, we investigate the properties of the product recovery system numerically. We assume $\Pr\{D(t) = 2\} = \Pr\{D(t) = 3\} = 0.5$ and $I_{max} = 10$. The cost parameters are set as follows:

$$c = 1, \quad h = 1, \quad b = 10, \quad \theta = 3, \quad \text{and} \quad \delta = 10$$

13.2.4.1 *Extreme Cases of the Systems*

We consider two extreme cases of systems. In the first case, the flow of the products is one way (Figure 13.2), and in the second case the product is circulated without disposal (Figure 13.3).

Case 1 in Figure 13.2 shows the traditional manufacturing system that has no recycling or remanufacturing process. This system can be viewed as a

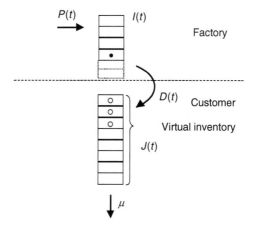

FIGURE 13.2
One-way flow model (case 1: $\lambda = 0, \mu = 1$).

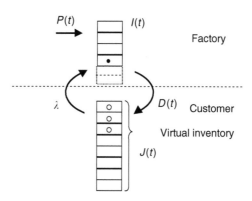

FIGURE 13.3
Circulated flow model without disposal (case 2: $\lambda = 1, \mu = 0$).

TABLE 13.1

The Expected Average Cost per Period in Each Case

	$\lambda = 0$	$\lambda = 1$
$\mu = 0$	—	17.5
$\mu = 1$	31.0	—

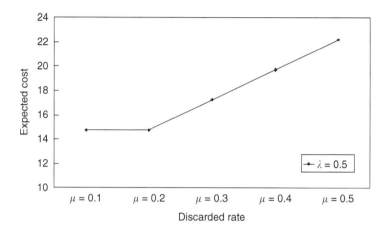

FIGURE 13.4
Effect of discarded rate.

typical model of the mass-consumed manufacturing system in the previous period. In contrast, case 2 in Figure 13.3 is the extremely advanced product recovery system in the remanufacturing environment. Table 13.1 lists out the expected average cost per period in each case.

We find that the advanced product recovery system performs better than the one-way model, which stands for the traditional manufacturing system without product return.

13.2.4.2 Optimal Policy

We study the effect of changing the recovery and discarded rates of the system on the total expected average cost and determine the optimal rates that minimize the expected average cost.

Figure 13.4 shows the behavior of the expected average cost per period with $\lambda = 0.5$, when discarded rate μ is varied from 0.1 to 0.5. As the discarded rate increases, the expected average cost also increases. This shows the effect of eliminating disposal in the case in which the proportion of the recovered products is constant in the remanufacturing environment.

Figure 13.5 shows the behavior of the expected average cost per period with $\mu = 0.5$ when recovery rate λ is varied. We can find that the optimal recovery rate is 0.2 under constant discarded rate. This value means that

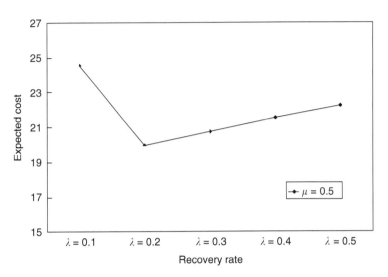

FIGURE 13.5
Effect of recovery rate.

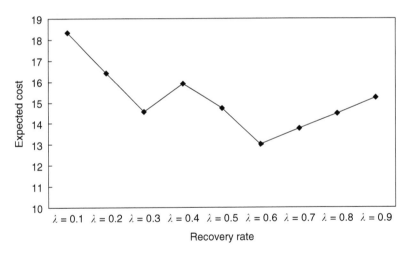

FIGURE 13.6
Behavior of minimum cost.

increasing the collected products for recovery is not optimal, and we should choose the optimal recovery rate carefully.

Figure 13.6 illustrates the behavior of the minimum expected average cost at each recovery rate. Each minimum expected average cost is obtained when the discarded rate is 0.1 at each recovery rate. We find that the optimal recovery rate is 0.6, and the optimal discarded rate is 0.1. This means that we can optimize the product recovery rate under the estimated discarded rate in the product recovery system.

13.2.5   Concluding Remarks

First, we proposed a new analytical approach to evaluate the product recovery system with stochastic variability, such as demand, in the remanufacturing environment. The management system was extended from traditional inventory management and formulated by a discrete-time MC model. Then, we calculated the exact total expected average cost per period. The product recovery system performed better than the traditional manufacturing system, which involved no product return. The properties of choosing the recovery rate and the discarded rate are illustrated. Finally, we determined the optimal recovery and discarded rates that minimize the total expected average cost per period.

13.3   Optimal Control of Environment-Conscious Manufacturing

13.3.1   Markov Decision Process

UMDP is a class of stochastic sequential processes in which the reward and the transition probability depend only on the current state and action of the system (Puterman 1994). A decision maker takes an action in each state. A policy is a set of actions, which the decision maker chooses for each state in advance. In the infinite time horizon problem, optimal policy exists in the stationary policy. The Markov decision process (MDP) models have gained recognition in diverse fields such as economics, communications, and transportation. Each model consists of states, actions, rewards, and transition probability. For example, in the field of engineering, the approach is used for controlling the production system (Ohno and Nakashima 1995). Choosing an action as a production quantity in a state generates rewards or costs and determines the state at the next decision epoch through a transition probability function. Then, we can obtain the optimal production policy, which minimizes the expected average cost per period in the optimal production control problem using policy iteration method. The policy iteration method is composed of policy improvement routine and value determination routine in which we have to solve the linear equation with the same number of dimensions as that of the states (Howard 1960).

13.3.2   UMDP Model with Product Life Cycle

Let us consider a single process that produces a single-item product. The finished products are stocked in the factory and phased out according to customer demand (Nakashima et al. 2004). Each product has its own remaining lifetime denoted by $i(i = M - 1, ..., 1)$ after it is sold. The remaining life time decreases one by one periodically. Traditional inventory management focuses only on the inventory in the factory. In the remanufacturing system, however, we should focus on the outdated products that are collected from the customers. Specifically, producers engaged in remanufacturing have to consider products in use as a part of future inventory. Here, products used by customers are regarded as virtual inventory. Managing the virtual inventory until products are collected and used in the remanufacturing process is as important as controlling the inventory at hand.

Remanufacturing preserves the identity of the product its or part and performs the required disassembly and refurbishing operations to bring the product to a desired level of quality with remanufacturing cost. In contrast, normal manufacturing uses new resources for production. The number of products by normal manufacturing at period t, $P(t)$, is chosen as an action k, that is, $k = P(t)$. Products are manufactured through normal manufacturing or through remanufacturing with parts taken back from customers. Both forms of production begin at the start of the period t, and all products are completed by the end of this period. At the end, customers buy new products. We assume that the number of finished products and the number of products bought by customers are $I(t)$ and $J_i(t)$ ($i = M - 1, ..., 1$), respectively. If backlog occurs, $I(t)$ is negative. Demand in successive periods $D(t)$ constitutes independent random variables with known identical distribution. When sold, products are remanufactured at the remanufacturing rate, λ_i ($i = M - 1, ..., 1$) with remanufacturing cost, and products in use are discarded at the discarded rate μ with out-of-date cost. It is supposed that $\lambda_i < 1$ and $\lambda_1 + \mu \leq 1$.

The state of the system is denoted by

$$s(t) = (I(t), J_{M-1}(t), ..., J_1(t)) \tag{13.6}$$

Figure 13.7 shows a remanufacturing system with a single product life cycle.

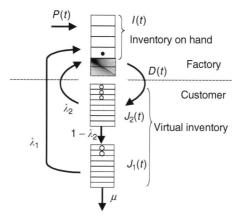

FIGURE 13.7
Product life cycle time model ($M = 2$).

The transition of the each inventory is given by

$$I(t + 1) = I(t) + k + \sum_{i=1}^{M-1} \lambda_i J_i(t) - D(t) \tag{13.7}$$

$$J_{M-1}(t + 1) = \min\{[I(t)]^+ + P(t) + \sum_{i=1}^{M-1} \lambda_i J_i(t), D(t) + [I(t)]^+\} \tag{13.8}$$

$$J_i(t + 1) = (1 - \lambda_{i+1}) J_{i+1}(t) \quad \text{for } i = M - 1, \ldots, 2 \tag{13.9}$$

and

$$J_1(t + 1) = (1 - \lambda_1 - \mu) J_1(t) + (1 - \lambda_2) J_2(t) \tag{13.10}$$

where $[x]^+ = \min\{0, x\}$. The action space in $s(t)$, $K(s(t))$, is defined by

$$K(s(t)) = \{0, \ldots, \max\{0, I_{\max} - I(t) - \sum_{i=1}^{M-1} \lambda_i J_i(t)\}\} \tag{13.11}$$

The transition probability of the system is given by

$$P_{s(t)s(t+1)} = \begin{cases} \Pr\{D(t) = d\} & \text{if } s(t + 1) = (I(t) + k + \sum_{i=1}^{M-1} \lambda_i J_i(t) - \\ & d, \ldots, (1 - \lambda_1 - \mu) J_1(t) + (1 - \lambda_2) J_2(t) \\ 0 & \text{otherwise} \end{cases} \tag{13.12}$$

The expected cost per period in state $s(t)$ under k, $r(s(t), k)$ is given by

$$r(s(t), k) = C_N k + \sum_{i=1}^{M-1} C_{Ri} \lambda_i J_i(t) + C_H[I(t)]^+ + C_B[-I(t)]^+ + C_O \mu J_1(t) \tag{13.13}$$

where the parameters are as follows:

C_N = Normal manufacturing cost of a new product
C_{Ri} = Remanufacturing cost of a product
C_H = Holding cost per unit
C_B = Backlog cost per unit
C_O = Out-of-date cost per unit

Let S and $|S|$ denote a set of all possible states and the total number of the states, respectively. The state s_n is numbered by s ($= 1 \ldots |S|$). An UMDP that minimizes the expected average cost per period, g, is formulated as shown in the following optimality equation:

$$g + v_s = \min_{k \in K(s)} \left\{ r(s, k) + \sum_{s' \in S} P_{ss'}(k) v_{s'} \right\} (s \in S) \tag{13.14}$$

where v_s denotes the relative value when the production system starts from state s (Howard 1960). An optimal production policy is determined as a set of k that minimizes the right-hand side of Equation 13.14 for each state s using policy iteration method (Howard 1960, Puterman 1994).

13.3.3 Computational Results

In this subsection, we show the numerical examples of controlling the reman-ufacturing system described in the previous subsection. The distribution of the demand is given by

$$\Pr\left[D_n = D - \frac{1}{2}Q + j\right] = \binom{Q}{j}\left(\frac{1}{2}\right)^Q, \quad (0 \le j \le Q) \tag{13.15}$$

where $D = 2$ and Q is an even number and variance (σ^2) is $Q/4$.

13.3.3.1 Optimal Policy

First, we show the optimal policy for the system with $M = 2$. The maximum number of parts inventory I_{max} is 10. The maximum number of backlog demand I_{min} is set as -5. The cost parameters are set as follows:

$$C_H = 1, \quad C_N = 2, \quad C_{R1} = 3, \quad C_B = 10, \quad \text{and} \quad C_O = 10$$

We can obtain an optimal control policy that minimizes the expected aver-age cost per period. It is assumed that the remanufacturing rate λ_1 is 0.2 and the discarded rate μ is 0.5. Table 13.2 shows the optimal control policy in case variance as 0.5. We find that the optimal number of normal production is restricted to the number of remanufacturing production.

We can find the optimal policy as a set of the numbers of production quan-tity in each state.

13.3.3.2 Behavior of the Minimum Cost

Figure 13.8 shows the behavior of the minimum cost under two types of variance. We find that as the remanufacturing rate increases, the minimum cost decreases in each case. It also illustrates the importance of production smoothing.

Next, we investigate the behavior of expected costs of the system with $M = 2$. It is assumed that $C_H = 1$, $C_N = 1$, $C_B = 10$, $C_O = 10$, $C_{R1} = 2$, and $\sigma^2 = 0.5$ and the discarded rate μ is 0.2. Figure 13.9 illustrates the behavior of the expected average cost under varying remanufacturing rates. As the rate increases, the expected cost tends to decrease.

13.3.4 Analysis of ECM Factors

In this subsection, we present the numerical results using design of experi-ments (Macclave et al. 1998) and investigate the effects of some of the factors on the system by analysis of variance. The distribution of the demand is

TABLE 13.2

Optimal Control Policy

State (I_2, I_1)	Action (k)	State (I_2, I_1)	Action (k)	State (I_2, I_1)	Action (k)	State (I_2, I_1)	Action (k)	State (I_2, I_1)	Action (k)	State (I_2, I_1)	Action (k)	State (I_2, I_1)	Action (k)
(−5,0)	7	(−4,7)	2	(−2,3)	3	(−1,10)	3	(1,6)	0	(3,2)	0	(4,9)	0
(−5,1)	6	(−4,8)	2	(−2,4)	3	(0,0)	3	(1,7)	0	(3,3)	0	(4,10)	0
(−5,2)	6	(−4,9)	1	(−2,5)	2	(0,1)	2	(1,8)	0	(3,4)	0	(5,0)	0
(−5,3)	5	(−4,10)	1	(−2,6)	1	(0,2)	2	(1,9)	0	(3,5)	0	(5,1)	0
(−5,4)	5	(−3,0)	6	(−2,7)	0	(0,3)	1	(1,10)	0	(3,6)	0	(5,2)	0
(−5,5)	4	(−3,1)	5	(−2,8)	0	(0,4)	1	(2,0)	1	(3,7)	1	(5,3)	0
(−5,6)	4	(−3,2)	4	(−2,9)	0	(0,5)	0	(2,1)	0	(3,8)	0	(5,4)	0
(−5,7)	3	(−3,3)	4	(−2,10)	0	(0,6)	0	(2,2)	0	(3,9)	0	(5,5)	0
(−5,8)	3	(−3,4)	3	(−1,0)	4	(0,7)	0	(2,3)	0	(3,10)	0	(5,6)	0
(−5,9)	2	(−3,5)	2	(−1,1)	3	(0,8)	0	(2,4)	0	(4,0)	0	(5,7)	0
(−5,10)	2	(−3,6)	2	(−1,2)	3	(0,9)	0	(2,5)	0	(4,1)	0	(5,8)	0
(−4,0)	6	(−3,7)	1	(−1,3)	2	(0,10)	0	(2,6)	0	(4,2)	0	(5,9)	0
(−4,1)	5	(−3,8)	1	(−1,4)	2	(1,0)	2	(2,7)	0	(4,3)	0	(5,10)	0
(−4,2)	5	(−3,9)	0	(−1,5)	1	(1,1)	1	(2,8)	0	(4,4)	0		
(−4,3)	4	(−3,10)	0	(−1,6)	1	(1,2)	1	(2,9)	0	(4,5)	0		
(−4,4)	4	(−2,0)	5	(−1,7)	0	(1,3)	0	(2,10)	0	(4,6)	0		
(−4,5)	3	(−2,1)	4	(−1,8)	0	(1,4)	0	(3,0)	0	(4,7)	0		
(−4,6)	3	(−2,2)	4	(−1,9)	0	(1,5)	0	(3,1)	0	(4,8)	0		

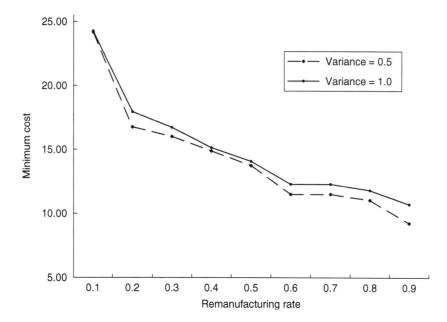

FIGURE 13.8
The behavior of minimum cost.

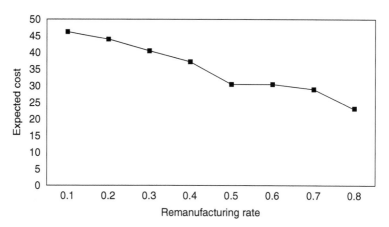

FIGURE 13.9
Expected average cost.

given by the same binomial distribution in the previous section with $D = 2$. The maximum numbers of actual inventory and virtual inventory are 5 and 10, respectively. The maximum number of backlog demand is set as -5. We assume that the remanufacturing rate is 0.6 and the discarded rate is 0.2.

We use $L_8(2^7)$ orthogonal array and take the five cost parameters defined in Equation 13.13 as the factors and the variance of demand. Each factor has two levels shown in Table 13.3.

TABLE 13.3

Factors and Levels

$A(C_N)$: $A_1=1$, $A_2=5$	$D(C_O)$: $D_1=10$, $D_2=30$
$B(C_H)$: $B_1=1$, $B_2=2$	$F(C_{R1})$: $F_1=2$, $F_2=8$
$C(C_B)$: $C_1=10$, $C_2=30$	$G(\sigma^2)$: $G_1=0.5$, $G_2=1.0$

TABLE 13.4

$L_8(2^7)$ Orthogonal Array and Numerical Results

Number	A	B	D	C	F	G	E	Total Cost
1	1	1	1	1	1	1	1	9.500
2	1	1	1	2	2	2	2	19.545
3	1	2	2	1	1	2	2	21.583
4	1	2	2	2	2	1	1	29.500
5	2	1	2	1	2	1	2	30.500
6	2	1	2	2	1	2	1	22.545
7	2	2	1	1	2	2	1	22.397
8	2	2	1	2	1	1	2	12.500

TABLE 13.5

Analysis of Variance

Factor	S	ϕ	V	F_0
$A(C_N)$	7.632	1	7.632	1764.903*
$B(C_H)$	1.892	1	1.892	437.394*
$C(C_B)$	0.002	1	864	0.350
$D(C_O)$	201.864	1	201	46,679.229**
$F(C_R)$	160.330	1	0.002	37,074.881**
$G(\sigma^2)$	2.071	1	160.330	478.810*
E	0.004	1	2.071	
Sum	373.795	7	0.004	

Note: The symbols * and ** mean significance if the significance level, α, is 0.05 and 0.01, respectively.

Table 13.4 shows the assignment and the results of the experiments.

Table 13.5 shows the result of performing analysis of variance on the data in Table 13.4. We identify the effects of normal manufacturing cost, holding cost, remanufacturing cost, out-of-date cost, and demand variance.

13.3.5 Concluding Remarks

We modeled the remanufacturing system as the UMDP with consideration for product lifetime. We optimized the remanufacturing system to obtain the optimal control policy that minimizes the expected cost per period. We also considered some scenarios under various conditions and discussed the effects of these factors on the remanufacturing system. The numerical investigation indicated remanufacturing management.

We also discussed cost management problem of the remanufacturing system with stochastic variability. Using design of experiments we considered some scenarios under various conditions and discussed the effects of cost factors on the remanufacturing system. The numerical investigation indicated the important factors in remanufacturing management.

References

Cohen M. A., Nahmias S., and Pierskalla W. P., 1980, A dynamic inventory system with recycling. *Naval Research Logistics Quarterly*, 27, 289–296.

Deleersnyder J. L., Hodgson T. J., Muller (-Malek) H., and O'Grady P. J., 1989, Kanban controlled pull systems: an analytic approach. *Management Science*, 35, 1079–1091.

Fleischmann M., Boemhof-Ruwaard J. M., Dekker R., Van Der Laan E., Van Nunen Jaee, and Van Wassenhove L. N., 1997, Quantitative models for reverse logistics: a review. *European Journal of Operational Research*, 103, 1–17.

Guide V. D. R., Srivastava R. M., and Spencer S., 1996, Are production systems ready for the green revolution? *Production and Inventory Management Journal*, 37, 70–76.

Gungor A. and Gupta S. M., 1999, Issues in environmentally conscious manufacturing and product recovery: a survey. *Computers and Industrial Engineering*, 36, 811–853.

Gupta S. M. and Taleb K. N., 1994, Scheduling disassembly. *International Journal of Production Research*, 32, 1857–1866.

Hoshino T., Yura K., and Hitomi K., 1995, Optimization analysis for recycle-oriented manufacturing systems. *International Journal of Production Research*, 33, 2069–2078.

Howard R. A., 1960, *Dynamic Programming and Markov Processes*. Cambridge, MA: MIT Press.

Inderfurth K., 1997, Simple optimal replenishment and disposal policies for a product recovery system with lead-times. *OR Spektrum*, 19, 111–122.

Kopicky R. J., Berg M. J., Legg L., Dasappa V., and Maggioni C., 1993, *Reuse and Recycling: Reverse Logistics Opportunities*. Oak Brook, IL: Council of Logistics Management.

Macclave J. T., Benson P. G., and Sincich T., 1998, *Statistics for Business and Economics*. Upper Saddle River, NJ: Prentice-Hall.

Muckstadt J. A. and Isaac M. H., 1981, An analysis of single item inventory systems with returns. *Naval Research Logistics Quarterly*, 28, 237–254.

Nakashima K., Arimitsu H., Nose T., and Kuriyama S., 2002, Analysis of a product recovery system. *International Journal of Production Research*, 40, 3849–3856.

Nakashima K., Arimitsu H., Nose T., and Kuriyama S., 2004, Cost analysis of a remanufacturing system. *Asia Pacific Management Review*, 9, 595–602.

Nasr N., 1997, Environmentally conscious manufacturing. *Careers and Engineer*, 26–27.

Ohno K. and Nakashima K., 1995, Optimality of a just-in-time production system. Singapore (World Scientific): Proceedings of APORS'94 (Selected Paper), pp. 390–398.

Puterman M. L., 1994. Discrete Stochastic Dynamic Programming. *Markov Decision Processes*. New York: Wiley.

Stock J. R., 1992, *Reverse Logistics*. Oak Brook, IL: Council of Logistics Management.

Van der Laan E. A. and Salomon M., 1997, Production planning and inventory control with remanufacturing and disposal. *European Journal of Operational Research*, 102, 264–278.

Wann D., 1996, *Deep Design: Pathways to a Livable Future*. Washington, D.C.: Island Press.

14

Disassembly and Reverse Logistics: The Case of the Computer Industry

K. Kathy Dhanda and Adrian Peters

CONTENTS

14.1   Introduction

The past couple of decades have witnessed a technological revolution. Dependence on electronic products has increased significantly, and most of the consumers in the developed economies own a computer and upgrade these systems every few years.

Indeed, electronic products are a fast growing portion of trash in the United States with 250 million computers becoming obsolete by 2005. In 2001, only 11% of the retired computers were recycled [1]. The average life span of a computer has shrunk from four or five years to about two years [2]. Americans are buying more computers than the people of any other nation, and, at present, more than half of American households have computers [3]. More than 50% of the computers turned in for recycling are in good working condition but are turned in to make way for the latest technology. One computer will become obsolete for every new one put on the market by the year 2005 [4]. In 2002, 63.3 million computer desktops will be taken out of service and 85% of these computers will end up in landfills [5].

Moore's law states that new microprocessors double in power every 18 months. One fortunate corollary is that computers are manufactured to be faster and cheaper as time goes along. However, the negative aspect of this law is that because of rapid innovations there is a glut of old machines that affluent societies do not know what to do with [6].

Obsolete computers and televisions generate millions of pounds of e-waste in the United States, and about 50–80% of the e-waste collected for recycling is exported. This export is propelled by the lack of environmental standards and the prevalence of cheap labor in Asia. Examples of e-waste recycling and disposal operations found in China, Pakistan, and India include open burning of plastic waste, dumping of acids in rivers, and exposure to toxic solders. The United States facilitates the export of toxic e-waste to developing countries rather than banning these exports under the principles of environmental justice. Although China has banned the import of e-waste, this ban has not been honored by the United States. With the electronics industry and the U.S. government not taking any initiative, consumers and local recyclers are left with very few sustainable options for e-waste, other than the trade of e-waste to lesser-developed countries [7].

14.2 E-Waste

E-waste or electronic waste consists of materials that are no longer usable or are computer scrap. E-waste comes from a range of electronic products, from computers to household appliances and cell phones. Large amounts of e-wastes end up in landfills in the United States or are exported to the developing countries. The main toxic agent is lead, which can leach into groundwater. The U.S. federal government has not enacted any laws banning e-waste from landfills, although a few states have imposed bans on cathode ray tube (CRT) monitors and televisions. On an average, there is 4–6 lbs of lead in CRT monitors. Circuit boards contain lead and other heavy metals such as arsenic. Half of the computer is plastic, and a large portion of this plastic is treated with toxic brominated flame-retardant chemicals in a process called doping [8].

E-waste is hazardous and contains numerous toxic substances such as lead and cadmium in the circuit boards, lead oxide and cadmium in the monitor CRTs, mercury in switches and flat screen monitors, and cadmium in computer batteries. The electronics industry is the fastest growing manufacturing industry and owing to the high rate of obsolescence, discarded electronic products or e-waste is the fastest growing waste stream in the developed countries [7].

14.2.1 How Much E-Waste Is Out There?

Twenty million computers became obsolete in the United States in 1998, and the overall e-waste volume was estimated to be 5–7 million tons [2]. These figures are projected to be higher today and rapidly growing. A study in Europe indicates that the volume of e-waste is increasing by 3–5% annually [9]. According to a projection, by 2001 about 41 million computers would have become obsolete in the United States [10]. In California more than 6000 computers become obsolete every day [4], and in Oregon and Washington about 1600 computers become obsolete every day [11]. According to experts, between 1997 and 2007 more than 500 million computers would have become obsolete in the United States [4].

Most of the e-waste comes from three major sectors: (1) individuals and small businesses; (2) large corporations, institutions, and governments; and (3) original equipment manufacturers (OEMs) [7]. A majority of this e-waste goes into storage. Three quarters of all the computers sold in the United States remain stockpiled and await disposal [2]. Some other studies indicate that the number of unused, stored computers is 315–680 million units [4]. On an average, a consumer has two to three obsolete computers in storage spaces. The value of these units drops down to 1–5% of the original cost of the equipment [12]. Old computers are worth very little. The 5 lbs of steel is worth $0.25, the CPU with gold tips and wiring is worth about $1.00, the motherboard with metal (gold, silver, and copper) connectors is worth about

$2.00, the cable approximately $0.09, the hard drive (15% aluminum) is worth around $0.10, and the monitor yoke (60% copper) approximately $0.80 [13]. So, computers are worth very little if not for the precious metals used. Recycling computers often entails subsidizing the recycler. For individuals, the cost incurred might be $10–$30. Hence, many consumers would rather throw away their old computers than pay this cost.

14.2.2   How Much E-Waste Is Exported

There are no definite statistics about how much e-waste is exported and where it is headed. One study conducted by the Graduate School of Industrial Administration at Carnegie Mellon indicated that 12.75 million computers would go to U.S. recyclers in 2002 [12]. Since industry sources estimate that 80% of this amount is sent to Asia, a tightly packed pile of computer waste of 1 square acre with a height of 674 ft (covers twice the height of the Statue of Liberty) can be conceptualized. Note that this estimate is only for one year and one country.

14.3   What Is Reverse Logistics?

Reverse logistics is a process in which a manufacturer accepts products from consumers for possible remanufacturing, recycling, reuse, or disposal [14]. Manufacturers also attempt to reduce the amount of materials used in the input process. Reverse logistics is much more than simple recycling since there is an emphasis on the reduction of materials used or on the remanufacture or the reuse of materials. In the traditional supply chain, the logistician managed the flow of products from the producer to the consumer. In reverse logistics, the reverse flow of the products from the consumer to the producer is managed.

Reverse logistics can extend the life cycle of a product and promote alternate use of resources that can be both cost-effective and eco-friendly [15]. Reverse logistics may help to increase profitability and productivity by reducing inputs [14]. BMW, DuPont, General Motors, and Hewlett Packard are some of the companies that have used reverse logistics in their operations [16].

14.3.1   Internal and External Drivers for Reverse Logistics

The following three primary intraorganizational activities impact reverse logistics:

- Sincere commitment to environmental issues
- Successful implementation of ethical standards
- Existence of policy entrepreneurs who are responsible for organizational adoption of an environment-friendly philosophy [17]

In addition, the reverse logistics activities of an organization are also directly affected by four environmental forces:

- Customers
- Suppliers
- Competitors
- Government agencies

14.3.2 Qualitative Models for Reverse Logistics

The reverse logistics hierarchy proposed by Kopicki et al. [18] and Carter and Ellram [16] states that resource reduction ought to be the ultimate goal. This resource reduction would include both the minimization of materials used in the product along with the minimization of waste and energy achieved through the design of more environmentally efficient products. After resource reduction has been achieved, the next aim is to reuse materials followed by recycling of as much waste as possible. Disposal, typically, is the last option employed, and even in this case incineration is preferable since some form of energy recovery is likely [7]. We modify this hierarchy as shown in Figure 14.1

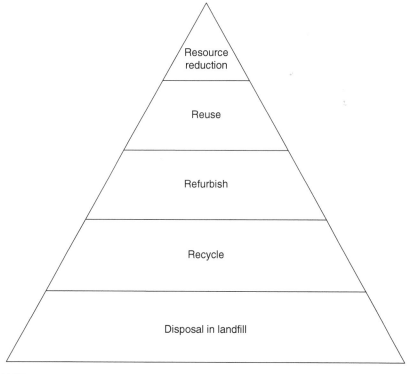

FIGURE 14.1
The modified reverse logistics hierarchy.

and apply each of the steps in this hierarchy to the computer industry. In this modification, we include another step called refurbishing, which fits between reuse and recycling, and combine the two disposal stages into one.

14.3.2.1 Resource Reduction

The major question is does the computer industry attempt to minimize materials used in the product? Alternatively, are there any attempts to minimize waste? Does the computer industry design for disassembly or design for environment? The answer in all these cases is in the negative.

Why do these problems exist? Computers are not designed for ease of recycling and, hence, the dismantling of computers is highly labor-intensive. The existence of toxic components is an added risk to recyclers. In addition, retrieving the valuable materials in e-waste is difficult since it is bound with plastics and mixed with other contaminants. Another issue is that the price of a computer does not include its end-of-life (EOL) costs. Hence, there is a lack of incentives to design for disassembly or design for environment. The only economically viable recycling that can take place is in a poor developing country, far from where this product was consumed and used. As long as U.S. recyclers compete with their Asian counterparts, it is highly unlikely that investment in infrastructure such as computer shredders and material separators for recycling of e-wastes will take place [7].

14.3.2.2 Computer Reuse

Computer reuse refers to the efforts aimed at keeping computers in operation by extending their life beyond three years. This reuse is direct second-hand use and made up a small percentage, about 3% in 1998, of the computers that had been discarded [8].

In noncommercial computer reuse, nonprofit and school-based programs do the refurbishing and computers are distributed to a particular population. Approximately 70% of this reuse serves schools and about 400–500 such programs exist in the United States with an average capacity of 200 computers per year. The largest program supplies about 10,000 computers. Some examples are (1) StRUT (students recycling used technology) in Portland, Phoenix, Silicon Valley, and Georgia; (2) Computers for Schools in Chicago; (3) the Computer Recycling Center (CRC) in the Bay Area; (4) Per Scholas in New York; and (5) Computers for Schools in Canada [8].

In commercial computer reuse, individuals or businesses can sell computers in auctions (eBay) or through local classifieds. The larger portion of computer re-use is the secondary wholesale market wherein "asset management" companies receive computers from large corporations or recyclers. These computers are then resold in the developing countries. In recycling electronic products, most of the profits accrue from reselling usable parts rather than crushing equipment to salvage metals, plastics, or glass [7].

Markets in developing countries have very cheap labor; in these markets, computers are bought, repaired at very a low cost, and resold for a profit. However, these older units also have a limited life span and end up as e-waste sooner or later [8].

Some large corporations such as HP Financial Services and IBM Global Asset Recovery Services have divisions for this work. Asset Recovery Center, Sysix, Micro Metallics, United Computer Exchange, and Waste Management Asset Management are other large secondary market companies [8].

14.3.2.3   Computer Refurbishing

This refers to the reconditioning of discarded computers to get them back in working order. Refurbishing is different from reuse since the computers are tested, repaired, and, in some cases, reinstalled with new software. Some examples of commercial refurbishers are IBM Refurbished, Dell Refurbished, and Nxtcycle. The noncommercial refurbishers are usually nonprofit or school-based companies. The refurbishers receive discarded computers that are then tested and repaired. In case a computer cannot be repaired, its usable parts are extracted. The nonworking equipment is sent to the recyclers for disposal. In general terms, it takes two to three computer donations to get a yield of one usable computer. A critical task is to clean out all the hard drives and install clean operating systems and software to make the computer functional. The cost of refurbishing is approximated to be $105 per computer, and this cost includes labor, parts, and disposal of e-wastes [19].

Some of the refurbishing is done in prison facilities. In the facilities, prisoners dismantle computer monitors, televisions, and other e-wastes. Since the prisoners are paid low wages for refurbishing activities, some domestic recyclers think that they will not able to compete as wages for refurbishing will be higher if done outside the prison. Others point out that the federally prescribed health and safety regulations of Occupational Safety and Health Administration (OSHA) will not be applicable to prisoners [8].

14.3.2.4   Computer Recycling

The break down of computer equipment to recover metals, plastic, and glass is termed as computer recycling. Recycling is a complex process since there are over a thousand different materials in a computer. In addition, diminishing amounts of precious metals such as gold, silver, platinum, and palladium have been used in computers in recent years. This fact combined with low commodity prices does not make computer recycling very profitable. For this reason, most of the computer recyclers are large commercial companies or government programs. The working parts and whole computers are also resold, and this area termed as "asset management or recovery" yields higher profits than smelting and shredding to recover materials. Noranda/ Micrometallics Corporation, Waste Management, IBM Credit Corporation,

MeTech International, Envirocycle, and UNICOR Federal Prison Industries are the largest recycling companies in this area [7].

Large corporations and OEMs have a much higher rate of recycling than individuals because of the application of the Environmental Protection Agency (EPA) regulations. These companies need to protect and hence destroy proprietary information, and some recyclers offer to clean the hard drives [7].

The literature fails to point out that in practice recycling actually involves export to a developing country. The recycling companies act as waste traders; figures indicate that 80% of the materials designated to be recycled is actually exported overseas and 90% of this material makes its way to China [20].

14.3.2.5 Disposal

Landfill and incineration: According to the EPA, more than 3.2 million tons of e-waste landed up in U.S. landfills in 1997. About 70% of the heavy metals found in landfills come from electronic discards and can contaminate groundwater [7].

A common assumption is that recycling is better than disposal in a landfill. However, if the recycling results in toxic waste exposures, open dumping, or burning of toxic residues, then this assumption does not hold true.

14.3.2.6 Export: The Solution to Recycling

The Basel Convention was created in 1989 to counter the unjust and unsustainable effects of free trade in toxic wastes. In 1994, the Basel Convention adopted a total ban on the export of all hazardous wastes from rich to poor countries for any reason, including recycling. Instead, the convention calls for all countries to reduce their exports of hazardous wastes to a minimum and contend with their waste problems within the national borders. However, United States is the only developed country that has not ratified the Basel Convention. The EPA admits that export is an integral part of U.S. e-waste disposal strategy. This policy of encouraging the export route under the green cloak of recycling stifles the innovation needed to solve the problem at source. As long as manufacturers can escape the true costs of disposal, they will not seek means to design better products. This aspect is especially critical since electronic products become obsolete rapidly and contain toxic materials [7].

Why are these e-wastes exported to Asia? There are three primary reasons for this phenomenon. First, labor costs are very low. In China, the costs are $1.50 per day versus the minimum wage of $5.50 per hour in the United States. Second, environmental regulations are lax. Even in the presence of legislation, the actual monitoring and enforcement does not always follow through. Finally, export of hazardous e-waste is legal in the United States [7].

Market forces state that toxic waste will run downhill on an economic path of least resistance. The toxic effluents of affluent societies will flood toward the world's poorest countries where labor is cheap and environmental regulations are slack. Free trade in hazardous waste leaves people in these countries with very few choices—poverty or exposure to toxins. In view of these facts, the Basel Convention was created in 1989. In 1994, the convention agreed to adopt a total ban on the export of hazardous wastes from rich to poor countries for any reason. The two compelling reasons to ban trade in hazardous wastes are as follows:

- *Downstream impacts.* This trade is environmentally damaging and fundamentally unjust since it victimizes the poor and burdens them with toxic exposure and environmental degradation. These victims had not benefited from industrialization in the first place.
- *Upstream impacts.* This trade enables waste generators to externalize their costs, thus creating a major disincentive for finding true solutions to the problem. As long as waste can be cheaply dumped elsewhere, there will never be any incentive to minimize hazardous waste at the source. Thus, the necessary innovation to design for the environment is forestalled [7].

14.3.2.7   Recycling or Trading

As stated earlier, most companies, which claim that they are recyclers of waste, actually engage in waste trading rather than waste recycling. Approximately 80% of e-wastes these companies collect is sent to Asia, and 90% of this material goes to China and the rest to India and Pakistan. In most of the cases, the recycler takes out the most valuable components of the product and sells these components to brokers. The rest of the material is generally broken down, sorted by waste (circuit boards, cables, and CRTs), and thrown into large cardboard boxes called gaylords. These materials are further sold to brokers who arrange the shipment to Asia. The warehouses in Asia (there are four such warehouses in the port of Nanhai, near Hong Kong) sell the most valuable parts in Asia. Alternatively, the e-waste broker might simply take the material in bulk and ship it to Asia without separation. Recycling or trading is an aggressive and competitive business, and circuit boards have the largest market among nonworking equipment because these boards are rich in precious metals such as gold, silver, palladium, and platinum [21].

Estimates indicate that it is 10 times cheaper to ship monitors to China than to recycle these monitors in the Unites States [7]. Computers are not designed for ease of recycling; hence, the dismantling of computers is a highly labor-intensive activity with an added risk of the presence of toxic components. Very few environmentally sound recycling facilities exist in the Unites States [7].

14.3.2.8   How Is This Export Rationalized?

Global standards. The stand of the U.S. regulatory body, EPA, is that the standards and operating procedures in developing countries must improve and this would justify the e-waste export to these countries. However, the reality is that these countries lack the infrastructure to monitor and maintain technology. Furthermore, the protection of worker and community rights is not guaranteed. Technology, by itself, is not enough to guarantee environmental and health protection [7].

Take back to Asia. There exists an argument that if electronic products are manufactured in Asia, then the waste from these electronic products must be exported back to Asia. The most toxic stages of the life cycle of electronic products—manufacturing and disposal—have been exported to the developing countries. Although cheap labor was first exploited in the production of the product, this fact cannot be used to justify a second exploitation, that is, the disposal of the product waste, especially when the consumption of the product has taken place in a developed country. Take-back programs do not imply physical transportation of the product to the country in which it was manufactured. Rather, take back must occur in the country of consumption and where the useful life of the product ends [7].

Export for reuse. Another argument is that obsolete computers need to be sent for reuse or refurbishing to Asia since this would add extra life to the product and would provide technology to the most needy. The reality is when these computers are sent to Asia, they end their life in Asia as well. The environmental and justice impacts would be the same as these would be if the computers were re-used or refurbished in the home country. However, given that the recycled parts end up in Asia in any case, the problem would simply be moved to Asia at a later date [7].

14.4   Disassembly Operations in China

In 2001, the Basel Action Network (BAN) conducted a field investigation in China to observe the recycling and disassembly operations of imported computers. This study was conducted in Guiyu, which is located west of Shantou city in the Chaozhou region of the greater Guangdong province. This area, made up of four small villages, has been transformed from a poor, rural, rice-growing community to a bustling e-waste processing center. The imported e-waste arrives into four large warehouses in the port of Nanhai and is trucked to the business centers in Guiyu. Hammer, chisel, screwdriver, and bare hands are used for the disassembly. Materials that are separated for recycling include materials containing copper, steel, plastic, aluminum, printer toner, and circuit boards [7].

14.4.1 Disassembly Details and the Hazards Involved

14.4.1.1 Toner Sweeping

The toner cartridges are taken apart in printer dismantling. A small amount of the residual toner is recycled, and the black plastic cartridge is discarded. The cartridges are opened with screwdrivers and paintbrushes are used to wipe the toner into a bucket. The workers engaged in this task do not wear any protective clothing, and this process creates clouds of toner, which are routinely inhaled by the workers [7].

14.4.1.2 Open Burning

The dismantled computers are dumped outside the town along a river. To recover copper, residents burn wires from the computer components. It is likely that the presence of PVC or brominated flame retardants contained in wire insulation leads to high levels of brominated and chlorinated dioxins and furans, both persistent organic pollutants. The inhabitants of the village are exposed to this pollution through airborne fumes, ash residue, and water waste in the fishponds adjacent to the river [7].

14.4.1.3 Cathode Ray Tube Cracking and Dumping

The popular claim is that the computer monitors are refurbished. However, the copper-laden yokes from the ends of the tubes are broken off. The CRTs are cracked and discarded. The yokes are sold to recover copper. The monitor glass, laden with lead, is dumped on open lands or into rivers. Guiyu, before becoming a computer-disposal site, was a rural, rice-growing community. At the time of the investigation, the irrigation canals were lined with broken monitor glass and plastic waste and bulldozers were brought in periodically to haul away this waste. Monitor glass qualifies as hazardous waste under the Basel Convention, and this glass has failed the leachate tests conducted by the EPA [7].

14.4.1.4 Circuit Board Recycling

Various components of the electronic circuit boards are recovered through this operation. The main approach to recycling a circuit board entails a desoldering process. The boards are placed on shallow grills that are heated. The wok-like grills contain a pool of molten lead–tin solder, and the circuit boards are placed in this solder and heated until the chips can be removed. These chips are then plucked with pliers and placed in buckets [7].

In another approach, the boards are slapped against a hard surface where the solder collects. Later, this solder is melted off and sold. The now-loosened chips are sorted out. The valuable chips are collected and sent for resale, and the remaining chips are sent to the acid chemical strippers for gold recovery. At times, the pins on the chips are straightened and dipped into fresh solder so that these pins can be reused in computer refabrication. After the desoldering process is complete, a laborer who uses wire clippers strips off the small capacitors and other less valuable components from the circuit board. After most of the components are picked off, the board is taken to a large-scale acid recovery operation where the remaining metals are recovered [7].

14.4.1.5 Acid Stripping of Chips

A process known as acid baths is used to recover precious metals from the circuit boards. The baths are aqua regia, which is a mixture of 25% pure nitric acid and 75% pure hydrochloric acid. This mixture is heated over small fires and then poured onto plastic tubs that contain computer chips. The workers swirl this mixture for hours and then add a chemical that precipitates the gold into the bottom of the tub. This gold is recovered, dried, and melted to a final form as a tiny bead of pure gold. On an environmental scale, this process is highly damaging. The acid stripping results in clouds of gases in the area. The aqua regia sludge is dumped on the riverbanks. In terms of protective clothing, the workers wear rubber boots and gloves. These workers do not wear any masks to prevent the inhalation of the fumes [7].

14.4.1.6 Plastic Chipping and Melting

Plastic is found in the computer casing, the monitor, and the keyboard. This plastic is chipped into small particles, and these particles are then sorted and separated by colors. Usually, children are employed for this task. The chips are placed in bags and sent to operations involving melting and extruding. The melting of plastic is done in rooms with little ventilation and no, if any, respiratory protection. A large proportion of e-waste plastic is deemed unrecyclable because of impurities or unmatched colors. Countless piles of plastic were observed in the landscape and near rivers [7].

14.4.1.7 Materials Dumped

A large portion of the imported e-waste material is not recycled but dumped in open areas such as fields and along waterways such as rivers, ponds, and wetlands. This dumped material consists of (1) leaded CRT glass; (2) burned and acid-reduced circuit boards; (3) toner cartridges; (4) dirty or mixed plastics; (5) residues from recycling operations, including ashes from open burning operations; and (6) acid baths and sludges [7].

14.4.2 Solutions to the Problem

This issue needs to be addressed at numerous levels—consumer, industry, and regulatory. So far, this problem has been ignored. Most consumers feel that dropping off a computer at a recycling station is an environmentally responsible course of action. However, most consumers are not aware that most of these materials end up abroad, dismantled, and disassembled manually. What are some of the approaches that can be followed to get around this problem?

14.4.2.1 Ban of Hazardous Waste

It is imperative for the United States to ratify the Basel Convention and ban the export of hazardous e-wastes that are found in computer monitors, computers, and circuit boards. The industrial world needs to manage its own hazardous wastes, and the developing countries need to be given tools and training to develop waste management strategies [7]. In the case of China, which has already ratified the Basel Convention, grassroots organizations within the country and international organizations need to play a proactive role. These organizations can exercise pressure tactics to make China abide by the convention and stop accepting these e-wastes, although the import of e-wastes has created jobs for Chinese citizens.

14.4.2.2 Extended Producer Responsibility

Under this approach, the producer has the responsibility to manage the entire life cycle of the product. In other words, the producer needs to take back the product and recycle or dispose it off in a responsible manner. In the absence of this responsibility, manufacturers are all too willing to externalize the environmental costs on to the consumers and, in the case of transnational pollution, to poorer, developing nations. Producers need to be encouraged to prevent pollution and reduce energy and resource in the product life cycle by changes in the design of the product and the process technology used. If the producers are financially responsible for EOL waste management, they would be encouraged to design products with more recycled materials and less hazardous materials [22].

The reasoning of extended producer responsibility (EPR) is based on the "polluter-pays" principle, first articulated in the German Ordinance on the Avoidance of Packaging Waste passed by the German government in 1993 [23]. EPR implies shared responsibilities in the product chain, although often the producer is in the best position, both technically and economically, to influence the rest of the product chain in reducing life cycle environmental impacts [7]. In the United States, there is a substantial amount of opposition to new take-back legislation among industry constituents. U.S. policy tends to prefer voluntary measures. However, these measures can be insufficient and unattainable when lacking economic incentives. Government intervention can provide for these economic incentives. Another approach would be to enact the "take-it-back" laws, wherein the producer is required to take back the product at the end of its life cycle. This approach is the ideal solution that would aim to close the loop of the product life cycle. The manufacturers of the product would need to work with the designers of the product to devise and implement sound recycling and reuse solutions [7].

14.4.2.3 Marketable Permits

A system of marketable permits may be the best way to protect the environment [24]. For example, developed countries could issue permits free of charge to existing e-waste polluters at some predetermined level of ecological degradation. Accordingly, if the United States generates 1 million tons of e-waste per year, then companies responsible for e-waste could receive permits that allow up to 1 million tons of pollution annually. Countries such as China and India that are victims of this biohazard would be allowed to participate and buy these permits to retire them. However, this approach violates international forums such as the U.N. Stockholm Conference, which requires polluting nations to assume responsibility for their actions and avoid damaging areas beyond their national borders.

14.4.2.3.1 Victim Pays Principle versus Polluter Pays Principle

In the case of marketable permits, the victims end up paying for the cleanup. This victimization is a matter of ethical concern. On transnational pollution issues, Baumol and Oates state that

> less-developed countries that choose uncontrolled domestic pollution as means to improve their economic position will voluntarily become the repository of world's dirty industries (25, p. 242).

These authors analyze the case of unidirectional transnational pollution theoretically to find a pareto-optimal solution. Interestingly, theory suggests that simply requiring the developed country to reduce e-waste would not be pareto-improving since there would be no benefits to that country. A mutual gain to both the countries requires the victim nation to make some payments to the developed country. Therefore, a pareto improvement implies a victim-pays principle!

The approach of EPR or "take-it-back" laws would fall in the line of the polluter-pays principle, wherein the manufacturer of the product would handle the costs of collection and disposal.

14.4.2.4 Design for Longevity, Upgradability, Repair, and Reuse

This approach states that the rapid obsolescence of the computer industry needs to be halted. The rapid advancement of technology has played a key role in the obsolescence of computers. A case could be made for the software and hardware engineers to create more flexible software and hardware systems that are upgradeable over time. For example, in the case of faster processors, a modular design could be incorporated to insert the newer, faster processor, rather than throwing out the entire computer or the motherboard [7].

14.4.2.5 Design for Recycling or Design for Disassembly

If a product needs to be retired, it needs to be designed in a manner that would ensure clear, safe, and efficient mechanisms for recovering the raw materials. There must be a preidentifiable recycling market for input materials. It would be useful to have a mechanism whereby the input material can be reconstituted and recycled. All the equipment components ought to be labeled as plastic or metal types. The product ought to be designed for rapid and easy dismantling or reduction to a usable form. Finally, warnings ought to be placed for any possible hazard in disassembly or recycling [7].

14.5 Conclusions and Future Work

In this chapter, we analyzed reverse logistics activities or the lack thereof in the computer industry. Numerous barriers exist to the recycling, reuse, refurbishing, or appropriate disposal of computers. The first challenge is the consumer mindset. Consumers in the United States are not as concerned about environmental issues as their counterparts in Europe. Although consumers in some pockets of the United States are involved in grassroots efforts to recycle or reuse computers, this phenomenon does not occur nationwide. Furthermore, there is a reluctance to pay for recycling, and most consumers state their preference for new products over the products made from recycled parts. Another challenge is political in nature. The Bush administration is pro-business and environmental concerns have taken a back seat as evidenced by the United States walking out of the Kyoto agreements. A small political presence is found within environmental concerns. This presence is evidenced by the popularity of some political candidates such as Ralph Nader. However, these voices are in the minority and are not loud enough to make a difference. EPA, the regulatory agency in the United States, has launched some initiatives to promote recycling but all these initiatives are in the pilot stage. Most of the initiatives are voluntary in nature, and companies take up these initiatives to be viewed as being "green" in nature. There is an absence of mandatory standards such as the directive on Waste from Electrical and Electronic Equipment proposed in the European Union. However, some leadership might emerge from the industry initiatives. If IBM, Dell, or any other major organization employs its reverse logistics practices as a market leadership strategy, then that company can pave the way for other companies to follow. Herein is an opportunity for

U.S. manufacturers to lead the way toward environmentally responsible practices in the disposal and the disassembly of computers.

Acknowledgments

The authors wish to thank BAN who prepared the report entitled "Exporting Harm: The High-Tech Trashing of Asia." This report includes a video that details the computer waste operations in China, India, and Pakistan.

References

1. Resource Conservation Challenge at http://www.epa.gov/epaoswer/osw/conserve/plugin/index.htm.
2. National Safety Council, Electronic Product Recovery and Recycling Base Report, 1999, p. 24.
3. U.S. Department of Labor, Labor of Bureau Statistics, Computer Ownership Up Sharply in the 90's at http://www.bls.gov/opub/ils/pdf/opbils31.pdf
4. Silicon Valley Toxics Coalition et al., Poison PCs and Toxic TVs: California's Biggest Environmental Crisis That You've Never Heard of, June 19, 2001, p. 8, at http://www.svtc.org/cleancc/pubs/poisonpc.htm. (hereinafter, Poison PCs).
5. Ten Tips for Donating a Computer, Exporting Harm, p. 5, at http://www.techsoup.org/products/recycle/10tips.cfm?
6. Manjoo F., http://www.salon.com/tech/feature/2002/09/23/antiglobal_geeks/print.html.
7. Exporting Harm, The High-Tech Trashing of Asia at http://www.ban.org/E-waste/technotrashfinalcomp.pdf, February 25, 2002, pp. 2, 6–8, 11–13, 15–22, 40–42.
8. Computer Recycling and Reuse Program, p. 8. Glossary at http://www.compumentor.org/recycle/glossary.html.
9. Arensman, R. Ready for recycling? Electronic business. *The Management Magazine for the Electronics Industry*, November, 2000.
10. Study conducted in 1999 by Stanford Resources, Inc. for the National Safety Council.
11. Northwest Products Stewardship Council. Government's saddled with electronic scrap. *Policymaker's Bulletin*; 1(1), p. 1, 2001.
12. Matthews, S. et al. Disposition and end-of-life options for personal computers, in Carnegie Mellon University Green Design Initiative Technical Report #97-10, p. 6, 1997, Appendix C.
13. Hamilton-Endicott, A. How Do You Junk Your Computer? *Time Magazine*, February 12, 2001.
14. Dowlatshahi, S. Developing a theory of reverse logistics. *Interfaces*; 30(3): 143–155, 2000.

15. Melbin, J.E. The never-ending cycle. *Distribution*; 94(11): 36–38, 1995.
16. Carter, C. and Ellram, L. Reverse logistics: A review of the literature and framework for future investigation. *Journal of Business Logistics*; 19(1): 85–102, 1998.
17. Stock, J.R. *Reverse Logistics*. Oak Brook, IL: Council of Logistics Management, 1992.
18. Kopicki, R., Berg, M.J., Legg, L., Dasappa, V. and Maggioni, C. *Reuse and Recycling—Reverse Logistics Opportunities*. Oak Brook, IL: Council of Logistics Management. 1993.
19. Geiselman, B. Electronics Recyclers Use of Prison Labor Makes Waves, Waste News, February 1, 2002 at http://www.wastenews.com.
20. Report for the United States Environmental Protection Agency, Region IX, Computers, E-Waste, and Product Stewardship: Is California Ready for the Challenge, p. 13, 2001.
21. US Environmental Protection Agency, Analysis of five community/consumer residential collections of end-of-life electronic and electrical equipment, November 24, 1998, at http://www.eeb.org/activities/waste/weee.htm.
22. Goldfine, S. Using Economic Incentives to Promote Environmentally Sound Business Practices: A Look at Germany's Experience with its Regulation on the Avoidance of Packaging Waste. GEO International Environmental Law Review. 1994.
23. Davis, G., Dillon, P., Fishbein, B. and Wilt, C. Extended Product Responsibility: A new principle for product-oriented pollution prevention. United States Environmental Protection Agency Office of Solid Waste, June 1997, at http://www.epa.gov/epr/pdfs/davis(1-4).
24. Dhanda, K.K. A market-based solution to acid rain: The case of the sulfur dioxide trading program. *Journal of Public Policy and Marketing*; 18(Fall): 258–264, 1997.
25. Baumol, W. and Oates, W. *The Theory of Environmental Policy*. Cambridge, UK: Cambridge University Press. 1993.

15

Evaluating Environment-Conscious Manufacturing Barriers with Interpretive Structural Modeling

Joseph Sarkis, Mohd. Asif Hasan, and Ravi Shankar

CONTENTS

15.1   Introduction

Environmentally-conscious manufacturing (ECM) and its practices have a variety of definitions and dimensions. ECM ranges from small, focused, and operational programs to broader, organization-wide strategic programs that have long-term implications for an organization. Sarkis (2001) has defined ECM within the broader context of corporate environmental management. Figure 15.1 summarizes the various elements and considerations within the ECM. This figure indicates the pervasiveness of the decision to integrate ECM and its many dimensions into an organization's culture, functions, and business processes.

The concept of ECM has been discussed in the literature, found in practice, and under research for almost two decades. Yet, its proactive adoption and pervasiveness are not as broad as expected. Proactive adoption means the implementation of the various aspects and elements of the ECM without coercion, specifically in response to compliance and regulations. Despite the many advantages of ECM, including possible "win-win" opportunities, competitive advantages, image enhancements, and long-term environmental benefits, numerous barriers exist for its implementation. Understanding

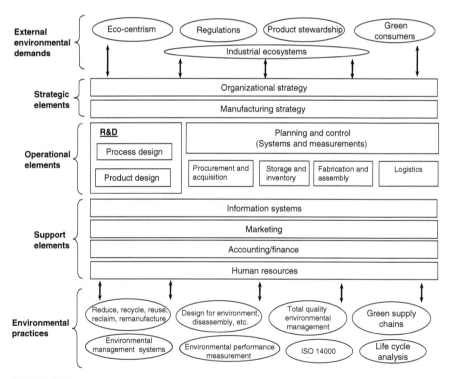

FIGURE 15.1
Strategic components of ECM.

these barriers and their interrelationships will help managers and organizations address or at least understand these issues so that they can make contingency plans.

To enable the managements of organizations to understand the barriers and their linkages, we introduce a planning tool to identify and analyze the barrier relationships. We call this technique interpretive structural modeling (ISM). ISM is a tool for "sense making" of complex situations and is very effective for planning exercises. In this chapter, we apply the mechanics of this tool to further understand the possible organizational ECM barriers and the interrelationships among these barriers. We outline some common organizational barriers and also provide an illustrative application of the methodology. The next stage of research would be to actually implement the methodology in a real setting or using an expert opinion. Our initial goal is to define ISM's applicability and possible relationships with other issues encountered in ECM. We begin with an overview of organizational barriers to ECM.

15.2 Barriers to ECM

Although we can identify a large group of possible barriers to ECM, we summarize only some major barriers based on observation and literature in the program management of strategic projects. We shall define these barriers from the organizational perspective in general.

Manufacturing can help organizations maintain competitive advantage as customer requirements evolve dynamically, especially with respect to environmental issues. A central issue facing organizations is the lack of appropriate roadmaps for adoption and implementation of ECM. As with most paths to organizational changes, there are numerous barriers and conflicts, causing the roadmap to become a vicious circle. Adoption and implementation of ECM principles in manufacturing organizations requires a systematic study of the various paths that may be followed along with a removal of barriers that would exist within these paths. These approaches would enable efficient and effective introduction of ECM practices. Our study is devoted to analyzing the barriers to ECM. Referring to the literature, we discern 11 barriers generally faced in the adoption of ECM. These barriers are summarized below.

15.2.1 Lack of Top Management Support and Commitment

Top management support is vital for the successful implementation of any strategic program (Hamel and Prahalad, 1989; Zhu et al., 2008). The top management has considerable ability to influence, support, and champion the formulation and deployment of environmental initiatives across the organization. In general, management support is a critical element in adoption and implementation of innovations, especially environmental systems, in an organization (Daily and Huang, 2001; Dechant and Altman, 1994). Frost and Egri (1991) emphasize that organizational innovations may not progress beyond the idea stage in the absence of dedicated champions for the cause. Carter et al. (1998) demonstrate that management support is especially useful for environmental practices such as ECM if such practices lie outside the domain of staff dealing with environmental concerns within the organization.

15.2.2   Resistance to Organizational Change

Cultural change is necessary to support the implementation of environmental programs in organizations (Kitazawa and Sarkis, 2000). ECM programs require significant organizational changes, including cultural change. New methods and structures may be necessary. For example, total quality environmental management (TQEM) emphasizes the use of performance measurement approaches and empowerment of people to complete environmental practices. This broader organizational perspective may also mean restructuring departments, especially environmental health and safety (EHS) departments. Organizations do not favor change due to the solidification of organizational routines through habituation and the risk of uncertainty associated with change (the fear of the unknown) (Clark, 1985; Hoffman and Bazerman, 2005).

15.2.3   Inappropriate Evaluation and Appraisal Approaches

Capital budgeting and investment justification tools primarily focus on the more tactical and short-term perspective of organizations (Hoffman and Bazerman, 2005). Payback, return on investment, and even discounted cashflow techniques typically favor short-term projects that have strong quantitative benefits. Even at the organizational (rather than programmatic) level, organizations use short-term performance measures such as stock price (Schmidheiny, 1996). ECM-type programs may require appraisal techniques that would incorporate intangible factors such as corporate image and social benefits of the decision (Presley and Sarkis, 1994). Environmental paybacks usually take much longer, some times generations, before they are realized. It is difficult to incorporate these factors into the more short-term financial techniques that consider discounting future benefits.

15.2.4   Lack of Methodologies and Processes to Enhance ECM

Although the field of ECM has witnessed significant growth, both theoretical and practical, over the past couple of decades, many tools and techniques still need to be developed for easier operational integration into organizations. For example, even rudimentary design for environment (DFE), environmental accounting, and other tools and processes for ECM are rarely implemented or integrated beyond the largest and the most forward-thinking organizations. Technology and tools for ECM practices such as demanu-facturing and disassembly are still in their infancy with much of the work carried out manually, and production and operations control tools are very limited (King and Burgess, 2005; Sarkis, 2003). Many organizations cannot afford the investment required by ECM methodologies and practices, and these organizations adopt such methodologies and practices only when they are coerced through regulatory measures or significant community pressures.

15.2.5 Difficulty in Integrating Life Cycle Analysis Elements into ECM

We mentioned the difficulty in integrating ECM processes into organizational practices and procedures. Within ECM, the life cycle analysis (LCA) linkage to ECM is an essential but difficult integration process (Gungor and Gupta, 1999). There is a significant uncertainty in LCA (Ross et al., 2002). Even the ISO 14040 (LCA standards) guidelines present cautionary statements about LCA. The major purpose of LCA is to determine the elements within and beyond the organization that would impact the environmental performance of products and processes. Unfortunately, the operational tools that can comprehensively apply and integrate LCA into most ECM settings are very limited, if at all existent. Helping organizations to think from this perspective requires significant organizational cultural change and further development of tools, information, and general business processes. The major elements of LCA are still not fully developed. The life cycle inventory is ever changing, the life cycle impact is still in early development, and the life cycle improvement elements are virtually nonexistent. Some researchers have looked at ways to mitigate these difficulties but essentially call for more development, enhancement, and study of LCA tools and methodologies (Udo de Haes et al., 2004).

15.2.6 Insufficient Training, Education, and Rewards Systems

Human and behavioral factors play a significant role in the organizational change and cultural change requirements for ECM development and implementation. Compared with traditional manufacturing environments, ECM has a number of characteristics that may alter workforce requirements. These characteristics include (i) closer interdependence among activities; (ii) different skill requirements, usually higher average skill levels on environmental issues and LCA; (iii) more immediate and costly consequences of poor practices; and (iv) continual change and development. The rewards system is related to the motivational explanation for a "misdirected attention" effect; organizations may offer incentives to and reward individuals and promote self-interested behavior that is at odds with wider environmental interests and goals (Tenbrunsel et al., 1997). A well-designed rewards system will help in ECM adoption. Yet, few organizations have financial incentives such as bonuses, incentives, or salaries tied to environmental performance

(Denton, 1999). In addition, communication through education and training of employees in ECM and broader environmental issues has typically been low on the priority list of organizations. Employee empowerment will greatly help in TQEM-type practices, but before employees can be given environmental responsibilities, training is required (Govindaraju and Daily, 2004).

15.2.7 Poor Design for Environment Interfaces

Organizations have difficulty with DFE integration despite the availability of DFE theoretical models and innovations (Handfield et al., 2001). This issue is important from organizational and cross-functional perspectives. DFE, which is related to LCA, is sometimes a difficult process to implement in organizations. Cooperation between at least four groups within an organization is required for the successful implementation of this program. Interfaces between marketing, engineering, manufacturing, and environmental staff are required. In addition to the organizational interface, technological interfaces are also required such that information and systems (design, manufacturing, and customer requirements) are all effectively integrated. Organizational support mechanisms, incentive systems, and review and audit processes are all required for effective integration of the DFE. The lack of appropriate environmental management systems, which can enable further integration of DFE with ECM, has also been found to be a major limitation (Ammenberg and Sundin, 2005).

15.2.8 Poor Partnership (Supply Chain) Formation and Management

To address rapidly changing environmental requirements and regulations, it is necessary to form long-term relationships across organizational boundaries (e.g., for regulations that require take back and the prevention of any materials in products such as the registration, evaluation, and authorization of chemicals (REACH) requirements in Europe (Brown, 2003)). Thus, supply chain and network relationships need to be managed efficiently and effectively for rapid and accurate information updating of products and materials, especially hazardous substance types. Financial and resource factors, evident in smaller firms, are possible reasons for some lost opportunities to improve environmental supply chain performance and data sharing. Some of the limitations of LCA are also evident in these circumstances. In addition, for various ECM practices, it is necessary to form partnerships with environmentally sound vendors (and customers). The necessary ECM evaluation tools and criteria as well as practices may not exist in many organizations (Sarkis, 2006).

15.2.9 Poor Incorporation of Environmental Measures into Decision-Making

The axiom "you can't manage what you can't measure" is applicable to ECM as well. Therefore, ECM requires appropriate measures for all types of environmental issues. These measures range from direct measures for waste generated and effluents emitted to measures for controlling toxicity levels of certain chemicals and processes and intangible measures such as creation of social image (Sarkis, 2001). Other authors have tried to define ECM precisely and to identify ways of measuring ECM as no simple metrics or indices currently exist for some environmental dimensions (Epstein and Roy, 2003). There is a need to identify and measure intangible environmental factors in enterprises. The other dimension of this issue is ways of measurement and capabilities of measurement systems, including systemic approach to develop, acquire, evaluate, and improve upon these measures. Adequate performance measurement systems, in general, are difficult organizational projects to implement. Development of environmental performance measurement systems for operational, employee, and project management purposes is rare and difficult to implement. The requirements of such systems can be quite extensive (Epstein and Roy, 2001). If environmental measures are not included in organizational and management performance, ECM measures would largely be confined to the environmental department.

15.2.10   Difficulties with Environmental Technology

Environmental technology can range from in-process, prevention technologies to end-of-pipe and interorganizational-type technologies. Some of these technologies may be traditional environment-based technologies such as dampers or filters or more advanced closed-loop systems that require integration with typical manufacturing processes. The difficulty with many of these technologies is that they are customized for a given plant or production line. Another aspect of environmental technologies is evident in remanufacturing and disassembly. Technologies for these reverse supply chain processes have not been fully developed due to the complexities involved in these processes. Currently, much of the processing in demanufacturing facilities is manual, and appropriate control technologies have either not been developed or implemented.

15.2.11   Limited Intraorganizational Cooperation

Intraorganizational collaboration is necessary to handle the organizational processes and limitations of DFE and LCA and other ECM practices. Organizational silos prevent multiple elements of organizations from creating and implementing effective strategies that cut across the organization. These silos are typically based on political divisions, and protective functional interests may blind organizations toward identifying the potential economic benefits of environmental initiatives (Hoffman and Bazerman, 2005; Lovins and Lovins, 1997). Environmental issues are prevalent throughout the organization, and ECM issues are also part of them. Environmental activities are considered the responsibility of the environmental staff. Expanding the relationships across functional areas, especially within manufacturing and engineering, requires sharing of environmental knowledge, technology, and practices.

15.3 Intrepretive Structural Modeling

ISM is a methodology used for identifying relationships among specific items, which define a problem or an issue. This methodology was first developed in the 1970s (Warfield, 1974; Sage, 1977) and provides a means by which relationship order can be imposed on a variety of interrelated factors. ISM can be used to analyze the barriers to ECM. This technique is intended to identify the interrelationships between the barriers and their levels. These barriers are categorized into autonomous, dependent, linkage, and independent barriers on the basis of their driving power and dependence.

ISM is an interactive group-learning process, whereby a set of different directly and indirectly related elements are structured into a comprehensive systemic model. This model portrays the structure of a complex set of issues in a carefully designed pattern, employing graphics as well as text. ISM helps to impose order and direction on the complex relationships among elements of a system (Sage, 1977). Since ISM provides a systemic view of the relationships among the barriers, the confounding effect of the barriers can be easily analyzed. This analysis would assist both manufacturers and researchers in exploring a path leading to the adoption of agile manufacturing in an efficient and effective manner.

ISM is interpretive because judgment of a respondent group determines whether and how variables (usually in textual mode) are related. ISM is structural because an overall structural relationship among the factors is extracted. Although this technique can be used as a group-learning tool, it can also be used individually to investigate various relationships (Jharkharia and Shankar, 2005). ISM generally incorporates the following steps (Ravi and Shankar, 2005):

Step 1: Variables affecting the system under consideration are listed, which can be objectives, actions, individuals, barriers, and enablers.

Step 2: A contextual relationship is established among variables identified in Step 1 with respect to which pairs of variables would be examined.

Step 3: A structural self-interaction matrix (SSIM) is developed for variables, which indicates pairwise relationships among variables of the system under consideration.

Step 4: A reachability matrix is developed from the SSIM, and the matrix is checked for transitivity. The transitivity of the contextual relation is a basic assumption made in ISM. The assumption is if a variable *A* is related to *B* and *B* is related to *C*, then *A* is necessarily related to *C*.

Step 5: The reachability matrix obtained in Step 4 is partitioned into different levels.

Step 6: Based on the relationships given in the reachability matrix, a directed graph is drawn and the transitive links are removed.

Step 7: The resultant digraph is converted into an ISM by replacing variable nodes with statements.

Step 8: The ISM model developed in Step 7 is reviewed to check for conceptual inconsistency and necessary modifications are made.

15.3.1 The Structural Self-Interaction Matrix and Data-Gathering Methodology

ISM methodology suggests the use of expert opinions based on various management techniques such as brainstorming, nominal grouping technique, and affinity diagramming in developing the contextual relationship among the variables (Ravi and Shankar, 2005). Typically, the approach would rely on expert or managerial opinion to arrive at the structure and the relationships of the barriers. We did not acquire expert opinion in this study since we focused on how the methodology can be used.

In the process of analyzing the barriers in developing SSIM, the following four symbols have been used to denote the direction of relationship between two barriers (*i, j*):

V—Barrier *i* influences the development of barrier *j*

A—Barrier *j* influences the development of barrier *i*

X—Barriers *i* and *j* influence the development of each other

O—Barriers *i* and *j* do not influence the development of each other

The following would explain the use of the symbols V, A, X, and O in SSIM (Table 15.1).

1. Barrier 1 influences the development of barrier 7 (V)
2. Barrier 11 influences the development of barrier 3 (A)
3. Barriers 4 and 5 influence the development of each other (X)
4. Barriers 1 and 2 do not influence each other (O)

The SSIM is developed on the basis of these contextual relationships (see Table 15.1). The number of pairwise comparison questions addressed for developing the SSIM are $(N(N-1)/2)$, where N is the number of factors.

TABLE 15.1

Structural Self-Interaction Matrix (SSIM)

Barriers	11	10	9	8	7	6	5	4	3	2
1. Inappropriate evaluation and appraisal approaches	V	O	V	V	V	V	V	V	V	O
2. Lack of methodologies and processes to enhance ECM	V	V	V	V	O	V	V	V	V	
3. Resistance to organizational change	A	A	V	X	A	O	X	O		
4. Difficulty in integrating LCA items into ECM	A	A	V	O	A	O	X			
5. Insufficient training, education, and rewards systems	A	A	V	O	A	O				
6. Difficulties with environmental technology	A	A	V	X	A					
7. Limited intraorganizational cooperation	V	O	V	V						
8. Poor partnership (supply chain) formation and management	A	A	V							
9. Poor design-for-environment (DFE) interfaces	A	A								
10. Poor incorporation of environmental measures into decision-making	V									
11. Lack of top management support and commitment										

15.3.2 The Reachability Matrix

The SSIM can now be converted into a binary matrix called the initial reachability matrix by substituting V, A, X, and O by 1 and 0 as per the following rules:

    i. If the (i, j) entry in the SSIM is V, the (i, j) entry in the reachability matrix becomes 1 and the (j, i) entry becomes 0.

    ii. If the (i, j) entry in the SSIM is A, the (i, j) entry in the reachability matrix becomes 0 and the (j, i) entry becomes 1.

    iii. If the (i, j) entry in the SSIM is X, the (i, j) entry in the reachability matrix becomes 1 and the (j, i) entry also becomes 1.

    iv. If the (i, j) entry in the SSIM is 0, the (i, j) entry in the reachability matrix becomes 0 and the (j, i) entry also becomes 0.

Following these simple rules, an initial reachability matrix for the barriers can be derived. This process also limits the number of pairwise comparison questions to $(N(N - 1)/2)$, where N is the number of factors. The final reachability matrix is obtained by incorporating the transitivities as enumerated in Step 4 of the ISM methodology. Table 15.2 shows the final reachability matrix with transitivities incorporated. This table also indicates the driving power and the dependence of each barrier. The driving power of a particular barrier is the total number of barriers (including itself) that it influences. Dependence is the total number of barriers (including itself) that may help in

TABLE 15.2

Final Reachability Matrix

Barriers	1	2	3	4	5	6	7	8	9	10	11	Driver Power
1. Inappropriate evaluation and appraisal approaches	1	0	1	1	1	1	1	1	1	0	1	9
2. Lack of methodologies and processes to enhance ECM	0	1	1	1	1	1	0	1	1	1	1	9
3. Resistance to organizational change	0	0	1	1	1	1	0	1	1	0	0	6
4. Difficulty in integrating LCA items into ECM	0	0	1	1	1	1	0	1	1	0	0	6
5. Insufficient training, education, and rewards systems	0	0	1	1	1	1	0	1	1	0	0	6
6. Difficulties with environmental technology	0	0	1	1	1	1	0	1	1	0	0	6
7. Limited intra-organizational cooperation	0	0	1	1	1	1	1	1	1	0	1	8
8. Poor partnership (supply chain) formation and management	0	0	1	1	1	1	0	1	1	0	0	6
9. Poor design-for-environment (DFE) interfaces	0	0	0	0	0	0	0	0	1	0	0	1
10. Poor incorporation of environmental measures into decision making	0	0	1	1	1	1	0	1	1	1	1	8
11. Lack of top management support and commitment	0	0	1	1	1	1	0	1	1	0	1	7
Dependence	1	1	10	10	10	10	2	10	11	2	5	72/72

influencing a barrier's development. These driving power and dependency values are used in the classification of barriers. The barriers are classified into four groups: autonomous, dependent, linkage, and independent (driver).

15.3.3 Level Partitions

The reachability and antecedent sets (Warfield, 1974) for each barrier are determined from the final reachability matrix. The reachability set for a particular variable consists of the variable itself and the other variables that it influences. The antecedent set consists of the variable itself and the other variables that may influence it. Subsequently, the common elements of the reachability and antecedent sets form the intersection sets, which are derived for all variables. The variable for which the reachability and the intersection sets are the same is assigned as the top-level variable in the ISM hierarchy, which would not help achieve any other variable above its own level.

After the top-level element is identified, this element is discarded from further hierarchical analysis (i.e., removing that element from all the different sets). For example, as seen in Table 15.3, "poor DFE interfaces" (barrier 9) is found at level I. Thus, poor DFE interfaces would be positioned at the top of the ISM model. This iterative procedure is continued until the levels of each variable are determined. The identified levels aid in building the digraph and the final model of ISM.

15.3.4 Formation of the ISM-Based Model

From the final reachability matrix (Table 15.3), the structural model is generated by means of vertices, nodes, and edges (Jharkharia and Shankar, 2005). A relationship between the barriers i and j is shown by a directional arrow from barrier i to barrier j. This graph is called a directed graph or digraph.

TABLE 15.3

Initial Reachability, Antecedent, and Intersection Sets

Barrier	Reachability Set	Antecedent Set	Intersection Set	Level
1	1,3,4,5,6,7,8,9,11	1	1	
2	2,3,4,5,6,8,9,10,11	2	2	
3	3,4,5,6,8,9	1,2,3,4,5,6,7,8,10,11	3,4,5,6,8	
4	3,4,5,6,8,9	1,2,3,4,5,6,7,8,10,11	3,4,5,6,8	
5	3,4,5,6,8,9	1,2,3,4,5,6,7,8,10,11	3,4,5,6,8	
6	3,4,5,6,8,9	1,2,3,4,5,6,7,8,10,11	3,4,5,6,8	
7	3,4,5,6,7,8,9,11	1,7	7	
8	3,4,5,6,8,9	1,2,3,4,5,6,7,8,10,11	3,4,5,6,8	
9	9	1,2,3,4,5,6,7,8,9,10,11	9	I
10	3,4,5,6,8,9,10,11	2,10	10	
11	3,4,5,6,8,9,11	1,2,7,10,11	11	

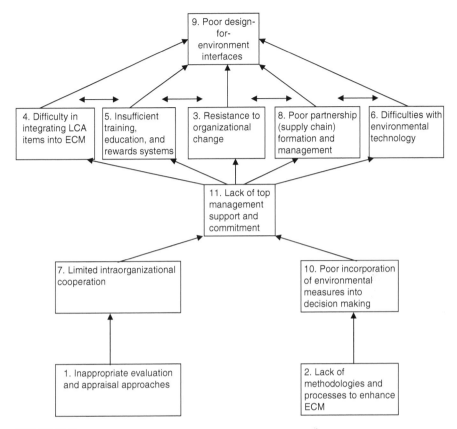

FIGURE 15.2
Final structural model of ECM barriers.

After removing the transitivities as described in the ISM methodology, the digraph is finally converted into an ISM as shown in Figure 15.2.

15.3.5 Classification of Barriers

The barriers may be classified into four clusters (Mandal and Deshmukh, 1994). The first cluster consists of autonomous barriers that have weak driving power and dependence. These barriers are relatively disconnected from the system and have relatively few links that may be strong; none of the barriers identified in this study fall within this cluster. The second cluster consists of the dependent barriers that have weak driver power but strong dependence. The third cluster is composed of linkage barriers that have strong driver power and also strong dependence. These barriers are unstable because any action on these barriers will have an effect on other barriers and also a repercussion on these barriers themselves. The fourth cluster includes independent barriers having strong driving power but weak dependence. Barriers with relatively strong driving power are key variables clustered into the category of independent or linkage barriers. We can observe that barrier 7 has a driver power value of 8 and a dependence value of 2 (as determined from Table 15.3). A graphical matrix may be shown but is left out of this chapter for brevity.

The objective behind the classification of the barriers is to analyze the driving power and the dependency of the barriers (Jharkharia and Shankar, 2005). In general, higher barrier driving powers mean that a larger number of barriers could be easily removed by its removal. Higher dependence values for a barrier necessitate a larger set of barriers to be addressed before its removal and a more likely success in the implementation of ECM. The classification of barriers into the four clusters helps identify the difficulty-removal potential of the barriers.

15.4   Conclusions and Future Research

In this chapter we have identified a number of barriers, organizational and technological, which may prevent organizations from considering ECM or pose difficulties in implementing ECM and its various practices. The 11 barriers identified in this chapter have significant overlaps and interrelationships that are sometimes difficult to observe. A more complete understanding of the barriers and their interrelationships, through a logical structure, will help managers to better prioritize and effectively target their resources. Eliminating some barriers may bring greater returns and payback than other barriers, from an ECM implementation perspective. This identification of barriers is the most important managerial implication of this chapter and would help management decision makers. Research implications also exist. Researchers can use this technique to more effectively iron out path relationships among various organizational factors. Research questions and various relationships can be related not only to adoption and implementation of ECM practices but their eventual success. For example, organizations may adopt and invest heavily in ECM practices, but with the existence of particular barriers the relative performance and success of these initiatives may be poor. As we saw in this example, albeit a conceptual example, the complexities in the relationships can become quite significant. Analytical modelers and prescriptively focused researchers may also benefit from this technique. Numerous decision models exist to help organizations effectively evaluate and implement ECM practices and general technologies. For example, the ISM technique may be applied with other approaches such as the analytical hierarchy process, which requires a decision structure to help determine the strength of relationships and for decision making. Simulation and system dynamics modeling may also be used to help identify how barriers to ECM in organizations will influence the various organizational and performance results.

The research and the managerial implications provide numerous directions for further research in this research stream. In addition to the research directions identified, this technique can be applied to a real-world setting to determine whether (1) the barriers are complete and (2) whether the relationships are similar to the relationships that exist here and found in the literature.

References

Ammenberg, J. and Sundin, E. (2005), Products in environmental management systems: drivers, barriers and experiences, *Journal of Cleaner Production*, 13(4), 405–415.

Brown, V.J. (2003), REACHing for chemical safety, *Environmental Health Perspectives*, 111(14), A767–A769.

Carter, C.R., Ellram, L.M., and Kathryn, L.M. (1998), Environmental purchasing: benchmarking our German counterparts. *International Journal of Purchasing and Materials Management*, 34(4), 28–38.

Clark, K. (1985), The interaction of design hierarchies and market concepts in technological evolution, *Research Policy*, 14(5), 235–251.

Daily, B.F. and Huang, S.C. (2001), Achieving sustainability through attention to human resource factors in environmental management, *International Journal of Operations and Production Management*, 21(12), 1539–1552.

Dechant, K. and Altman, B. (1994), Environmental leadership: from compliance to competitive advantage, *Academy of Management Executive*, 8(3), 7–27.

Denton, D.K. (1999), Employee involvement, pollution control, and pieces to the puzzle, *Environmental Management and Health*, 10(2), 105–111.

Epstein, M.J. and Roy, M.J. (2001), Sustainability in action: identifying and measuring the key performance indicators, *Long Range Planning*, 34(5), 585–604.

Epstein, M.J. and Roy, M.J. (2003), Improving sustainability performance: specifying, implementing and measuring key principles, *Journal of General Management*, 29(1), 15–31.

Frost, P.J. and Egri, C.P. (1991), The political process of innovation. *Research in Organizational Behavior*, 13, 229–295.

Govindaraju, N. and Daily, B.F. (2004), Motivating employees for environmental improvement, *Industrial Management & Data Systems*, 104(4), 364–372.

Gungor, A. and Gupta, S. (1999), Issues in environmentally conscious manufacturing and product recovery, *Computers & Industrial Engineering*, 36, 811–853.

Hamel, G. and Prahalad, C.K. (1989), Strategic intent, *Harvard Business Review*, 67, 63–76.

Handfield, R.B., Melnyk, S.A., Calantone, R.J., and Curkovic, S. (2001), Integrating environmental concerns into the design process: the gap between theory and practice, *IEEE Transactions on Engineering Management*, 48, 2, 189–208.

Hoffman, A.J. and Bazerman, M.H. (2005), Changing environmental practice: understanding and overcoming the organizational and psychological barriers, Harvard NOM Working Paper No. 05-04; Harvard Business School Working Paper No. 05-043.

Jharkharia, S. and Shankar, R. (2005), IT-enablement of supply chains: understanding the barriers, *Journal of Enterprise Information Management*, 18(1), 11–27.

King, A.M. and Burgess, S.C. (2005), The development of a remanufacturing platform design: a strategic response to the directive on waste electrical and electronic equipment, *Proceedings of the Institution of Mechanical Engineers, Part B: Journal of Engineering Manufacture*, 219(8) 623–631.

Kitazawa, S. and Sarkis, J. (2000), The relationship between ISO 14000 and continuous source reduction, *International Journal of Operations and Production Management*, 20(2), 225–248.

Lovins, A. and Lovins, H. (1997), *Climate: Making Sense and Making Money*. Snowmass, CO: Rocky Mountain Institute.

Mandal, A. and Deshmukh, S.G. (1994), Vendor selection using interpretive structural modeling (ISM), *International Journal of Operations and Production Management*, 14(6), 52–59.

Presley, A. and Sarkis, J. (1994), An activity based strategic justification methodology for ECM technology. *International Journal of Environmentally Conscious Design and Manufacturing*, 3(3), 5–17.

Ravi, V. and Shankar, R. (2005), Analysis of interactions among the barriers of reverse logistics, *Technological Forecasting & Social Change*, 72(8), 1011–1029.

Ross, S., Evans, D., and Webber, M. (2002), How LCA studies deal with uncertainty, *International Journal of Life Cycle Assessment*, 7(1), 47–52.

Sage, A.P. (1977), *Interpretive Structural Modeling: Methodology for Large-Scale Systems*. New York: McGraw-Hill, pp. 91–164.

Sarkis, J. (2001), Manufacturing's role in corporate environmental sustainability: concerns for the new millennium, *International Journal of Operations and Production Management*, 21(5/6), 666–686.

Sarkis, J. (2003), Operations of a computer equipment resource recovery facility. In Williams, E. (ed.), *Computers and the Environment*. Tokyo: United Nations University, Kluwer Publishers, Chapter 12, pp. 231–251.

Sarkis, J., (ed.), (2006), *Greening the Supply Chain*. Berlin: Springer.

Schmidheiny, S. (1996), *Financing Change*. Cambridge, MA: MIT Press.

Tenbrunsel, A.E., Wade-Benzoni, K.A., Messick, D.M., and Bazerman, M.H. (1997), The dysfunctional effects of standards on environmental attitudes and choices. In Bazerman, M.B., Messick, D.M., Tenbrunsel, A.E., and Wade-Benzoni, K.A. (eds.), *Psychological Perspectives to Environmental and Ethical Issues*. San Francisco, CA: New Lexington Books.

Udo de Haes, H.A., Heijungs, R., Suh, S., and Huppes, G. (2004), Three strategies to overcome the limitations of life-cycle assessment, *Journal of Industrial Ecology*, 8(3), 19–32.

Warfield, J.W. (1974), Developing interconnected matrices in structural modeling, *IEEE Transcript on Systems, Men and Cybernetics*, 4(1), 51–81.

Zhu, Q., Sarkis, J., Cordeiro, J.J., and Lai, K.-H. (2008), Firm level correlates of emergent green supply chain management practices in the Chinese context, *OMEGA*, 36(4), 577–591.

Index

Printed and bound by CPI Group (UK) Ltd, Croydon, CR0 4YY

28/10/2024

01780122-0001